The Sitwells

By the same author

GONE TO TIMBUCTOO
BLUEBIRD AND THE DEAD LAKE
THE PERSUASION INDUSTRY
(with Graham Turner)
THE LIFE OF IAN FLEMING
ARENA
THE PROFESSION OF VIOLENCE
THE AUTOBIOGRAPHY OF JAMES BOND
EDWARD THE RAKE

The Sitwells

A FAMILY'S BIOGRAPHY

John Pearson

A Harvest/HBJ Book

Harcourt Brace Jovanovich

New York and London

Published in England under the title *Façades*

Printed in the United States of America

Library of Congress Cataloging in Publication Data

Pearson, John, 1930–
The Sitwells.

(A Harvest/HBJ book)
Originally published under title: Façades.
Bibliography: p.
Includes index.
1. Sitwell, Edith, Dame, 1887–1964—Biography.
2. Sitwell, Osbert, Sir, bart., 1892–1969—Biography.
3. Sitwell, Sacheverell, Sir, bart., 1897– —Biography.
4. Authors, English—20th century—Biography.
5. Sitwell family—Biography. I. Title.
[PR6037.I79P4 1980] 820'.9'00912 [B] 80-14371
ISBN 0-15-682676-3

First Harvest/HBJ edition 1980

A B C D E F G H I J

For Lynette with love

Contents

List of Illustrations

The publishers are grateful to the *Sunday Times* for help in finding illustrations.

Introduction

In a sense, this biography began when I trod on Sir Osbert Sitwell's toe at the big post-war exhibition of Mexican Art at the Tate Gallery in London. I was a schoolboy at the time, and this was my first physical contact with a literary celebrity. I recognised him instantly from his photographs in the press – *Great Morning* had only recently appeared – and from his gleaming brass bust by Dobson, which in those days was the brightest object in the section of the gallery devoted to modern British art. But as I stammered an apology, I was horrified by the look of anguish on the face of this distinguished writer. Too young to comprehend the agonies of gout from which he was tormented at the time, I remained puzzled for a long time afterwards by this image of such extraordinary suffering. It haunted me as I meandered through the bland and stately pages of his autobiography and gave me, I imagine, my first inkling that for all its massive length, the book revealed only what its author wanted to appear. I sensed a mystery, but had no idea what it really was, or why.

The mystery deepened for me as I read the other Sitwells. All three of them shared what Edith somewhat regally described as 'the remote air of a legend', and the legend puzzled me. What was it, and why did they evidently feel the need of it? How had this trio managed to become a legend early in their lifetime, and what lay behind it? Like the great majority of the Sitwells' readers and admirers, I had no idea, and finally forgot about it.

Then in the summer of 1973 the mystery suddenly revived, for me at any rate, when my friend, the late Leonard Russell of the *Sunday Times*, suggested Osbert Sitwell as the subject of a long article for the paper. I was in Florence at the time. Sir Osbert had been dead for just over five years, and by a strange coincidence, soon after Leonard's suggestion arrived, I heard some Italian friends discussing the famous Sitwell castle at Montegufoni which I had once read about in the pages of *Great Morning*. It was for sale. Apparently Sir Osbert's heir could not afford to run it, and there were rumours that an Italian was in process of acquiring it. Gossip inevitably followed in the best Florentine manner – chiefly about the price and how the deal was progressing – but also about the various members of the family who had lived there. One of the Italians knew someone who had visited Sir Osbert at the house not long before he died, and someone else remembered the eccentric Sir George Sitwell there before the war.

'And does anybody live there now?' I asked.

'Good Heavens, no! The place is empty and neglected. Why don't you go and have a look if you're so interested.'

I did, a few days later, driving some twenty kilometres up the historic Via Volterrana from the edge of Florence and finding the enormous castle just beyond the little village of Montagnana, like some Tibetan lamasery set on its private hill. It was forbidding and at first sight seemed entirely closed up, but it was not completely uninhabited. As I drove up the drive a small man in dark glasses stopped my car and asked me what I wanted. I explained, and he replied in English that it was impossible to see the castle as it had just been sold, but that if I was really interested in Sir Osbert Sitwell, there was a chance that he could help me. He introduced himself. His name was Frank Magro, and he had been the great man's secretary and companion for the last years of his life. He had a small flat in the castle and hospitably invited me up to have a drink. He talked about the last years Sir Osbert spent here in the castle as a virtual recluse, and the dramas that ensued upon his death; and as he talked I was back with the mystery that had always puzzled me about the Sitwells. Why should the author of *Great Morning* have ended up like this?

There was no simple answer. For as I soon found out, the truth about Sir Osbert's retirement and death was inextricably bound up with much that lay behind the complex Sitwell legends of the past. Not for nothing had Dame Edith's best-known work been called *Façade*, and I decided there and then to find what really lay behind the various façades Sir Osbert and Dame Edith had assiduously erected in their lifetime.

I discussed this with one of Sir Osbert's most devoted friends, Sir Harold Acton. He gave me a number of invaluable clues, but also warned me of the enormity and hazards of the task – the vastness of the Sitwells' literary output, the range of their friends and enemies, and certain matters – here he smiled with ineffable discretion – which were bound to be 'shall we say, slightly touchy?'

While still in Florence I saw the remaining few who had seen Sir Osbert just before he died, the former British Consul, Christopher Pirie Gordon, the head of the British Institute and founder of the Oxford Sitwell Society, Ian Greenlees, and the priest who saw him on his death-bed and conducted the funeral service, the Rev. George Church, vicar of the Anglican Church of St Mark's. Then I returned to London to begin what would turn out to be practically four years' obsessive work.

Here I must put on record the invaluable assistance of the many other people, including friends – and enemies – of the Sitwells who helped me: the Dowager Lady Aberconway, A. Alvarez, Lorna Andrade, Sir Frederick Ashton, Francis Bamford, Sir Cecil Beaton, Sir Basil Blackwell, Hugh Burnett, Father Philip Caraman, Major J. W. Chandos-Pole, Lord Clark, Serafina Clarke, Cyril Connolly, Lady Diana Cooper, Father Martin D'Arcy, Alan Dent, Rache Lovat Dickson, Geoffrey Elborn, Constantine FitzGibbon, James and Minnie Fosburgh, Philip Frere, Sir John Gielgud, Martin Gilbert, Michael Goodwin, Geoffrey Gorer, Geoffrey Grigson, Sir Alec Guinness, Robert Heber-Percy, Peter Hollingworth, Viva King, Lincoln Kirstein, Professor Desmond Laurence, Dr Frank Leavis, John Lehmann, Cole Lesley, Jack Lindsay, Lady Lindsay, Nancy Luscher, Harold Macmillan, Dan Maloney, Delveo Molesworth, Dr Patrick Moore, Sheridan Morley, Angus Morrison, Raymond Mortimer, Beverley Nichols, John Piper, Alan Pryce-Jones, Peter Quennell, Kathleen Raine, John Russell, Elizabeth Salter, Hugo Southern, John Sparrow, Stephen Spender, Mrs Adrian Stokes, Claire Sutton, Julian Symons, Herbert van Thal, Sir William Walton, Glenway Wescott, Dame Rebecca West, Monroe Wheeler, Gillian Widdicombe.

David Hardman gave me an invaluable collection of early press material on the Sitwells, Sybille Bedford and Maurice Richardson gave me sage advice on the Sitwells' relationships with Aldous Huxley, Anne Olivier Bell provided quotations from Virginia Woolf's diaries while they were yet unpublished, and that most generous of London booksellers Anthony Rota actually lent me several early Sitwell rarities.

I must also thank the Sitwell family. It was clear from the beginning that none of us wished this to be an 'authorised' biography, but I received most generous help from Sir Sacheverell and Lady Sitwell, who invited me to Weston and answered my interminable questions with infinite courtesy and patience. So did their son, Reresby, and his wife Penelope, who let me stay at Renishaw while their second son and daughter-in-law, Francis and Susanna, were an invaluable source of information on Dame Edith. To Francis Sitwell, Dame Edith's literary executor, I am particularly indebted for permission to quote from her works. I am all too well aware that this is not necessarily the book that they would all have liked, but far from hindering me in any way, they have given me much encouragement and sympathy.

I need to say a word here of what I am very conscious of as a weakness of this book – the lack of attention which I seem to pay to Sir Sacheverell, compared with his two siblings. But biographers depend upon the material they get. Sir Osbert and Dame Edith loved the limelight and wrote interminable letters. Sir Sacheverell loved privacy and wrote books. I have respected his desire for privacy as much as possible and have left a fuller assessment of his considerable literary achievement to Denys Sutton, who will be dealing with it in a forthcoming book.

Frank Magro has kindly made available to me the papers that are still in his possession, including his own diary, and I am also indebted to him for permission to quote from Sir Osbert's published and unpublished writings.

The unpublished material I have drawn on comes from various sources, and I must thank the librarians and owners of Sitwell correspondence for the kindness they have shown me. Foremost among these is the great collection of twentieth century literary source material at the Humanities Research Center at the University of Texas at Austin. They own by far the greater proportion of material on Dame Edith and Sir Osbert – their manuscripts, and their correspondence, including Dame Edith's letters to Allen Tanner, and to her family and friends, and also Sir Osbert's entire correspondence to David Horner. Other material is held by the Manuscript Department of the British Museum and the New York Public Library.

Lincoln Kirstein allowed me to consult his correspondence from Sir Osbert; and Elizabeth Salter, Lorna Andrade, Ann Charlton, John Sparrow have also made available papers in their possession.

I must give heartfelt thanks to my publishers – particularly to Alan Maclean, who made available to me the long and detailed correspondence between the Sitwells and his firm and encouraged and sustained me through my work; to my agent, Deborah Rogers; my research assistant, Christine Zwart; and to the most patient and meticulous of editors, Richard Garnett, who has saved me from I shudder to think what literary pitfalls. Those into which I have still contrived to fall are entirely of my own devising.

CHAPTER ONE

The Youngest Baronet in England

1860–1886

ONE DAY in 1949, when he was still a young and tender playwright, an admiring Tennessee Williams met Edith Sitwell over tea. Six feet tall, with one of the most memorable profiles of the century, Edith was being hailed – and was behaving – as the greatest female poet of her time. That autumn her lecture tour on modern poetry had brought her standing ovations across the length and breadth of the United States. Her reading of *Façade*, to William Walton's music, at the New York Museum of Modern Art had been a triumph. She was very grand and very gothic, looking as if she regretted not having worn her crown that afternoon.

Was it true, asked Mr Williams, cautiously, that she really was of royal descent?

'I,' she replied, drawing herself up and staring at him as Elizabeth Tudor might well have stared at some upstart commoner who doubted her paternity, '*I* am a Plantagenet.'*

Her brother Osbert, who was with her in New York that afternoon, was also quietly enjoying great success. Since the publication of the first volumes of his long and long-awaited autobiography, *Left Hand, Right Hand!*, America had given him the status of an instant literary celebrity. He was a poet on his own account, a master of English prose in the grandest manner, and for more than thirty years had been a figure of influence and controversy, with close friends and dedicated enemies among the greatest names in European art and high society.

* The Plantagenet connection was in fact extremely distant. Edith was indeed descended from John of Gaunt and Catherine Swynford, as was Elizabeth Tudor. But Gaunt and his lady were a fertile couple with fertile progeny, and their accredited descendants are now legion.

Although less flamboyant than his sister, he too appeared to have a definite royal aura of his own. A quarter of a century before, Gertrude Stein described him being like the uncle of a king, and now at fifty-seven the description seemed to fit him more than ever. He occasionally referred to his own heavy Hanoverian features, and rarely refuted the discreet assumption that in some stylishly illicit way he was descended from King George IV, while in profile he uncannily resembled George III.

But the true fame of the Sitwells rested, as it always had, on the reputation of the literary trio which they originally formed together with their younger brother Sacheverell, and which had become a legendary part of the artistic history of the twenties. With total single-mindedness – and total loyalty to one another – they had created what was almost an artistic movement of their own, writing their poetry, their novels and their travel books, battling for the modern movement in the arts against their enemy, the philistine, scourging the middle classes, publicising one another and their followers and friends, until they had seemed to be the most effective British shock-troops of the *avant-garde*, who with their wit, their taste, their instinct for publicity, were rivalling the staider denizens of Bloomsbury as the artistic leaders of their time.

During the twenties, in her poem, 'Colonel Fantock', Edith already had proclaimed, 'We all have the remote air of a legend'. Now in late middle age it was as legendary creatures that all three of them appeared before their public, despite Sacheverell's retiring disposition and the intensely private life he led, and it was no coincidence that their best-known work was still *Façade*. Their battles and their stunts, the lives they led, their work itself, had always held an element of fantasy, which had built up across the years, and Edith's grand Elizabethan role was only one of the façades which she and Osbert consciously maintained to keep the inquisitive at bay.

To get behind these layers of protective myth, one must begin with Renishaw Hall in Derbyshire, the grim yet beautiful ancestral home of the Sitwells surrounded by some seven thousand acres of Sitwell land on the edge of Sheffield. For Renishaw provides the background to the legendary history of the Sitwells. It remains the symbol of their all-important ancestry, and it was once the setting for the series of excruciating melodramas and family disasters which plagued their youth, and helped to launch them as a unique trio pledged to work together in their life as dedicated artists.

The one survivor of these happenings, Sacheverell, has described Renishaw as a house of tragic memories, and it is with these memories of Renishaw, tragic and otherwise, that their story starts.

Osbert described the Derbyshire countryside he knew so well as:

that country of abrupt cliffs, wooded valleys, and hanging woods, with its dark-leaved old trees and gigantic plumes of smoke rearing themselves like serpents in the still and watery air . . . this land of two populations, the white, rustic, old, stationary population that labours by day in the open air of the countryside, and the shifting masses of black-faced miners who work, day and night, under the earth.

But the Sitwells were at Renishaw long before the coal-miners. The first mention of them is of one Simon Cytewel, who in the year 1301 was named as son and heir of a certain Walter de Boys, an adventurous spirit who had died that year on pilgrimage to the Holy Land.

During the centuries that followed the family developed in the usual way of minor country gentry. The Sitwells were sharp, circumspect and shrewd in the manner of the old-style Derbyshire squires. They prospered unspectacularly but steadily. They added to their lands by careful marriages – Robert Sytwell of Stavely Netherthorpe was already one of the six richest men in Derbyshire in 1588 – but it was industry and trade that were ultimately to give the Sitwells their first real taste of eminence and wealth and lay the foundations of the present house at Renishaw.

Early in the seventeenth century the enterprising royalist, George Sitwell, turned from agriculture to iron-founding, and with the new wealth from his iron-works at Eckington built the first gabled manor-house at nearby Renishaw. Sheffield was already growing as an industrial centre just six miles from George Sitwell's gates, and by the end of that century the Sitwells were enjoying the prosaic but valuable distinction of being the world's largest manufacturers of iron nails: one tenth of the country's iron trade passed through their extremely competent hands.

The peace and plenty of the eighteenth century brought polish, social opportunity and even a touch of learning to the family – as with similar rich country gentry throughout the land – and the squires of Renishaw, while still living off the profits of the iron trade, were starting to collect their books and pictures, improving their estates and adding to their lands.

Politically they were Whigs, but they kept clear of London politics – and clear too of the perilous extravagance that titles and fast living bring. Not until the turn of the century, and the succession of a new heir with the resounding name of Sitwell Sitwell, did the family succumb at last to such temptations. The new heir was not a Sitwell by birth. He was born Sitwell Hurt in 1769, son of one Francis Hurt by his first wife, Mary Warneford. In 1776, when Francis inherited Renishaw on the death of his maternal uncle, William Sitwell – a bachelor, and the last male heir of the old line of the family – he changed his name from Hurt to Sitwell, thus giving his young son his name of Sitwell Sitwell. (In the 1950s when

Osbert Sitwell was seeming over-sensitive, Evelyn Waugh suggested he should revert to the old family name and call himself Hurt Hurt.)

In fact that strange duplication of the Sitwell name was most appropriate, for it expressed the need the new heir evidently felt to enhance it. In 1791 his father had inherited a considerable fortune from his cousin Samuel Phipps, and soon after this had sold the iron-works as well. Therefore, by Sitwell Sitwell's time, the family was out of trade and had accrued estates of more than three thousand acres and a fortune of over half a million pounds. The change marked one more step in the family's ascent. The Whig connection had been conveniently dropped, and Sitwell Sitwell flourished as an exuberant, ambitious, horse-racing, fox-hunting Regency Tory.

He was an energetic, somewhat reckless gentleman, famous throughout Derbyshire for the splendour of his fighting-cocks and the breeding of his pack of harriers that once gave chase to an escaped tiger from a menagerie in Sheffield. His energy led him to a mania for building, and he was soon intent on changing old George Sitwell's simple manor-house into a mansion which could almost hold its own with any of the great houses in the district.

He had a certain flair for architecture. A splendid stable block was built to house his thoroughbreds. He personally designed a gothic temple and an archway on the drive. A fine new dining-room was added to the house, and then in 1808 both Renishaw and the Sitwells were embellished further. Sitwell Sitwell held a ball in honour of the Prince Regent. The house received a grand new ballroom; the host received a baronetcy. Then, three years later, at the age of forty-one, Sir Sitwell Sitwell died.

But this non-Sitwell, who had so enhanced the Sitwell name, had also left his heir, young George, the second baronet, a dangerous legacy in the expensive life-style he had taught him. George lived as extravagantly as his father, but less shrewdly. He married outside the neighbourhood – a Miss Tait from Edinburgh. The Taits of Harviestoun had long been a distinguished legal family in Scotland, but there was also quite a horde of dangerously dependent low-church Scottish relatives, most of them all too eager to enjoy whatever hospitality Renishaw could offer. With his wife's encouragement, Sir George shot eagerly in Scotland. He hunted no less passionately round Renishaw, and while the rents from agriculture were steadily declining in the aftermath of the Napoleonic wars, he fought an expensive local election in the Tory interest, permitted his solicitor to rob him of a good part of his fortune and, with unerring instinct, invested what remained in a bank in Sheffield. In 1848 the bank collapsed. The second baronet expired soon after.

Under his son, Sir Reresby Sitwell, the doom of the Sitwells seemed assured. Rents went on falling, debts went on mounting, and much of the estate and contents of the house were sold. The great house, gloomy now and stripped of all but a few remaining pictures and pieces of ancient furniture, was finally closed up. Sir Reresby came to live in London, and in 1862 he too expired. His title, and the empty house in Derbyshire, passed to his two-year-old son George.

In the year 1864 a small boy and his nurse were travelling in a train, and the inevitable kindly passenger sitting opposite inquired, after the manner of the day, 'And, little boy, pray tell me, who are you?'

'I,' came the answer, 'am George Sitwell, baronet. I am four years old and the youngest baronet in England.'

With this, his first recorded statement – so reminiscent in its way of many of his daughter Edith's public utterances – the fourth baronet began to make his presence felt. He evidently needed to, since the childhood of this orphaned, all but dispossessed new owner of Renishaw was far from happy, being oppressed by relatives, by straitened circumstances, and by insistent Evangelical religion.

His mother – formerly a Hely-Hutchinson – was as frugal and devout as earlier Sitwells had been profligate. She seems to have possessed a loving nature, a sound business head and a will of iron; while her firm religious faith was further buttressed by the beliefs of her dead husband's Tait relations, in particular by that distinguished Bishop Tait who in 1869 found fame on his appointment as Archbishop of Canterbury. It was in the household of his great-uncle, the Archbishop, at Lambeth Palace, that young Sir George inevitably spent a great part of his boyhood years.

The Lambeth of Archbishop Tait was a very different world from Renishaw, but the young baronet was soon exhibiting a true Derbyshireman's sturdiness of mind by totally rejecting almost everything the great man stood for. As George matured, he made it plain that prayer-meetings, psalms, the inspiration of the mission field, were not for him. As he told his son, Osbert, 'when I was three or four, I came to the conclusion that I was too young to understand such things properly, and so I had better reserve my judgement until I was old enough to form an opinion of my own'. And when he grew up he had other matters on his mind.

Even as a boy he showed that he was interested not in God but gardens, not in angels but in architecture, not in the litany but literature. He possessed taste, intelligence and will-power to an extraordinary degree; by his early teens he also had developed considerable self-sufficiency and a profound conviction of his own infallibility – as one would have required to become, as George did now, a practising agnostic in the heart of Lambeth Palace.

What is so interesting in the form that this reaction took is that it made George Sitwell proof against the passions of both sides of his family. Unlike the godly Taits, he had no time for good works or for piety; and, unlike the rakish Sitwells, he had no interest in sport or drink or gambling. In short, he represented a strange cancelling-out of his antagonistic ancestry. For he had absorbed the negative qualities of precisely those beliefs he seemed to be rejecting. When he threw out the low-church piety of the Taits, he carefully retained what Osbert called their curious, incongruous puritanism, which would always seek to cramp any side of life which seemed to him pleasant. When he abjured the rollicking manner of the Derbyshire squires, he clung to a passionate concern for families and land and ancestry. This would become his true religion.

To make these contradictions in his character the more extreme, it seemed that, whatever he believed, he believed in double measure. No one could have been more against the demon drink than this young atheistic puritan; or more obsessed with families and genealogy than this non-huntin'-shootin'-fishin' baronet; or more fearfully conventional than this uncomfortable young man who had no time for people outside his family, and who would tell his son that to have friends was 'such a mistake'.

None of this made poor George particularly endearing – even in the bloom of youth. Tall, pale, intense, with watery eyes and perfect, slightly frigid manners, he lived by his beliefs and seems to have had no friends outside the adoring circle of his family. The adjective his daughter used to describe him was 'insipid' – 'the insipidity being largely the result of blinking, with pale eyelids'. But this would seem inaccurate as well as most unkind. The word 'insipid' surely conveys a sense of weakness: Sir George was never weak.

Despite his manner and those watery eyes, there were two features which must certainly have rendered him attractive after a fashion. One was a certain subterranean but quite definite sense of fun, which took, as it was bound to do, a distinctive form. He had a quirky, esoteric sense of humour which in later life became so totally absorbed within his general eccentricity that it is often difficult to tell how much was humour and how much downright dottiness. But there was something undeniably appealing about someone who could spend his time at Eton inventing a toothbrush which played 'Annie Laurie', and a revolver for shooting wasps (predecessors, these, of countless subsequent inventions, including an elaborate lying-down desk, a car with a bed in it, and a rectangular synthetic egg for sportsmen, travellers and big-game hunters).

His other quality, which must have been attractive once he was in his early twenties, was his money. For, thanks to his mother's resolute frugality – and to the timely discovery of coal in Renishaw Park – the

Sitwells, by the mid-1870s, were finally beginning to enjoy their own small economic miracle. The seams of coal beneath their lands were rich, and Lady Sitwell, with the Lord behind her, was adroit enough to get extremely beneficial terms from the mining company who were so anxious to exploit them. The money was impeccably invested. Fresh land was bought. Before long, Renishaw was beginning to be restored to something like its former glory.

When young Sir George went up to Christ Church, Oxford, as a gentleman commoner in 1880, he was already wealthy in his own right. Thanks to a twist of fate (or, as his mother would have said, to the workings of the Lord), and thanks, too, to strict adherence to the puritan beliefs of the Archbishop, the Sitwells had managed to achieve what all ruined aristocrats dream of. They had recovered power and lands and wealth and lived down the disgrace of an improvident father. By the 1880s Lady Sitwell had bought a house, Wood End, in Scarborough, and for the summer months she and her son were living in Renishaw.

One cannot over-emphasise the importance of this dream come true on young George Sitwell. For this uncanny change of fortune really provides the key, not only to so much of his own subsequent unhappiness and eccentricity, but also to the storms and bitterness that raged within his family.

On the surface his good fortune was essentially a mid-Victorian morality tale straight from the improving pages of Samuel Smiles – frugality and faith and business-sense had brought this all-but-disinherited young orphan back from a straitened and unhappy childhood to the ancestral splendours of his true inheritance. But behind this lay a terrible uncertainty. How easy it would be to lose it all again! And it is now that one begins to see those two exaggerated sides of his strange personality starting to assert themselves in that weird clash of opposites and contradictions which never failed to exasperate his children, and fuelled the legends that grew up around him.

For on one side there were stacked the elements that made his life worth living – his baronetcy, his fine estate at Renishaw, his ancestry and above all his rapidly acquired fortune. All these assets gave him grounds for confidence and help to explain that curious high-handedness and wayward optimism which would express themselves in the great gardens he laid out at Renishaw, the castle that he bought on a sudden impulse in Italy, and the whole life of wilful self-indulgence that he conducted to the day he died. This side of him was typical of a certain type of cultivated, rich Victorian, whose wealth and birth and background gave them an almost godlike confidence to exercise their faintest whim with total selfishness, and an inspired disregard for everyday reality. This side of Sir George produced what Cyril Connolly described as the 'monolithic egotism' of that dominating, wilful figure who in extreme old age reminded his daughter of Cesare Borgia.

But there was really nothing carefree in these acts of high extravagance, and one can see that always lurking in the background lay Sir George's puritanism and the one extreme emotion which was to dog him until the day he died – straightforward fear, the dreadful fear of following his father and his grandfather and once more losing everything that made his life attractive – fortune, estates and even Renishaw. Despite his present wealth, his childhood had been lived under the shadow of his ruined family. The shadow never left him.

The tug of these antagonistic forces must have done much to make him the lonely and unpredictable character that he became. He was obsessively concerned with everything pertaining to his family, since through a kindly turn of fate his family and ancestry had together proved his true salvation: and the success and continuity of his family formed the one credible defence this fearful young agnostic could rely on for the future. There was no guarantee from God – since God did not exist – nor yet from friends and relatives: Sir George could not forget that one of the contributory causes of the original Sitwell ruin had been the horde of cronies and penurious Scottish kinsfolk who had battened on the Sitwell hospitality at Renishaw.

This profound distrust of other people seems to have ruled his life. Neither at Eton nor at Oxford did he indulge in the dangerous luxury of friendship – nor, needless to say, did he fall for the hideous temptations of the flesh. Instead he appears to have sailed through his time at Christ Church in beleaguered isolation from the rich, fornicating 'fast set' in his college who were still setting the fashionable tone of Oxford with their antics. By nature he was ambitious, but he wavered between two interests which he apparently felt proper to a gentleman – politics and literature.

Thanks to his mother's purchase of Wood End, her house in Scarborough, he had a stake now in that fashionable Yorkshire seaside town, and in 1882 he was adopted there as local candidate in the Conservative interest for the next election. Two years later, at the age of twenty-four, he was returned to Parliament. While he was there – from 1885 to 1886, and 1892 to 1895 – he was to be a conscientious – if not particularly inspired – backbencher, a good constituency member, and a sound orthodox Conservative.

During his career in Parliament he would raise his voice on four occasions: once in 1886 to ask the Secretary of State for War, Mr Campbell-Bannerman, 'whether it is the intention of the Government to build artillery barracks at Scarborough, and if so at what date?'; again in 1893 to question the Foreign Secretary about the imprisonment without trial of the Archbishops of Marash and Zeitoun; and for a third time that same year to ask the Postmaster-General when telegraphic communication was to be restored with the South Cliff, Scarborough. He also introduced a private member's bill, against the opposition of the Sabbatarians, intended to make it possible for sacred music to be performed on Sundays with money being taken at the door.

His literary activities proved more extensive. Aged twenty, he admitted that what he would really like to do with his life would be to live in Nuremberg, study genealogy and collect books. As this was patently impossible, he felt his name and talents would be best employed at the head of some suitable literary journal. As might have been predicted, there was a certain oddity about his choice.

In the 1880s the *Saturday Review* was publishing the early essays of that budding socialist George Bernard Shaw, while its editor was the unscrupulous rogue-elephant of late Victorian morality and letters, Frank Harris. Neither fact deterred Sir George from putting up sufficient money to acquire for himself the resounding title of the chairman of the *Saturday Review*. His literary credentials thus established, his literary career could start.

It was inevitable that it should reflect his obsession with his ancestry and his inherited position in the scheme of things. His passion, more than ever, was for genealogy; and it was now that he began those loving and laborious researches into the archives of his family and distant forebears which were to be his real life's work and which he thought would bring him fame and immortality.

His first book, *The Barons of Pulford*, was to appear in 1889, 'Printed and sold by Sir George Sitwell at his Press in Scarborough'. It was an indigestible collection of local charters and assize rolls dating from the eleventh and twelfth centuries, and the indomitable Sir George described his working methods in a splendid conclusion to the Preface: 'Copies of early charters and other MSS. in the British Museum have been mostly taken by myself, but in a few cases, by an expert, whose work I have corrected.'

Much of Sir George's life was to be spent correcting experts. According to his secretary, Francis Bamford, his chief criticism of his children's books was that he could have written them so much better himself. But although today *The Barons of Pulford* appears as a somewhat cranky essay on the distant origins of the Sitwell family, Sir George, with his passionate conviction of always knowing best, clearly envisaged it as something rather grandiose. As he went on to outline in his Author's Preface, he was intent on challenging 'the great authority of Dr Stubbs and Professor

Freeman, and to give a new and revolutionary theory on the subject of the English palatinates'.

Alas, his challenge went unheeded; and the great theories of both Stubbs and Freeman on the palatinates remain to this day undented by Sir George and his revolutionary theories. At the same time, his dearly purchased chairmanship of the *Saturday Review* was hardly more successful. Certainly no editor as irrepressible as the outrageous Harris would pay a great deal of attention to the chairmanly advice of anyone as strange and distant as this pale young baronet, however certain he might be that he knew best. And certainly Sir George appears to have made no appreciable impression on the contents or the running of the journal during his period as chairman. As might have been predicted, he soon tired of his powerless position.

He used to claim that his real reason for his resignation was that he discovered that Frank Harris had been using his editorial position as a way of blackmailing a 'celebrated public man'. Sir George was typically imprecise over the identity of this famous man, and what the story underlines is Sir George's ineffectualness outside the orbit of his family: one normally expects an editor accused of blackmail to depart and not the accusing chairman of the board.

During his early twenties, young Sir George made one other real attempt to make a name for himself. This time the effect was more in character with his professed agnosticism – and more successful. This was the celebrated showing up of a fake medium during a so-called seance at the British National Association of Spiritualists in 1880. Here he was very much the man of action, and rather brave.

Like all his children, Sir George was fascinated by the whole subject of psychic phenomena, but at this time, when spiritualism was in its heyday, his interest was emphatically that of a highly critical non-believer. For him spiritualism, with its extravagant claims of scientific proof of after-life, was one more instance of the great fraud of religious superstition from which he suffered since childhood.

He must have planned his escapade quite carefully, choosing a meeting held by a famous medium, whose speciality was to produce the ghostly presence of her spirit guide, 'Marie', into the darkened hall. Sir George became suspicious when he heard the rustling of the medium's stays, and when 'Marie' appeared he bounded from the audience and grasped the spirit by an all too solid arm. When the lights went up and the uproar had subsided, it was revealed that 'Marie' was none other than the agile lady suitably attired in a shift of ectoplasmic white.

This was an achievement of which Sir George was always proud. The

Evening Standard of the day, describing the affair, called it 'the heaviest blow which has yet been dealt to the ridiculous system of imposture known as Spiritualism', and until his death Sir George's *Who's Who* entry would include the curious battle honour – 'captured a spirit at the headquarters of the Spiritualists, London, 1880'.

But while this must have bolstered his profound conviction of invariably 'knowing best', Sir George's spirit-catching was an isolated happening. He travelled widely in these early years – chiefly through Italy, which he already loved, and where he had now begun to study both the gardens and the architecture. In Moscow he was curiously impressed by the coronation of the Tsar.

But, despite his travels, and despite those self-inflicted 'duties' which brought him regularly to London – attendance at the House and genealogical research in the British Museum Reading Room – Sir George's passionate concern lay more than ever with his family, his ancestors and his inheritance. His disappointing lack of real achievement outside his family turned him increasingly to Renishaw, where he was busy planning the new gardens in accordance with his own elaborate and expert principles. The house was now recovering from the years of closure and neglect, but to complete the true dynastic picture he envisaged for the house there was still one important item lacking there – a wife, to be the lady of the manor and to provide those heirs and progeny who alone could guarantee the future of the Sitwells and thus consolidate that whole dynastic world that meant so much to him.

He obviously planned this very carefully, as he planned everything in life; for the same extreme genealogical passion that had inspired *The Barons of Pulford* also decreed his strategy for marriage.

'Oh, Osbert,' exclaimed Margot Asquith about Sir George, 'what a look in the eye! *Cold* as *ice*!' And there was something glacial about his courtship. His was a curiously insulated nature. His fear of friendship, far from being a defence, had left him dangerously ignorant of human weaknesses – his own included – just as his puritanism had deprived him of quite elementary knowledge about women or the explosive workings of the human heart.

Instead, as he embarked on his romance, he was concerned with those essentially non-human rules that he employed in garden architecture and in tracing pedigrees – symmetry, a fine effect, and the baroque splendour of a noble name. Love he apparently ignored – or possibly just took for granted, as he would naturally assume his shrubs would grow when planted in some great design. But then, he always was convinced that he knew best, so why should love be any different?

The young lady finally selected by Sir George for his exercise in dynastic architecture was ravishingly beautiful – and barely seventeen – though it is doubtful whether either fact appealed to him as much as that she was the daughter of a famous peer and grand-daughter of a duke, and so would bring the maximum enhancement of the Sitwell name. But had he paused for just a moment to regard her as a human being rather than as an entry on a family tree, he must have noticed also that she was hardly suited as the life companion of a person of his tastes and temperament. Behind her great brown eyes and Burne-Jones profile, young Ida Denison possessed a nature that was uncontrolled, unformed and virtually uneducated. Indeed, this high-born schoolgirl lacked the faintest interest in those tastes and rarefied pursuits he lived for. Where he was serious, she was frivolous; where he was mean she was unthinkingly extravagant; where he appeared the coldest of cold fish, she was a passionate young woman with a fiery temperament.

But none of this apparently concerned Sir George, for Lady Ida Denison was the daughter of Lord Londesborough, her mother was a Somerset, a daughter of the Duke of Beaufort, and the Beauforts (as none knew better than Sir George) could claim a straight line of descent from the Plantagenets. 'Duke of Beaufort, Marquess and Earl of Worcester, Lord Botetourt, Lord Herbert of Herbert and Lord Herbert of Raglan, Chepstow and Gower, Hereditary Keeper of Raglan Castle . . .' – the titles borne by his bride's grandfather must have sounded like a fanfare of trumpets to George's medievally attuned ears, and as he studied her descent from John of Gaunt and Katherine Swynford (whose sister was the wife of Geoffrey Chaucer) he, more than anybody, would have realised just how colourful and splendid were the threads about to be interwoven with the more homespun Sitwell family tapestry. Even the fact that Ida's great-grandmother was a sister to the Duke of Wellington must have suggested here a useful strain of common sense and practicality to enhance the Sitwell genes. Further, Sir George would probably have hoped his unborn sons would naturally inherit that addiction to all forms of field sports which had distinguished the Somersets for the previous three centuries – activities at which he too would doubtless have excelled if only his life had not already been so crammed with time-consuming and important duties. With ancestors like these around, how could he pause to notice that by nature and by nurture Lady Ida was quite clearly born to bring about that very nemesis he feared?

Since the distinction of young Ida's blood was of such key importance to her future husband (and, for that matter, to their children also) we must pause briefly to examine who the Londesboroughs were. Although much grander and more obviously wealthy than the Sitwells, the actual Londesborough line was relatively *parvenu.* Four generations back the male line was founded by a self-made banker, one Joseph Denison of Leeds, who left a very humble family and trudged his way to London in the 1740s. By constant work and scraping he amassed an enormous fortune in the City, which was to form the basis of the Londesboroughs' wealth. By his second wife, Elizabeth (a Southwark hatter's daughter), he had three children. His only son and heir was William Joseph Denison who carried on the banking business, added to the fortune, sat as M.P. for Kingston upon Hull, and died childless in 1849. Joseph Denison also had two daughters: Maria, who married Sir Robert Lawley, later created Lord Wenlock, and Elizabeth, who in 1794 married Henry, third Baron Conyngham.

This was the lady who as Marchioness of Conyngham (her husband was created Marquess in 1816) achieved her special place in history as the last, the fattest and possibly the most rapacious mistress of King George IV, by ousting Lady Hertford from the royal favour early in 1820. Rowlandson certainly exaggerated the more grotesque features of these ageing lovers, but from contemporary accounts it seems that they were not a pretty pair. It was Lord Glenbervie who remarked that if they both loosened their stays their stomachs would have touched the floor. Despite – or possibly because of – this, the King was certainly in love with her and showered her with gifts until he died. She was considered to have acted with propriety by returning such of the Crown Jewels as he had given her, but the Conyngham family seat, Slains Castle, Co. Meath, is still full of furniture 'borrowed' from Windsor during this period. As the King was fifty-eight and her ladyship fifty-one when the liaison started, and the first Baron Londesborough was already fifteen, there is no question of Hanoverian blood having entered the family.

It was this son, Lord Albert (pronounced Awbert) Conyngham, who inherited old William Joseph's millions when the banker died in 1849. As something of a mark of gratitude, Lord Albert decided to assume the arms and name of Denison, and a year later he was raised to the peerage as the first Baron Londesborough. He died in 1860, leaving the Londesborough barony, two million pounds in stocks and shares, and a rent-roll of a hundred thousand pounds a year to his son Henry, then aged twenty-six. This was Ida's father – and George Sitwell's future father-in-law. He was to become Earl of Londesborough in 1887, and it was he who steered the Londesboroughs to their mid-Victorian zenith as a family.

As well as being so immensely rich, Lord Londesborough was a popular and well-known public figure, a 'heavy swell' in the language of the time, and custom-built for the fast set around the Prince of Wales. His wealth and style were matched by his extravagance. He was good-looking, dashing, an impressive dresser and a great driver of a four-in-hand. His tastes coincided with his Prince's – particularly where actresses and horses were concerned – and in 1870 he nearly gained a melancholy place in history when he invited Albert Edward as his weekend guest to Londesborough Lodge, near Scarborough. The Londesborough drainage failed to match the Londesborough hospitality; the royal party rapidly contracted typhoid (the hardier Londesboroughs, presumably immune to the local germs, did not); and Albert Edward very nearly died.

One might have thought that this impressive, wayward gentleman – 'the best whip in the country', as the *Tatler* called him – would have looked askance at young George Sitwell as a son-in-law. Their backgrounds and their tastes could hardly have been more different. Surely his lordship must have seen that this intellectual, pale young puritan was the last man on earth to make his daughter happy? But for him, as for Sir George, such commonplace emotions as mere happiness scarcely came into it. Ida was a daughter – and a second one at that. And then, George Sitwell was the local Member and a sound Conservative – whilst the Sitwells were a reasonable county family. On top of that, Sir George at twenty-six was quite rich: apart from Renishaw, his mineral rights and his mother's house at Scarborough, there was property in Rotherham, and another house, Long Itchington, near Rugby.

By now such mundane matters as mere wealth were starting to be important to the Londesboroughs, for his Lordship was not quite as rich as he had been. The racehorses, the actresses, the hordes of servants* had started to erode even his brilliant fortune, and as many found around this time, the Prince of Wales was an expensive friend to have. As a measure of miserable economy, which depressed him greatly, Londesborough would shortly move his London house from Berkeley Square to Grosvenor Square. With such a cold wind blowing, little Ida might be better off as Lady Sitwell than she could reasonably hope to be by sitting on at home and waiting for a better offer.

And so the marriage preparations went ahead, with a resolute Sir George, as Osbert would describe him, 'plotting of detail, of each move in the countless games in which he was engaged'. For the whole business of the marriage was essentially a matter to be settled between him and his intended father-in-law: his bride's feelings scarcely seemed to matter, although it does seem fairly clear that she despised him from the start.

* One of the Earl's more amiable, if hazardous, habits had been to issue cheque-books to his servants on which they could draw quite freely for expenses without having to bother him. He was also reputed to have lost £30,000 in backing the Covent Garden production of a spectacular French extravaganza called *Babil and Bijou*.

'A baronet,' she would mutter crushingly to her children in the years ahead, 'is the lowest thing on God's earth.' But the feelings of a seventeen-year-old girl could hardly be allowed to interfere with something as important as the linking of the Sitwells and the Londesboroughs; and in November 1886, in some magnificence, George Sitwell, baronet, and Ida Denison, the second daughter of Lord Londesborough, were made man and wife at St George's, Hanover Square. Early in January they returned to Scarborough from their honeymoon. Bells rang, bonfires blazed, and the Scarborough lifeboat crew drew their carriage in a loyal procession to their house.

But before this, just a few days after the wedding, the new Lady Sitwell had scampered from her marriage-bed in horror – and run home pathetically to mother. She was returned to her husband by her family (she had her duties to perform, and Dukes' grand-daughters don't behave like shop-girls). Nine months later her first child was born, and christened Edith Louisa.

CHAPTER TWO

'My Childhood was Hell!'

1887–1909

Sir George, as his son Osbert put it later, had won the booby-prize of fatherhood by the mere fact of having had a daughter. One can picture his exasperation. After such deep-laid preparations to enhance the Sitwell line and produce an heir whose quarterings he was uniquely fitted to appreciate, he had – Edith. Fate had no right to practise such an ill-judged trick upon him, and he was so put out he even got the date wrong when registering the birth and had to change it later.

Edith herself would never forgive her parents for her childhood, and all her life she hinted at some hideous event, some unforgivable betrayal, deep in the black recesses of her earliest years, accounting for the sense of tragedy and loss that echoed on throughout her life and poetry. But what exactly *had* occurred? She never really said, and the stray hints and anecdotes she did let drop about her early childhood are often suspect: certainly some of the stories that she told, particularly in old age, seem to have been fantasies, or complex embroideries on truth, which she constructed to express some deeper bitterness about her parents.

But there is no need for excessive subtlety to understand this bitterness, when all the circumstances point one way. Edith was obviously unloved at birth. Her parents' marriage was itself painfully devoid of love – an eccentric father lost in dreams of primogeniture, an emotional child-mother compelled to consummate a union with a man whom she despised – and Edith the all too speedy outcome of it all. Such a child could either heal the rift between her parents, or make it worse. Edith was destined from the start to do the latter.

Had she but been a boy, all would have been quite different, and Sir George would have prized his first-born as proof that his dynastic plans were working out. Had she been pretty, she might still have brought a flicker of affection to those cold, pale eyes. But Edith was not a pretty baby.

As for her eighteen-year-old mother, it is not strange that she resented and ignored this ill-favoured infant female who had been forced upon her by the insistence of her parents and the embraces of a man she did not love. Significantly, Edith's birth was actually brought on by a violent row between her mother and her grandmother, Lady Londesborough; and Lady Ida tried to regard her as George's child rather than her own. Later she would often tell her that, of course, Edith was not as well-born as she was: Edith's father was a mere baronet, hers was an earl. (Much later, Harold Acton heard Lady Ida say, 'of course the two boys get their brains from me, but where Edith gets hers from, I simply wouldn't know'.)

Such were the feelings of this dissatisfied and highly temperamental girl when faced with her own small female version of Sir George, nor was it in Ida's nature to keep her feelings to herself. Her rages were quite frightening, and tantrums and hysterics seem to have formed the most memorable part of the infant Edith's contact with her mother – interspersed with long periods when she was left entirely with the servants, and particularly with her beloved nurse-maid, Davis.

It was not in Edith's nature to accept this passively. Soon she was learning from her parents, responding to her mother's tantrums with even greater tantrums and to her father's icy self-possession with an extraordinary childish self-possession of her own. If they could be intolerable, so could she. 'You were an exceedingly violent child,' her mother told her later, and she herself admitted, in a moment of rare candour, that she 'must have been a most exasperating child'.

For, as the months passed, relations between Edith and her parents seem to have got worse rather than better. She told the poet Jack Lindsay that 'hardly a day passed in her childhood when her mother did not threaten to throw her out of the window or to commit suicide'. As for Sir George, his attitude towards his daughter seems to have crystallised now into cold distaste – particularly as her behaviour became as unfortunate as her appearance.

Edith's wretched very early childhood clearly damaged her: she was to bear the mental scars of an unwanted child for life. The underlying self-doubts, the desperate shyness, the sense of being really both undesired and unloved would never leave her – nor would her morbid touchiness and her profound mistrust of human beings for the pain they could inflict on her. When she wrote later of her infant self, that she possessed 'the eyes of someone who has witnessed and foretold all the tragedy of the world', she was not being entirely fanciful.

But at the same time the various indignities inflicted on this tiny child served to force on precocious qualities of survival: that quick-witted and determined battler of later years, so ready to administer a verbal 'slap' for any hint of an impertinence, made its appearance early on in life. Like Sir George, she was capable of extraordinary dignity and self-possession when the occasion called for it. Some of the earliest family tales about her show this clearly and reveal a formidable personality quite frightening in a three-year-old.

Osbert, for instance, has described the visit of an unsuspecting family friend called Rita to the Sitwells, and offers an intriguing glimpse, not just of Edith, but also of young Lady Ida's attitude towards this alarming child whom fate had wished on her:

> My mother rang the bell and told the footman to send for Davis and Miss Edith. She did not want to see her, for the child, whom sometimes she really hated, always filled her with fear: for the little creature possessed a striking personal dignity. . . .

Edith and her beloved Davis duly appeared, to be presented to the hapless Rita, who in her turn asked the solemn little girl what she would like to be when she grew up. She should have known better than engage in such banalities with the young, and Edith's reply must rank as the most effective put-down from the childhood tales of almost any writer.

'A genius!' said Edith.

A somewhat more endearing case of Edith's self-assertion was recorded at the age of four, this time with a doughtier opponent than Rita. Edith was seated in her perambulator when Mrs Patrick Campbell called, and had the temerity to call the infant Edith 'Baby'. Edith hit her.

What is most evident from all the stories of Edith's early childhood is that even then she plainly had two separate personalities. Left to herself – or with those she loved and felt at ease with – she was clearly sensitive and loving to a degree. Her Sitwell grandmother, by now a touching old lady living at Scarborough at her house, Hay Brow, and dedicating her remaining years and wealth to charity for the fallen women of the town* has described how Edith sang hymns and danced for her. Indeed, Lady Sitwell always referred to Edith, now a plump and round-faced little girl, as 'poor little E.'.

The servants loved her too – particularly her nurse, Davis, who seems to have given her the first regular affection that she ever knew. And then there is the story that she liked to tell of the first time she fell in love – again aged four – with a tame peacock in the grounds of Renishaw.

* In 1886 Lady Sitwell had made over Wood End, Scarborough, to Sir George – who immediately spent a fortune trying to reconstruct the gardens – and retired to Hay Brow, a small house on the outskirts of Scarborough with a large, untidy garden. She also bought and financed a house in the town as a refuge for reformed prostitutes.

Symbolically, the romance ended with the intrusion of her parents – this time her father – who decided from on high that it was time that Edith's favourite peacock had a mate; the peacock left her. 'It was', she wrote, 'my first experience of faithlessness.'

It was her parents and her parents' world that always seemed to be the threat she had to fight against; and when she fought, 'poor little E.' could become a mortal terror.

For Edith the crowning insult and indignity of childhood occurred on 6 December 1892 when she was five. Her brother Osbert was born, and the new baby's sex made him from the start the successful and beloved child that she had never been. Sir George had his longed-for heir at last, Ida had a son to lavish her extravagant affections on, and Edith, 'relegated to a second place in the nursery', could now be discreetly overlooked. She was, as she put it, 'in disgrace for being a female', and she retaliated in the only dignified way left to her. She ran away.

When finally brought back, she seems to have shown no jealousy of the new baby, which is strange. Perhaps this was because he did not threaten her directly, and his helplessness and smallness brought out the gentle side of her own nature.

For the new baby – he was christened Francis Osbert Sacheverell Sitwell – there could have been no greater contrast than between his start in life and Edith's five years earlier. The family nightmare seemed to have subsided, and Osbert was welcomed just as Edith had been rejected. Sir George was as thrilled as his uncomfortable nature let him be; for not only had the Sitwells an undoubted heir, but they were linked for ever now with the Beauforts, the Plantagenets and all those branches of the upper aristocracy that this fanatical hypergamist had worked and planned for in his marriage.

Even Osbert, who came to hate him in the years ahead, admits that Sir George did demonstrate much love towards him when he was a child – for, as he says, 'My father considered sons, especially elder sons when they were small, as a valuable extension of his personality'.

Ida's reaction was more straightforward. She adored him. At birth he was prettier than Edith, and from the start quite clearly had a more accommodating nature. Osbert, with his later passion for ascribing almost everything in life to biological inheritance, used to put down this difference to the fact that he had inherited more of the pacific Londesborough temperament, whereas his sister drew more heavily upon the Sitwell genes. More to the point, his birth was notably devoid of all those rages and resentments which had surrounded Edith's entry into the world, so that his sunny temperament continued to bring out the spoiling, doting side of Lady Ida, making his early childhood a period of unusual happiness. He adored his mother in return, and in his autobiography he has left a nostalgic picture of almost Proustian intensity, as he describes the way his adored young mother, 'so beautiful in her light-coloured evening dresses, pale pink or yellow', would come to the nursery at night to give him one last treasured and forbidden kiss. 'With her, as she walked out of the door, she took the last reminder of all the light that the day had held.' Like Proust's father, Sir George did not approve of such excessive demonstrations of affection, and did his best to stop them.

This love that both the Sitwell parents felt for their son must have brought them closer – if only temporarily – and there is a clear impression now of the gloom lifting slightly from the mock-gothic battlements of Renishaw. Ida was in her middle twenties – no longer the hysterical young girl thrust from the school-room into marriage. Sir George as well was finally recovering from the setbacks of those early years. Indeed, these years of Osbert's early childhood were probably the happiest his father ever knew. Not only had he sired an heir, but his estates were starting to reveal the imprint of his personality. The gardens at Renishaw were progressing. The flower-beds, which he abhorred, had given way to the elaborate formal garden which he had planned on the best Italian model; the statues were in place; and work would soon begin on the ornamental lake in the park. At Scarborough there were the gardens of Wood End to put in order, and his affairs were flourishing – even by *his* pessimistic standards – for during the mining recession of the nineties, when many pits went bankrupt, the Sitwell mining interests flourished, thanks to the richness of their coal deposits and the success of the companies who rented them. During this period one can catch a glimpse of the potential paradise behind the gloom and shadows that obscured so much of Edith's childhood, for the Sitwells really were extremely blessed by fortune. During the summer months they lived at Renishaw, and Renishaw *could* be a place of magic. There were the endless corridors and empty rooms for children to explore, the gardens to escape to, the Wilderness and the great park to wander in. There were the servants who befriended them and told them the legends and the folk-tales of this part of Derbyshire. The countryside appeared enchanted then, with distant hills and far-off palaces like Bolsover and Hardwick shimmering in the summer haze. When night came the enchantment entered Renishaw itself, with oil-light and candle-light, and ghosts and shadows. According to Sacheverell there were at least twelve ghosts resident at Renishaw about this time, including one who hung above the main front door and scared the postman, and another, the ghost of Henry, the last of the Sacheverells, who was drowned in the river, and who kissed pretty women as they slept.

Then in the winter there was Scarborough, with its electric light and children's parties, the excitement of the winter storms at sea, and gentle, devoted 'Granny Sitty' to be visited at Hay Brow.

This legendary side of the Sitwell children's childhood seems like a microcosm of the golden age of late Victorian upper-class country life, but it was flawed rather as Edith's own internal life was flawed. Sheffield lay just beyond the park: its furnaces lit up the night and covered Sir George's statuary with smuts, and the coal-mines were a constant presence just beyond the confines of the park.

More than ever after Osbert's birth, Edith's strange double-life continued. Thanks partly to her shyness, partly to her position as Sir George's daughter, she had scant contact still with other children of her age. 'I remember thinking, when I was a schoolgirl, how delightful it must be to enjoy oneself. I wished definitely to enjoy myself. I thought it would be rather fun.'

Instead, for the most part, she was kept apart from other children like some pale princess immured in her eccentric father's palace on the hill. She had her beautiful domain through which to wander, her dark house in the wood, the servants and her ancestral ghosts to keep her company. All this was the stuff of that 'air of a legend' she created for herself, where she could escape and where she felt happiest.

For she still needed to escape from the loveless, disapproving world of contact with her parents. The rumpuses continued, her self-confidence declined under what she called her mother's 'bullying', her nervousness grew worse.

'It was worse when I was good,' she said later. 'I was simply frozen with nerves'; and now began the cruellest form of self-doubt that can afflict a sensitive young girl. She became convinced that she was hideous.

One will never know how real the slights and cruelties inflicted by her parents were – and how much she imagined them – but certainly Sir George, with his frigid manner and clear-cut sense of what was beautiful, was not the man to bolster an unusual-looking daughter's pride in her appearance. Edith used to tell of how as a girl she was once sitting in a railway compartment, opposite her father who was reading *The Times*. Suddenly he lowered the paper, shuddered visibly, then raised *The Times* again so that he would not have to see her – all this without a word spoken.

In later life she made a lot of what she christened her 'Bastille', which she was forced to wear in early adolescence by her parents. This was a complicated patent orthopaedic brace and corset which was prescribed to 'correct' a distinct curvature of the spine and the weak ankles which were developing with puberty. She also had to wear some sort of facial brace at night to straighten out the hawk-like Sitwell nose. Sir George apparently had a thing about female noses.

All this rectifying hardware must have been degrading – and probably downright painful for the wearer. But, on the other hand, such treatment was very much in vogue at this period; leg-irons and various ingenious orthopaedic braces were recommended by advanced medical opinion as ways of combating actual physical deformity, and for Sir George and Lady Ida to have taken Edith to a London specialist may have been ill-advised but can hardly be used as proof of cruelty.

But for Edith her 'Bastille' became exactly that. It was a hated symbol of oppression, a degrading prison, just as the Paris Bastille had been. Had she believed her parents loved her it would have been quite different; as she could not believe this, the 'Bastille' was just one further part of some hideous conspiracy. In her imagination these uncomfortable clamps and braces, even the sinister Mr Stout the specialist – who 'looked like a statuette constructed of margarine, then frozen so stiff that no warmth, either from the outer world or human feeling, could begin to melt it' – epitomised the restricting pain and insult that the outside world, as well as her unloving parents, could impose upon her precious privacy.

Her sense of outrage now was so intense that she complained about it all her life, but her worst affliction was still the certainty that, in contrast with her beautiful young mother, her father found her hideous. During the bitter years of her extreme old age she would tell friends and interviewing journalists of how Sir George had 'mocked' her nose, her looks, the curvature of her back. She told the American novelist Glenway Wescott that as an adolescent girl she was taken walking by her mother through Eckington Woods and was made to wear a veil like some leper daughter, just in case some villager should see how hideous she was.

This must be Edith's rebellious memory of some hated head-dress which her mother made her wear, like any well-brought-up young lady of the time, when she appeared in public. But the mere fact that she could believe herself an outcast shows just how much she suffered from the feeling that she was cut off, 'different' from ordinary humanity, and that her father (and hence all other men) could hardly fail to find her physically repulsive too.

But she was learning new ways to defend herself. In public, and with her parents and her parents' friends, she had refined her methods of aggression since the time when she had actually hit Mrs Patrick Campbell. Even in childhood she was already using her extremely sharp intelligence – her power with words, and in particular her sense of wit and ridicule – to protect herself and be revenged on anyone who slighted her. She could be dangerous as well as dignified, as her parents' friend Mrs George Keppel evidently realised when she was brought in to read young Edith a lesson after an upset with her parents.

'George and Ida,' that wise and very worldly mistress of the King is said to have exclaimed, 'always remember that you never know what a young girl may become!'

But for the most part Edith's favoured strategy was not attack but to retreat – into the secure recesses of her rich imagination where she could build herself an ideal world like any miserable 'slum child' that she always said she envied. There was, however, one important difference between Edith's dream-world and that of a poorer child. Hers could be lovingly created from her own surroundings, which already seemed to have a fairy-tale quality of their own – the splendour of that countryside, the magic of her father's woods and gardens, the servants' legends and the household ghosts, and finally those kinsfolk of hers – distant Macbeth, the old Plantagenet kings of England, rapacious Lady Conyngham, Elizabeth I. Sir George with his ancestral passion could not forget such characters. Neither could Edith. 'To me, as a child, glory was everywhere,' she wrote.

She was an autodidact; her early education, in theory presided over by a succession of whey-faced governesses, was very largely of her own devising, and this too allowed her to pursue whatever interested her. What learning she acquired was very much her own. She always claimed that she could read and write at three (another trait that she apparently inherited from Sir George), and this retiring, difficult young girl 'with her hair of a shallow gold that was almost a polar green', and the white 'Plantagenet' face that Osbert would remember, was soon scribbling in notebooks, filling her boredom and her loneliness with nursery rhymes and fairy stories which were her first attempts at verse.

Early on she decided she must be a 'changeling', a child left by the fairies in exchange for a human baby, and it was as a self-created other-worldly being that she began writing of this legendary world that she inhabited.

Meanwhile, the whole family of 'George and Ida' was growing up in outwardly enviable, even opulent, surroundings. Victoria was in the last years of her reign and on the surface Sir George Sitwell and his family complacently reflected the splendours of their time – the regiments of gardeners at Renishaw, substantial mineral royalties, inherited possessions in four counties, a rented house in London to cover Sir George's periods in Parliament. (It was in this house, now demolished, at 3 Arlington

Street, just round the corner from the Ritz, that Osbert was born during a time when Parliament was sitting.)

The Sitwells outwardly conformed to the activities of their relatives and the surrounding gentry, and at first glance gave an impression of a normal, rich, conventional family. But this was very much a matter of appearances, especially with Sir George, who more than ever now began to live his own eccentric, isolated life. He never rode or hunted, never drank, and never really hobnobbed with the local gentry. Like Osbert's old nurse, they thought him 'too clever for a gentleman', a view that offered more than a grain of truth. Like Edith, he was bound up in his extraordinary dreams and multifarious ambitions, and had innumerable projects to keep him occupied – his new book on the fourteenth-century Sitwell relatives, the Hurts; his celebrated history of the two-pronged fork; a learned article on Eckington and the Sitwells.

But though he was such an odd, isolated man (and getting odder and more isolated with every year that passed) he always was to be concerned with building the façade of strict conventionality with which he could confront the world. It was as if those childhood fears of his were best assuaged by acting out the part he felt the Sitwells ought to play. He had his place in Parliament, his noble wife, and then on 15 November 1897 he even had a second son. He had him named in honour of the ancestral Sacheverell ghosts that he believed in.

Sacheverell was beautiful and much beloved – so much so that, even more than with Osbert, his early childhood pointed up the contrast with the way their parents treated Edith. As the baby of the family he enjoyed a charmed position and was quite free from shyness at this time. One of his early habits was to invite passing strangers in the streets of Scarborough home to meet his parents. Because of the difficulty of pronouncing his distinguished name, everybody called him 'Sachie'. He was intelligent and unusually receptive from the start; Sir George inevitably made him something of a favourite.

Snapshots show him as a pretty child with golden curls, standing in the Crescent at Scarborough or posed on a rocking-horse beside his mother. Where Edith had a peacock as a pet, he was given a pet lamb.

This sunny early childhood he led makes his own accounts of early youth appear to contradict Edith's black reminiscences of Renishaw. He writes that his earliest memories of his mother are of how easily she laughed, and talks of her gaiety and powers of mimicry. She was 'tall and thin, and dark, and beautiful, with straight Grecian nose, small mouth, dark brown eyes, and little shell-like ear, set close to her head'.

One gets a clear impression of the picture that Sir George desired to offer to the world, of himself in the bosom of his united family, from the group portrait he commissioned in 1900 from the American portrait painter John Singer Sargent. Today the painting hangs at Renishaw, exactly where Sir George first placed it, in pride of place over the Renishaw Commode. Sir George is painted very grand and confident indeed in riding gear, while Lady Ida, in a ball-gown and a hat, arranges anemones in a silver bowl. Edith wears a scarlet dress, and the two small boys are playing in the foreground with their mother's pug.

Sir George – still knowing best – apparently decreed how his family would pose and what they should wear, and was by all accounts delighted with the result. One can see why. For there in that crowded, slightly vulgar canvas the Sitwell family appears like the epitome of some haughty, fashionably rich, Edwardian county family, and it must have reassured Sir George that Sargent could depict them like the rest of those Edwardian grandees who patronised his work, and who, according to Osbert's theory, loved him because with all his merits he showed them to be rich.

But as one looks a little closer at the painting it is by no means as straightforward or as confident as first appears; not for nothing was Sargent the sly deflater of these Edwardians he painted.

In the first place, as Edith pointed out, few of the details that Sir George insisted on really make much sense. He never rode a horse, and Lady Ida never arranged flowers, least of all when in a ball-dress with a hat. The picture is, in short, a not particularly convincing tableau, complete with fancy dress, of the way Sir George wanted the world outside to see him.

But there is something even more revealing in the painting. The more one looks at it, the more one sees that it is dominated not, as one might imagine, by Sir George, nor by his handsome, flower-arranging spouse, but by the most unexpected member of the family – despised and rejected Edith. According to Osbert's account of the painting of the picture during the sittings, Sir George callously drew Sargent's attention to the fact that Edith's nose was slightly crooked 'and hoped that he would emphasise this flaw'. ('This request much incensed Sargent, obviously a very kind and considerate man; and he showed plainly that he regarded this as no way in which to speak of her personal aspect in front of a very shy and supersensitive child of eleven.') Edith was most upset, and to console her Sargent made a point of straightening her nose upon the canvas, and giving Sir George's nose a decided skew. He did more than this, and with extraordinary perception made the tall girl with the scarlet dress (who was actually thirteen) the centre of the painting. Sargent was the first real artist she had met, and she got on well with him. There is certainly no sign of shyness in the way he painted her, nor of her conscious misery about her looks. Instead she stares unblinkingly towards the painter, oblivious alike of her weirdly garbed parents by her side and of the curiously contrived surroundings of the studio. Here one is looking at a powerful young presence, already very much aware of who she is and how she wishes to present herself to life. It was a useful lesson on the power of artists and how they can transform life with their own vision of reality.

As for the two younger children, they are still loved and safe within their idyllic childhood. For Sacheverell, only two, the idyll will continue for some time to come. But for young master Osbert the first of life's unpleasant shocks is just round the corner.

Osbert's childhood was to be a lesson in the dangers of trusting overmuch in those one loves. The first to teach him this was Lady Ida. He began by worshipping her extravagantly – this much is evident, both from the first childish letters that he wrote her, and from the way he described her in his autobiography. But Lady Ida was a dangerous person for a small boy to worship. He wrote of her 'strange temperament, her kindness, indulgence, and furious, sudden rages'. 'The first time she lost her temper with me . . .' he wrote later, 'the whole world temporarily assumed a more tragic tone.' Before long he too, favourite though he may have been, was, like Edith, feeling the rough edge of Lady Ida's rages. Unhappy and unbalanced lady that she was, she would spoil him one moment, then round upon him angrily the next. She must have taught him early on how love can bring betrayal.

As for Sir George, young Osbert seems to have admired rather than loved him, since he was the most intelligent person he knew, but otherwise Osbert, like Edith, must have led a curiously cut-off existence as a child. Like Edith, he was left a lot with servants, and like her made little contact with other children. One of his major dreads about the return from Renishaw to Scarborough each autumn was having to cope with children of his own age when the parties started. Osbert seemed to have inherited his father's solitary temperament, his closeness with his family, his strange uneasiness with his contemporaries. For all their wealth and ancestry, the Sitwells were a family apart.

It was this curious apartness that must have made the first outside event of Osbert's life such an uncomfortable shock. Aged nine, he was sent to school. Until now he had been taught by a governess, Miss King-Church, and it is not surprising that this sensitive and molly-coddled mother's boy should have had a particularly hard time of it in the Scarborough day school which Sir George decided would be good for him. Writing about the whole experience forty years later in his autobiography, Osbert composed a fearful picture of the bullying that he endured from the 'jostling, screaming mob' of middle-class boys and local tradesmen's sons, who blacked his unsuspecting eyes and mocked him on account of his father's baronetcy. This was, he wrote, his 'first encounter with victorious British Democracy', and an important early lesson in the contemporary class war. It was here, he suggests, that he received his first crucial impressions of the dismal nature of democracy, and the distressing, hearty, boring features of the ascendant middle classes (or, as he liked to call them later, 'burgess orders'). The school itself, he wrote, was nothing more than a 'miniature model prison, essentially middle-class, with all the middle-class snobbery and love of averageness'.

This is as may be, but the fact is that he stayed at the Scarborough school for no more than a few months before going on to Ludgrove School, New Barnet, in the beginning of 1903. Ludgrove was a preparatory boarding school for Eton, and the earliest letters he sent home from here show no particular aversion to the school itself, the other boys, or even to those hearty games he so objected to in later life. On 14 January 1903 he was writing to his mother of how he had 'kicked about my football in my football boots', and a week later was imploring her, 'Do come and see me *soon*. It is rather fun here. Four new boys came here as well as myself.' As for the other boys, he specially asked his mother to tell Edith that 'the tiny tiresome toads aren't such beastly bothering bores'.

Apart from this, most of the references in these early letters home from Ludgrove are to his family, for he was evidently missing them. 'I thought you might like a present of a tame rat or mouse so I am going to send you one,' he told Lady Ida cheerfully, knowing how terrified she was of rodents. 'If Granny Sitty sends another parcel, please no figs as they are not as good as they might have been,' he wrote peremptorily in another letter, and there was always a special note for Edith, or 'Dish' as she was known in the family.

But in these letters that he wrote from school early in 1903 his only real source of worry seems to be Sir George. 'I wrote to Father yesterday,' he told Lady Ida on 14 January, and then on the 25th, 'I am sorry to say I am at the bottom of the school, I am afraid Father will be so disapoitted about it.'

Then on 1 February he was again inquiring of Lady Ida, 'What about writing to Father? I don't know his address what do you think I better do.' A week later he was again asking for news about Sir George, and then at the end of March he was asking gloomily, 'Have you heard [from] Father lately? He has not written to me for a long time.'

At the beginning of his remarkable essay on *The Making of Gardens*, Sir George explains that it was 'written during a period of broken health, when slowly recovering from the effects of overwork'. Outrageous hypochondriac that he was by now (the book appeared in 1909) – and passionate proclaimer of his own self-sacrifice – this reads like one of those continual plaints about his health by which he tried to make his family – and in this case the reader too – feel as sorry for him as he felt for himself.

But now behind this all too typical Georgian moan there lay for once a certain residue of truth. The plea of overwork, of course, is nonsense. No one was more devotedly self-indulgent than Sir George. Whatever overwork he suffered from was self-inflicted, for Sir George only worked – or overworked – at subjects that appealed to him, and which he painfully pursued with meticulous obsession.

But since Sargent had depicted that curious façade of the Sitwell family, cracks had started to appear. Sir George's state of irrational anxiety – his 'anguish', as Edith called it – which he had grown up with had got slowly worse. She has described his manic walking round and round the house, his pacing of the corridors at night. He seems to have become more solitary than ever. The tensions in the family increased. And then, during the Christmas holidays of 1902, Osbert overheard the governess and tutor at Wood End 'remarking how extraordinary it was to have a nervous breakdown just because you could not get your way in everything'. Only later did he realise that they were referring to Sir George.

But what did this strange remark portend? And what did George Sitwell want that caused him so much nervous damage?

The root of all Sir George's troubles lay in his childhood: his orphaned state, mistrust of all outsiders, concern at losing once again the family possessions – these were what caused his 'anguish'. But his *Angst* had recently been fed from other sources. One was apparently a sense of failure as a politician, the feeling that, despite his own acute awareness of his talents and of knowing so much better than the party leaders, he had not made the impact at Westminster that he should have done. Then he was also having trouble with his literary work. His essay on the origins of the gentleman – a topic dear to him from which he was hoping for considerable *réclame*, was proving harder to elucidate than he imagined.

But his profoundest source of misery was his marriage. He was now being forced to pay – in every sense – for that glittering alliance he had so carefully constructed with the Londesboroughs. That vulgar element he had so carefully ignored in all his calculations – simple human nature – had finally rebelled. His wife was ruining him. With every year that passed, Ida's extravagance was growing worse.

Wild spending – and reckless generosity – had become a habit. It was a habit very much in keeping with her romantic nature. It also presumably alleviated the boredom of her marriage, and kept the horde of cheerful, grasping friends that she attracted loyal and happy. But it was also, whether she intended this or not, the surest way she had of getting her revenge upon her husband. For George was not merely mean. He had the true Edwardian obsession over money. It was a matter for intense emotion, since its possession meant security and power, its loss catastrophe. Far rather she had been unfaithful than she had squandered this precious substance of his being.

'But, Ida,' he exclaimed despairingly, 'one simply cannot overspend one's stipend *every* year!'

'But why not, George!' she artlessly replied, remembering that cool two million which Lord Londesborough had already nearly finished with so little effort. 'Father *always* does.'

Here the real aristocrat was speaking. Against an attitude like this Sir George, with his old-maidish, middle-class concern with petty-fogging, everyday economy, was really powerless. Also, she bought and entertained on credit, something which, as daughter of an earl and wife of a well-known baronet, she could always do. So George had no earthly way of stopping her. It must have been slow torture for him, especially when she started gambling too. That dread reality of ruin which had so haunted him in childhood had suddenly returned to Renishaw. The family façade which Sargent painted was collapsing and Sir George began to do the same.

The form his breakdown took appears as an attempt to opt out from a situation he could not control. He simply fled – first from his family to his various houses where he would stay a night and then, with manic restlessness, move on to the next. Then, just before Christmas of 1902, in near despair, he finally appealed for help to his Scarborough doctor, who recommended what he really needed – change, instant relief from his responsibilities, or, as he put it, 'convalescence'. Medical advice was always sacrosanct to Sir George. If his health really did depend on his departure, Renishaw, Wood End, the family would have to take care of themselves, and Ida must be left to her own extravagant devices. It was his duty to himself to go where he was happiest. He chose Italy. He needed somebody to keep him company, and since he had no friends the doctor finally agreed to go with him himself. Once in Italy, and away from Lady Ida, Sir George's recovery was spectacular: in no time at all he had started travelling and studying the Renaissance gardens of the whole peninsula (something he had longed to do systematically for years). He had escaped at last – not just from worry over Lady Ida but also from himself.

His 'breakdown' was the perfect pretext his anxious nature needed to excuse a life of self-indulgence, and luckily his wealth was more than adequate to bear the strain. The anguished, self-denying, family-obsessed Dr Jekyll in Sir George had at last discovered how he could let himself become a garden-visiting, architecture-studying, Italy-loving Mr Hyde. This was a discovery of key importance, for, as Osbert writes, Sir George's trips to Italy were to form the mainspring of his life during the next ten years.

Henceforth he would always be 'the dear invalid', as his devoted mother, old Lady Sitwell, used to call him, and it was as 'the dear invalid' that he would always treat himself. He would have been most hurt at the suggestion that his breakdown had been cured. In one form or another it went on for years, for it was as a very lively invalid indeed that his relentless egomania had found the role it needed. He had a clear and overriding duty to do only those things which would abet his all-important peace of mind. Travel was beneficial – so whatever the rigours and expense and inconvenience to others, travel he must. The worries of his family were bad for him, and so they, equally, must be avoided.

Sir George had made his own pre-Laingian discovery of the value of neurosis as both refuge and defence against a situation which he deemed impossible. Soon he was doing something more, and using it as means of actual self-expression. The part of the 'dear invalid' was to become his new façade and fantasy, and with its help he now proceeded to construct the final version of the fully fledged eccentric that he finally became; the drawbridge to his ivory tower came clanking up.

Not just his family but the servants too were drawn into this elaborate and deadly serious charade. 'Poor Sir George, he really is an hero for his bed,' his butler, the enormous Henry Moat, remarked. 'I have known him often being *tired* of laying in bed, get up to rest, and after he had rested get back into bed like a martyr.' He took his meals increasingly alone: people, especially his family, upset him and wrought chaos with his gastric juices. He announced that he was never, under any circumstances, to be contradicted: the results on his nervous system would be too horrible to contemplate.

As a traveller in Italy Sir George was original as well as indefatigable. At the beginning of this century, when the Italianism of the Anglo-Saxons was still ruled by shades of Ruskin and the Brownings, most cultured English tourists would see Venice, visit Rome and then end up in Florence where, in that grey-stone city with its English tea-rooms, they felt instantly at home. Not so Sir George.

Just as he once reacted against the conventional Christianity of his uncle the Archbishop, so he was now intent on showing his firm independence against the accepted cultural pieties of his age. Armed with his air-cushion, hamper (with the ubiquitous cold chicken for his lunch), mosquito-net, spyglass, notebooks, and large green umbrella to protect his pale complexion from the sun, the dear invalid was soon off on safari to the south. The doctor had returned to Scarborough: his place was taken permanently now by Henry Moat, and Henry Moat with his sturdy Yorkshire common sense, his humour and his loyalty towards 'The Great White Chief' was all he needed. For a man who thought that he was suffering from such appalling health, Sir George was a resilient traveller, braving the food, the heat, the hardships of the road, and even the threat of bandits to see what he had set his heart on. (Osbert has given a memorable picture of him sitting oblivious of his surroundings in a low, bug-infested doss-house in the south, watched by an audience of fascinated villagers as Henry Moat solemnly set up his elaborate mosquito-net above his bed.)

At a time when Baedeker was advising that 'there is little of interest to attract the traveller South of Rome', Sir George was carefully observing Naples, and then passing on to those southern forgotten cities – Lecce and Sicilian Noto – which were such masterpieces of the unfashionable baroque. Ever the original, Sir George appreciated them. He also appreciated the gardens, the deserted palaces, and, of course, the solitude. Culturally and personally he was as he liked to be – alone.

In fact, though, he *did* sometimes take his family abroad with him on his travels – when it suited him, or when there was no alternative. He and Lady Ida had always been keen travellers. In her mid-forties Edith would write to a friend, 'I can still smell the mimosa trees and fields of narcissi at Cannes which I smelt when I was four.' According to Osbert, her earliest memories were of this holiday. She recalled 'walking along, under a feathery avenue of palm-trees and gazing up at a very blue sky. The place was deserted, when suddenly a carriage appeared, with a little old woman rolling about in it. "The Queen," her nurse said: and my sister dropped, as best she could, a solitary curtsey, to which Victoria ceremoniously responded with an inclination of her hand.' Then at the beginning of 1902 – the year of the fateful breakdown – Sir George had already taken her and Lady Ida off to Naples for a holiday. Among Edith's unpublished papers there is a brief early impression of the trip. She doesn't seem to have enjoyed herself. Indeed, Italy as a whole never came to mean as much to her as it did to either of her brothers. She had no real taste for pictures, no interest in architecture, no feeling when it came to it for things Italian. Instead one gets a firm impression that these fraught and serious expeditions with her parents helped turn her against the things her father valued for the remainder of her life.

On this occasion she describes how she and Lady Ida trudged their way miserably across the lava at the summit of Vesuvius, and how Sir George embarrassed her by typically refusing to pay the porters until they had brought them back in safety to the railway station.

Whereas her brothers – Sacheverell in particular – would remember Naples for its churches, baroque palaces and memories of King Bomba, Edith was struck by simpler happenings – the wretched children in the street 'looking for insects in each others' hair and laughing in the sun', the bed-bugs at the hotel and the chaos at the fish-market on New Year's Day. At fifteen with her button boots and hair in pigtails, Edith quite clearly failed to enjoy the place. So, one gathers, did her mother.

By a coincidence, that same year ten-year-old Osbert also had his first experience of Italy. *His* reaction was quite different. His period at the hated Scarborough day school ended when he had pleurisy and the Sitwells' doctor, young Dr Dawson,* had recommended somewhere warm for him to convalesce. He was taken to San Remo on the Italian Riviera, and his response was everything his father could have wanted. As he wrote later, 'the very first morning that I woke on Italian soil, I realised that Italy was my second country, the complement and perfect contrast to my own'. Before long, and thanks entirely to Sir George, Osbert would have his own extraordinary stake in this land of his adoption.

Half way between Florence and the mysterious Etruscan city of Volterra, the ancient road, the Volterrana, switchbacks its way across the loveliest countryside in Italy. During the early autumn of 1909 Sir George was driving through it in a motor-car, when suddenly the car broke down.

Beside the roadway there were two antique stone lions guarding the overgrown entrance to a path that led away between the olive trees. Sir George inquired where it went. To a castle, he was told. Its name was Montegufoni – the hill of the screech-owls. If the *barone* was interested he might care to see it, as it was for sale. Sir George *was* interested, and when he saw it he was most impressed; so much so that he decided he would buy it there and then.

To this day local tradition has it that the breakdown of Sir George Sitwell's car was not the accident it seemed: and as one of the Italians he was driving with finally received the customary five per cent which an intermediary in Italy demands at the conclusion of a deal, it would appear that the rich baronet was being taken for more than one sort of ride when his car came jolting down the hill from Montagnana. But even so, who-

* Dr Dawson of Scarborough, partly through Lord Londesborough's patronage, became one of the most successful physicians of his age, and as Lord Dawson of Penn ended his career as George VI's doctor.

ever stage-managed the affair deserved his money, for he showed extraordinary understanding of this strange *Inglese* with the cold eyes and the frosty manner. Only a rich man of supreme self-confidence and more than a touch of Anglo-Saxon madness could ever have considered taking on that veritable herd of trumpeting white elephants which made up the Castello di Montegufoni.

It was almost twice the size of Renishaw – for in the manner of great medieval Tuscan castles it had proliferated endlessly across the centuries. It was originally the seat of the great Acciaiuoli family, steel merchants (as their name implies), bankers, and allies of the early Medici, whose fortunes suffered from lending money to Edward III of England. It had five separate courtyards, three massive terraces, a chapel, an impressive *limonaia* (where more than two hundred ancient lemon trees were placed for protection from the frost each winter), an extraordinary grotto, over a hundred rooms (there were originally seven separate houses there until they were knocked into one in the fifteenth century), a hall somewhat larger than the great hall at Hampton Court, and a central bell-tower modelled on the campanile of the Palazzo Vecchio in Florence.*

It was not merely very large. It was also in a state of near collapse. As Sir George wrote cheerfully to Osbert, 'the drains can't be wrong as there aren't any.' And during the previous twenty years or so, during which it had been left officially unoccupied, squatters had moved in. Sir George described the place as 'a rookery of poor families who lived by agriculture and straw-plaiting, there being 297 inhabitants of the castle all told'. They had their cattle in the great hall, their goats in the *salone*, their children everywhere.

Far from discouraging Sir George, all this appears to have excited him. Like the great gardener whose work he so despised, he could detect here 'capabilities of improvement', a challenge, a life's work. Also, there was something in the atmosphere of this great, medieval ruin of a house which must have spoken to the author of *The Barons of Pulford* and *The Social History of Medieval Eckington*. The house was haunted. It had an air of authentic gothic gloom, strangely reminiscent of the gloom of Renishaw, and in an uncanny way the house was ready to become a strange extension of Sir George's own medieval mind – towered, turreted and fiercely walled, with secret passages, resplendent courtyards, intricately linking sets of extraordinary rooms. 'The air of forlorn grandeur is very attractive,' he wrote, 'and this I hope to keep.'

* This tower was built by Donato Acciaiuoli in 1386. Legend has it that he had made a solemn vow to live always underneath the shadow of the great Florentine *campanile*. Donato's brother, Neri, inherited the family Duchy of Corinth, and in 1392 became the Duke of Athens, a title which the Acciaiuoli held until 1461, when the Turks finally over-ran Athens, and the last Acciaiuoli Duke disappeared, according to Gibbon, 'to end his days in an ignominious position in a Turkish harem'.

The castle also had those Acciaiuoli ancestors. One was a cardinal, others had been Dukes of Athens – all could in a sense be annexed to the Sitwell name by right of purchase, just as the Plantagenets had been annexed by marriage.*

This must have been the baronet's excuse for what would have seemed for anybody else an egregious folly and extravagance – especially as he claimed to be near bankruptcy with Lady Ida's debts about this time. The actual property cost four thousand pounds – not in itself a great sum, even in those days – but with the certainty of thousands more requiring to be spent to make it habitable. None of this appeared to worry him, since he managed to regard the purchase as a serious investment and enhancement of the Sitwell patrimony. As if to emphasise this aspect of his acquisition, he had it registered in the name of his son and heir, young Osbert, who thus at seventeen suddenly became, in strict legality if not in actual fact, lord of a thirteenth-century castle in the Tuscan hills.

* Sir George was delighted when he discovered that the Acciaiuoli were, like the seventeenth-century Sitwells, connected with the iron trade – and still more delighted when he learned that they shared a similar heraldic device: lions rampant.

CHAPTER THREE

A Closed Corporation

1903–1912

As SOLUTIONS to that Chinese puzzle of his life, Sir George's breakdown, his long flights to Italy and his adoption of the role of hypochondriac eccentric were all ingenious – but only partially successful. They may have saved his sanity, but they did nothing to improve the situation in his family, relations with his wife, or his financial worries back in England. Nor could they possibly avert the fate that now began to stalk the Sitwells. Rather the reverse.

While he was still actually present in the family – as he had been until the moment of his breakdown – he had maintained a semblance, even if that was all it was, of order and conventional control at Renishaw. The children had all been comparatively young, his rents and mineral royalties had guaranteed continuing prosperity and paid for Lady Ida's debts, the whole conventional façade of Renishaw had been maintained.

But with his breakdown and its aftermath, whatever balance and restraint the family had once enjoyed were lost, particularly where Lady Ida was concerned. While he was off in Italy, her ladyship was unrestrained and plainly very bored. As might have been expected, she reacted in the way that she knew best – with jollier parties, longer trips to London, more acquaintances and more and more extravagance. Left on her own at Renishaw or Scarborough now for months on end, she played the profligate Edwardian lady to the hilt. She rarely rose till noon – until then she stayed in bed reading the same light novel or newspaper over and over again to while away 'the nullity of her days', as Edith put it.

Her afternoons and evenings would be spent with the friends she found who were always willing to consume her food and drink and keep her happy. She kept *them* happy simply by spending: dresses for herself and dresses for the ladies, presents for everyone, and loans for anyone in trouble. She was by nature as generous as Sir George was mean. If anyone admired a ring, a brooch, a piece of furniture, she gave it instantly. She was high-spirited, gregarious and could be wonderfully amusing as a mimic. She gambled more and more. She shared her father's easy-going attitude to cheque-books. She started drinking. Then, when her money started to run dry, she borrowed.

Osbert was soon to find himself drawn into the troubles that were brewing now between his parents. In Italy Sir George was free from worry, but whenever he returned his *Angst* immediately grew worse. Since his breakdown his attempt to cut himself off from the world obliterated any lingering sense of proportion to his fears. He knew how Ida was over-spending, but there seemed nothing he could do, and in this hopeless situation all his family must have seemed a source of bitterness and worry.

Edith records that since her twelfth birthday – when he gave her a tortoiseshell cigarette-case – relations with her father had actually improved. He had even made suggestions how she might enjoy herself, such as by playing the cello or exercising on the parallel bars. 'Nothin' a young man likes', he would murmur sagely, 'so much as a girl who's good at the parallel bars.' But in the aftermath of his breakdown he turned against her once again – and soon began to do the same with Osbert.

Ostensibly the cause of this was Osbert's disappointing work at school. He had remained around the bottom of the form throughout his time at Ludgrove and when he moved on to Eton in the autumn of 1905 there was no improvement – rather the reverse, for, as he wrote later, he soon discovered that 'Public schools are to private schools as lunatic asylums to mental homes: larger and less comfortable'.

Eton, that 'wasteful, antiquated, rather beautiful machine', bored and depressed him, and there is no cause to doubt his celebrated private verdict on the place – 'I liked Eton, except in the following respects: for work and games, for boys and masters.'

But why did he find the school so totally unsympathetic? There was none of that 'middle-class snobbery and love of averageness' which had depressed him so at Scarborough: he was free from the burgess orders, and nobody would bully him for being the offspring of a baronet. And yet he remained a lonely, isolated boy, more than ever an outsider here among his own class, just as Sir George had been. He was miserable, learned little, and did badly, so badly that in later life he never showed the faintest sign of interest in or affection for his school. When asked to contribute as an Old Etonian to a fund for some new playing-fields, he did send Provost James a cheque, on condition that the new fields would not be used for compulsory games. His cheque was returned.

The reason for his misery at Eton was that he clearly found the upper classes *en masse* as depressing as his hated middle classes, and what he most objected to at Eton was the conformism demanded of its members. That 'antiquated, rather beautiful machine' had, as he saw it, one remorseless aim:

> so many boys turned out from it each year, standardised and to type, amused by the same jokes, in arms at the same challenges, with no intellectual zeal that could be knocked – or rather, bored – out of them, with no ideas, and few ideals save a valuable if restricted sense of duty, together with a natural love of pleasure, which they have never been taught to differentiate from happiness.

The English public schools, he insisted later, were simply 'large-scale factories of that boring and emasculate commodity, the English Gentleman'; and in adolescence Osbert was far more like Sir George than he admitted, sensitive, solitary and bookish. If Sir George was 'too clever for a gentleman', Osbert was too much of an individual to be a good Etonian.

He refused to be confirmed. He refused, as much as it was in his power, to join in any sort of games. As for learning, it was 'at Eton, that I learnt the first lesson requisite for a comfortable life in this age . . . to regard the possession of any intelligence as a guilty secret between Man and his Maker'.

But the school did leave its mark upon him. As he wrote about the hero of his novel *The Man Who Lost Himself* (another old Etonian), when the machinery of Eton failed to 'standardise' a boy, its effect was 'to make more emphatic a boy's idiosyncrasies'. To survive he needed to dissimulate them – particularly if they included a sensitive nature and a love of art. This 'constant need of disguise' became, he said, a habit. It was a habit that soon became second nature, and stayed with him for life.

Osbert's failure at Eton, coming on top of all the other Sitwell worries, seems to have upset Sir George to an extraordinary degree – so much so that it caused the first real rift between him and his eldest son. For in his own neurotic way Sir George discerned in Osbert the emergence of precisely those distressing tendencies which were bringing him such trouble now with Lady Ida – bad handwriting (Ida's was dreadful too), a tendency to unseemly laughter, incompetence with figures, a love of the theatre, and incipient extravagance – in short, the whole irresponsible disposition of the Londesboroughs. Had Sir George not been quite such a passionate believer in heredity, his worries about Osbert might have been easier to bear. He would have realised his son would change with time; nurture as well as nature would begin to play its part.

But here, in Osbert, Sir George must have discerned a fresh genetic retribution, thwarting his anxious plans for Renishaw. It was dangerous enough to have a wife squandering the Sitwell patrimony. What hope remained for the family if his son and heir continued with the process?

Sir George began increasingly to disapprove of Osbert, and Osbert reacted, first with disappointment, then with growing opposition of his own. By his fifteenth birthday father-and-son relations had already become distinctly strained, and it was then that Osbert made his first attempt to retaliate:

> I remember that, as a schoolboy on holiday, when my father had been particularly disagreeable, I used always to go into the garden to tend a rhododendron that carried a purple blossom of a particularly obtrusive and fiery appearance which he could see from his study window, and which greatly offended his eye.

It was a feeble gesture – but an important one. For the more difficult Sir George became, the more Osbert's mood of silent mutiny inevitably increased, confirming his contemptuous rejection of authority, of the values of the despised boys and masters at his school, and of the whole hostile and oppressive world it seemed that he inhabited. Under that bland, habitual 'disguise' he spoke of, Osbert was in personal revolt against the conventional Edwardian world which Sir George and the conformist members of the upper classes set such store by.

Paradoxically, the origins of this revolt began, like almost everything in Osbert's life, with Sir George's influence. That sense of apartness and isolation which Osbert picked up from his father helped him define the omnipresent enemy around him. He was to christen them 'The Golden Horde'.

It was a good phrase which echoes Byron's *Don Juan*:

> Society is now one polish'd horde,
> Formed of two mighty tribes, the *Bores* and *Bored*

And he felt that it perfectly described the mindless, sporting local gentry and county aristocracy he met in Derbyshire; the red-faced hunting men, the genial drinkers, the Edwardian practical jokers who were always so suspicious of Sir George. Despite the riding dress he wore in the Sargent portrait, Sir George could never really be at ease in such society. Nor could Osbert. Like Sir George, he was too fastidious for the rough friendship of his peers in Derbyshire; like him, he detested horses, hunting and all sorts of so-called rural sports, and now at Eton he was once again confronted with the '*Bores* and *Bored*'. The Golden Horde was more intrusive – and more complex – than he had suspected. There was the bullying conformist element at school for whom any real intelligence had to remain that 'guilty secret between Man and his Maker'. There was also the group he called 'The Fun Brigade', the cheery gang of spongers, hedonists and nit-wits who were battening upon his mother in Sir George's absence and plainly leading her towards disaster.

These groups that Osbert so resented and disliked had certain things in common. They were all firmly anti-intellectual and conformist, inartistic, intolerant and dominating to a degree. They may have included stray outsiders from the middle classes, but they essentially represented the all-powerful elements in Osbert's own cherished social class which he could never stomach.

Sir George pretended to accept certain of these elements when it suited him, if only to create the conventional façade he wanted. His real inclinations lay elsewhere – chiefly in Italy and in his multifarious writings and researches. But Osbert was more rebellious than Sir George and possessed no such refuge. (It is perhaps significant that, unlike his son, Sir George had never had to battle with a dominating father during adolescence.) And one of the most important factors now in Osbert's adolescent mutiny was that it gave him common cause with Edith.

She was a seasoned ally. She, after all, had been battling with her parents almost from the day that she was born. Like Osbert, she was a solitary soul who loathed the Golden Horde in all its guises; and she, unlike young Osbert, had her defences charged and ready for the fray. Now, as their parents' troubles cast ever gloomier shadows over Renishaw, Osbert and Edith could unite. From her he could learn the martial arts for dealing with their parents – how to use ridicule and wit against them, how to avenge an insult, and how to project a steely presence to confront their mother's rages or their father's persecutions. She had a still more valuable resource which Osbert could soon share with her – the private world of books and poetry, which was her refuge from the world around her.

Sacheverell describes her at the beginning of this period: 'my earliest memories evoke the figure of a young girl, thirteen or fourteen years old, tall and thin, and already copying out reams of poetry into her note-books'.

For poetry had become a secret cause for Edith, something to put against the cello and the 'Bastille' and the whole regime her parents were still trying to impose on her. Poetry could offer a forbidden liberation from her miseries – she has described how she first read Pope's 'Rape of the Lock' by candle-light beneath the bedclothes. And poets could become the allies and the heroes that she needed. After Pope she fell in love with

Swinburne – and, typically, her admiration for his poetry gave her an excuse for just the sort of gesture her dramatic nature craved. Accompanied by her puzzled, dimly disapproving maid, she made a forbidden journey to the Isle of Wight to offer a libation of milk and honey on the poet's grave. Shy though she was with strangers, Edith already could act out a role like this to great effect. A more impressive demonstration of her poetic sensibility came when her parents took her to hear Sousa conducting his own brass band, and she was promptly sick in the Royal Albert Hall.

With 'art for art's sake' in the air, these years before the First World War were a great period for aesthetically inclined young ladies to espouse the cause of art, and Edith was fortunate enough to meet several artists in the flesh and to discover that, unlike her parents' friends and disapproving local notables, they could be kind and interested in her and put her instantly at ease. Sargent had done this. So did Walter Sickert. That handsome, high-bohemian painter was for a time the lover of her one 'artistic' aunt, Mrs George Swinton. It was through her that Edith met him, and she was captivated by the charm that Sickert always could dispense to shy young ladies, so much so that some forty years later she would write to Denton Welch about an article that he had written on the painter: 'I can't tell you how kind he was to me when I was seventeen, and trembled with shyness when spoken to!'

But throughout her teens the most important intellectual and artistic ally Edith had was undoubtedly her new governess, Helen Rootham, who, late in 1903, had taken the place of Miss King-Church. This intense young woman, with her 'dark hair and eyes of almost Spanish vehemence', was an unlikely person to have turned up at Renishaw as governess, and in later years few of the Sitwells' friends felt particularly at ease with her. Harold Acton found her 'rather dreary and impossible, the essential hysterical spinster', and there were times when even Edith found it hard to cope with her possessiveness and touchiness. But in her early years at Renishaw young Miss Rootham proved a most exciting, liberating influence – particularly for Edith. A cross between Miss Havisham and one of Bernard Shaw's 'new women', Helen Rootham was far better educated than was expected of an ordinary Edwardian governess. She was intensely musical: her uncle, Dr Rootham, had taught music at Cambridge, and Osbert said that she was 'perhaps the finest woman pianist it has ever been my good fortune to hear'. She had lived for several years in France, where she became a passionate admirer of Verlaine and Baudelaire, and that most enthralling of poetic prodigies, Rimbaud. She had already started to translate his poems (her translation of 'Les Illuminations', published with Edith's help, would finally be set to music by

Benjamin Britten) and inevitably introduced Edith to the world of French symbolist poetry.

But for all the children the importance of Miss Rootham in the house went far beyond her literary influence. 'She was the first person we had ever met who had an *artist*'s respect for the arts, that particular way of regarding them as all-important – much more important than wars or cataclysms, or even the joys of humanity,' wrote Osbert. And, also, here at last Edith had an older friend she could confide in, a loyal supporter and a guide for the artistic world she wanted. If Edith had lacked confidence and direction, determined Helen Rootham could supply them both, confirming her incipient belief that the only calling worthy of her nature was to become a poet.

From her late teens Edith possessed another ally too – Sacheverell. Although he was more emotionally attached to both his parents than either Edith or Osbert, this had done nothing to affect his deep devotion to his older siblings – and to Edith in particular. Sacheverell was still unscarred by life. He was precocious and extremely clever. In him Edith found the perfect pupil and disciple for her poetry, and before long he was writing on his own account. As he says modestly, 'When she wasn't writing poetry herself, she was reading it to me and encouraging me to write, so that poetry appeared a natural part of life. In those days Edith would have made anyone a poet.'

At the same time, all Sir George's troubles and the atmosphere within the Sitwell home had strengthened the devotion which the three children felt for one another and, as Osbert put it, 'had developed to the highest degree a sense of mutual confidence and interdependence among my brother, my sister and myself. . . . We formed a closed corporation.'

Now, thanks to Edith's influence, this 'closed corporation' was beginning to exhibit similarities with another group of writers who developed in the isolation of a nearby unhappy family – the Brontës. Renishaw, as Edith used to say, is 'near the *Wuthering Heights* country', and, like the Brontë children, the young Sitwells were beginning to unite in a poetic world of imagination to offset the darkening colours of the real world around them.

Her poetry apart, Edith had now reached a state of limbo on the edge of womanhood. Normally a well-brought-up young lady of her class and her position in society would have swum automatically into the waters of the London season, to presentation as a debutante at court, and then on to the fecund haven of a suitable marriage. But this was clearly not for her. By her eighteenth birthday Edith was convinced that her appearance, if nothing else, ruled out entirely any such conventional solution to her life. Certainly she was unusual – six feet tall, pale-faced, lank-haired and with that contentious nose of hers and deep-set Londesborough eyes, her physical appearance seemed to underline the fact that she was a being totally apart.

But many stranger-looking women than the youthful Edith Sitwell have been courted, married and desired. As the Sargent portrait showed – and Cecil Beaton's photographs of her in her thirties emphasised – she did possess a haunting beauty of her own. And certainly she showed no lack of impetus to fall in love. Indeed, around her eighteenth birthday it appears she did exactly that – with a good-looking, eligible, but sadly unresponsive young Guards officer she met at Renishaw. (Elizabeth Salter believes, from what Edith told her, that this anonymous young man was the inspiration for her first published poem, 'Drowned Suns'.)

The trouble was her paralysing shyness – particularly when faced with any ordinary male – and this in its turn derived from her lifelong certainty that no ordinary male, as both her parents must have told her countless times, could ever possibly desire her.

Again this must have helped convince her that she had no place in that upper-class conventional society to which she nominally belonged. As a child she had believed herself a changeling. She was a changeling still – a poet – and she began to emphasise the fact. She could not run away – where would she run to, how could she survive? She could at least refuse to play the part in the debutante and marriage-market world that was expected of the daughter of a baronet.

Her mother evidently had ideas that she should dress in something 'suitable' and make a start of the local balls. Edith submitted – briefly:

> In my first ball dress of white tulle, lightly spangled and streaming with water-lilies, with my face remorselessly 'softened' by my hair being frizzed and then pulled down over my nose, I resembled a caricature of the Fairy Queen in pantomime.

Enough of that! To her mother's horror she went off and bought herself a long, black velvet dress. 'I knew I was right to look different from other girls,' she said, 'because I was different.' Or as she put it later, 'If one is a greyhound, why try to look like a Pekingese?'

Black velvet must have helped her self-respect, but it did nothing to resolve the problem of her life. What could she do? She seemed to be unmarriageable. She had no money of her own. She was totally impractical, and as far as anyone could see had no interests beyond poetry and music. As her twenty-first birthday loomed, she continued her prematurely spinstered life at Renishaw and Scarborough with Helen Rootham still for company, and her parents always there to plague her. She became steadily more bored and discontented with her lot.

Meanwhile, the century was ticking on, with storm clouds gathering not only over Europe but over Renishaw as well. Osbert described the two years 1910 and 1911 as 'the peak years, both of my father's irritability and of my own misery'. They coincided with Osbert's leaving Eton. He was eighteen and, despite his lack of obvious success, he was rapidly developing an attitude and sensitivity to life that were all his own. At Eton, behind that carefully maintained 'disguise', he had been reading everything he could – Dickens, the plays of Strindberg, George Meredith and above all Samuel Butler's *The Way of all Flesh*, that wonderfully subversive Victorian indictment of families and fatherhood. (When Osbert gave it to Sir George to read the baronet spent the next three days in bed.) And now, with Osbert leaving Eton, Sir George began to act exactly like a Samuel Butler father.

Osbert had been counting on following Sir George up to Christ Church, Oxford, and pleaded hard to be allowed to go. Sir George thought otherwise. Relations had been worsening of late. Writing some thirty years later, Osbert referred to that 'mutual diffidence upon which, I fully realise, a sound father-and-son relationship must rest'. Sir George would have found the whole idea of diffidence difficult to understand – particularly towards his son. Who better than his son to bear the weight of his didactic inspiration? Who else to benefit from those gems of worldly wisdom which he, Sir George, had painfully acquired? He was now in his early fifties, and age had made him more – not less – convinced that he was right in almost everything, and Osbert simply *had* to be corrected.

With Osbert home from school, and suddenly at the most painful period of late male adolescence, he and his father clearly got on one another's nerves. Hardly surprisingly, Osbert remarked that he was bored.

'I never *allow* myself to feel bored,' replied Sir George.

'That's just the difference between us,' Osbert answered. 'I never allow *myself* to bore other people!'

With remarks like this Osbert soon found himself in 'continual disgrace at home'. He suffered, for despite his rudeness to Sir George he still respected him. Soon, as he wrote,

> My self-respect had entirely perished: for of what use was I, if my father so little esteemed me? After all, he was the most intelligent and learned person of his generation who was within my range. . . . But my feeling for him must, within the space of a year or two, have very much altered, until . . . I would feel ill for an entire day at the sight of his handwriting on an envelope.

Osbert would later blame much of his subsequent abrasiveness upon this period. As he wrote in his novel *The Man Who Lost Himself*, 'a gift for over-sharp sayings in any man is usually the sign that it has been forced upon him, a token of nervous exacerbation caused by constant, wilful and vituperative misunderstanding'.

The climax of this 'constant, wilful and vituperative misunderstanding' came in the summertime of 1912.

Despite his 'anguish' and his hypochondria, Sir George was undoubtedly a wily old opponent. Osbert was still extremely gullible and Sir George played on this when he suggested that, instead of Oxford, Osbert should spend six months at a military 'crammer' in Camberley, preparing for the Sandhurst examination. For anyone of Osbert's temperament the idea of a military career was ludicrous. His only explanation for agreeing to it was that as 'I was always being told my faults . . . I agreed with my father in thinking that I was ill-qualified for any profession. . . . And it was precisely this sense of weakness that led me into captivity for a while'.

To begin with it was benign captivity – certainly far more agreeable than Eton. He was to describe the Camberley 'crammer' in some detail in a short story, *Happy Endings*, and the picture that he gives is of a run-down, ineffective, rather touching school for the sons of gentlemen and officers. The teaching staff were dingy, anachronistic characters, and the whole place smelled of musty dreams of long-gone military splendours. No one objected to Osbert's reading what he liked. Nobody criticised him, and away from Sir George he seems to have rapidly recovered his self-confidence. One thing he did learn at the crammer – the army was emphatically not for him, for he hated discipline and heartily disliked communal life as he had known it at school. In theory, though, Osbert had no difficulty if he decided not to be a soldier. All he had to do was to make sure he failed the Sandhurst examination which Sir George wanted him to take.

For once he had underestimated Sir George. The baronet must have got wind of Osbert's plans, for he thwarted them with ingenuity. Without a word to Osbert, he arranged to have him instantly gazetted into the Yeomanry, through a friendly general. Yeomanry officers had no need to go to Sandhurst, so that Osbert opened *The Times* one morning to discover that he had overnight attained the rank of Second Lieutenant. Worse still, as an officer in the Yeomanry he had been in turn seconded to the Eleventh Hussars – a crack cavalry regiment stationed at Aldershot.

There was, it seems, no possible way out. To refuse to go could be legally construed as mutiny. Cunning Sir George had deftly managed to 'lasso' him and, on his favourite principle that it was always good for people to do what they thoroughly disliked, he was determined that the cavalry would make a man of him.

For Second Lieutenant F. O. S. Sitwell the Aldershot day began at crack of dawn on horseback in the regimental riding-school learning 'a whole repertory of helpful circus tricks *à la Russe*' from a bellowing instructor. It ended around midnight – if he was lucky – watching his brother officers drinking themselves beneath the table, and feeling as if he had suffered 'banishment to Siberia'. On the eve of the First World War most cavalry officers were still intent on rigidly maintaining just those qualities which had brought disaster in the Crimea. They solemnly believed that they were the cream of the officer corps; and, as Osbert wrote of one of them, 'a routine of port and a fall on his head once a week from horseback kept him in that state of chronic, numb confusion which was then the aim of every cavalry officer'.

Apart from the physical discomfort, finding himself among such company was for Osbert rather worse than going back to Eton. He was tormented by two dreadful characters – Major Gowk and Major Fribble-Sadler. Clearly they squashed Lieutenant Sitwell from his first evening in the mess. He had the temerity to talk. They made it plain that brash young officers did no such thing.

Later he could be philosophical about the cavalry: 'No life shared with savages of the Congo or with the web-footed natives of New Guinea could have been more strange,' he wrote, and it was a useful way of learning at first hand 'to comprehend the mentality of the members of any primitive tribe'.

More useful still, for a potential poet and lover of the arts, his fellow-officers demonstrated 'the immense negative energy, if I may so term it, of the opposition' and 'the virulence of the feeling existing between those who hate beauty and those who love it'.

This was a cruel and a crucial lesson. Previously he had shied away from the Golden Horde: now he was suffering at their hands, and seeing them as philistines incarnate. But with his sufferings his greatest shock of all came when he realised the full extent of what he termed Sir George's 'treachery'. From now on he felt that he could never trust him – nor forgive him, for that matter. Henceforward Osbert would regard him not as a member of the Golden Horde – he could never possibly be that – but as a sort of Quisling who had betrayed him to the enemy.

For Osbert these unhappy months at Aldershot were the time when he became aware of the dichotomy which would obsess him for the remainder of his life – the chasm which divided life as it was from life as he longed for it to be. Here he was, banished to 'Siberia'; but not so far away, in London, there were art, society, the joy of civilised behaviour. From the columns of *The Times* he knew that the new show of Augustus John had opened at the Leicester Galleries, that the Italian Opera was at Covent Garden and the important second post-impressionist exhibition of Roger Fry was just about to start.

But all this lay in a far-off world, and by midsummer it was clear that the sheer misery of life had become too much for Osbert. He cracked – and, like Sir George exactly ten years earlier, Osbert now had a nervous breakdown of his own.

It would always remain a mysterious affair. There is no record of the form the breakdown took, beyond the fact that in late August Lieutenant Sitwell was granted compassionate leave of absence from his regiment on medical grounds, and he never enlarged upon it. He confined his version of events to his unhappiness in the Hussars, and how he finally fled to Florence to beg Sir George – who held the all-important strings to his allowance – to help him leave the hated cavalry.

In his autobiography the account he gives of this Italian journey is a romantic one. After the gloom of Aldershot and London – 'this year the fog had already coffined the city for the winter' – Florence appeared to him a sort of miracle, a city of instant happiness in which all his troubles lifted. He seems to have even found Sir George in a receptive mood. He and Lady Ida were both staying at the Grand Hotel, and when Osbert met him the baronet was sitting swathed in mosquito-netting, receiving telegrams in code from Naples on the progress of the cholera epidemic in that southern city. The weather in Tuscany was perfect and Sir George took Osbert for his first glimpse of the castle at Montegufoni which he was now enthusiastically restoring.

Osbert's account of the visit reveals his father at his happiest – as he generally was in Italy – with the enormous Henry Moat unpacking Sir George's inevitable cold chicken on the terrace, and the *contadino*, Angelo Masti, bringing them out red wine from the castle, figs, *pecorino* cheese and truffled macaroni to complete their picnic.

In such surroundings even Sir George could not refuse his son's request to leave the cavalry – at least to the extent of sanctioning his transfer to the Brigade of Guards; and Osbert, knowing that this would mean an end to Aldershot and horses, seems to have been satisfied.

Suddenly, after months of wretchedness, Italy was showing him a double freedom – freedom from the cavalry and the wider freedom to enjoy the splendour of Montegufoni and the Tuscan autumn with 'its pyramids of figs and peaches, its dust, its roses and oleanders, and brown, baked hills, spiky here and there with dark cypresses, and its terraces hung with opalescent, smoky bunches of white grapes'.

It was a golden picture, and Osbert's future seemed assured. But it was a deceptive happiness. During those autumn days in Florence disaster and disgrace began to stalk the Sitwells. Even Osbert was involved by now, and there was more to his breakdown than his misery at Aldershot.

CHAPTER FOUR

A 'Yorkshire Scandal'

1911–1914

THE TROUBLE had started twelve months earlier, when Osbert was still at Camberley. One weekend, back at Renishaw – during one of Sir George's frequent absences – his mother had confided in him on the subject of her debts. Debt was a constant problem now for Lady Ida, who had been overspending on her personal income of £500 a year from the day she married. Sir George had been settling for her each year – but unenthusiastically. Indeed, a few years back there had been quite a crisis in the family when he had positively refused that honour – only to perform it after Ida's brother, Lord Londesborough, had gallantly stepped in to offer to do so himself. Even so, there had been quite a rumpus. Ida had had to promise never to overspend again, and finally Sir George had settled all her debts by shrewdly purchasing life insurance on her person with her marriage settlement.

But now, as she explained to Osbert, despite her promises and protestations at the time, she was in debt again, more seriously in debt than she had ever been before – over two thousand pounds. What could be done? After her previous promises she could not possibly ask Sir George for help again. Osbert, who had been having trouble with his father over *his* overspending recently, saw her point. However, he had an alternative. He had a friend at Camberley, a boy called Martin who was entering a rifle regiment and who appeared to know all about borrowing money. Why not ask his advice?

Martin was invited up to Renishaw, and proved forthcoming. It would be simplicity itself for somebody like Lady Ida to borrow all she needed, and there would be no need for Sir George to be involved. He had a friend, who was a financial wizard, a gentleman and a literary figure, who was accustomed to assisting people of good family with their money problems. He would arrange an introduction. It duly followed, and shortly afterwards Julian Osgood Field, the *éminence noire* of the Sitwells, arrived for a weekend at Renishaw.

Field was not physically engaging. He was then in his sixties, and Osbert has graphically described his appearance – his 'stunted, stooping, paunchy body . . . carried a heavy head . . . with a beak like that of an octopus, which spiritually he so much resembled, and a small imperial and moustache that were dyed, as was his hair, a total and unnatural black'. With his striped trousers, pointed beard and pouchy yellow face, Field bore a marked resemblance to that unsavoury monarch Napoleon III, whom he had met in Paris.

Field had known many other famous people in his day. He came of a prominent New York family; his father had been Assistant Secretary of the United States Treasury; and Field himself had gone to Harrow and to Merton College, Oxford. Jowett had been impressed with him. Edith's great hero, Swinburne, had stayed with him, and in the seventies and eighties, when he had moved to Paris, he got to know an impressive list of literary celebrities including Victor Hugo and de Maupassant. The book of sketches which he published under the title *Aut Diabolus, Aut Nihil* remains a useful source of Parisian characters of the period. Using the pseudonym 'X.L'. he also wrote a book of Russian sketches and ghosted the memoirs of the Duke of Hamilton.

All of this naturally appealed to Osbert. For Lady Ida what was more important was that this plausible and fascinating little man seemed to be offering immediate escape from all her troubles. With her distinguished name and family connections she would have no problems raising the money she required. Sir George need never know, and Field assured her that she would never have to suffer the indignity of dealing with sordid money-lenders personally. He would be honoured to do this for her and arrange everything with absolute discretion.

During the next few weeks – it was just before Osbert's 'lassoing' into the Hussars – Field had had several further meetings with Lady Ida. In her impulsive way she was soon trusting him implicitly – especially when he produced the scheme that he had promised. He knew a lady by the name of Dobbs. She was eccentric and extremely rich. According to one rumour she owned half Regent Street. She would be willing to lend Lady Ida £6,000 at interest of five per cent. She was lonely and desired to enter good society: perhaps it might be tactful now if Lady Ida asked her up to Renishaw for Christmas. As for repayment of the loan, Field carefully explained that she could do this by taking out further life insurance, just as Sir George had done. There would be no problem, since she was borrowing enough to pay the premiums for several years to come.

Ida was impressed, but said that she did not need £6,000. Her debts were only £2,000. Field airily replied that it was always best to borrow more than you needed while you could, and added casually that there would of course be certain incidentals. There were his expenses and those of his agent, a man by the name of Herbert who was arranging everything with Miss Dobbs. Also, of course, Lady Ida would not get quite all of the £6,000. Miss Dobbs would 'back' a bill – guarantee a loan over a period, say six months, so as to give him time to find a money-lender who would 'discount' the bill, paying in cash a certain sum on the strength of it. This would be less than the £6,000 (in fact it turned out to be £4,250) and Lady Ida would have time then to arrange her life assurance so that repayments would be met.

As Osbert wrote later, Ida 'had been married from the schoolroom. Money had no meaning for her.' But even a reasonably bright child of ten ought to have smelt this particular rat for what it was. Lady Ida didn't. Feckless and desperate to avoid still further trouble with Sir George, she would clutch anything by now, and Field was skilful at his business. Since 1901, when he went bankrupt and was imprisoned for forging a contract with a publisher, he had found this sort of shady dealing with the improvident well-to-do more lucrative than literature, and had perfected a technique combining money-lending, fraud and blackmail in varying proportions. Once Lady Ida scrawled her all but illegible signature to Miss Dobbs's bills, Field had her in his grasp.

Early in 1912 the two bills for £6,000 were discounted by a London money-lender by the name of Owles, and the money paid to Field. £1,000 went immediately to Herbert, and Field explained that he had to pay his own expenses too. Out of the £4,250 Owles had lent, £200 seems to have found its way to Lady Ida.

When she complained, Field started to reveal the character more of a blackmailer than a money-lender by telling her that if she was not satisfied she ought to tell her husband what had happened. When she replied that she would do anything but that, Field told her that the only way of avoiding this, and paying for her life insurance, would be by borrowing some more. For somebody like Lady Ida with so many friends and so much social influence it would be easy. All she would have to do would be to find somebody of wealth to back a further bill: once she had found a wealthy backer, he, Field, would do the rest.

But how could she persuade anyone to guarantee a further loan for her?

Field replied that she must know countless rich, ambitious women only too ready to pay something for the privilege of being taken up by a daughter of Lord Londesborough: bankers' wives with daughters needing to be presented at Court, or sons who wished to enter an exclusive London club. Such things were done continually, but if this was impossible, what about Osbert? He was a devoted son, and a fashionable regiment such as the Hussars must have a number of young officers of wealthy families who would be willing to guarantee a bill for Lady Sitwell – in return for immediate hard cash. Surely to help his mother Osbert would not refuse to mention this to one or two of them?

Either through panic or stupidity – or both – Ida apparently agreed, and by the spring of 1912 had found herself involved in a spiral of accelerating debt and deception. For Field she must have seemed the victim of a lifetime. She followed his advice implicitly – and wrote extremely compromising letters to him regularly, detailing everything she had done. Field kept the letters carefully, knowing that the moment would ultimately come when he would need them – as indeed it did.

> Oh dear! [she wrote in one of them] Have just had a letter from Lady —— saying she does not see her way to signing the paper. I think it is the money-lender. What is to be done?

Shortly afterwards she was writing once again – this time about the women to be pushed into society.

> I am rather sensitive about getting hold of people in a great hurry. Is it not possible to get hold of this woman you told me about? Of course, I will do everything in my power to get her into the society she requires. That lies in the hollow of my hands. . . . Now as to telegrams, I think it would be advisable to address them to Miss —— as there would be no chance of them getting into the hands of my family.

But despite this Lady Ida seems to have had problems with her candidates.

> It seems impossible [she wrote] to make the woman understand that our transaction is absolutely sound and straight, doesn't it? So do try to get me a man as a second backer. Of course I could push ladies into good society if only they would have faith, but somehow they do not seem to believe my word.

Because of this, she turned again to Osbert. Martin, his original friend from Camberley, was persuaded to guarantee a bill for a small amount. There was another subaltern called Glass who seemed inclined to do the same, but there were difficulties with him. As she wrote to Field, 'I find Glass is only 20. My boy thinks if he were with him for a few days he could get hold of him, and he quite hopes, if he joins the 11th Hussars, there must be some boy he can get hold of.'

Apart from Martin, who joined his regiment later in 1912, Osbert was fortunately unable to 'get hold of' anybody else. After the Dobbs loan Field's best catch turned out to be a further bill, this time for £4,000 which Lady Ida managed to persuade an improvident young Yorkshire cricketer called Wilson to guarantee for her – for an outright payment of £300, and a promise to persuade her brother, Londesborough, to propose him for the Marlborough Club.

What Field knew – and Lady Ida never understood – was that with every further loan she took the nearer she approached inevitable disaster. Only a minute fraction of the loans ever reached her (the rest lined Field's pockets) and by that fateful summer of 1912, when Osbert was in Aldershot, the reckoning had begun. Ida had been unable to arrange the life insurance she had originally planned to pay off the £6,000 she now owed to the money-lender Owles; and since the bill was due, Owles was now pressing for repayment. So were the other money-lenders who had paid out short-term loans on bills backed by Martin and by Wilson. By the end of that July the pressure was increasing. Field engineered a two-month extension on the Dobbs loan, but Wilson was compelled to mortgage his reversion for £4,000 to meet his guarantee, and Martin was writing frantically to Osbert that because Lady Ida could not meet her debts the money-lenders were now threatening to tell his parents and his colonel. One can understand the pressure that this put on Osbert – on top of his other troubles at the time – and it was now that he finally had his mysterious 'breakdown', and fled from Aldershot. His mother, by this time in Florence with Sir George, seems to have temporarily escaped the worst of the repercussions. Ironically, although she was now liable for debts approaching £12,000, she was worse off for cash than ever, and that autumn, out in Florence, was having to borrow off her servants – her personal maid and the butler, Henry Moat. Yet even now she trusted Field, and wrote imploring him for aid as late as that October. 'Will you be very kind and let me know when I have some money? I must pay that spiteful maid, who has a very bitter tongue about me. . . .' And she added in another letter, 'I owe my butler £125, the brute, I want to get out of his clutches.'

When he received these letters Field, who was a past-master in the gruesome art of exploiting such affairs, must have realised that the crucial moment had arrived. It was time for outright blackmail.

Not long before, Ida had written to Field: 'I need hardly say, if the worst comes to the worst, my brother, Lord Londesborough and my husband, Sir George Sitwell, whose income is about £15,000 a year, would pay at once such a debt of honour.' But would they? It was up to Field to make sure they did, and, by stressing every aspect of the scandal, to extract still further money from the family. But everything depended on Sir George.

Writing about this period in his autobiography, Osbert gives the firm impression that Sir George had no idea of what was happening until he broke the news to him. Even then, when Osbert told him 'Mother has got into the hands of a moneylender,' he makes Sir George reply, 'I never heard such nonsense: if she were, *I* should have known about it!'

In point of fact, however, it is quite clear that by this time Sir George *did* know a good deal of what was going on. As he lay shrouded in mosquito-netting in the Grand Hotel the baronet was by no means as ignorant or insulated as he appeared to Osbert. His solicitor, Sir Henry White, had already written to him that Owles, the money-lender, was clamouring for payment.* When Ida left Florence for England soon after Osbert came, it was to face the first of the melancholy legal actions over the Dobbs bills, and it was also now that news arrived that Wilson's trustee, a Major R. B. Turton of Kildale Hall, Yorkshire, outraged at what had happened, was demanding full repayment of the £4,000. Failing this, there was going to be what Field described as 'a public Yorkshire scandal'.

Indeed, the more one learns of the events that followed, the more one is compelled to the conclusion that Sir George was shrewdly keeping his own counsel and planning, Chinese fashion, how to avert or lessen the disaster.

On this occasion he was by no means the ineffectual, comic character Osbert painted. Certainly he treated Osbert well. One does not know how much Osbert actually told him of the attempts that he had made to 'get hold of' various young officers in the mess at Aldershot to back his mother's bills – or of the army scandal which was threatening with Lieutenant Martin. But Sir George had undoubtedly learned enough by now to understand the cause of Osbert's breakdown. This was one reason why he agreed to Osbert's change of regiment, and it was Sir George who promptly settled Martin's debts, and made it possible for Osbert to join the Grenadiers without the threat of scandal hanging over him.

Once he was back in England that November, Sir George also swiftly set about clearing up the remainder of the mess his wife had made. He contacted the Turtons and the great 'Yorkshire scandal' was averted by paying off young Wilson's debts and so restoring him to his inheritance. Other debts were settled too. In all Sir George paid out something approaching £5,000 on Lady Ida's debts. But he made one ominous exception. He emphatically refused repayment of the £6,000 still due to the money-lender Owles. He had his own good reasons for refusing this, for, as he soon found out, the arrangement of this loan was the one

* Sir George Sitwell made this plain in his Old Bailey evidence in March 1915:

Mr. Muir (cross-examining): When did you first know that Miss Dobbs was being sued upon the acceptances she had given for your wife?

Sir George: My solicitor, Sir Henry White, had a communication from the moneylender's solicitor on the subject.

That was some time in August, 1912? – I think it was early in September.

transaction where Field could actually be proved to have acted quite illegally. When Field had obtained for Lady Ida her two-month extension on the loan, he had told Owles that Miss Dobbs, the guarantor, had written her agreement to the delay. She had in fact done no such thing. Had Sir George paid he would have been condoning Field's behaviour. The matter would have been closed. Field would have escaped scot-free with something like £10,000 before he picked another victim. To prevent this – and expose Field for what he was – Sir George was now prepared to fight. On the advice of his solicitor, he soon retained Sir George Lewis, perhaps the shrewdest and most celebrated legal brain in Britain, to represent him.

Up to this point Sir George had acted admirably, coping with the crisis with a coolness few other husbands similarly placed could muster. Where he apparently failed – and this is not surprising – was in restoring any sort of calm and happiness to his unfortunate family. Indeed, it is now that their suffering began in earnest. All the wretchedness of the past, the loneliness and anguish of that unhappy marriage were revived. The Sitwell home became a place of misery.

Osbert was out of it. As a new subaltern in the Grenadiers he can have only felt relief when he was placed on duty at the Tower of London that Christmas. Fifteen-year-old Sacheverell came home from Eton and was shocked by what he found. On Boxing Day he wrote to Osbert from Wood End thanking him for various books he had sent for Christmas, and reporting that their mother was extremely ill – far more than he had ever realised. She had had what he described as 'clots of blood and congestion of the brain' and had, he said, nearly lost her reason. She had only just started coming down to dinner in the evenings. She was drinking too, and the atmosphere within the house was worse than ever. Sir George had been telling him of the extent of her debts – over £11,000 – and he could not see what on earth could happen. Desperate with worry, he appealed to his 'darling Osbert' somehow to arrange to meet him before he went back to school, either at Scarborough or at Renishaw a few days later. He needed somebody to talk to, and only Osbert could offer him the calmness and the understanding that he needed. The two brothers always had been close, and this latest crisis was to make them more dependent still upon one another.

It is interesting that in the letter he wrote to Osbert Sacheverell made no reference to Edith. Unlike her brothers, she had been at home with both her parents, enduring the hysterics and the accusations which had followed their return from Italy. It is clear that she felt little sympathy with either of them, and she blamed her mother's 'sheer stupidity' for what she called the 'sordid tragedy' she had called down upon the family.

Had Edith had a saintly or forgiving nature she might presumably have felt a touch of pity for her mother's misery. But by now Edith was twenty-five and still living out her unsatisfactory, put-upon spinster's life between Renishaw and Scarborough. Since earliest childhood she had been suffering the rages of this sad, neurotic woman, and she was shocked to see that even now her mother seemed to show no sign of real awareness, let alone remorse, for what had happened. 'All *she* wanted', as she wrote impatiently, 'was to return to her silly daily life of bridge and watching the golfers on the golf-course.' Far from being grateful to Sir George, Ida was blaming him for not paying all her debts immediately, and in what Osbert called 'the atmosphere of hysterical violence that prevailed' the conditions of Edith's life became intolerable. It was impossible for her to write her poetry in that 'state of absolute bondage' to her parents in which she still existed. Edith herself more tersely stated that she finally 'had to leave home owing to my mother's conduct and habits'.

But leaving home was easier said than done. Edith could deal with violations of her dignity or personal integrity. Her outraged presence could be most impressive. But she was totally impractical. Harold Acton, who remembers seeing her when her father brought her out to Florence when he was a boy, recalls her even then as 'utterly unworldly, nervous and shy and quite incapable of asserting herself in any practical way'. Later she reminded him of a Firbank heroine, like the young lady in *Caprice* who went to London and was killed by a mousetrap.

But there was somebody at Renishaw who was intensely practical – her former governess and friend, the dark-haired Helen Rootham. She was what Edith needed, a committed ally. She believed in the same cause as Edith – poetry. And she supported Edith through this time of crisis, telling her that she would have to make a break from her appalling parents if she was ever to fulfil herself either as an artist or a human being.

Then in the spring of 1913 Miss Rootham's arguments received fresh weight. Edith's first poem was published. Thanks to the talent-spotting skill of Richard Jennings, a journalist and influential Fleet Street literary figure, 'Drowned Suns' was published in the London *Daily Mirror*.* Edith was paid two pounds and had the extraordinary excitement of suddenly emerging from her private world into the bold publicity of print. More important still, in Richard Jennings and the unlikely medium of the *Mirror* she had found what she required – an editor to back her and

* Edith was always grateful to the *Daily Mirror* and had a soft spot for the low-brow paper all her life, even when it backed the Labour government. Just before her death she said: 'I have a great feeling of gratitude' towards the *Mirror*: 'it published my first poem – in 1913 – when no other paper or publisher would have anything to do with my work.'

a platform for her poetry. During the next three years the *Mirror* was to publish eleven of her poems, including her 'Love in Autumn', 'Serenade', and 'Water Music'.

This was the spur she needed. Late in the summer of 1913, coolly ignoring Lady Ida's accusations that she was ungrateful, heartless and completely mad, she packed her bags. After a tussle with Sir George she sold a small diamond pendant she had inherited. Encouraged and sustained by Helen Rootham, she took the train for London, where the two women found themselves lodgings in a Bayswater boarding-house kept by one Miss Fussell. They had little money – Edith had £100 a year from a legacy – and sustained themselves on soup and beans and buns. But Edith had finally escaped. From now on she could dedicate her life impractically and undistractedly to poetry.

Although Osbert was safely in the Grenadiers, and most of the nightmares of the year before were now apparently behind him, there was one way in which his mother's debts still plagued him – through what seemed to be the obsessive meanness of his father. The act of paying Lady Ida's debts had made Sir George more obsessional than ever for economies – at least where his dependents were concerned – since he was now convinced that all that stood between the Sitwells and disaster was his own rigorous control of the family's finances. Not unnaturally, perhaps, Lady Ida was kept all but penniless – which hardly added to the cheerfulness at Renishaw that year – and Sacheverell at Eton was always short of pocket-money. Just before Christmas 1913 he was apologising miserably to Osbert for being unable to buy him anything for Christmas. But the chief object now of Sir George's meticulous concern with money and extravagance was Osbert. Osbert, after all, had been mixed up with Lady Ida's debts. Sir George, as we have seen, was already worried that his eldest son was tainted with the Londesborough temperament: the only way of saving him from fresh catastrophe was constant vigilance.

If Osbert really did feel physically ill at the sight of his father's writing on an envelope, he must have suffered almost constant nausea in 1913, as successive letters, penned in immaculate black copper-plate (which was in itself so potent an expression of his father's character) arrived at the Guards mess for Lieutenant Sitwell.

The first of them came on 4 February, just eight weeks after Osbert joined the Grenadiers. Sir George was already worried about whether Osbert was living within his allowance – and taking no chances.

Your allowance will be £700, but I am arranging to keep £240 of this to the end of the year to meet bills incurred in the year and payable at the end of it – tailor, hosier, hatter, etc. etc. . . . You hardly realise the amount of labour incurred checking your old bills, obtaining particulars about your account and making arrangements in order to keep you.... It has cost me hours of thought.

This was undoubtedly quite true, and Osbert is misleading in his autobiography when he presents his father's desperate concern with his finances and accounts as part of some perverse and self-indulgent game with which he chose to persecute his son. For anguished Sir George, the attempt to keep Osbert solvent was profoundly serious. What he wanted was not just economy but a complete change in Osbert's attitude to life – as he showed all too clearly in the letter that he wrote one month later.

I am not satisfied with your letter. What I want is some indication that you are sorry about them [Osbert's debts], that it is a mean thing to spend so much on oneself, especially on things one engrosses, such as food, clothes, cigarettes, and that you realise what a serious thing getting into debt is. £340 seems no great sum, but what about the thousands it must cost you in death duties because I have lost confidence in you and dare not hand over property until at least time has re-established confidence.

The only way that Osbert could really 're-establish confidence' was by a thorough change of heart: Sir George went on to list a few ways he could start. Firstly (and here Sir George revealed his constant fear of the weaknesses of the Londesboroughs), Osbert must promise solemnly never to book theatre tickets. Then he must send Sir George a detailed estimate of all his personal expenses for the next twelve months – and how and by how much he proposed to curb them. Next he required a run-down on his son's expenditure on cigarettes, since such small details 'indicate a person's frame of mind' and help in 'exposing the consequences that could follow'. Sir George concluded this extraordinary letter with a warning which, more than anything, revealed *his* state of mind and the real nature of his fears:

Who could have thought that Lord L. would have got through £2 million and have been on the point of bankruptcy, that another person would have been on the point of several frauds, that Butler would have been cut out of his inheritance, that Cayley would have been robbed by his friends, and all because of a tendency in youth to be generous with themselves.

Osbert's reply was evidently spirited. (He was concluding his apprenticeship in the lifelong art of dealing with Sir George's criticisms.) He was sorry for the pain the subject of his supposed extravagance had caused Sir George – but pointedly refused the penitential diet of humble pie Sir George demanded. This brought an instant counterblast from Sir George, which is such a classic in its way and offers such an insight into the turmoil in his mind that it deserves quotation here in full:

My dear Osbert,

I was very glad to get your letter. Of course this has been very painful to both of us. As regards the debts, the sum, though a real trouble to me this year, is not very great. If I had no previous experience [of such behaviour] I would not feel it so much. But I am sick and sore to the soul from 26 years of the same experience, and I recognise in this the same ideas, part of the same cycle. If I could have felt you were in your heart sorry and would be more careful in the future, I would have paid without a word, as I expected to be able to do. But the conclusion has been forced on me that it was only the regret of good manners, that you still maintain that self-denial of any kind is a mistake and that you have no feelings of shame at allowing yourself what I do not think it right to allow myself. You write that restrictions of any kind are a mistake and refused on one ground or another every single sacrifice, however small, which I proposed to keep this year's expenditure within the £700.

You would not give up your servants, would not promise not to book theatre tickets, nor to smoke less expensive cigarettes. In fact, no reduction of expenditure will you make. And you seemed to be under the impression that the money to pay the old bills was falling from the sky: to be precise, I suppose the idea was that you and I were to join in 'doing' your next heir of that sum.

This difference between us has been growing like an evil tree until it has now led to a break of the old relations. I dislike all that set of ideas entirely. I think that men who spend profusely on their bodies are really swine. One ought to be self-denying with one's expenses except where the mind is concerned.

Loyalty to your regiment will oblige you to dress well . . . but there should be some feeling of shame, not only at spending on clothes, cigarettes, etc., much more than I do, but at having to be self-indulgent at all. . . . You should be self-denying in small things, must realise that you must make some return to the world, must work and take trouble without grumbling, must not complain about boredom when you have a single dull day. Then I shall be able to feel real respect for your character,

ever my dearest boy,
your loving father,
GEORGE R. SITWELL.

Throughout this uncomfortable correspondence one detects not only Sir George's passionate concern with money, heirs and the well-being of the Sitwell dynasty, but also something of the effect all this was having now on Osbert. For it was Osbert's passionate rejection of almost every tenet of Sir George's personal philosophy, expounded in these arguments on money, that was to form the basis of his own beliefs – and fix his character for life. Indeed, Osbert's private programme for the future can be mapped out by taking the opposite position of almost every point Sir George advanced for his salvation. No one would ever be a more convinced antagonist of self-denial in his private life than Osbert Sitwell. Few men in his time paid more attention to their dress and their appearance, were more in love with the theatre, or made such a successful way of life from elevated self-indulgence. He always drank the best wine, ate the richest food and smoked the most expensive cigarettes. (One of Osbert's favourite remarks about finance, which would have incensed Sir George, was, 'Take care of the *pounds*, and the pence can take care of themselves.') Restrictions never ceased to be anathema to him, and few heirs to a distinguished name showed less concern about *their* heirs and the future of their inheritance.

But most important was the part this battle played in crystallising Osbert's opposition to anything that even hinted of paternal authority. His mind was being formed around a passionate rejection of Sir George's image – and almost all he stood for.

Sacheverell was far more tolerant of both his parents. The atmosphere at home obviously upset and irked him. Home for the Christmas holidays in 1913, he wrote to Osbert to complain about the world of pantomime in which both his parents seemed to pass their days.

But while he was clearly worried and depressed by the gloomy antics of his parents, particularly when his mother, as he wrote to Osbert, seemed to turn against him for no reason, he was still partially protected against the more desperate aspects of the situation. He was upset by what had happened, but felt nothing like Osbert's and Edith's bitterness. He, after all, was not receiving lengthy missives on the subject of his debts, and could afford to see the funny side of Sir George's foibles, who would, for instance, spend twenty minutes telling him his patent system for taking the soap out of the eyes when washing – something which it had taken *him* twenty extremely painful years to learn.

Another important difference between his adolescence and Osbert's was that Eton was proving nothing like the purgatory it had been for his elder brother. He played no part, of course, in any of the heartier side of Eton life, and seems to have remained fastidiously aloof from as much of the more humdrum items of the regular school syllabus as he could decently avoid. But he possessed great charm, had no difficulty making friends with a few select spirits, and was already showing signs of such precocious style and sensitivity that he was accorded the status of an accepted dilettante and potential aesthete in the school. His unpronounceable first name proved a problem. One master called him 'Savonarola', but to everybody else he was simply 'Sachie', and a master who appreciated the tall, shy boy's originality was the celebrated 'Tuppy' Headlam. It was largely thanks to Headlam that he was soon enjoying that very special status which Eton even then occasionally offered the unusual boy. He had the privilege of reading undisturbed in the gallery of the school library, and thanks to this was soon developing an unusual range of esoteric learning. Edith may have awakened his lifelong love of poetry, but Sir George deserves the credit for sparking off his younger son's extraordinary interests in painting and architecture.

'Even while I was really very young,' he says, 'my father was always talking to me about the paintings and the buildings he had seen, particularly in southern Italy,' and it was Sir George who took him first to Venice. With Osbert he was soon to develop a passionate interest in opera and ballet, and by thirteen or so he had already worked out, he claims, an ambitious programme to which he would dedicate his life. He would be a poet – with Edith for a sister it was inconceivable to think of any other way of life. But he would also spend his life pursuing art and beauty, seeing for himself the greatest masterpieces of every school and period, studying the lives of poets, painters, architects, musicians, and discovering the artistic peaks of the human spirit.

His taste already was omnivorous. C. W. Beaumont, the ballet critic and historian, remembered him from 1913, entering his famous bookshop at 85 Charing Cross Road with Osbert, and knowledgeably ordering the most *recherché* books on Russian ballet. At the same time he was also studying seventeenth- and eighteenth-century English poets for himself, and exhibiting a precocious response to what was new and vital in the arts. At sixteen he was excited by the work of Sickert, of Augustus John, of Epstein and of Wyndham Lewis, and when the Italian futurist Marinetti arrived in London, causing something of a stir among the *avant-garde* with his violent manifesto 'Art and Literature', the sixteen-year-old Etonian who had seen his paintings at the Sackville Gallery thought it quite natural to start a correspondence about modern art with this interesting Italian who believed the time had come to blow up picture galleries and bomb Venice from the air so as to destroy the stultifying influence of the past.

One evening in the spring of 1914 Lady Diana Manners, the future Lady Diana Cooper, was dining at Lady Cunard's, who told her that she had placed her next to a Guardee. Lady Diana, who was still as innocent as she was beautiful, inquired what a Guardee was. Guardee, replied Lady Cunard with elaborate foreboding, was a slang term for 'Guards Officer', a member of the upper crust of the officer corps, a dedicated militarist, the British equivalent of the Junkers. Suitably impressed and rather nervous, the gorgeous Diana met her English Junker – a tall, fair-haired, impressive-looking, silent man – and marched in to dinner on his arm. Only when they were seated did he address a word to her.

'Tell me,' he inquired, 'what do you think about Stravinsky?'

Such was her first encounter with a live Guards officer – and with Lieutenant Osbert Sitwell of the Grenadiers.

By then he was twenty-one, witty, impeccably dressed, and to understand the contrast now between this suave young military music-lover and the unhappy rebel with the nervous breakdown of just eighteen months before one must appreciate the social difference that existed then between the cavalry and the Brigade of Guards.

Whereas the cavalry were essentially the Golden Horde on horseback, the Guards were officered by young and leisured aristocrats of fashion. They were a curious relic from the past, one of the last repositories of that essentially eighteenth-century concept of soldiering as a polite and fashionable activity. Since the main peacetime task of the Guards was to protect the royal palaces, the officers were in close proximity to the Crown. For the most part they were rich – and, compared with line regiments, extraordinarily privileged. The life of a Guards officer of this period has been described as one of alternate ritual and idleness, Captains got four months' leave a year, majors five, and colonels six. As foot-soldiers they were not subject to the skull-thickening influence of the horse; as young men of leisure they could have independence, a dandy's attitude to life, and an extraordinary degree of tolerance (provided that the officer was not a cad, a bounder or a manifest outsider). The Brigade was, in short, a most exclusive London-based club for the sons of the very rich, and as such it immediately suited young Lieutenant Sitwell.

One of the first questions he was asked in the mess at Wellington Barracks was the inevitable one a guardsman would address to a former member of the Hussars. Did he like horses? Osbert replied idly that he really did prefer giraffes – they had such a beautiful line. This was a remark which, far from showing how unfitted he was for a military career, would have enhanced his reputation as something of a wit and man of style. There were no Gowks and Fribble-Sadlers in the Brigade.

It is no coincidence that the most memorable chapters in his autobiography were to be the ones in which he lovingly recalled his life as a subaltern in London on the eve of war. For this was to be his university, and for all his subsequent attempts at pacificism, for all the battles he would one day fight against the military mind, the fact was that he enjoyed this pre-war service in the Guards. It put its stamp on him for life.

He always had the regular soldier's keen regard for punctuality and discipline (he used to boast that he had never missed a train in his life). He learned a fine Edwardian awareness of his appearance – a foretaste of the famous dandy of the twenties – the splendid suits and shirts, the Charvet neckties, and the nice ritual of the hair trimmed every fortnight by Mr Trumper of Curzon Street. And it was now too that he began developing what was to be his later manner – measured, urbane, and more than a shade forbidding when he wanted it to be.

But for Osbert the real importance of these months he spent in London in the Guards – from November 1912 to the outbreak of war in 1914 – was that, like his glimpse of Italy, it gave him a taste of a certain sort of freedom; the freedom of a London which was still a glittering, full-blooded and artistic capital. True, the old king was dead, but twenty-year-old Osbert, after the boredom and distraction of the months before, was being offered his own favoured seat for the last few circuits of the Edwardian roundabout.

For him the London of this time would always be a golden period – a lost domain which nobody could possibly bring back, an exile's city like the St Petersburg of Diaghilev and Nabokov, bright in his memory from this time before the wreckers came and the philistine took over.

It was a period of conscious happiness. As he wrote of his spell in Wellington Barracks, 'I still recall the time I spent there as the first period of my life that I enjoyed with a full sweep, and fortunately I recognised my happiness while it lasted'. It was now too that he formed an ineradicable and lifelong taste for what he called 'the silly pleasures of the rich'.

For as a young member of the Brigade – and with the assistance of his various relatives in London – he suddenly began to find himself leading an insider's life in the higher reaches of society, safely away from the attentions of the Golden Horde. He had never been popular before. At school and in the cavalry he had been lonely, wretched, a confirmed outsider. Now, almost overnight, all this had changed, and he was being introduced to that world of hostesses and London drawing-rooms which he was to make so much his own until they vanished with the second war of 1939. There was his Aunt Londesborough, back from the Delhi Durbar, her small, marble-lined mansion in Green Street, Mayfair, now crammed with 'large presentation portraits of jewelled and dusky potentates'; there were his Beaufort cousins and his Swinton cousins, and, as Sir George Sitwell's son, he was expected to make an occasional appearance too at Lambeth Palace.

He was eminently presentable: in an age that valued tall men he was six foot three, he dressed immaculately (his suits were made by Pode and Rede, his shirts by Hodgkinsons of Pall Mall), his profile was distinguished, his manners were above reproach. Appearances apart, it was soon evident that the turn of phrase and knack of sharp rejoinder which he had picked up in his encounters with Sir George were useful social assets. Already he possessed a certain wit. He could amuse and flatter too – older women in particular. This was a talent which would never fail him, and in those Mayfair evenings on the eve of war it brought him much that made his life agreeable – rich food and good champagne, smart talk, fresh invitations and further contact with the eminent, the fashionable, the rich. London society was ruled by a small group of ladies of a certain age – he called them the *Vieillesse Dorée*, 'belles of the Edwardian summer' who 'survived in a kind of September splendour': and he became a protégé of several of them now. One was his parents' old friend, Mrs Keppel. Thanks to her friendship with the late King Edward and her adroit investment, on the advice of Sir Ernest Cassel, in Canadian stocks and shares, she was by now a grand and very wealthy lady. Osbert found her house in Grosvenor Street 'one of the most remarkable' in London. With its thick carpets, lacquer cabinets and masses of expensive porcelain, it was still something of a temple of high Edwardian luxury – just as Mrs George herself remained the out-and-out epitome of the Edwardian *grande dame*. Osbert – like Sir George – admired her enormously, particularly relishing her good-natured common sense, her lack of humbug, and the air of stately notoriety that naturally surrounds the one-time mistress of a king. He seems to have preferred her to her daughter, Violet, although he guardedly admitted that he found that passionate young lady 'cosmopolitan and exotic from her earliest years, with a vivid intelligence, a quick eye for character' and 'irresistible gift of mimicry'. For all her liveliness and love of poetry and art, frivolous Violet must have been something of a trial to Osbert – as she continued to be at intervals for the remainder of his life. When her adored Vita Sackville-West married Harold Nicolson in October 1913, and the Keppel parents were insisting that it was time for her to think of marriage too, Violet suddenly announced that she was going to marry Osbert Sitwell. He knew nothing about this until she made the startling announcement, and he was considerably embarrassed – as she presumably intended him to be. But although the whole incident was little more than another of young Violet's 'jokes', she often referred to Osbert in years to come as her 'ex-fiancé'. Osbert was not particularly amused.

A less compromising friendship was with that erratic social dynamo, Margot Asquith, wife of the prime minister. Thanks to her the bland young Grenadier became a frequent visitor to Downing Street, 'then a centre of such an abundance and intensity of life as it had not seen for a hundred years and is never likely to see again'.

But it was the company of even grander, older ladies that he most enjoyed, particularly when they sported something of the atmosphere, the life, the gossip of an age gone by. One of these was witty old Lady Brougham, now in her seventies, and married to the nephew of the great Lord Chancellor, still with her large house in Chesham Place and her memories of European royalty of fifty years before. Another was the extraordinary old Lady Sackville, the Pepita of V. Sackville-West's biography, mistress of Knole, both 'peasant and aristocrat' by birth, who several times invited him to dine *à deux* with her amid the oriental splendours of her London house.

As for the male world of this high Edwardian London that he loved, he had a favoured glimpse of it not only through his regimental life but also through his automatic membership as a young Guardsman of that remaining relic of the old King's rakish Marlborough House Set, the Marlborough Club. Osbert described the way he joined the Marlborough Club in a brief fragment he dictated a few years before his death but never published.

When I joined the Marlborough I was too young for any of the members to know me well enough to blackball me when my name came up for election. So inexperienced was I in the ways of the world that I entered it for the first time as I would have entered a house. Finding the door shut I rang the bell and was greeted by Warwick, the grumpy but kindly hall porter, saying to me: 'I suppose you know, sir, that that was the fire-bell you were ringing.'

Here, as in some lovingly maintained menagerie, he gazed upon the creatures and the habitat of Edward's reign: indeed, that pleasant monarch's shade still seemed to haunt the place – 'the smoke of Havana cigars still lay on the air', there was 'the icy crack of billiard balls' from the two tables on the ground floor that were eternally in use, and in the hot summer afternoons he used to see old members of the club sitting in heavy leather armchairs by the open window and 'drinking that mysterious drink of the nineties, hock and seltzer'.

Here too he saw many of the famous faces of the old King's cronies – 'the squat, blue-chinned figure of the Marquis de Soveral, filling every inch of his clothes, the chirpy, genial Sidney Guille', and the gigantic figure of Lord Chaplin, once known as 'the Squire of England', and now 'so large (he was said to weigh thirty stone) that at first one would think that one must be looking at a statue'.

He had his younger friends of course – his fellow Guardsman, the future Field-Marshal Earl Alexander of Tunis (in those days still a dandified young subaltern with a taste for gossip and malacca canes), Duff Cooper, Edward Horner, Raymond Asquith, young men he was later to describe as being 'sure of themselves . . . full of vigour, with a wit all their own which I take, in essence and origin, to have been a legacy from the old Whig Society'. They were the absolute antithesis of the Golden Horde.

Enjoyment was the aim of life, and gaiety and high spirits the links that bound them, high spirits that could find continual outlet in the theatre and the music hall (he has described how he had to hurry once from a reception at Lambeth Palace to be in time to see a play called *Swat That Fly!*) and in weekending, parties and in ballroom dancing. Somewhat surprisingly, Osbert was almost all his life an enthusiastic dancer. At the age of four he was taught the hornpipe on the sands of Scarborough by an old sea captain, and at twenty he was instructed in the tango by the inextinguishable Mrs Hwfa Williams, 'a woman already for three generations famous for her chic'.

But despite this round of pleasure, it seems clear that Osbert was already viewing this society he had finally discovered with premature nostalgia. He already loved the past more than the present, already sensed that little of this Edwardian afterglow could possibly endure. His own enjoyment of it rested on the merest fluke – had Sir George not agreed to his transfer from the Grenadiers he would have missed it; if Sir George cancelled his allowance it would all be over.

More important still, his London life formed an extraordinary contrast with his life at home. He has described how he and Sacheverell and Edith had always suffered from a 'terrible duality' which had taken such a heavy toll of their nerves, 'the apparently prosperous, traditional life stretched over and disguising the frenetic disputes, the rages and cold hardness'. The rages were, of course, his mother's, while Sir George provided the 'cold hardness'.

But never had this 'terrible duality' been so extreme or so bizarre as now. On one side lay his brilliant life in London – his new-found happiness, his friendships with the rich, the ease of his acceptance in the greatest houses in the land: and on the other side lay everything embodied in his parents – scandal and pettiness, financial nagging, constant scenes, and that entire black history of unhappiness and lack of logic which he had suffered from until he joined the Guards.

The contrast was too great for any compromise, and just as Edith had rejected everything their parents stood for, his inclination was to do the same.

For Osbert it was not as easy as it was for Edith. Edith could effectively sever contact and live on her soup and buns in Bayswater – Osbert could not. His way of life required Sir George's £700 a year (and probably a good deal more) and it was this accursed dependence on his father that made his situation so uncomfortable. However much he wanted to reject his parents, and the whole wretched life they represented, he could not do so. Sir George still kept the power to end his cherished London life; because of this Osbert was still shackled to this 'terrible duality' of living in his two antagonistic worlds.

How could he reconcile his life in London with the situation back at Renishaw? He did so in the only way he knew – by making his father the embodiment of this uncomfortable absurdity from which he suffered. For as a comic character Sir George lost any moral hold he had upon his son. Osbert refused to see him merely as eccentric: he was ridiculous, and those anxious letters urging Osbert on to self-denial and a total change of life could be treated simply as a joke; so could Sir George's building plans, his gardening, his passion for the family, his trips with Henry Moat to Italy, and finally the whole depressing business of his desperate concern with Osbert's overspending and with Lady Ida's debts.

There was one final element in Osbert's developing philosophy during these golden months in London on the eve of war – his profound belief in the sacredness of art. It had been implanted partly by Sir George and partly by Edith and Sacheverell. From Sir George he had absorbed the attitude of a potential connoisseur of painting and of architecture, while from Edith he had picked up something very different – her romantic attitude towards the function and the sacred calling of the artist.

Through her dedication to the poet's role, Edith had found the cause which elevated her above the misery of life at Renishaw and now sustained her in her meagre life in London too. For her and for Helen Rootham, their great hero, the creative artist, was a being set apart – a Swinburne or a Rimbaud or a Baudelaire – and for Osbert too the artist was now appearing like a 'priest, prophet and law-giver, as well as interpreter: the being who enabled men to see and feel and pointed out to them the way'.

It was all rather high-flown and very much in line with the still fashionable, post-Wildean, art-for-art's-sake theories of the day. Unlike Edith, Osbert could not see himself as a potential artist for, as he says, he was ill-educated, badly read, and totally lacked confidence in any powers he did possess. (A blank-verse drama about Jezebel, based all too obviously on Oscar Wilde's *Salome*, which he had written at fifteen had not survived Edith's stringent disapproval.) But his enthusiasm for the artist as a hero naturally led this twenty-year-old Grenadier in eager quest for these priestly beings. It was not all that difficult for him to meet them, for this was the great age of the artist as celebrity; and as he made the rounds of London's fashionable drawing-rooms Osbert encountered many of these legendary creatures. At Lady Speyer's house in Grosvenor Street he shook hands with Claude Debussy. (Osbert was so overcome at meeting the great man that, as he admitted later, 'I recall my emotion, I remember clearly Madame Debussy's face – but that is all!') At Lady Cunard's he chatted more memorably with Delius, got to know George Moore, and heard Richard Strauss in person playing pianoforte selections out of *Der Rosenkavalier*. At the Swintons' he renewed acquaintanceship with Walter Sickert. And at Hill in Essex, the 'Edwardian arcady' belonging to the lady he described as 'the Kitchener of hostesses', Mrs Charles Hunter, he met the painter Tonks, heard anecdotes about Henry James, and first encountered the amiable Robert Ross, once the friend of Oscar Wilde and now of almost everyone who mattered in literary London.

From these encounters Osbert learned an important lesson for the future – that however sacred the artist's calling, there are few actual artists, however famous, who will not respond to good food, well-directed flattery and smart society.

As for appreciating art itself, Osbert was, as he admits himself, a slow developer. During these months in London he began forming an intelligent but cautious taste in modern painting, joined the Burlington Fine Arts Society, attempted unsuccessfully to persuade Sir George to buy some drawings by Augustus John and was predictably impressed by Roger Fry's important second exhibition of French post-Impressionist art in 1912. But his most deeply felt experience of art during this time came not from painting – and still less from literature and poetry. While still at Aldershot in June 1912, he had booked a single ticket for himself at Covent Garden. It is some measure of the intellectual exile he was suffering that he had no idea what to expect from the performance by the Diaghilev company of Stravinsky's *Firebird*. At this period ballet performances were interspersed between the Italian operas of the summer season, and Osbert had booked his seat on spec, not realising the considerable success Russian ballet was enjoying at the time.

As it turned out, that evening helped to shape his life. Karsavina, as the Firebird – 'the greatest female dancer that Europe had seen for a century', as he was later to describe her – was at the summit of her career. Stravinsky's music he had never heard before. Bakst's décor was a revelation. The entire performance was one of those overwhelming, new experiences that occur once in a lifetime, and for Osbert *Firebird* came as a potent demonstration of something he had never realised before, the power of art to supersede reality, and to create its world of fantasy and joy. He also found a private symbolism in the ballet which appealed to him. The evil wizard, played that night by the great Cechetti, was defeated by one feather plucked from the Firebird's breast. This seemed to tell him, as he wrote later, that 'The raging of the old tyrant, and his sycophantic cronies and dependents, *could* be faced' – a lesson which imparted hope in his struggles with Sir George.

So impressed was Osbert by the ballet that he decided there and then that his greatest hope of defeating his own powers of evil, and his old raging tyrant of a father, was through art. 'Now I knew where I stood,' he wrote, 'I would be, for as long as I lived, on the side of the arts.' The fact that he felt he had no skill himself as an artist did not matter. If he could not be an artist, at least he could dedicate himself to art, and pledge himself to defend the artist in his battles with his enemies, the philistines. Art needed champions as well as exponents, and that evening, he says, he decided 'I would support the artist in every controversy, on every occasion. . . . And in my bones I felt that this opportunity would most frequently come my way. . . .'

It was a role – although a strange one at the time, especially as he knew few artists who could need his help. The only ones who did were 'two most rare, but as yet unfledged artists, my own brother and sister'.

CHAPTER FIVE

Disasters

1914–1915

The troubles of the Sitwells, like some mysterious malady, had been temporarily alleviated by the discovery of Lady Ida's debts, but they were far from cured. Her ladyship was still a nervous wreck, unrepentant and impossible, spending the greater part of every day in bed, and blaming Sir George for what had happened. He, poor worried man, was still tormented by his chronic fears – fears of financial ruin, fears about Osbert's character, and fears of that unspecified disaster which had long haunted him. Letters about Osbert's overspending still sputtered back and forth between Renishaw and London. (At one stage Sir George was calculating that his son would soon be £20,000 in debt, and was threatening to make him leave the Guards.) The presence of the money-lender, Field, remained a black threat in the shadows.

And gradually the sickness of the family grew worse again, reaching a sort of crisis during the long hot summer of 1914.

Edith, with rare good sense, had been managing to stay aloof from what was happening at home, and had moved on from Miss Fussell's Bayswater establishment into a rented fifth-floor flat in Pembridge Mansions, Moscow Road, which was to provide her with a London base for eighteen years to come. Moscow Road, Bayswater, was hardly an address at which one would expect to find a Sitwell, and Pembridge Mansions were an unlovely and unheated red-brick block constructed, like so many others in the district, to house the respectable London lower middle classes of the 1870s. She and Helen Rootham had to share three barely furnished rooms (Sir George had offered some discarded curtains from Renishaw which Edith had refused) for which they paid twenty-five shillings a week. However dominating Helen may have been – she was still very much the governess with Edith and would soon show social pretensions as a singer – Edith was intensely grateful to her. Edith had her £100 a year to live on, Miss Rootham rather less, but they appear to have been extraordinarily happy. Part of their happiness came from sheer relief at being away from home, and Edith loved London. Although the poetry that she was writing now was heavy with country images – 'sparkling geese' and nutmeg groves, and satyrs horned as a summer moon – she was by inclination a profoundly urban person.

After her absolute bondage with its atmosphere of hysterical violence for so long at Renishaw, London meant freedom and excitement and escape. The country, she wrote cheerfully to Richard Jennings soon after she arrived at Moscow Road, she personally considered a depraved taste, although she knew some people did enjoy it. 'My parents, for instance, are perfectly happy posturing as lords of creation to an audience of cows. I am not.'

Instead, she and Helen Rootham had embarked on the great adventure of becoming poets. Miss Rootham was still translating Rimbaud and Edith had had two more poems published in the *Daily Mirror* in January 1914 – 'Song: When Daisies and White Celandine' and 'From an Attic Window' in which she lovingly described the urban landscape from her bedroom window:

> Houses red as cat's meat, steam
> Beautiful as in a dream.

Edith's crammed notebooks from this period show something of the energy and seriousness with which she was pursuing her great calling. She was writing with some fluency – poems which were a curious mélange of nursery rhyme and doggerel and country characters, lifted above the commonplace by her sense of fun and isolated lines of sudden beauty. She was working hard. Just as she had virtually educated herself through omnivorous – and eccentric – reading, so she was now engaged in a conscious process of making herself a poet, and by July that year she evidently felt she had sufficient publishable work to make her first slim volume. 'Sir,' she wrote to the publisher Elkin Matthews in much the same tone of voice in which Sir George addressed his private printer, 'I am intending at some time to have a small volume of verse published. I shall probably have it done in paper cover form, like Mr Gibson's "Fires". There will be about thirty pages. Would you kindly let me know about what the cost would be?'

Either her confidence – or, more likely, her finances – failed her and the project came to nothing* but until the summer of 1914 her new life away

* In January 1917 she was again in correspondence with Elkin Matthews, saying, 'I do not want to pay for my book. I would be obliged if you would return my M.S. as soon as possible as I have found a publisher for them.' Several of her early poems were in fact published by Basil Blackwell in *Clowns' Houses* in June 1918.

from home was full of hope and high ambition for the future. Sacheverell's life was optimistic too. However worried and depressed he may have been by what was happening to his parents, he was naturally high-spirited, and in his scholarly pursuits, his love of art and poetry, his natural sympathy with artists he seemed to have already settled for a distinguished life as a scholarly connoisseur, the sort of life, in fact, which Sir George himself once dreamed of. At Eton he had his own room, and thanks to Headlam was permitted to read exactly as he pleased. One of his closest friends was Roger Norton, a precocious mathematician whose family was on the fringe of Bloomsbury, and it was through him that he first heard of people like Clive Bell and Roger Fry. Meanwhile, with Osbert, he was beginning to enjoy the artistic life of London, and it was Osbert who first introduced him to Diaghilev, and took him to the Russian Ballet. Osbert himself had finally met the great impresario at the Keppels', and Sacheverell too had become an instant enthusiast of the Ballets Russes.

As for the family, it was Osbert – outwardly the most favoured and fortunate of the trio – who felt the first disturbing tremors of what was in the wind; and as usual in his life these tremors emanated from Sir George. The baronet, now fifty-four and sporting a newly grown red pointed beard that gave him an uncomfortable resemblance to the Kaiser, had begun the year with one of his periodic trips to Italy. March saw him busily engaged at Montegufoni, manfully supervising an erratic army of Italian masons, plasterers and plumbers in heroic efforts to restore the castle. In Florence the English colony were agog about him and kept asking, who was this rich baronet who was being fleeced by his extraordinary hangers-on? But Sir George was apparently oblivious of the stir that he was causing and in exuberant mood he wrote to Osbert describing in detail a successful buying trip he had just made to Venice and Milan with his 'adviser', Professore Bracciaforte. In Milan he had bought a *cinquecento* sideboard, in Padua a batch of *seicento* flower-pots; but his prize acquisition was a bed. Since his conversion to the need for lying down as much as possible, Sir George had understandably become obsessed with beds, and this latest one, he hastened to explain, was 'the most glorious thing I ever saw: old gold with here and there a little browning, and some eight feet high to the top of the finials over the twisted pillars'. With loving detail he proceeded to describe its finer features, the gilded bosses in the form of foliaged vases and the panaches of plumes at each corner. The one thing he did not reveal was the price, but he clearly failed to see the irony of such a letter. In Sir George's eyes all this expenditure in Italy was justified *because it was not going on himself*. Like the large sums which he had also recently been spending on an artificial lake at Renishaw, or on the golf course which was being laid out in the park, it was a capital investment for the future, and an enhancement of the precious Sitwell patrimony.

This was a distinction which was lost on Osbert, particularly when Sir George returned and finally exploded over Osbert's debts. By regimental standards they were not excessive, but in Sir George's eyes they proved conclusively that his son and heir was heartlessly determined to avoid that all-important change of attitude on which so much depended. Osbert's accounts were scrutinised as if they held some dreadful secret to his character. Money was being squandered on theatre tickets, on Sullivan and Powell's best Turkish cigarettes, on restaurants and entertaining friends. The last straw was apparently an entry marked 'To one pair Laced Pyjamas', which, as Osbert vainly tried to explain, meant that the pyjama jacket was frogged across the chest instead of buttoning in the usual way.

Sir George pretended not to understand, and, as Osbert wrote, 'for the rest of his life I used to hear him, from time to time, confiding in acquaintances, "As my son *insists* on wearing *lace* pyjamas, Lady Ida and myself are obliged to economise. . . . I should never have dreamt of wearing lace pyjamas myself at that age!"' This was the sort of old man's joke, half serious and half playful, with which Sir George quite patently enjoyed needling his son, but there was nothing playful in the action that he took early that summer of 1914. Sir George's foot descended. Osbert's indulgent London life could not be permitted to continue. Sir George would pay his debts this once; then Osbert would have to leave the Grenadiers for good. Since his father held the purse-strings, Osbert had no alternative to accepting his father's ultimatum. Sir George had originally forced him into the army against his will – now he was forcing him out in the same way, and 'with a feeling of the most profound depression' Osbert bade farewell to his beloved London on 20 July to take up civilian life.

Sir George had arranged a suitably corrective job for him – at the Town Clerk's office in Scarborough. Here he would have to be methodical, self-disciplined and diligent. He would have no temptations to extravagance – and Sir George could firmly keep an eye on him.

It is hard to exaggerate the sense of bitterness and degradation this meant for Osbert. This was Sir George's most high-handed exercise of paternal power so far, and for Osbert it was unforgivable. Ridicule was no defence, for it was not much help to go on regarding his father as an interesting old buffoon. He was too powerful for that. Osbert was twenty-one. He had enjoyed the best society in London and been treated as an equal by the wife of the prime minister and by the one-time mistress of a king: for eighteen brilliant months he had started to fulfil himself: for the first time in his life he had known freedom, friendship and success: and now, by the decision of his father, all this was finished, and he was to be thrust back into a clerk's existence and the miserable family life he loathed.

It was an extraordinary trick of fate that made Osbert's expulsion from his earthly paradise of London coincide so neatly with its extinction by the outbreak of war. On 22 July, after saying his farewells in his regiment, he arrived back to find the family assembled for the customary high-summer gathering at Renishaw. Sacheverell was home from Eton, Edith from Moscow Road, and Sir George was as usual anxious to maintain the elaborate façade of a united, happy and distinguished family. But behind this show of Sitwell solidarity it would have been hard to find a family less united. Edith and Sacheverell were both indignant at Sir George's treatment of their brother; their mother was again distracted by the ceaseless worry of her debts. For once again – as in his attitude to Osbert – Sir George was determined to be utterly inflexible. At whatever cost, Field must be exposed for what he was, and everything was set now for a legal action by Sir George alleging fraud and misappropriation of the large sums Field had borrowed in Lady Ida's name. Sir George Lewis had advised him that this was the only way to bring Field to justice – and also to clear Lady Ida's honour. Nor surprisingly, she was dreading the ordeal.

And so it was that Europe and the Sitwell family were both poised that summer on the edge of an abyss. In both cases there was an unnatural show of calm and outward prosperity, as if that stately, well-fed world would naturally go on for ever. 'Never had Europe seen such mounds of peaches, figs, nectarines and strawberries . . .,' wrote Osbert later. 'Champagne bottles stood stacked on the sideboards.'

But even as he savoured the symbolic luxury he found at Renishaw, Osbert already sensed an ominous foreboding. 'It seemed,' he wrote, 'as if the whole earth were waiting, waiting for bells to sound or a clock to strike. . . . And, being of a superstitious turn of mind, I was unhappy when on the first of August . . . I counted the lacquered clock on the stairs outside my bedroom strike thirteen at midnight.'

There were other portents – though whether for the world or for the Sitwells it was hard to tell. Mark Kirkby, the Renishaw woodman, had seen a comet fall in Eckington Woods, and the Archduke Ferdinand had died at Sarajevo. On 3 August Osbert was summoned back to London to rejoin the regiment he thought that he had left for good. On 3 August he was among the crowd who saw 'the Lord Mayor's gilded coach roll trundling up the Mall . . . to Buckingham Palace; a sure sign of the imminence of catastrophe that was coming'. That evening he listened as the crowd 'roared for its own death' outside Buckingham Palace. On 4 August the Great War started. Too late, a distracted Lady Ida scrawled her son a letter:*

* Lady Ida dated the letter 4 July 1914, but it must surely have been written a month later.

My Darling Boy,
I miss you too too dreadfully it is all too awful – Darling I could not stand seeing you leave and I walked to where I was when you left. I pray God you won't have to go I shall be distracted. My darling if I have ever been hard or cruel or unkind please forgive me and realise I adore you and if it was in my power I would do anything in the world for you.

For Osbert and Edith and Sacheverell the Great War would always be the Great Catastrophe, lying 'across the years like a wound that never heals'; and in many ways their lives and work would always stand as an affirmation against the horror and the deprivation of the war – and against the ideas they believed had made it possible. The source of this attitude was primarily Osbert: he was the only one of them to see active service, and he was to be the one most closely linked with the main group of anti-war poets led by Sassoon and Graves and Wilfred Owen which had emerged by 1917. But even from the start Osbert's reaction to the war was far from typical of other poets of his generation: his first-hand experience of the fighting would also be unusual; so would the way in which his ideas developed from the experience of war.

By training and profession he was a regular soldier, and he rejoined the Grenadiers with none of that lemming-like excitement with which so many war-starved young civilians rushed off to meet the slaughter in the north of France. His fastidious nature was congenitally opposed to mass emotions; and, questions of character apart, he had already sat through sufficient lectures on the coming war by plainly incompetent very senior officers to have his doubts about the great adventure which the London crowds were now applauding. As a trained soldier he would have turned a very frosty eye on Rupert Brooke and his

> Now, God be thanked Who has matched us with His hour,
> And caught our youth, and wakened us from sleeping,

Osbert did not want 'wakening'; he did not believe in God, and what had 'caught his youth' had been those eighteen precious months in his beloved pre-war London. So, from the very start, the war appeared to him as an ending, not a challenging beginning. True, it had saved him from Scarborough and Sir George, but it had also cancelled at a stroke the London life he loved: later he wrote of the abruptness with which the glittering city of lights ended on 4 August 1914, and a new London in transition came into being. Even as this was happening his pessimistic nature was already sensing in this war a further element within that lurking doom which had so long been threatening his family and the world he loved.

One might have thought that, even so, as a young professional soldier Osbert might just have welcomed war as a chance to exercise his craft;

but, as he makes quite clear, he really had scant aptitude for military business. He felt no need to prove himself in battle; he loathed discomfort, was an appalling shot, lacked any real group feeling for his regiment, and never learned to read a map. He was a peacetime soldier; but now, since war had come, he was prepared to do his duty – whatever it might be.

Yet even now the continuing disaster in his family seemed to be competing with the wider catastrophe of the war: early that November, while Osbert was in Chelsea Barracks preparing to embark for France, his mother was appearing in the King's Bench Division for the start of her long-awaited action against Julian Field.

As it turned out, the actual trial was nothing like the grim ordeal which Lady Ida evidently feared. It was a brief, one-day hearing. Her counsel made a brisk, efficient speech alleging fraud and breach of duty against Field in that he had misappropriated £7,775 borrowed in Lady Ida's name. Field appeared but, to what must have been her ladyship's profound relief, made no attempt to contest his guilt. The jury gave their verdict against him, and judgement was given in Lady Ida's favour for her lost £7,775, and the costs of the action. As was claimed at the time, Lady Ida's honour had been vindicated and Field exposed for the fraud he was.

But even at the trial it was clear that this was all distinctly academic, and it was not too hard to see why Field had not bothered to defend himself. As the evidence revealed, he was an undischarged bankrupt. His own debts were enormous, and there was not the faintest chance of Lady Ida or Sir George ever recovering the money they had been awarded by the court. Worse still, the money-lender Owles had not been repaid the money borrowed from him, nearly three years earlier, in Lady Ida's name. Until he was the miserable affair would never end; and until it did there was still the very real danger of the whole scandal being published. That first autumn of the war the fear of this still hovered like a black cloud over Renishaw.

By a strange coincidence, just before Osbert left for France in December, Sir George and Lady Ida in Scarborough had both felt the power of the enemy when three cruisers from the German battle fleet steamed out of the morning mist and shelled the town. While Sir George sheltered in the cellar, and Lady Ida remained resolutely in bed, fragments of shells struck their house, Wood End. According to Osbert, Sir George would always be convinced that he had been one of the prime targets of the Germans, and that he had endured a worse bombardment than any that his son was to face at the front.

Osbert's reaction to the fighting he experienced in France was less dramatic than Sir George's. He felt no excitement or surprise, but simply the depressing feeling that he had somehow been through it before. The actual danger of the front line did not worry him: sudden death was almost the least of his concerns. What did oppress him was the monotony, the 'excess of bleak boredom and grey discomfort', and the loss of friends. The landscape of the front with its 'flying fountains of dead earth, the broken trees and mud' appeared like a vision of some dreadful limbo in which he was perpetually lost; the front line formed 'a monochromatic geographical entity of its own, floating, cloudlike, across a continent'.

It was a nightmare land which corresponded with a nightmare which had haunted him for years. It had begun in earliest childhood when the cry of the rag-and-bone man at Scarborough crystallised his night-time fears of the destruction of the precious world around him, a world that was linked with that overwhelming love his mother gave him, but which was threatened by her rages and by Sir George's cold, forbidding presence. The threat had grown through childhood. His father's breakdown and his mother's instability had nourished it: later the grim world of his unhappy schooldays had given it still further substance. It was a loveless, lonely world of grey monotony and hopelessness, the fearful, introverted world of an unhappy schoolboy, bullied and oppressed by those around him; and during these early months in northern France Osbert's immediate reaction was to compare his active service life with life at school. In the front line, 'At least there were no masters, matrons, or compulsory games. The discomfort was, at times, perhaps a little greater, the food, though tinned, perhaps a little more palatable.' The thought struck him that, perhaps, his schooldays had been specially arranged to prepare him for the ordeal of war, and that through the hated system of private and public schools the children of the upper classes had been 'taught to bear with composure a high degree of physical hardship and spiritual misery, while enclosed in an atmosphere of utmost frustration'.

By the first Christmas of the war it was quite clear that there would be yet one further trial for Lady Ida. The money-lender Owles had died, but his heirs were clamouring for payment of their debt. Either they got their money or they would go to law. Would it never end? Sacheverell wrote despairingly to Osbert. Lady Ida was primarily concerned at what a further trial would reveal, and trying frantically to make Sir George settle with the Owles estate. Sir George stayed true as ever to his character. Since judgement had been pronounced against Field in open court, that was an end to it. Field had been shown up as a criminal. He had obtained the money from the money-lender. If anyone should now be pressured for repayment, it should be Field and not the Sitwells. On this Sir George was adamant.

Stuck in the freezing Flanders mud in early 1915, Osbert was increasingly oppressed by the grey wretchedness around him and by the complementary wretchedness at Renishaw. The family wrote regularly, so that he knew exactly what was happening; his mother was hysterical and sick, his brother desperate with worry, and the continual shadow of the threatened trial had now 'grown to such dimensions that it stretched across the whole day as well as night', darkening the whole horizon.

One can see how in such circumstances the two events – the growing disaster of the war and the inexorable disaster of his parents – should have become part of the same recurrent nightmare, and how a single villain seemed to inhabit it. Was it by chance that the Kaiser and Sir George looked uncannily alike, shared the same birthday, albeit a year apart, and that Sir George was now professing himself a keen admirer of the German emperor? Was it by chance that for as long as Osbert could remember it had always been Sir George who had been plaguing him with his continual talk of duty and self-denial – and using his power to inflict boredom and unhappiness on him? Wasn't this war with its hideous monotony and sacrifice – and old men's exhortations to the young for duty and self-denial – an obvious extension of the war he had been fighting now for years with his father and his father's allies?

If Osbert had any lingering doubts about his father's perfidy, they were effectively dispelled in the anxious weeks that followed. As the finest troops of the British professional army – the same army which the generals had maintained would bring home victory by Christmas – attempted to make one decisive thrust against the encircling Germans, Osbert found himself fighting a never-ending battle against cold and sleeplessness and mud and an all but invisible enemy. And back in England the long campaign against Lady Ida and her debts was reaching its own dénouement.

On one side stood inflexible Sir George and his advisers: on the other the impatient representatives of the Owles estate. The money-lenders had by now informed Sir George that they wanted justice – either they received repayment of their £6,000 plus interest, or they would prosecute everyone involved, Field, Herbert the money-lender's tout – and Lady Ida. It would not be a pleasant case. Since Lady Ida had already secured judgement against Field, they now had no alternative but to allege a far more serious matter than simple non-payment of a loan. The charge would be one of conspiring to cheat and defraud Miss Dobbs when Field and Lady Ida originally persuaded her to guarantee the bill.

There is always an element of bluff in cases such as these, and Sir George evidently decided he would call it now, and damn the consequences. It is uncertain if he knew the full extent of the evidence against his wife. The worst of it lay in those letters she had written to Field, and Field undoubtedly tried using them to persuade Sir George to end the case by paying off the debts. Sir George, inflexible as ever, refused to have anything to do with a man he regarded as a blackmailing scoundrel. At this point Field would appear to have carried out his threat to send on the letters to the prosecution. Presumably he thought that, if Sir George refused to deal with him, he would have to listen to the lawyers of the Owles estate once they explained the evidence against her ladyship.

Here Field, like many others in their time, failed to reckon with Sir George's infinite obtuseness. The case was fixed for Monday, 9 March at the Central Criminal Court; and the moment was suddenly past when Sir George could do anything to stop proceedings, even had he ever wanted to.

Over one thing Field was right – his warning of the scandal that the case would bring. The rights and wrongs of Lady Ida's situation were hardly here or there. By letting the case occur at all Sir George had committed the one unspeakable crime of the old Edwardian society: the indignity of what King Edward used to call 'public laundering of dirty linen'. It was an unspoken rule that the upper orders never publicly exposed their faults before the lower orders; but if they did so they were naturally fair game for everything that followed. During the second week of March the Sitwell case soon stole the headlines from the U-boat war and the latest news from France. 'Earl's sister at the Old Bailey' shouted the *Daily Mirror*; 'Sidelights on Society' replied the *News of the World*.

The press were to get their money's worth. The judge, Mr Justice Darling, was an egregious old legal martinet, who enjoyed showing off his wit in court and loved publicity.* And Sir George and Lady Ida, whatever their feelings for each other now, put on a great show of dignity and solidarity. Ida, 'tall and slight', as one reporter ventured to describe her, maintained an impeccable demeanour through the whole ordeal, and her clothes were much remarked on at the time. So, for that matter, were Sir George's, as he sat impassively throughout the trial in what must have been one of the last private frock-coats ever worn in London.

Despite their dignity it proved a wretched case, with the prosecution taking pains to drag out everything it could against Lady Ida, and largely ignoring Field's part in the affair. Miss Dobbs was called, and long

* He specialised in a certain brand of false naïveté, and on one occasion intervened when counsel referred to Charlie Chaplin.

'Who,' he inquired, 'is this Mr Chaplin?'

'My Lord,' came the reply, 'he is the Darling of the Halls,' a two-edged reply that contained more than a grain of truth.

passages from Lady Ida's letters were read in court, however seemingly irrelevant. The judge, quite clearly out to show that he was not overawed by rank or title, pointedly made no effort to direct the prosecution's case to the points at issue, and by the end of her seven-hour examination in the witness-box Lady Ida's reputation was in ruins. Great play was made of passages she wrote to Field saying she would get 'her boy', who was just entering the Hussars, to try to 'get hold of' other young officers to back her bills. She tried explaining that she used the phrase 'get hold of' simply to mean 'to meet', but was not particularly convincing. Nor did she fare much better when she attempted to explain her remarks to Field about 'getting women into Society'. By the time she left the witness-box the jury must have been convinced that tall Lady Ida in her 'long black cloak of military cut and a black Cossack cap' was a ruthless, scheming woman who needed to be taught a lesson.

There was one other member of the family who was compelled to undergo the agony of sitting through the trial. Because of those unguarded letters Lady Ida wrote to Field, and the references to the way she hoped her 'boy' could 'get hold of' his young fellow-officers to guarantee her bills, Osbert was inevitably involved in the whole scandal too. His name was mentioned in the evidence in court, and since he might have been required by his mother's defence he was dragged back from France and forced to endure the whole miserable performance.

Apart from his feelings for his mother, it was a galling and degrading situation as he sat in court. For here he was, a lieutenant in the Guards, having to listen to the imputations of the prosecution of what sounded like his own dishonourable behaviour. He had no chance to answer back or to explain himself without going into the witness-box himself, and causing further trouble to his mother. All he could do was to sit there and endure it all.

But one can easily imagine how he felt against the man who had allowed this whole disaster to occur – Sir George; and when he wrote about this in his autobiography a quarter of a century later, it is the one place in the book where his real hatred of Sir George breaks through. While the trial was on he describes Sir George acting with macabre insensitivity – particularly to him. According to his account, Sir George insisted that his son, on his return from France, should sleep in a distant bedroom in their London house because he had read that front-line troops were verminous; and as the trial proceeded Sir George made no reference to the ordeal his children and his unhappy wife were facing. Instead he insisted on enlivening 'on several occasions luncheons and dinners of the most fearful family anguish by discoursing on *The History of The Fork*'. As a result, wrote

Osbert, throughout this whole ghastly period of the trial the house seemed 'to be full of a highly coloured, evil levity of spirit'.

One could interpret this as Sir George's way of trying to maintain the family's morale and a façade of dignity against disaster; but Osbert clearly did not think so. For Osbert there was no possible excuse for what Sir George had done; with his meanness and his coldness and his canting talk of self-denial, he had permitted this unspeakable disaster and disgrace to befall his family.

On 14 March the full extent of the disaster was revealed, when Mr Justice Darling pronounced sentence. Thanks to his admitted guilt and his criminal record, Field received three years' imprisonment. Herbert was acquitted with a warning. And Lady Ida – to expressions of the judge's infinite regret – was sentenced to three months' imprisonment. 'I can only say', he added, 'that if it were not for the state of your health, it would be more severe.'

'Lady Ida', wrote the reporter from the local *Sheffield Daily Telegraph*, 'took the sentence very calmly and walked firmly from the dock to the cells below.'

CHAPTER SIX

'Pictures, Palladio, and Palaces'

1915–1918

IN THAT first shock of the disaster and disgrace, as Lady Ida settled into Holloway, the remainder of the family – including Osbert, who was given leave – withdrew to Renishaw. The sentence had been totally unexpected, and everyone was stunned by its severity. The worst hit was probably the one who had always been the most protected, seventeen-year-old Sacheverell. He was still at Eton, and had no idea of what was happening until he saw a headline in the *Daily Express* announcing his mother's prison sentence. The headmaster immediately gave him leave of absence from the school – something rarely done at Eton in those days, even for death or sickness in the family – and in a state of numb despair he travelled up to Renishaw.

He was so overcome with the horror of the event that, as he says, his mind 'blanked out completely on arrival', and he has no memory of what ensued within that gloomy house. Edith never talked about it either, and Osbert confined his own account of what he called 'this, our private calamity' to a description of the great gale that struck Renishaw the night that they arrived. As usual with the Sitwells, nature was providing a symbolic background to the dire event. 'Never', wrote Osbert, 'have I known such storms as those which now battered the old house, until it seemed alone on its tableland in a world of fury.'

But, strangely, it appears that even now the three children made no overt attack upon Sir George for what had happened. They were still frightened of him in the flesh, and the Sitwells had a habit of avoiding head-on confrontations with each other. Instead, at this time of suffering and crisis, the courtesies were carefully observed, the old façade of the united family painfully maintained.

On the surface it would soon appear as if everyone was acting admirably. Scandal has rarely worried the English aristocracy for long, and once those initial storms at Renishaw subsided it might have seemed as if the air was clearer now than it had been for years – even for Lady Ida. Whatever her suffering and disgrace, she had atoned for her stupidity and was free at last from years of blackmail and financial pressure – and Sir George's miserable nagging over her extravagance. Predictably, she bore the next ten weeks with dignity and fortitude, taking the respect of wardresses for granted, calmly declining to be visited by the Archbishop of Canterbury on the grounds that it might prove mutually embarrassing, and briskly turning down a visit from some upstart neighbour's wife by saying that since she had never entertained the woman in her house she could not do so in her cell.

For Sir George too it must have been a vast relief to have the whole affair behind him. There is no evidence of remorse – or even mild regret – for his part in the affair. There was a lot of criticism, naturally, of this rich man who allowed his wife to go to prison rather than pay her debts, but if it ever reached him – which is doubtful – he had a lifetime's practice in the art of always knowing best, and could console himself with the knowledge that he had not betrayed his principles. He had resisted blackmail, Field was in prison, and as for Ida, Sir George was a great believer in life's little lessons. If she was cured of her extravagance at last, all might yet work out for the best. Certainly he was not entirely heartless over her sufferings. At the beginning of May he was writing anxiously to one of her old friends, begging her to come and stay at Renishaw when Ida returned from prison. 'Your influence with her has always been so good,' he wrote, 'and I dread the first few weeks, especially if she has no friend with her.' He also spent a lot of time and trouble petitioning the Home Secretary to revoke the sentence on the grounds that Lady Ida Sitwell never understood figures.

Much of the fascination of the older Sitwells comes from the way in which their dramas and absurdities seem like an acting-out of the obsessions of the society in which they lived. Sir George and Lady Ida had long appeared as parodies of different Edwardian upper-class attitudes – Sir George with his crippling concern with birth and money, his hypocrisy, his lunatic omniscience, and the continual power-games he played with his dependents; and Lady Ida with her extraordinary stupidity, her self-indulgent sense of 'fun', and the disastrous belief that everything could ultimately be fixed and work out for the best.

Now at this point of high disaster they had performed their ultimate symbolic role for all their children, and at a blow the catastrophe finally had freed the trio from their parents and their parents' world and the restrictions of their class. It also freed them from Sir George's domination. In all their arguments till now he had always been able to fall back on his claim that he alone knew what was best for the welfare of the family: now that his always knowing best had ended with this family disgrace, his real authority had gone.

It naturally took some time for the full implications of this dramatic change to work out for the children. Osbert would finally rejoin his Grenadiers in France, Edith return to Bayswater and the unlikely job that she had taken with the War Supplies Depot in Kensington, and Sacheverell go back to Eton for the beginning of the summer half.

But the immediate effect was to cement all three of them into their 'closed corporation'. Since childhood they had relied on one another; now they closed ranks against the world to replace their parents and protect each other from the shame of the disaster. For years to come the only people they would trust entirely would be one another.

Osbert took the first initiative. He has described how on his return to London he attended a sergeants'-mess concert at Chelsea Barracks and heard a tactless comedian crack a joke about Lady Sitwell, which was received in stony silence by the audience. This show of loyalty encouraged him. Several years later he would admit to a friend that he 'decided there and then to brazen things out. Rather than hang my head, I would go out to all the parties and theatres and be seen everywhere in London just to make it plain that I had nothing to be personally ashamed of.'

This was very much in character, and one detects here the beginnings of his later passion for publicity, but there was more to it than this. In the first place there was a clear desire by all three Sitwells to make a name and reputation to live down the disgraced name of their parents, as if emphasising that henceforth the name Sitwell had nothing to do with Sir George and Lady Ida. (Edith carried this to the extreme of refusing to name her parents in her early *Who's Who* entries, which read: 'Sitwell, Edith; *b.* Scarborough; *sister* of Osbert Sitwell and Sacheverell Sitwell.' Sacheverell also refused to mention either parent in his early entries.) And this in turn, as Sacheverell says, 'was to tie the three of us together, two brothers and a sister, in our determination to live and leave a mark of some sort or kind'.

Apart from Osbert's social witness, Edith had shown already how this could be done – by publishing her poetry. To her unforgiving eyes 'the sordid tragedy in which my mother was involved', was already mirroring something of the horror of the war itself with its life-denying misery and was 'a dwarfish imitation of the universe of mud and flies'. Her response to it was to submit her first small book of poetry to the young Basil

Blackwell at Oxford for publication, and the book appeared early that autumn in a paper cover, priced sixpence.

This first book of Edith's took its title from one of the poems that she wrote that spring. It was called 'The Mother' and is a violent and macabre piece of writing in which a woman tells of how her son, presumably meant to represent the young men fighting in the war, became a murderer and killed her in the process. But in the last verse the mother compassionately forgives him from the grave:

He did no sin. But cold blind earth
The body was that gave him birth.
All mine, all mine the sin; the love
I bore him was not deep enough.

This poem, with its heavy-handed imagery, was hardly typical of her later style; not surprisingly, she omitted it from her *Collected Poems*, but, as she told Geoffrey Gorer, the mother of her poem, with her compassion and her understanding for her child, was an idealisation of the sort of mother she had always wanted for herself, and never found in Lady Ida. There was no such ambiguity in the second poem in her book. 'The Drunkard' graphically described how she entered the bedroom of a drunken woman:

But yet she does not sleep. Her eyes
Still watch in wide surprise

The thirsty knife that pitied her;
But those lids never stir,

Though creeping Fear still gnaws like pain
The hollow of her brain.

She must have some sly plan, the cheat,
To lie so still. The beat

That once throbbed like a muffled drum
With fear to hear me come

Now never sounds when I creep nigh.
O! She was always sly.

And if to spite her, I dared steal
Behind her bed, and feel

With fumbling fingers for her heart . . .
Ere I could touch the smart,

Once more wild shriek on shriek would tear
The dumb and shuddering air . . .

As well as coinciding with the publication of Edith's first bitter little book of verse, this time of shock and misery following the family disgrace saw the beginning of Sacheverell's emergence as a poet. For some time now Edith had been encouraging him to write. She had great faith in his lyric gifts – and in her power to help him realise them; and he acknowledges her all-important influence. 'Perhaps in my case I was as much a pupil of my sister as any poet ever has been of another.' With Edith publishing *The Mother*, that summer he began to write what he has called his own first real poetry, largely inspired by her.

But in a way the most important and surprising influence of all upon the direction of the trio during the crucial summer of 1915 was Osbert's.

Late that July he was under canvas in a camp near Marlow waiting for embarkation. Marlow is not far from Eton, and, with school ending, Sacheverell impulsively decided he must somehow see his brother before going back to Renishaw, and so took a room at a hotel near the river.

It was a time of deep emotion for them both. Once back at Renishaw, Sacheverell would be with his parents for the first time since Lady Ida's return from prison, and by then Osbert would be back in France. The slaughter was increasing at the front; Sacheverell had no certainty of ever seeing his 'Darling Osbert' again.

During their first day at Marlow they walked for miles together, talking of everything except the war, of the whole world of art that seemed all but lost for ever, and especially of poetry. It was an emotional time; Sacheverell says that he was almost 'hysterical' with grief and anxiety and it was now that he decided he would commit himself to the poet's calling, just as Edith had already done; for now that he was discovering his talents he could envisage poetry, and his role as poet, as the one credible alternative to the risks and obscenities of war, to the family disgrace, to Edith's 'universe of mud and flies'.

Osbert was more than sympathetic, making it plain that if they both survived the war he would do his best to help his brother. They would work together, live together and, in effect, Osbert would take the place of their disgraced and futile father.

As if to show that this was no idle promise, the next night Osbert took Sacheverell to London, and out to dinner – to the Eiffel Tower in Percy Street.

The Eiffel Tower has a unique place in the cultural archaeology of the twentieth century. During the twenties and the thirties it became a centre for the flusher members of that so-called 'Higher Bohemia' who made 'Fitzrovia', this area between Bloomsbury and Soho, so much their own. During its somewhat seedy decadence on the eve of the Second World War, it would become a home from home for Dylan Thomas and Augustus John and still later it was Ernest Hemingway's favourite London restaurant; today it has disappeared entirely under the glossy trappings of a new and re-named restaurant. But in 1915 the Eiffel Tower had just been discovered by a number of the more intelligent and adventurous 'artistic' children of the rich whom Osbert had already met in their parents' houses during his socialite–Guardee period. These included Beerbohm Tree's rebellious daughter Iris – already a distinctly *outrée* student at the Slade – Lady Diana Manners and Lady Cunard's nineteen-year-old daughter Nancy, for whom the Eiffel Tower became 'our carnal-spiritual home'.

Quite why it did so is a little difficult to understand. The food was expensive and not always very good. Stulik, the fat Austrian proprietor, spoke little English and, despite his frequent hints of illegitimate descent from the Emperor Franz Josef, was hardly an exciting or particularly endearing *patron*. The décor was confined to ferns in large brass pots. But Osbert knew what he was doing when he took his brother there to dine, for the Eiffel Tower already represented something important for the future. Its windows looked down Charlotte Street, where Walter Sickert had his studio. Wyndham Lewis had recently painted one of his Vorticist murals in the private dining-room upstairs and Augustus John with his red beard and roving eye had already helped to give the place what Michael Holroyd calls 'something of the Café Royal atmosphere, but on a scale much simplified'.

The world of the grand artistic drawing-rooms which Osbert had so much enjoyed was hit by the war, and the Eiffel Tower was offering a foretaste of the freer, racier café society which would largely take its place as the common ground between society and art when the war was over.

Sacheverell was painfully shy in those days and found himself tongue-tied in the face of Miss Cunard's stark white complexion and outrageous talk. But Osbert, polished, witty Osbert, was very much at home here with his friends, and in his company Sacheverell could glimpse something of that world of art and artists which he and his brother had so fervently discussed the day before. For the schoolboy poet this was a miraculous escape from the family disgrace, the gloom of Renishaw and the holocaust of war. It was, he says, 'extraordinarily exciting'.

Within hours of their dinner at the Eiffel Tower the brothers parted, Sacheverell for Renishaw, Osbert for northern France, where he began his second tour of duty at the Front. This was to be a very different matter from his first. During the first spell in the Flanders trenches, depressed by his mother's coming trial, he had stoically endured the mud, the boredom and the misery. Now he was waiting in reserve for the big offensive which would go down in history as the Battle of Loos, and he had the time and opportunity to observe the war for the military disaster that it was. His brightest friends were being killed. The Desboroughs' rumbustious poet son, Julian Grenfell, had died that May of head wounds just a few weeks after writing 'Into Battle'. Then in October Osbert's close friend from the Grenadiers, Ivo Charteris – Lady Diana Cooper called him 'an enchanting child-soldier' – had followed him.

But more disturbing still was the grim experience of these days before the battle. As part of the reserve battalion Osbert spent his time with his men as they moved up, enduring all the chaos that went on behind the lines. From time to time they would be chivvied on by staff officers, talking about some mythical great victory in the offing. For Osbert their 'clipped, fox-terrier-like phrases' could only bring back memories of Gowk and Fribble-Sadler and the cavalry mess at Aldershot. Then on the day before the battle there was one more example of the British military mind – an enthusiastic general who harangued the troops and boasted of the secrecy of the British plans. 'The Germans haven't begun to get an idea of it!' he assured them.

The general's voice was curiously familiar, and as the battle rolled on in the days ahead he could not fail to see a similarity between the generals and Sir George: the same brand of optimistic yet fallacious logic, the same way of bravely staying safely in the rear, the same glib reliance on the young to do the fighting – and the same ultimate catastrophe. Loos was a battle of outrageous blunders, of missed opportunities and wasted lives. A major cause of British failure was a mistake which Osbert witnessed – Sir John French's delay in bringing up the reserves – and when his battalion reached the battlefield Osbert would see the hideous results. Despite the general's talk of secrecy, the Germans had been waiting for the British. Before long 'the bodies of friends and enemies lay, curious crumpled shapes, swollen and stiff in the long yellow grass under the chicory flowers'.

Loos was an unforgettable example of something Osbert knew already – the terrible disasters which ensued from these powerful elders who were so certain they knew best. At home his mother went to prison. Here there were twenty thousand British dead.

Loos was Osbert's final battle. When it was over he was promoted captain, given a company, and throughout the early spring of 1916 he remained in Flanders, while the generals argued and the shattered British army painfully regrouped.

Then in April he contracted blood-poisoning from an injured foot. It was serious enough for several weeks in hospital, and at this point in the war officers were still being sent home for convalescence. So he spent several weeks that spring at Renishaw with a trained nurse looking after him.

If in his time away his memories of Sir George had softened, these weeks were quite enough to banish any sentimental feelings for his father. The war had made him more preposterous than ever. The nurse – and the implication that Osbert's state of health could actually be worse than his – upset Sir George considerably, particularly since his war-work, growing potatoes for the troops, was so much more demanding and important than Osbert seemed to realise.* There was still trouble over money too, since Sir George had for a period cut off his son's allowance on the grounds that he could not spend money at the front. Sir George was also up in arms about *The Mother*. 'Edith's poems', he kept repeating, 'make *me* look ridiculous!' It was with a sense of great relief that a restored Osbert found himself ordered back to Chelsea Barracks at the end of May.

With Osbert unexpectedly in London, the ambitions of the three Sitwells to 'leave a mark of some sort or kind' were able to proceed much faster than seemed possible when Osbert and Sacheverell made their plans beside the Thames at Marlow ten months earlier. Also, since then their prospects had received an unexpected boost. Osbert had made an important self-discovery. Stuck in a billet outside Ypres during a rest-period earlier that year, 'some instinct, and a combination of feelings not hitherto experienced, united to drive me to paper' and, 'astonished' at his sudden ability to concentrate his thoughts, he started to compose some verse. The result was a poem he entitled 'Babel'.

The war with its heightening of the emotions was a great creator of sudden poets, and 'Babel' would not stand out today from the mass of now forgotten 'soldier poetry' that found its way into print. A solemn and distinctly baleful hymn on the destruction left by war, it seems to echo the rhythm and the feeling of the schoolboy's favourite battle poem, Thomas Campbell's 'Hohenlinden':

> Deep sunk in sin, this tragic star
> Sinks deeper still, and wages war
> Against itself; strewn all the seas
> With victims of a world disease
> – And we are left to drink the lees
> Of Babel's direful prophecy.

But the act of writing 'Babel' meant more for Osbert than it would

* Sir George took his agriculture seriously, farming his three thousand acres with considerable profit and success. He formed a company to manage the financial side of the business, and for tax purposes Osbert was made a director. This explains his celebrated *Who's Who* entry for the war: 'Fought in Flanders and farmed with father.'

have done for some otherwise obscure military rhymester. In the first place, Osbert had his invaluable 'closed corporation' back in England to seize on his efforts and promote them. In this case it was Edith's advice that brought 'Babel' to the attention of that same poetic talent-scout who had originally ensured her publication in the *Daily Mirror*. Clever Richard Jennings must have seen from the beginning that Osbert's poetic future lay some way up-market from his sister's: on 11 May 'Babel' was published in the London *Times*.

For Osbert this was naturally a great encouragement. In those days it was quite the thing for smart young soldier poets to be published in this way. Only a week or two earlier there had been a great response when *The Times* published Julian Grenfell's 'Into Battle'. 'Babel' was something of an answer to its warlike sentiments, and Osbert was always to be proud of his literary debut in what he called 'an organ of such national and international celebrity'.

But much as he enjoyed publicity, there was something more important still for Osbert in this whole event. He had not merely written a poem; he had become a poet. Instead of that humble relegation to the role of 'champion' of the arts and artists which he had once envisaged for himself, he had a chance now to become one of the elect and to be linked more closely still with Edith and Sacheverell.

Had that been all there was to it, Osbert's conversion to the artistic life would have been mildly ridiculous, a self-indulgent art-for-art's-sake posturing like many a Wildean nineties aesthete before him. As it was, there was certainly a whiff of that around the trio from the start, with their self-proclaimed artistic earnestness, their cultivated sensitivity, and their extraordinary skill at self-promotion. But they would rapidly progress beyond this. Osbert fulsomely proclaimed that, 'From the moment of my beginning to write, my life, even in the middle of war, found a purpose.' It was a purpose which he would share increasingly with Edith and Sacheverell.

Paul Fussell in his remarkable book *The Great War and Modern Memory* has written of how the war finally produced a whole 'generation of bright young men at war with their elders'. Osbert could with justice claim to have been the first of them and certainly the first to put this warfare into words.

The major poets of the war – Blunden, Sassoon, Graves and Owen – were all soldier-civilians whose formative experience of war came later – in the long slaughter of the Somme and in the dragging months of horror that went on until the Armistice. Their writings all reflect a first-hand vision of this holocaust, and it was this, and the nearness and constancy of death, the comradeship of the trenches, the revelation of a crucified humanity, that filled their minds.

This sort of feeling is strangely lacking in all Osbert's writing on the war. Apart from a few sparse pages on the fighting in his autobiography, he made curiously little use of his actual experiences at the front. One gets the feeling that, even when under fire, his mind was elsewhere; and for all three Sitwells the war meant something much more complex and involved than the straightforward horror of the battlefield.

In a sense all three of them had been at war for years – with their parents, with the Golden Horde and the philistine, with all those malevolent black forces which had been threatening their comfort and their ideal world of beauty. And just as the disaster of the war had coincided with the disaster of Lady Ida's trials and disgrace, so did the war itself appear as the culmination and extension of this private conflict. For them the greatest enemy was not the German army but those attitudes of mind which had blighted their own lives and helped produce this international calamity – the blindness and glib hypocrisy of the old, the games-playing optimism of the young, the herd-instinct with its patriotic call to duty, and all the Edwardian absurdities their parents represented.

In essence, what all three Sitwells wanted was quite simple, sane and selfish. Edith had her burgeoning ambition to become a major poet, Sacheverell his desire for a private world of poetry and scholarly appreciation of the arts, and Osbert a longing for the elevated social round, good talk, rich food, artistic and receptive friends, foreign travel and the Russian ballet.

Their aims, like their characters, were quite separate, but in the middle of the war they interlocked as they increasingly appreciated what they had lost. Instead of this golden world, their friends were being slaughtered, their mother was disgraced and their father was increasingly impossible. The generals and the politicians were bungling the war and killing off the young, the arts were in eclipse, and any moment now Osbert could be ordered back to France; Sacheverell within a few months' time would follow him.

Such was the situation which they faced. The 'purpose' they had found was to do something positive about it all.

This 'purpose' was none too evident at first as Osbert settled into London. By far the most sociable of the trio, his appetite for people had been sharpened by his time in Flanders and the spell at Renishaw, and, since he was back in Chelsea until the medical board pronounced him fit enough for active service, he took advantage of his opportunities. The social diaries of the period offer stray glimpses of the young captain in the Grenadiers making the most of the rich country house and London life which still went on in grotesque contrast with the grim world of the battlefields across the Channel. Lady Cynthia Asquith records in her diary meeting him on several occasions now – dining in style and some hilarity at the Ritz, with Lady Horner and Duff Cooper and his Lady Diana at the next table, visiting the Gilbert Russells in their new house in Audley Square, and playing poker until three-thirty in the morning with Lady Diana and Lord Stanley after an 'oyster, baby potato, and champagne orgy' with Venetia Montagu. Lady Cynthia found him 'quite agreeable to talk to' but one thing struck her even then as odd: 'He told me he read every night of his life till three or four in the morning.'

Colonel Repington, the military critic of the *Morning Post*, described how he met young Captain Sitwell at Watlington Park, the Keppels' country house in Oxfordshire. It sounds like an echo of the old weekend parties Osbert had so enjoyed before the war. Both Keppel parents were there, and their daughters Violet and Sonia, Lady Wemyss, Lord Ilchester, Osbert's old tango teacher, the stone-deaf but redoubtable Mrs Hwfa Williams, and Harold Nicolson and his wife Vita. Osbert had brought 'another young fellow in the Guards'. It was, the colonel said, a 'gay party', with tennis in the grounds, hide-and-seek on a wet English Sunday morning, and 'a lot of Bridge'. Then on Monday morning the colonel returned to London with Osbert, Harold Nicolson and the young officer 'and instead of discussing soldiering, we talked of nothing but pictures, Palladio, and palaces. They were all three extraordinarily well-informed, and knew Italy well.'

The colonel was quite put out to hear such topics from a Grenadier, but for Osbert palaces and pictures mattered far more than soldiering. Poor Colonel Repington would have been still more disturbed had he known that Captain Sitwell now believed quite firmly that the massacre should stop 'if the fabric of European civilisation was to survive', and that such feelings were making him dangerously intolerant of the sort of people he had been meeting that weekend. The old, fashionable society he once enjoyed was starting to upset him with its complacency and bland acceptance of the slaughter. Instead, as a neo-pacifist, he was inclining now to the company of those he felt most at home with – artists and writers and the sort of fashionably artistic young he dined with at the Eiffel Tower.

With them talk of 'pictures, Palladio, and palaces' was more than mere idle gossiping about the arts. Osbert, the Wildean aesthete, was starting his own aesthete's revolt against the war and the society that bore the guilt for making it. Art and the ideal of artistic pleasure had become a sort of shibboleth, a bond of union between those who were for 'civilisation' and against the generals.

The most important personnel in this revolt were naturally his brother and his sister. Family loyalty apart, they were the two 'artists' he was closest to. They shared one another's views on everything that mattered and believed passionately in one another's talents. But there were others too. To start with, there was Helen Rootham: she was a poet, a translator of French poetry and almost a member of the family. Then there was the Eiffel Tower group, particularly that extraordinarily liberated pair, Nancy Cunard and Iris Tree. Both were poets, both had violently rejected their rich parents' world and what they stood for, and both were emphatically against the war. And Osbert had other socially acceptable friends who would make useful allies: Victor Perowne, a literary young man he had known at Eton, young Arnold James, and 'Bimbo' Wyndham Tennant, Lord Glenconner's son and a fellow Grenadier.

Apart from Helen Rootham they were primarily Osbert's friends, and the idea of publishing a small anthology of all their poetry seems to have evolved spontaneously that summer. The idea almost certainly originated from the energetic Nancy Cunard, who wrote the poem *Wheels* which was to give the magazine its title.* But it was Edith who took the idea up and introduced it to her Oxford publisher, Basil Blackwell, as something of a counterblast to Edward Marsh's *Georgian Poetry*. Blackwell agreed to back it and meet all the publishing expenses. By the end of 1916 *Wheels: An Anthology of Verse* was in the bookshops.

In years to come this first wartime edition of *Wheels* would become part of the early legend of the Sitwells. In a symposium for Edith's seventieth birthday held by the *Sunday Times* in 1957, Raymond Mortimer, for instance, spoke of the excitement of its appearance. 'It was immense, a manifesto, like the first Communist manifesto. It was a juggernaut, "Wheels".'

But today it requires some imagination to understand the fuss, for *Wheels* was the slenderest of slim anthologies. None of the verse is particularly memorable, and most of it is best forgotten. Edith's 'The Mother' and 'The Drunkard' are included, and it would have taken a perceptive critic to detect the seeds of her later poetry in such inflated offerings as 'A Lamentation' from a mercifully uncompleted verse play, 'Saul', or her poem on 'Thaïs in Heaven':

* There is some confusion about who actually began *Wheels*. Nina Hamnett, who was around at the time, but was writing long afterwards, wrote: 'Nancy Cunard, who was often at the Eiffel Tower, started a poetry magazine called *Wheels*. Three young poets called Sitwell wrote for it.' Edith was to edit all the subsequent issues, and implied in later life that she had edited the first one.

I'm curious now to know if love
Is really heaven – where *you* rove.
Your kind of love . . . or mine, Thaïs?

Osbert's poems give even less hint of the real talents he possessed, or the way they would develop. Indeed, in the brief few months since he discovered his poetic muse at Ypres, it must have run away with him. There are eleven poems. Only one of them, his '20th Century Harlequinade', is memorable with its picture of Fate, 'malign dotard, weary from his days, too old for memory, yet craving pleasure,' – a malevolent old harlequin out of the *commedia dell' arte* who decides the time has come to break up the pantomime of life. Sacheverell's first published poem, his 'Li-Tai-Pé Drinks and Drowns', stands out as something totally apart from anything by Osbert or by Edith, a meticulously conveyed *chinoiserie* picture of the Chinese drunkard who drowns as he tries to catch the moon. It is extraordinarily polished and complete for this precocious nineteen-year-old poet.

As for *Wheels* being something of a manifesto, as *The Times Literary Supplement* remarked in a condescendingly polite review, this 'little troupe of nine singers . . . have not as a body very much in common'. Helen Rootham contributed three short enigmatic translations of prose poems by her beloved Rimbaud, Nancy Cunard offered the idea 'that all our thoughts are wheels rolling forever through the painted world', and Wyndham Tennant's 'Song', with its opening lines:

How shall I tell you of the freedom of the Downs?
You who love the dusty life and durance of great towns,
And think the only flowers that please embroider ladies' gowns,
How shall I tell you?

could easily have been written by one of the country-loving Sussex bards that all the Sitwells later ridiculed so savagely.

The most that can be said of this first number of *Wheels* was that it was a start – for the Sitwells. At the time it was generally seen by the press and by their friends as one more 'Society anthology', as the *Evening Standard* called it, and without the impetus and aid of Osbert's social world it could hardly have come into existence. But once *Wheels* was born the situation changed. The Sitwells had ambitions and a sense of loyalty to one another which the other poets lacked, and gentle Edith was more than a match for both Miss Cunard and Miss Tree.

It was now that the Sitwells began to show that they were no ordinary writers. Sacheverell says simply, 'If you are two brothers and a sister and you all write poetry, it is extremely hard not to be lumped together as a group.' But the Sitwells were an extraordinary group, and their effect was something more than the sum total of the work of three loving, dedicated poets. From the beginning they were a self-contained literary and artistic movement in themselves, and as with any movement they were inevitably concerned with power. They needed contracts, allies and publicity – an entire machine – and it was in creating this, rather than in their earliest work, that they revealed the full force of their powers.

Here the main influence was Osbert's. Robert Ross sometimes wrote to him as 'Dear Impatient Poet', and it was this impatience, this eagerness for recognition and success, that drove him now. There were other motives too, and it would not always be too clear exactly which was which. One was his righteous anger with the war; another was his passionate ambition for both Edith and Sacheverell; another his unusual appetite for personal publicity; and yet another was a strong aggressive streak, a quality he traced back to his battling forebears, and which had prompted that decision to become a 'champion' of the arts against the philistine.

But such feelings were not confined to Osbert. Edith had them too. So far her poverty and sex had largely kept them hidden in the obscurity of Moscow Road, but her aggressiveness was every bit as sharp as Osbert's, her longing for publicity, now that it was roused, was even greater. Sacheverell too had great ambitions. His temperament was gentler than the others'. He had no wish for notoriety, but he was still much influenced by Marinetti, and dreamed of doing 'something of great importance in the arts' as well as simply writing poetry. He was already seeing himself as some sort of scholarly impresario when the war ended, discovering fellow geniuses, encouraging their work, and promoting ballets, exhibitions and performances of modern music. Certainly no member of the trio was content to sit patiently and wait while their poetry produced its own rewards.

Even in the middle of the war the artistic life in London was a full-time occupation. Osbert was strategically placed in Chelsea: his duties left him time for a determined social life, and he now started to reveal extraordinary energy and flair as he began to build the network of those influential friends and contacts on whom much depended.

It was a role that suited Osbert to perfection. Those pre-war drawing-rooms of the Asquiths, Keppels, Cunards and Swintons had already given him a useful grounding in the tricky art of meeting and impressing men of letters. Since then he seems to have refined his manner, and at twenty-four this convalescent Chelsea Guardsman recently returned from Flanders was an impressive personality. Witty, urbane, apparently well-read and an amusing anecdotist, he had two qualities to which all but the most self-confident men of letters are vulnerable – social acceptability and a flattering belief in art. His manners were, of course, impeccable, his turn-out was above reproach, and there were few established writers, let alone mere journalists and editors, who would turn down an invitation to the Marlborough or the St James's from this beguiling great-grandson of a duke.

Osbert quite genuinely loved to meet his artists and his men of letters; instead of being stage-struck he was art-struck, and he was quite eclectic in his tastes. Despite his youthfulness and championship of the young, he had the sense to cultivate much older men, particularly the now middle-aged survivors from the world of Oscar Wilde.

The high priest of the Wildean cult and guardian of the sacred memory was the former art editor of the *Morning Post*, Robert Ross. He had first met Wilde in 1886, and later became his lover, a position which he held until dislodged by his lifelong enemy, Lord Alfred Douglas, five years later. But Ross stayed true to Wilde, even after his disgrace, meeting him in France after his release from Reading gaol, and sitting by his death-bed with the only other faithful follower, Max Beerbohm's friend, ugly, witty Reggie Turner. Since then poor Robbie's looks and hair had gone, and with his heavy-lidded eyes, small Chinese moustache and noble brow he seemed to suggest a personality of foreign origin, perhaps a chain-smoking impresario from Russia or from Germany, a milder Max Reinhardt. His *Morning Post* articles had made him an influential art critic, but the one great mission of Robert Ross's life remained the reinstatement of Wilde's artistic reputation.

Osbert, as we have seen, had first met Ross through his 'Kitchener of hostesses', Mrs Charles Hunter, at her luxurious Essex house just before the war, but it was only now that he began to know him well. One can but guess at the full extent of Ross's influence on Osbert: it was a subject over which he was always guarded. In *Noble Essences* we have a vivid but discreet account of Ross's famous first-floor chambers at No. 40 Half-Moon Street, which possessed, so its author assures us, 'the same genial glow' as the owner's personality, with their dull gold wallpaper, their antique shelves of inscribed first editions, and the 'panel by Giovanni di Paolo of St Fabian and St Sebastian, now in the National Gallery'.

How much of Ross's sexual tastes Osbert already shared is hypothetical. One thing the Wilde case had instilled in his survivors was the profound need for discretion, and Osbert must have learned this early on from Ross. What is beyond dispute is that Ross was a man of extraordinary charm and magnetism – when he died Siegfried Sassoon addressed him as, 'O heart of hearts, O friend of friends!' – and that at the late-night sessions in Half-Moon Street, which Osbert so enjoyed, Osbert could make common cause with Ross's literary friends in an atmosphere of Wildean aestheticism and hatred of the war. Ross was pro-German by inclination, and anti-war in practice. More practically, he was an influential literary contact man and maker of reputations, with friends everywhere in publishing and journalism, a kindly old homosexual godfather in the literary Mafia of the day, whom Osbert always treated with immense respect. He made sure to introduce Ross to both Edith and Sacheverell, and when Blackwell reprinted *Wheels* in 1917 Edith sent 'Dear Mr Ross' a copy, 'from us all, with grateful thanks for all your kindness and encouragement'.

It was through Ross that Osbert also came to know that irascible mandarin of English letters, Edmund Gosse. A less *rusé* character than Ross, he was the walking image of the literary establishment. He had been Librarian of the House of Lords before the war, and was to be chief literary critic of the *Sunday Times* until his death in 1928; he was an artful, powerful old snob whose thundering reviews could still make or break a writer's reputation, and who looked, in Lord Clark's memorable phrase, 'like a cross between an old tom-cat and an octogenarian pirate'. Ross carefully arranged a dinner at the R.A.C. so that Osbert could meet him, and Edmund Gosse was hooked. If there was one thing he admired more than an aristocrat, it was a literary aristocrat; and though for years now Gosse had been the most uxorious of husbands, he too as a young man had known Oscar Wilde and would not have been entirely immune to the delicate male flattery of an Osbert Sitwell.*

Soon he was calling Osbert 'Ossie', and when he was introduced to Edith and Sacheverell, Gosse the social snob, if not Gosse the demanding literary critic, was soon captivated by the trio. The 'deleterious' contributors to *Wheels* were soon being invited to dine at the Gosse's house in Hanover Terrace, and it was on one of these occasions that Edith showed that she was not a Sitwell for nothing. An air-raid was in progress and Osbert had already arrived when Gosse's parlour-maid entered to announce, 'Miss Sitwell has telephoned. She sends her compliments, but says she refuses to be an Aunt Sally for the Germans, so she is not coming to dinner.' This was a remark Gosse and the Sitwells treasured.

Outside the Wildean connection Osbert's attempts to muster literary allies were less successful. Through Sickert he had got to know Roger Fry and had even bought a table from the Omega Workshop, but the rest of Bloomsbury seem to have viewed the Sitwells at this time with slightly bored indifference. Plainly an Osbert had no place among these high-flying Cambridge intellectuals, and it is not entirely surprising that, though he organised a party for Bertrand Russell on the night he was released from prison – after six months inside for pacifism – he never met the great philosopher again.

* Gosse had a sublimated, almost mystical excitement over handsome young poets till he died. The great 'discovery' of his life was Rupert Brooke, and he wrote rapturously about the young Sassoon and the Keatsian Robert Nichols. In his youth, as Phyllis Grosskurth has revealed in her biography of his friend J. A. Symonds, Gosse's 'tendencies' were less restrained.

He seems to have had scant success with H. G. Wells, whom he admired considerably, and although he dined with Bernard Shaw and laughed at the great man's jokes, the friendship went no further. However, Osbert did make one other important conquest among his distinguished elders. After Shaw and Wells, the most influential man of letters in the capital was Arnold Bennett, although at this time he was more publicist than novelist, helping to run the Ministry of Information under Lord Beaverbrook. Again it was Osbert's literary godfather, Robert Ross, who arranged the introduction – this time amid the pseudo-Roman splendours of the Reform Club – and a considerable friendship soon developed. On the face of it, it was an improbable relationship: the man who wrote *Clayhanger* was not likely to respond to the pretensions of an aesthetic captain in the Grenadiers. But Osbert amused and flattered him from the start, and Bennett was quite shrewd enough to recognise the energy and seriousness behind that bland, well-dressed exterior. Ross must have helped behind the scenes as well, for within a week of the meeting at the Reform Bennett was offering to back Osbert in a venture he had already set his heart on – his own literary magazine – and introduced him to his friend, Frank Swinnerton at Chatto & Windus. The project came to nothing, but the friendship with Bennett prospered, particularly once Osbert introduced him to the other members of the trio. Bennett, like many self-made men, had a sharp eye for the up-and-coming. He liked the aristocracy. He had a plain man's taste for what was new in poetry and art, and Edith in particular had the good sense to play up to him. Soon he was 'Uncle Arnold', and the Sitwells would be grateful for his avuncular support.

The remainder of the war became a time of furious activity for all three Sitwells as they pursued the serious business of establishing their name and their credentials. Indeed, Osbert writes of the 'fury' with which he 'applied myself to life'.

Sacheverell had left Eton and followed his brother with a commission into the Grenadiers, where he was stationed for a while at Chelsea Barracks. He was as tall as Osbert now, but slenderer, and lacked the Edwardian dandy's glitter and dominating presence of his brother. Instead, he had what Osbert has described as the gold complexion and reddish curls of a young Greek shepherd and everything about him made him seem a person to be cherished and protected – his shyness and appealing *gaucherie*, his cleverness, his hidden charm. He was still suffering from unhappiness about his mother and from the horror of having just lost several close school-friends in battle. One of his nightmares was that Osbert would soon be ordered back to France and follow them. But for Osbert and for Edith the real dread was that Sacheverell would be the one who would have to go. He was an unconvincing soldier, and after three years of slaughter England had become a place of doomed young geniuses. About their young brother's genius there seemed to neither of them any further doubt. Throughout that long winter of 1917 he had been working on some of the poems which would make up his first book, *The People's Palace*, which Blackwell would publish in the summer of 1918. The clarity and the assurance of these poems from a boy of nineteen was quite awe-inspiring.

But since Sacheverell was so sensitive and so retiring, this seemed to place a further obligation on his brother and his sister to establish the Sitwell name.

The task seemed pressing, for in 1917, and despite her poems in the *Daily Mirror* and in *Wheels*, Edith with her thirtieth birthday looming was still totally unknown, so unknown that Cynthia Asquith noted in her diary for March how she had arrived too early for the Gilbert Russells' wedding reception and 'found only a sister of Osbert Sitwell waiting'.

Edith had no intention of remaining 'only a sister of Osbert Sitwell' for much longer, and one can feel something of the energy with which she was now pursuing fame and fortune in a letter written just a few months later to Lady Ottoline Morrell by another obscure poet, Aldous Huxley:

> Then I rush to meet yet another unknown figure – the editress of *Wheels*, Miss Edith Sitwell, who is passionately anxious for me to contribute to her horrible production. These Wheelites take themselves seriously: I never believed it possible! While I sit in the Isola Bella, naïvely drinking in the flattery of the ridiculous Sitwell, in dart Carrington and Barbara, borrow half-a-crown from me, and whirl out again. What a life!

As he wrote this, Huxley must have been feeling an acute need to forestall any jealousy from the possessive Lady Ottoline that a protégé of hers was daring to have dealings with a rival patroness by making Edith seem as subfusc and absurd as possible. But even so the letter is revealing, as it shows Edith's overriding eagerness to get the poets that she wanted now for *Wheels*, and the anxiety and flattery with which she chased them. By now – with Nancy Cunard firmly off the scene – Edith had established herself as the sole and undoubted editor of her own anthology. And, however much Huxley might write it off as a 'horrible production', he was quite shrewd enough to realise that both Edith and *Wheels* might have their uses.

Edith undoubtedly saw *Wheels* as her best hope of getting her name before the public. She always had a taste for titles, and from now on she was firmly 'Edith Sitwell, Editor of *Wheels*.' This was how she asked Basil Blackwell to publicise her next book of poems, *Clowns' Houses*, which he brought out in the summer of 1918; for as she realised, an editor with her own anthology had far more power and status than a struggling poetess, however dedicated.

Again, though, it was Osbert who was having most success in getting the Sitwells talked about, and it was now that he gave the first hint of that real flair for publicity which was to bring them so much notoriety during the years ahead. When Blackwell had reprinted the first *Wheels* early in 1917, it was his idea to publish at the back the press notices, bad as well as good, which the first edition had inspired. He was smart enough to realise that even for poets there is no such thing as bad publicity.

More important was the new preface this edition sported. It was in verse and was entitled 'In Bad Taste'. Although unsigned, it was by Osbert, and was the first example of an as yet unsuspected skill which he possessed. It was his first attempt to write poetic satire. Here he had suddenly escaped from the turgid imagery and rhymes of those earliest poems on the horror of the war. Instead, with a dandy's elegance and wit, he went into the attack himself.

At last he was pointing to the real enemy, the enemy he had suffered from for years and which had grown with war into what he called the 'platitudinous multitude', the mass of the insensitive, complacent public who were expecting the young men to fight for them. But his critique was subtler than that and once again derived directly from his own experience. The unforgivable thing about this 'platitudinous multitude' was that they were led by safe old men trying to tell the young men who were facing death not to be morbid, not to be critical, above all not to think. Instead they should be happy with the simple life.

Then suddenly one recognises a familiar voice promising the young:

> Then would you grow to a malign old age,
> Watching your sons a-cricket on the green
> And hear your daughter's cello in the dusk.
> These are the joys the future holds in store.

It is, of course, Sir George, who has become for Osbert a sort of mouthpiece for the hypocrisies and vices of the older generation. It is the Sir George who tried to turn his daughter from her poetry to the cello, and who had wanted Osbert to play cricket like a conventional, sports-loving member of the Golden Horde. It is Sir George attempting to fob his children off with a sentimental talk about the past. But as Osbert points out as the poem ends, this talk of the unquestioning love of the simple life is horribly beside the point at a time when 'Moloch, God of Blood' is devouring young men by the million. As the awful god rides by, Osbert pictures him holding in his hand 'a fingered treatise on Simplicity'.

Osbert's first satire coincided with the tragic worsening of the war in 1917, and his meeting with a poet who was still experiencing the horror of the front at first hand, a relation of the man he once fagged for at Eton, the immensely wealthy Philip Sassoon's thirty-year-old cousin, Siegfried. As a pre-war dilettante poet whose unusual mother had once edited the *Sunday Times*, Sassoon had already got to know most of the London literary establishment that Osbert knew, including Gosse and Edward Marsh, but yet again it was the invaluable Ross who effected the introduction at one of his late-night gatherings at Half-Moon Street.

But now Sassoon had been through the Somme and Passchendaele, been disillusioned by the fighting, and had begun what Blunden called his splendid war on war with his poetry. Collected under the title *The Old Huntsman*, the poems form the war's most vivid, bitter depiction of the agony of the ordinary soldier and the horror of the Front. They also manage to express the soldier's tired disgust at the mindless patriotic cant of the civilians who were still cheerfully insisting that the war must continue at all costs.

Sassoon was an imposing figure now, good-looking, tall, and passionately committed to his soldier's pacifism. He was to write that 'the man who really endured the war at its worst was everlastingly differentiated from everyone except his fellow soldiers', and from the moment that he met him Osbert felt this bond between them.

More important still, Sassoon the heroic fighting man was acting on his beliefs. Late in the spring of 1917 he decided he would fight no more, and in a celebrated gesture threw away his Military Cross ribbon and tried to get himself court-martialled by sending his commanding officer what he called 'A Soldier's Declaration' – *Non Serviam*. Thanks to the intervention of another poet, Robert Graves, Sassoon was denied his military martyrdom, and was sent instead to the mental sanatorium of Craiglockhart in Scotland.

But by then Sassoon had already made a profound impression on Osbert. He strengthened and confirmed his pacifism – to the point where Osbert now agreed that the war must be ended by almost any means. He also seems to have convinced him that the time had come for soldiers to speak out. There was perhaps no point in Osbert making the sort of gesture that Sassoon had done (one can imagine the reaction to a captain in the Grenadiers who refused to go on soldiering in Chelsea Barracks). Instead, Osbert joined Sassoon in the propaganda poetry he wrote against the war and all in England who supported it.

Their targets were essentially the same – the profiteers, the generals, the complacent patriotic public and the glib journalists and politicians who still egged them on against the enemy. But from the start Osbert proved himself the more effective adversary. Sassoon wrote about the suffering of war, and when he turned his bitterness against the public back in England it was too brutal and extreme. In 'Blighters' he described with relish how a tank might crush some warmongering audience to death in a patriotic music-hall; in 'Fight to a Finish' the troops turned their bayonets on the hated 'yellow journalists' of the popular press.

Osbert's attacks were subtler, and more effective, as in the summer of 1917, under the self-proclaiming pseudonym of 'Miles' – Soldier – he began publishing a series of satirical poems, first in the *Spectator* and then in the more radical *Nation*. (The introduction to H. W. Massingham, the editor, was yet another benefit which Osbert owed to Ross.) In 'Rhapsode', published in October 1917, he chose a target which would have considerable significance for the future – those who always wanted poets to prettify the dreadful facts of war.

> We know you now – and what you wish to be told:
> That the larks are singing in the trenches,
> That the fruit trees will again blossom in the spring,
> That Youth is always happy; . . .

Instead, he proclaims:

> We shall sing to you
> Of the men who have been trampled
> To death in the circus of Flanders; . . .
>
> You hope that we shall tell you that they found their happiness in fighting,
> Or that they died with a song on their lips,
> Or that we shall use the old familiar phrases
> With which your paid servants please you in the Press:
> But we are poets,
> And we shall tell the truth.

It was a proud and an important boast. It was also distinctly arrogant for the author of a handful of not particularly distinguished poems to be claiming the high poetic role of bearing witness to the truth, along with the other war poets of his generation.

By the autumn of 1917 the direction of the Sitwells was becoming clearer. Osbert, with faithful Robbie Ross behind him and Sassoon for inspiration, seemed to have taken up position as a committed controversialist, ready and willing to do instant battle against a number of opponents – who included warmongers, fathers, press-lords, generals, philistines, the middle classes and the church. Politically his loyalties were none too clear – beyond the fact that he was overridingly opposed to any politician who desired to prolong the war – but his passion and his energy were undeniable. So were Edith's as she continued to conduct her role as the editor of

Wheels. By August 1917 she had finally persuaded Aldous Huxley to contribute, and the letter that he wrote his brother now from Garsington, announcing this, offers a cynical but perceptive outsider's view of the trio. Referring to *Wheels* he writes:

> The folk who run it are a family called Sitwell, alias Shufflebottom, one sister and two brothers, Edith, Osbert and Sacheverell – isn't that superb – each of them larger and whiter than the other. I like Edith, but Ozzy and Sachy are still rather too large to swallow. Their great object is to REBEL, which sounds quite charming; only one finds that the steps which they are prepared to take, the lengths they will go are so small as to be hardly perceptible to the naked eye. But they are so earnest and humble . . . these dear solid people who have suddenly discovered intellect and begin to get drunk on it . . . it is a charming type.

This condescending attitude towards the Sitwells would long be typical of many of Huxley's Bloomsbury friends, although in Huxley's case it did not stop him taking full advantage of these 'dear solid people'. When *Wheels, Second Cycle* appeared that December Huxley, with nine of his poems included, was the most heavily represented 'Wheelite' of them all.

The second *Wheels* was, as Huxley had predicted, 'quite a bright production' and a considerable improvement upon the first. It also marked Edith out as something of a poetic leader of those who disagreed with the homespun poetry which was being published in Sir Edward Marsh's *Anthology of Georgian Poetry*, precisely the poetry which Osbert had attacked for pretending that the larks were singing in the trenches. But although the *Pall Mall Gazette* felt that all the poems in *Wheels* had one thing in common, being 'conceived in morbid eccentricity and executed in fierce factitious gloom', this was still the nearest that they came to any unity of style or attitude. Clearly no school of poetry was forming yet around the Sitwells. Two of the contributors, Sherard Vines and Iris Tree, read embarrassingly today, and Huxley's jaunty verses hardly seem worth the effort Edith put into getting them. More even than with the first *Wheels*, the main interest in *Wheels, Second Cycle* was as a family anthology for the Sitwells and a shop-window for their work in progress.

Osbert's most impressive contribution was another preface 'In Bad Taste', signed this time and entitled 'Armchair'. This was the most polished of his satires on the old men who still longed to run the war, and once again it pilloried Sir George:

> All through the long spring evenings, when the sun
> Pursues its primrose path towards the hills,
> If fine, I'd plant potatoes on the lawn;
> If wet, write anxious letters to the Press.

Sacheverell in a long poem, 'Soliloquy and Speech from The Mayor of Murcia', revealed his extraordinary talent for using poetry to create the authentic landscape of a dream. As for Edith, these early poems still show her curiously divided. She has completed four more pages of 'Saul' – 'You should have stabbed my womb, Saul, my son Saul' – but she also offers two poems which at last show the way her talents will develop. One is 'The Satyr in the Periwig', a country fantasy half nursery-rhyme, half surrealist soliloquy, where for the first time she exhibits her gift for creating lines of a haunting, other-worldly magic:

Plums sunburnt as the King of Spain
Or gold-cheeked as a Nubian,

The other memorable poem was her short conversation piece, 'The County Calls':

I saw the County Families
Advance, and sit, and take their teas;
I saw the County gaze askance
At my thin insignificance,

While little thoughts like fishes glide
Beneath their eyes' pale glassy tide:
They said: 'Poor thing! we must be nice!'
They said: '*We know your father!*' twice.

That December of 1917 also witnessed a curious step forward in the Sitwells' gathering power and reputation – the historic verse reading given at Lady Colefax's house. In years to come the unfortunate Sibyl Colefax would become one of the Sitwells' favourite butts, 'old Coalbox', the beady-eyed original of Osbert's mythical 'Lady Flinteye', the voracious, lion-hunting hostess and terror of the London drawing-rooms. But that day all three Sitwells held her drawing-room in awe as they entered it to face the most important moment so far in their career as a poetic trio. Huxley, a performer there himself, described the scene in another letter to his brother Julian:

Gosse in the chair – the bloodiest little old man I have ever seen – dear Robbie Ross stage-managing, Bob Nichols thrusting himself to the fore as the leader of us young bards (*bards* was the sort of thing Gosse called us) and myself, Viola Tree, a girl called McLeod and troops of Shufflebottoms, alias Sitwells bringing up the rear: last and best, Eliot. But oh – what a performance: Eliot and I were the only people who had any dignity: Bob Nichols raved and screamed and hooted and moaned his filthy war poems like a Lyceum Villain who hasn't learned how to act . . . the Shufflebottoms were respectable but terribly nervous.

The fastidious Huxley may have found the whole event distasteful: one of the reasons he thought Gosse so frightful was because that irascible, toadying old gentleman chose to dress down Eliot, the American bank-clerk, in front of the smart society audience for arriving late. But for the Sitwells it was a profoundly serious affair. Gosse and Ross between them were presenting their three protégés to intelligent society as poets of accredited and noticeable promise. Irene Macleod read poems by Sassoon, who was unable to attend, having returned dramatically to France. Robert Nichols – another favourite of Gosse's – had inherited the mantle of Rupert Brooke, and after three weeks' service at the Front had become the most famous war poet of his day. And it was at the dinner given for the poets by another of the organisers, Madame Vandervelde, the socially influential wife of a Belgian politician, that the Sitwells had already met T. S. Eliot.

Huxley's tart view of the proceedings does not appear to have been shared by the audience, who were gratifyingly enthusiastic. Cynthia Asquith found it 'very moving'.

The almost simultaneous publication of *Wheels, Second Cycle* with the Colefax reading seems to have put the Sitwells firmly on the poetic map of London: and Edith and Osbert both proved wonderfully adroit at making sure they stayed there. 'In those days', Sacheverell says, 'you had to get up very early in the morning to catch my brother and my sister out.'

Even at this early stage they made it plain that they were not merely poets but potential allies, leaders, friends and aristocratic supporters of other poets. Their dedication, not just to poetry, but to the life of poetry, was all-consuming and embracing, and however nervous they may have seemed at Lady Colefax's there is no sign of nervousness about the way they plunged into the fray.

Despite the sparseness of their published work, they had other assets to attract their fellow-poets. By now Osbert had somehow persuaded Sir George to pay for a short lease on a small house in Swan Walk, Chelsea. It was a pretty house, near to the river, overlooking the old Physic Garden with its mulberry trees, and it enabled Osbert to do something he was always good at – play host to a small but sympathetic dinner-party of hungry intellectuals. The food, the wine, the anecdotes were better than most poets were accustomed to. Nor were his dinner-parties confined to poets. Osbert was always careful to keep his social contacts in adroit repair: within a few weeks of the Colefax reading, Lady Colefax and her overshadowed husband, Arthur, were dining at Swan Walk. Edith was hostess, and, although she thought their influential guest 'looked more like a wrought-iron railing than usual', the evening, like almost all the Swan Walk gatherings, was a great success.

Apart from one reference in a note from Edith that 'Osbert is route-marching', there is no sign that Osbert's military duties now impinged upon his social and poetic ones. Nor, for that matter, did Edith's daily duties at the Kensington War Supplies Depot, where she continued working almost to the very end of the war.

Indeed, they suddenly appear to have become a distinctly formidable pair. With Sacheverell based at Aldershot, Edith and Osbert seem to have spent more of their time together, and Huxley's letters strike a new note of admiration and respect for them now. There is no more talk about the Shufflebottoms. Osbert had influential friends throughout society, and, like it or not, such friendships had their uses; and Edith is no longer the wan, bird-like poetess high on her lonely perch in Moscow Road. Now she has tasted her first real success, the Editor of *Wheels* is out to conquer London.

One of her earliest acquisitions at this time was Robert Nichols, with whom she was soon conducting a determined literary flirtation. Later in life, when she had become more discerning about the poetry of the poets that she patronised – and when Nichols had declined from grace – she would be sadly scathing over his *Ardours and Endurances*, which had made his name. And, true, the unhappy man was finally shown up as something of a charlatan. But in these last months of the war he was still a hero, the suffering soldier-poet personified, with his 'flame-opal ring, his wide-brimmed hat, his flapping arms' and what the devoted Gosse had called 'his mournful grandeur' in repose. For her he was 'Prometheus', and when he told her that he feared that he was losing inspiration, she urged him in an eager letter to 'remember that your poetry is what you are doing for humanity: nothing else you could possibly give would be so valuable. I cannot *bear* to hear you say your poetry will never be the same again.'

She invited him for tea and then for drinks at nine o'clock at Pembridge Mansions – 'Dinner is also impossible owing to the primitive nature of the cooking.' She also asked him round for 'supper' at Swan Walk to celebrate Osbert's birthday, adding that Osbert was still anxious to discuss setting up his magazine. The failure of the attempt to start a publication backed by Arnold Bennett seems to have made them keener than ever now to start a journal in opposition to the 'closed shop' of the established Georgian poets of the day where young writers could 'vent their opinions or publish the writings of those whom they admired'.

Edith was passionately up in arms about it all and clearly determined to push Osbert forward and bring Nichols in for additional support. 'We *will* have our magazine, and nothing will interfere with that, and Osbert is with us about it. We will discuss it all on Sunday.'

In this same letter that she wrote to Nichols, she also showed her mettle over the question of the press reviews for *Wheels, Second Cycle*. Here from the start she exhibited that forthright shamelessness with which she would treat the press for the remainder of her life. 'Meanwhile,' she wrote, 'I am trembling about the *Morning Post* and "Wheels". You are possessed of great tact. Couldn't you manage to review it? There is nothing like asking for a thing straight out and so I have.' Nichols, like many a good man after him, did as he was told.

Edith's pursuit of Robert Nichols is instructive, since it formed the pattern of so many similar literary seductions during the years ahead, and explains a great deal about the successes and the tribulations of her private life. It was emphatically not carnal: Sacheverell speaks about her 'strangely virginal quality which was bound up with her poetry'. She was extraordinarily 'proper' in the way of well-brought-up Victorian young ladies of her day; whenever anyone arrived at Moscow Road for tea, Miss Rootham would invariably be there to act as chaperon. And although Edith was beginning to be admired now for her own extraordinary brand of beauty, it had a quality which excluded fleshy familiarity. She could be wonderfully amusing with those she liked and trusted – usually, it is true, at somebody else's expense. But there was something in her manner and appearance which made it plain exactly where familiarity stopped – and 'impertinence' began.

This was not, as has been gratuitously suggested, because she was a lesbian. Life would probably have been much easier for her if she had been. But from the moment that Sir George turned up his nose at her appearance and her mother told her that she could never possibly attract a husband, Edith had been building a protective personality. Her poetry was part of it, a sublimation and defence of her romantic nature: hence its passionate importance in her life, and hence the uncomfortably involved emotional affairs which now increasingly surrounded it.

With the romantic-seeming Robert Nichols she was soon artlessly exposing her poetic life. There were her *ennuis*: 'I have the sort of headache which untidily empties one's head and then lines it with thick yellow flannel. Do come back soon.' There were her powerful ambitions, and always there were the dreadful people who would try to thwart them. For the remainder of her life Edith would never stop complaining of these private enemies of hers, this regiment of what she came to call her 'lunatics and bores'. All the Sitwells had their enemies – Sir George, the Golden Horde, the philistine, the 'Little Man'. In the letters that she wrote to Nichols, they were currently her 'virtuous female cousins like very large omnibuses on very small wheels, who live in drawing rooms like railway carriages', and who had been pursuing her for weeks and preventing her from writing.

At the same time she was showing deep and sisterly solicitude for Nichols, flattering him outrageously – 'Whatever happens you *must* get the Hawthornden Prize. If only you knew how much I believe in you: I believe there is nothing you cannot accomplish,' – and showing an older woman's sympathy for the trouble he was having with his neurotic wife – 'Why should a fiery-hearted creature like you have to suffer so?'

As a result, Nichols inevitably became a dedicated member of the suddenly expanding circle of young poets who visited Swan Walk, and were starting to regard the Sitwells as poetic equals and potential patrons. Some like Sherard Vines and the Icelandic poet and playwright Haraldur Hamur – known as 'Iceland' – are today totally forgotten. But there were others who are not. One was T. S. Eliot. The author of 'Prufrock' had few friends in London. He was living at the time with his uncomfortable but devoted wife Vivienne in a small flat off the Edgware Road, and after the ordeal of the Colefax reading all three Sitwells went out of their way to befriend them.

> I remember many afternoons in 1918 and 1919 [Osbert wrote in an unpublished memoir of the poet] when I waited with my brother and my sister to have tea with the Eliots near the Marble Arch or Oxford Street, in dank London teashops, seemingly papered or panelled in their own damp tea-leaves. Vivienne would be the first to arrive, and then Tom would come from his bank by underground, and join us in consuming hot tea and muffins. Tea drinking is not a habit of Americans and I often thought that for him it possessed an exotic charm that for us it lacked.

There was one other poet who suddenly swung into the orbit of the Sitwells during the last months of the war. In July 1918 Sassoon wrote to Osbert a short note from France. He was dispirited. 'I get so damned tired. My feeling for the troops is the only thing that makes life endurable.' After asking Osbert to be sure to send his 'benediction' to his sister – whom he had still not met – he went on to ask, 'Have you met Wilfred Owen, my little friend, whose verses were in the *Nation* recently? He is so nice and shy and fervent about poetry, which he is quite good at, and will do *very well* some day.'

The answer to Sassoon's head-prefect's note was that he should have known that Osbert had already taken his 'little friend' firmly under his protective wing. The previous September Owen had arrived in London from Craiglockhart where he had met and apparently fallen in love with Sassoon, who had given him an introduction to the inevitable Robbie Ross. Gentle Robbie dutifully arranged a bed for Owen at Half-Moon Street, gave him the customary dinner at the Reform, and that same evening insisted that Osbert came to meet him back in his rooms. Interestingly, on the telephone he had already asked Osbert to be careful 'not to frighten him'.

Osbert found him shy, looking younger than his age – he was twenty-four – but with the 'easy supple good manners of the sensitive'. He must have avoided the forbidding manner with which a captain in the Grenadiers might treat a mere lieutenant in the South Wales Borderers, for the two of them got on, and during the following few months, now that Sassoon was back in France, Osbert took care to meet Lieutenant Owen whenever he came to London.

Osbert behaved to him as one war poet to another. Their opposition to the war was, Osbert wrote, like the 'force with which faith had knitted together the early Christians', and they began exchanging poems as they wrote them. It was one of Osbert's – a satire published in the *Nation* in which he wrote about the calm satisfaction with which Clemenceau might have reacted to the news of Christ's crucifixion – which brought back in reply one of the most moving letters Owen ever wrote:

> For 14 hours yesterday I was at work – teaching Christ to lift his cross by numbers, and how to adjust his crown; and not to imagine he thirst until after the last halt; I attended his Supper to see that there were no complaints; and inspected his feet that they should be worthy of the nails. I see to it that he is dumb and stands at attention before his accusers. With a piece of silver I buy him every day, and with maps I make him familiar with the topography of Golgotha.

It was in the summer of 1918 that Osbert saw the last of Owen, who was in London *en route* for France. It was a fine, hot, summer day. Sassoon was back, and he and Osbert met Owen and took him to hear their friend Violet Gordon Woodhouse play the clavichord, making 'the afternoon stand out as an oasis in the desert of war'. Afterwards the three poets had raspberries and cream at Swan Walk and then sat for a while beneath the mulberry trees in the Physic Garden before Owen left to catch his train. On 4 November he was killed in action; a week later the war ended.

CHAPTER SEVEN

The Machine Starts Up

1918–1919

FEW LONDONERS celebrated Armistice Night with more distinction and panache than the Sitwells. They began at Swan Walk by giving dinner to Diaghilev and Massine, both of whom were currently the toast of fashionable London. Diaghilev had recently brought back his Ballets Russes to London, having spent most of the war in exile and obscurity in Spain. Now with Massine and Lydia Lopokova as principal dancers, and with completely new productions like *Le Tricorne, La Boutique Fantasque*, and Cocteau's *Parade*, the great impresario had once again become the midwife to the rebirth of his ballet as it achieved what Osbert called its 'second Golden Age'.

Now, too, it had suddenly become the rage with London's intellectuals, particularly with Bloomsbury. During its pre-war London seasons most of the so-called intelligentsia had shunned it, leaving it to society and to what Osbert called 'the old kid-glove-and-tiara audience of Covent Garden and Drury Lane'. Since Osbert and Sacheverell had been such enthusiastic members of those audiences, and since they had got to know Diaghilev himself, they were now in a tactical position to upstage Bloomsbury over their new-found admiration for the Ballets Russes. They knew far more about the subject than these *arriviste* balletomanes from Garsington and Mecklenburgh Square, and being Sitwells they were more than ready to make the most of the situation.

True, Lady Ottoline Morrell had made a bold bid to become an unofficial patron of the ballet, with her enthusiastic invitations, and since Diaghilev was vulnerable to wealthy titled ladies she had achieved a level of success; but even this was working to the Sitwells' clear advantage. As far as the ballet was concerned, Bloomsbury could not ignore them any longer, and when the Sitwells threw a full-scale party for the Russian Ballet at Swan Walk on 12 October, Bloomsbury arrived in force. Indeed, it was here that Maynard Keynes first met Lopokova, and started the most unlikely union of all between Bloomsbury and the Ballets Russes.

It was apparently pure chance that the Sitwells' dinner invitation to their friend Diaghilev happened to coincide with Armistice Night, but it was splendidly appropriate. For, more than anyone, Diaghilev represented what they wanted from the peace.

Dinner over, they all walked through Chelsea hardly daring to believe that the nightmare of the war was over and watching as 'the great impresario, bear-like in his fur-coat, gazed with an air of melancholy exhaustion at the crowds'. Unlike the cheering masses in the streets, they had their private celebration to attend, for as the guardians of Diaghilev that night they were invited to the Adelphi where in Monty Shearman's spacious Adam rooms 'the dark flower of Bloomsbury' had assembled to welcome the ending of the war.

It was an improbable but an historic gathering, for here, *en fête*, were Ottoline Morrell, like 'a rather over-life-size Infanta of Spain', Clive Bell and Roger Fry, Mark Gertler, Maynard Keynes and Lydia Lopokova, as well as a 'sub-rout of high-mathematicians and low-psychologists, a tangle of lesser painters and writers'. The D. H. Lawrences were there. Almost everyone was dancing, even the gangling, myopic Lytton Strachey who brought himself to jig around the room with the 'amiable debility of someone pleasantly awaking from a trance' – everyone, that is, except the great Diaghilev whose role in life was making other people dance.

For the three Sitwells that unlikely evening with the high priests of Bloomsbury may have marked the end of all hostilities in France, but it also marked the opening of several other wars which they would wage themselves during the months ahead.

Edith gave her reaction to the Armistice in a letter to Robert Nichols. 'Isn't it inconceivable that men are no longer being butchered? It has made one's nerve go all to pieces, this sudden relief from the intolerable pain. I cried and cried and cried.' Sacheverell felt stunned relief, but Osbert's mood was more aggressive. He was 'unhappy and furious at the waste of lives and years, and frantically glad, on the other hand, that the struggle was over'. This contradictory attitude of Osbert's – extreme happiness and violent anger – is important, for it explains a great deal of the zest which galvanised the trio into action now. It also helps explain the form that action took.

Even the idea of making up for the lost years of the war was not as simple as it sounds, for all three Sitwells had a complex and absorbing

The Sargent portrait: Edith, Sir George, Lady Ida, Sacheverell, Osbert

Lady Ida and Edith

Sir George coming out of the Old Bailey, 12 May 1915

Sitwell family group, by Cecil Beaton

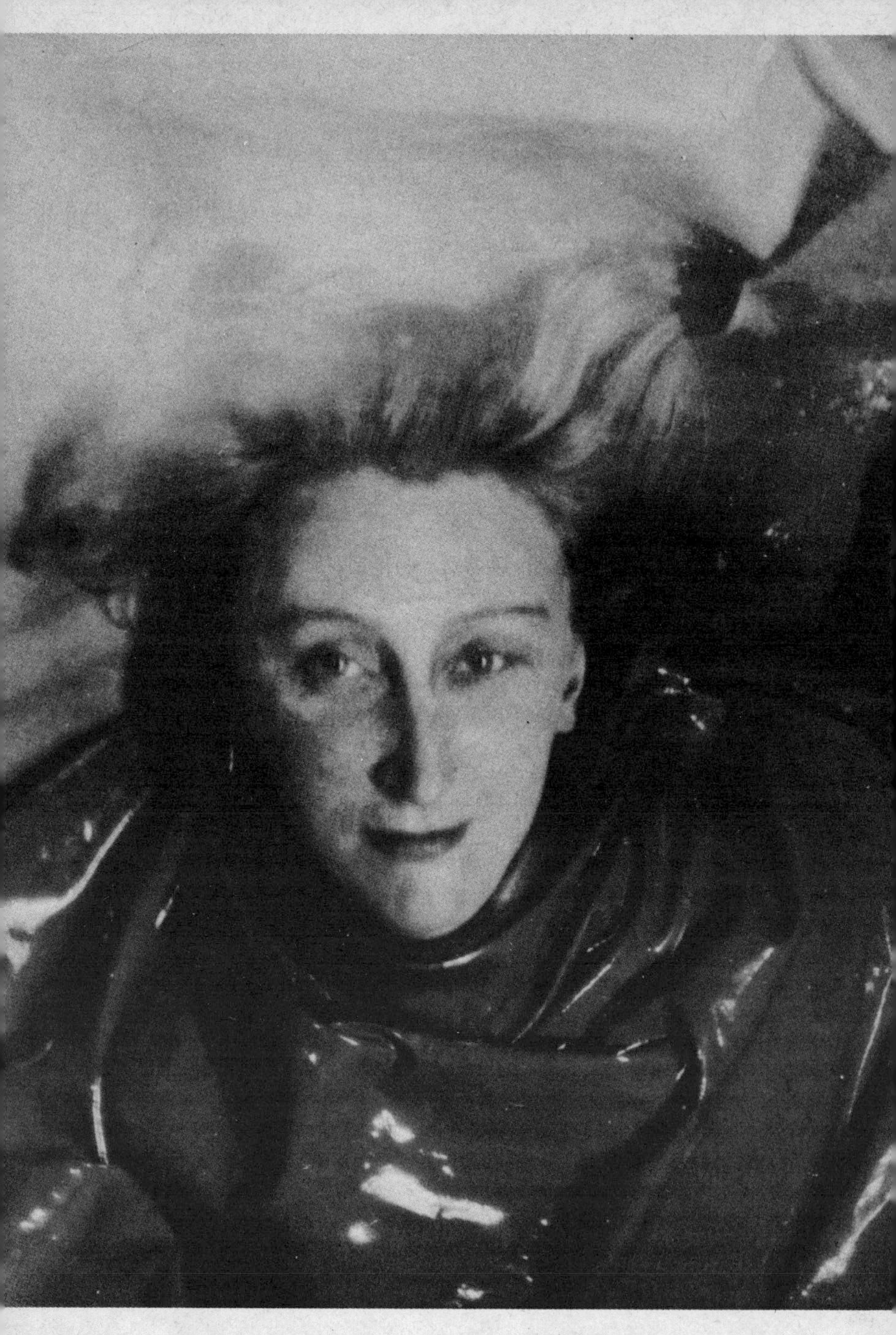

Edith, 'entirely beautiful, a most wonderful aesthetic object',
by Cecil Beaton

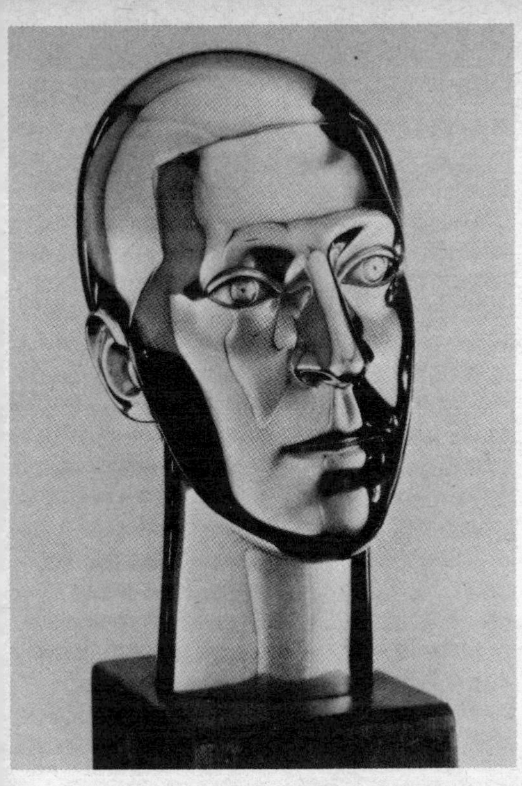

Osbert, by Frank Dobson 'authentic and magnificent . . . as loud as the massed bands of the Guards'

Sacheverell, drawn by Wyndham Lewis in competition with Robert McAlmon (*see page* 177)

Edith by Wyndham Lewis

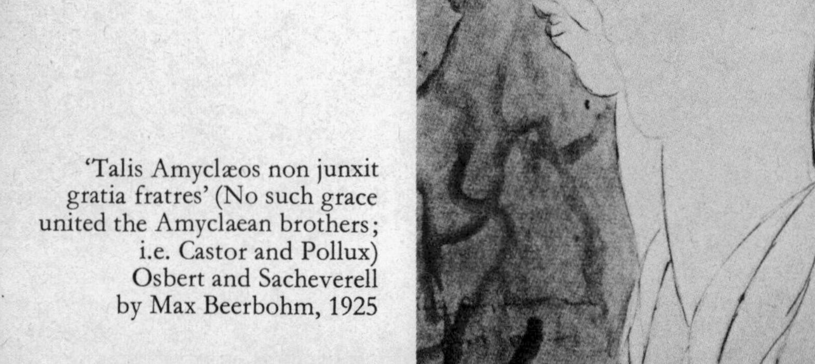

'Talis Amyclæos non junxit gratia fratres' (No such grace united the Amyclaean brothers; i.e. Castor and Pollux) Osbert and Sacheverell by Max Beerbohm, 1925

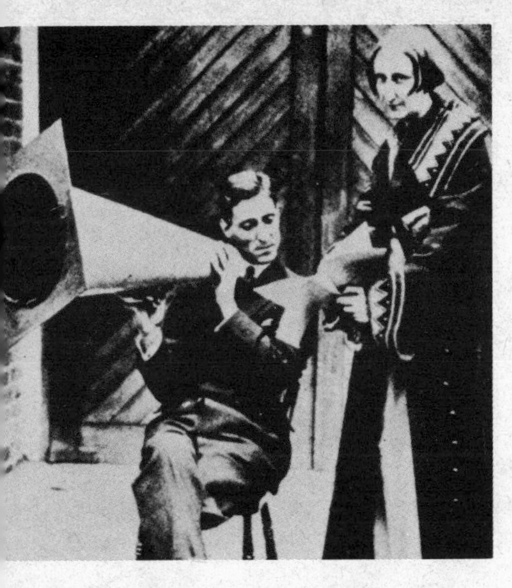

Edith rehearsing an unidentified actor in the Chenil Galleries *Façade*, April 1926

William Walton and Sacheverell at Renishaw, 1924

The Swiss Family Whittlebot (Maisie Gay with Tubby Edlin and the Brothers Childs)

Noel Coward with Joyce Carey (who also acted Hernia Whittlebot)

Osbert and Edith in Carlyle Square

The bowl of press-cuttings in Carlyle Square

The three Sitwells with the cast of *All at Sea*

sense of loss: indeed, the range and richness of their unfulfilled ambitions was suddenly their major asset.

Inevitably some of their ambitions varied. Edith was still willing to subordinate almost everything to her dream of becoming a major poet. Osbert was longing to resume his social life. Sacheverell had his private world of scholarship and learned appreciation of the arts. But in an uncanny way these ambitions seemed to interlock: all three considered they were primarily poets, all three were anxious to attract like-minded allies in the arts, and all three seriously believed in their artistic mission.

This sense of mission drew great strength from that second emotion Osbert felt now that the war had ended – his bitterness and rage. The targets of his wartime satires had not disappeared, nor had the attitudes they represented. As the war ended he wrote a poem which was a sort of battle hymn for the Sitwells and their followers as they formed up to face the peace. It had a suitably rousing title – 'How Shall we Rise to Greet the Dawn?' – and the answer that the poem gave sounds like an angry aesthete's call to arms:

> We must create and fashion a new God –
> A God of power, of beauty, and of strength –
> Created painfully, cruelly,
> Labouring from the revulsion in men's minds . . .
> We must create and fashion a new God –

But what sort of God? It was a question they would have to answer not only for their friends but for themselves, for the days were over when it was enough to be opposed to warmongers, racketeers and fathers. It was a question, too, which plainly worried young Herbert Read who met them a few days before the Armistice.

Read was a more or less self-educated Yorkshireman, a courageous soldier and an idealistic socialist more influenced by William Morris than by Marx. A year earlier he and Frank Rutter, a one-time director of the Leeds City Art Gallery, had founded a small quarterly review called *Art and Letters*, which he had helped edit from the trenches. Several of his comrades had contributed, and one gets an idea of Read's high-mindedness from a sentence in the preface that he wrote about them: 'Engaged, as their duty bids, on harrowing work of destruction, they exhort their elders at home never to lose sight of the supreme importance of creative art.'

Osbert himself could hardly have put it better, and in October 1918, when the future of *Art and Letters* was in doubt, Read was in favour of his co-editor's suggestion that they should sell it to the Sitwells. Read described his first meeting with Osbert and Sacheverell at the Café Royal to discuss it all.

Osbert and Sachie turned up. They are sons of Sir George Sitwell – aristocratic, wealthy, officers in the Guards, Oxford University and what not. But *also* furious socialists, good poets (Sachie very good) and very young (about my own age). They are crammed full of enthusiasm for the future and it is with them that I can imagine myself being associated a good deal in the future.

The inaccuracies of this first impression are in themselves revealing, but there was evidently much excitement that day amid the gilt and plush of the Café Royal, now that the Sitwells had their long-desired artistic quarterly within their grasp. Arnold Bennett had stuck by his promise to back them financially, Rutter was in favour, and Osbert clearly did his best to reassure Read that they were firmly on his side.

Doubts followed. Two days later, after a second meeting, Read was again confiding to his diary:

The Sitwells are too comfortable and perhaps there is a lot of pose in their revolt. But they are my generation whereas Lewis's is the generation before and it is with the Sitwells that I must throw in my lot to a large extent. . . . I am meeting Osbert Sitwell on Thursday to discuss the future.

That reference to 'Lewis' almost certainly explains Read's second thoughts about his 'furious socialist' Guardees. For Wyndham Lewis, painter, publicist and one-time editor of *Blast*, was one of Read's heroes at the time, and Lewis had recently been seeing quite a lot of all the Sitwells. They seemed to share many of his ideas, professed great admiration for his painting, and had already entertained him at Swan Walk. But Lewis was a suspicious friend who had his doubts over what he called the Sitwells' 'special brand of rich-man's gilded bolshevism', doubts which he must have passed to Read.

This makes it doubly interesting that Read decided even so to 'throw in my lot' with the Sitwells, because they had youth and the future on their side. Osbert and Read became joint literary editors of a reconstituted *Art and Letters*, and by November Read was noting optimistically that he had 'secured the invaluable co-operation of Lewis and Eliot' along with a 'jolly open-faced' young English poet called Richard Aldington. As well as the Sitwells' contributions, there would be poems by Huxley and Sassoon. Read's own poems would appear, and the only doubt was over his friend Ezra Pound, who 'withholds his "concurrence" from *Art and Letters* for two issues and then he will "see"'.

The war had inevitably left a vacuum in the artistic life of London which the Sitwells were both ready and equipped to fill. On the one hand was the ferment peace was bringing to its poets, painters, intellectuals: and on the other was the dislocation of the old artistic groupings. In poetry the Georgian Group, though still extant, had grown flabby and discredited among the younger generation. The pre-war world of the artistic drawing-rooms which Osbert loved had been disrupted too: some individual hostesses like Emerald Cunard and Sibyl Colefax would keep their private courts, but their appeal was limited to food and conversation. Even that unifying and supportive role which Robbie Ross had long provided ended with the war: that autumn, unexpectedly, he died, leaving the Sitwells and those other poets who had enjoyed his pacifying company to fend for – and feud among – themselves.

This open situation gave the Sitwells their big chance to build what Cyril Connolly has called 'their own alternative Bloomsbury', and that autumn one can already see the process starting. Osbert and Edith were both avid organisers and inviters, with an uncanny nose for talent. They had inherited their mother's taste for entertaining, along with that indefinable attraction of the Edwardian upper classes – style; their snob appeal was undeniable and, as Herbert Read reluctantly admitted to himself, they had the sense of youth and the future being on their side. There was a whiff of power about them too, as they discussed the downfall of their enemies and the advancement of their friends. They had a small but useful armoury of allies in the press, and with *Art and Letters* added now to *Wheels* they were becoming patrons on their own account.

Their poetry was just becoming fashionable as well – at least, with other poets. Edith had finally abandoned 'Saul' and poetic melodrama for a poetry which began to match her talents and the life-style which the Sitwells were presenting to the world. It was no longer gushing, passionate, naïve; instead she seemed to have created something quite new in her book *Clowns' Houses* and in the poems she contributed to *Wheels, Third Cycle*, which came out in January 1919. Here for the first time she had allowed her wit and actual tone of voice to infect her poetry, and shifted from the strident realism of her earlier poems into a world of fantasy and elegance and colour, a world of blackamoors and pantaloons and satyrs, all very much in tune with that decorative *commedia dell' arte* world her brothers loved.

Osbert's poems too, particularly his satires, which were still appearing in the *Nation* signed by 'Miles', served to remind his friends that for him at least the war was not over, and that his rage and despair and contempt for those who would not see continued.

But it was Sacheverell who was emerging as the most extraordinary poet of them all, a twenty-one-year-old prodigy whose recent book, *The People's Palace*, had led to Aldous Huxley's hailing him as *le Rimbaud de nos jours*.

During that first winter of the peace the Sitwells already had a lot to offer anyone who interested them.

In the tight, snobbish little world of artistic London the social life was crucially important, and Edith and Osbert both enjoyed the role of social freebooters. Sacheverell was still nominally tied to Aldershot. ('Qu'est-ce que c'est, cette Aldershot – c'est une femme?' asked Diaghilev as Sacheverell rushed off to catch the last train back to barracks.) But Moscow Road and Swan Walk were beginning to attract a glittering clientele. Indeed, the two addresses were almost complementary. Some poets such as Ezra Pound and T. S. Eliot were most at home with tea and buns and Edith's dedicated conversation against the spartan background of her flat. Other more ponderous figures such as Bennett, Gosse or Lytton Strachey could hardly be expected to brave Bayswater, five flights of stairs, and Edith's buns. For them Swan Walk and possibly some good champagne were more in keeping.

Both Edith and Osbert were extremely busy now as they began to organise their social and artistic lives – so, for that matter, was Helen Rootham, who in October 1918 was considerably involved with them in organising an impressive charity concert in aid of the Serbian refugees. All three of them were anxious to combine the *avant-garde* with the socially acceptable.

'Besides Royalty, Helen wants poets,' Edith wrote to Robert Nichols, begging him to come. 'The list of patrons includes everyone from Princess Beatrice and all the Ambassadors down to Richard Nevinson and the three Sitwells; and there will be a first performance of two of Mr Van Dieren's works.'

According to Edith, this first attempt at actual patronage was 'an enormous success', and, as she wrote afterwards to Robert Nichols (who apparently failed to materialise), the only jarring note in the whole evening came from Ezra Pound, 'who made a bit of an ass of himself about the Van Dieren Music which Helen produced. The man,' she added, 'really is intolerable.'

Their other poets knew for the most part how to mind their manners in the presence of their social betters, and the Sitwells managed to attract an impressive range of allies. Some were bohemians like Sickert's friend, the scatter-brained and generous painter, Nina Hamnett, or the one-legged nature poet and Welsh 'Super-Tramp', W. H. Davies. There were the young composers like Van Dieren and the rumbustious Philip Heseltine. There were the war poets like Sassoon and Robert Graves. And there were even Bloomsburyites of a more cavalier persuasion like Clive Bell and Roger Fry, who were attracted to the Sitwells' hospitality, or talk, or liveliness, or earnestness, or sense of fun. For the Sitwells held a multiple attraction – anger and frivolity, conspiracy and cosmopolitanism,

Osbert's anecdotes and satire, Edith's poetic passion, Sacheverell's charm and gentleness and air of genius.

The autumn of 1918 saw another interesting development – this time in the Sitwells' politics. Until now it had been hard to know exactly where they stood. During the war their politics had been dominated by their pacifism. Osbert had been unwaveringly opposed to any politician who wanted to continue with the fighting. Edith, though more concerned with poetry than Parliament, agreed with him. Sacheverell was more consciously left-wing. Towards the end of his time at Eton he was, he says, 'extremely Red', and had refused a fag. Trotsky joined Marinetti in his pantheon of private heroes, and when the Communists began the Russian Revolution, stopped the fighting on the eastern front and then began to infiltrate the German army, Osbert and Sacheverell applauded. As Osbert wrote, 'my brother and I admired the Bolsheviks for putting a stop so quickly to a war that had become for the Russian troops, left without arms, increasingly a shambles, and for defeating Germany in so novel a manner'.

This did not mean, however, that they would welcome Communism in Britain. Quite the contrary. It was invaluable for Russians and for Germans, but for Britain Bolshevism, Osbert said, 'appalled me'. This held a certain logic – at least on a short-term basis – but it gave substance to the impression Huxley had already formed that there was something phoney in their appearance as rebels, or, as Herbert Read had put it, that there was 'a lot of pose in their revolt'. This was the basis too for Lewis's complaints about their 'rich-man's gilded bolshevism'.

So what *were* the Sitwells' politics as the war ended? Osbert gave an answer – of a sort – when he decided he would stand for parliament as a Liberal and contest the family seat at Scarborough – held by Sir George's Conservative successor, Sir Gervase Beckett – during the 1918 General Election.

Re-reading Osbert's election address and the speeches he delivered to the burgesses of Scarborough it is a little hard to take him seriously as an orthodox Liberal. As Edith wrote to Robert Nichols, 'we are all a bit vague about our principles, but *ever* so staunch, which is the great thing. Osbert will have to "preach and prance" to country gentlemen gaping like fish with red and yellow waist-coats, red gills and no chins. Ugh! It is dreadful to have to pretend that they are alive! Such a mockery!'

Osbert had simply chosen Liberalism as the least obnoxious party label to give him a chance to enter parliament and present his views on the two great issues which obsessed him – the prevention of another war, and the rebuilding of a new and just society. He did not receive – or want – the notorious 'coupon' of approbation from that warmonger Lloyd George.

Instead he chose to stand as a 'Squiffite' – a follower of the already discredited and lost leader of the old-style Liberals, Asquith – counting upon the support the Sitwell name would naturally attract in Scarborough, and on the tacit help of the Labour Party. Neither was sufficient. Lloyd George's reputation as the man who had won the war dominated the election. The 'Squiffites' and the independent Liberals were annihilated by the politicians Stanley Baldwin called the hard-faced men who had done well out of the war. At Scarborough the Conservative polled 11,764 votes, Osbert a respectable but inadequate 7,994.

For Osbert it was a depressing and important episode. He had not been well at the beginning of the hustings: Sir George dogged him with maddening advice until Lady Ida told the baronet he looked unwell and packed him off to bed; but worst of all was the experience of actually meeting the electorate. Osbert had little taste or talent for political campaigning. He was a stiff and uninspiring public speaker, and although he dutifully preached and pranced and kissed the Scarborough babies, he never seems to have communicated much of his bitterness and idealism to the electors. Here he was Captain Sitwell, and reading the bland platitudes of his election address it is hard to believe that they were written by the ironic, bitter pen of 'Miles'. Perhaps he should have tried to use 'How Shall we Rise to Greet the Dawn?' as an electoral address instead. Perhaps, occasionally, 'Miles' should have appeared on the platform instead of Captain Sitwell. He never risked it, and his defeat convinced him once and for all of what he had long suspected – the hopeless lack of vision of the common herd, the almost criminal complacency of the 'burgess orders', and the depressing, terrible inadequacy of democracy. Although he continued as the half-hearted champion of Scarborough's Liberals for the following three years, this one experience of active politics inevitably confirmed his anger and contempt for those who would not see.

Further irritations and distractions followed during these early months of peace, first and not least of them the family Christmas at Wood End. Whatever the children felt about their parents, they dutifully foregathered for the festive season at Scarborough. Sir George and Lady Ida seemed to have come to terms with life now in their own strange ways. Ida's days of extravagance and anguish were behind her and, as Edith put it, 'Time was for her but an empty round between the night and night, a repetition of sad nothingness.' She would still invite stray friends and acquaintances she met to Renishaw, still stay in bed till noon, still mock her husband's wilder absurdities when he wasn't looking. But she was curiously free of bitterness about the way he had allowed her once to go to prison. She was ageing, and one young visitor to Renishaw would soon find it 'difficult to associate her with the bizarre adventures and tragic misadventures that had once distracted her existence'.

As for Sir George, that dedicated hypochondriac had now achieved an almost timeless state of health and selfishness, having spent the war, as Edith put it, preoccupied with 'his sole war-work – that of nourishing himself – "Keeping in good condition for whatever may turn up"'. By now he had earned himself the nickname which would stay with him until he died. During an argument over an inadequate tip, a cockney taxi-driver had said, 'After the war, Ginger, I'll get even with you!' and drove off leaving Sir George distinctly puzzled at this strange mode of address. He had asked Osbert about it. What was this odd term of familiarity? Could it have been a reference to the colour of his beard? Osbert seized on it gleefully. From now on Sir George was 'Ginger' – and Sir George and Lady Ida would be known collectively as 'the Gingers'.

Edith adopted the name delightedly.

> Ginger is too marvellously fantastic [she wrote to Robert Nichols], so Italian Comedy, I expect him to break into tail-feathers at any moment. Whenever he looks at me, he bursts into floods of melodious tears, – my personal appearance not having come up to his expectations. At one moment I used to play the piano, but had to give it up as jellified females would melt over me; Ginger weeps over this: 'Oh, my darling, every girl should have *some* talent or other!' This is a very favourite device for it swipes my poetry one in the eye without the possibility of retaliation on my part.

It was all very funny, and the Gingers would keep their children in anecdotes for years, but actual contact must have been uncomfortable, and it was now that Osbert started that elaborate game of hide-and-seek with his father that would soon see him dodging him through London and the south of Italy. For while Sir George was still convinced that he knew best, his children saw him as a growing menace to their work. This fear had already started to oppress them as they did their best – or worst – to celebrate what Edith called 'this bloody festival' of family Christmas at Wood End. By Boxing Day the days there had already started to become 'all raw edges tinged with irritabilities', and she could hardly wait to return to Moscow Road and beloved London.

It was not just the Gingers who had upset their daughter. On Christmas Eve the *New Statesman* had reviewed *Wheels, Third Cycle* and had remarked: 'Miss Sitwell's talent is accomplished, charming and rather trivial. . . . She can write *Fêtes Galantes* and perverted nursery rhymes as well as any poet alive. That is all! It is not much but it is something definite.'

Edith was never one for patiently enduring criticism, and the words rankled through those tedious, exasperating days of Christmas. Finally, in a letter to Robert Nichols, she exploded.

Damn them, – oh damn them! If they only knew the amount of concentration I put into these things, the amount of hard work and the frayed nerves it entails. They grumble because they say women will try to write like men and can't – then if a woman tries to invent a female poetry, and uses every feminine characteristic for the making of it, she is called trivial. It has made me *furious*, *not* because it is myself, but because it is so unjust.

Osbert was soon suffering more serious afflictions. Soon after Armistice Day he had caught influenza: now it returned again, and in the first new year of peace he became a victim of what he called 'the great plague, Spanish Influenza'. Indeed, he nearly joined the ranks of the twenty-one million who died of it, and for over a month was confined to King Edward VII's Hospital for Officers on Millbank.

Even there his social life continued after a fashion, and Aldous Huxley and Lytton Strachey made a determined point of visiting him (Strachey reminded Osbert of 'a rather satyr-like Father Time' – Huxley expanded on the sex-life of the octopus). But Osbert's recovery in the Influenza Ward was slow. While he was there he says that he contracted the illness three times. His heart was affected, and there would be a further legacy as well, although it would be many years before he would be forced to recognise it for the dreadfulness it was.

Sacheverell escaped influenza: he escaped the Grenadiers just after Christmas and went up to Oxford, doing what Osbert had so longed to do before the war when Sir George had 'lassoed' him into the Hussars instead. He even joined Sir George's college, Christ Church.

It was a time of returning servicemen. Sassoon was in Oxford. So were Roy Campbell and the Canadian poet Frank Prewitt. Robert Graves would soon be at St John's, Maurice Bowra at New College, but Sacheverell says he found the whole place 'something of a desert'. Oxford was freezing and he remembers the food as 'simply execrable'. Organised learning did not suit him: Oxford had little that could match the artistic life around his brother and his sister: dons struck *le Rimbuad de nos jours* as tedious beyond belief.

'Mr Sitwell,' said one of them, commenting on an essay, 'you write like Ouida.'

'That,' he replied, 'was the intention.'

'By now,' he says, 'I had become terribly ambitious to do anything and everything possible for and about the arts. As well as writing poetry I had ideas of becoming a sort of artistic impresario, mounting exhibitions, commissioning books and paintings, organising concerts and discovering geniuses.' Despite its drawbacks, Oxford did enable him to start to realise the last of these hazardous ambitions.

The first of his 'discoveries' was the disinterring of the author Ronald Firbank from all but total seclusion of his rooms opposite Magdalen. Edith had already introduced her brothers to Firbank's novel *Vainglory* and some time in February 1919, when Osbert was in Oxford visiting Sacheverell, both brothers called on this extraordinary recluse. According to Osbert, they were the first human beings – apart from his charlady and the guard on the train to London – to whom Firbank had communicated for the past two years. More than this, they actually persuaded him to read a chapter of the novel he was writing to an audience of nine after a private dinner party. With his exaggerated nervousness, and his voice 'which contained in it the strangled notes of a curate's first sermon', Firbank's reading was a dubious success, but Osbert and Sacheverell were delighted with his work. They were even more entranced (for a while, at least) by his outrageous and extraordinary presence. For, as Wyndham Lewis noted, Firbank was 'the reincarnation of all the Nineties – Oscar Wilde, Pater, Beardsley, Dowson and all rolled into one and served up with a *sauce creole*', and was tailor-made to be something of a cult figure for the Sitwells – and the twenties. That spring Osbert used his position as editor of *Art and Letters* to publish an extract from *Valmouth*, while for Sacheverell friendship with Firbank must have provided some relief from the cultural desert of the university.

Oxford spring term of 1919 witnessed a still more enterprising piece of genius-spotting by Sacheverell. The youngest undergraduate at Christ Church at the time was a tall, shy, Oldham music scholar by the name of Walton. Sacheverell heard him play part of a piano quartet he was writing, and despite the quality of the playing – young Walton was an atrocious pianist – became convinced he had the authentic seeds of genius within him. Several influential Oxford personalities had already reached the same conclusion, including Dr Strong, the Dean of Christ Church, and Dr Henry Ley, the college organist. They were urging him to continue studying composition in London at the Royal College of Music. Sacheverell quite calmly disagreed with them, and almost there and then persuaded Osbert that they had a duty to support young Walton if only to protect him from the scourge of academic teaching.

It was a courageous decision. They had little money, they were neither of them musicians, and the musical establishment of Oxford was convinced that Walton ought to train at the Royal College. Against this, Osbert and Sacheverell were acting in the high-handed and expansive style of eighteenth-century patrons. Sacheverell coolly argued that, at all costs, Walton must be saved from the influence of Sir Charles Villiers Stanford who was still teaching at the College. To prevent this and allow the young composer to develop as his genius dictated, he and Osbert would in effect adopt him, take him to live with them, and treat him as an extra member of the family.

During the months that followed something of a tug-of-war developed over William Walton, a tug-of-war which Walton's father finally insisted that his son should settle for himself. He chose the Sitwells – and was to share their lives for the next fourteen years, a period to which he now looks back with gratitude.

'If it hadn't been for them, I'd either have ended up like Stanford, or would have been a clerk in some Midlands bank with an interest in music. Life would have been a very great deal duller.'

For the Sitwells dullness was the great enemy of life. During the war they had endured it quite long enough, and now with the first spring of the peace Sacheverell and Osbert both had a chance to flee from it. It was a flight which soon became a voyage of exploration.

Osbert departed first. After his illness he was still pale and shaky, and aroused the sympathy of Mrs Ronnie Greville. Mrs Greville – or 'Aunt Maggie' as Osbert usually referred to her – was the first of the devoted band of rich and grand society ladies of a certain age who adopted Osbert. The daughter of a Scottish brewer – 'I'd rather be a beeress than a peeress,' she growled on one occasion – and a favourite hostess of King Edward in his prime, she was a woman who aroused strong and often contradictory reactions. Cecil Beaton found her 'a galumphing, greedy, snobbish old toad who watered at the chops at the sight of royalty . . . and did nothing for anybody but the rich', and Chips Channon wrote that there was 'no one on earth so skilfully malicious as old Maggie . Sacheverell thought her 'sheer hell and extremely boring', but Osbert loved her. He also loved her houses – one at Charles Street, Mayfair, and the other Polesden Lacey. Here he enjoyed the side of her which was summed up in her remark, 'one uses up *so* many red carpets in a season'. For Mrs Ronnie was a thriving relic of that *mondaine*, Edwardian society which he had known as a young Guards officer. When she suggested taking him to convalesce at Monte Carlo, he gracefully accepted.

It was, of course, extraordinarily proper. She was a solid lady in her childless late fifties and with the face of 'a small Chinese idol with eyes that blinked'. Osbert was twenty-six, attentive and amusing with that entirely asexual manner he adopted now with women, and which was earning him a reassuring reputation as 'a thoroughly tame cat'.

It appears to have been a very happy trip. They gambled – modestly, at her expense. They stayed less modestly in the Hôtel de Paris – again at her expense. And for Osbert, the plaster shores of the French Riviera, with its vacuous, rose-pink, middle-class tidiness, appeared suddenly a sort of paradise. It was four years since he had seen the Mediterranean and he had 'forgotten, in those fifty months of darkness, the sumptuous plenitude of Italian light'.

Edwardian society, the light of Italy – after the nightmare of the war the past was truly coming back, but the most important period of this short holiday was still to come. For after Monte Carlo Osbert went on to Biarritz, that other shrine of all Edwardians abroad, and here he waited for Sacheverell. Sacheverell arrived from Paris, bubbling with excitement. In the French capital he had seen 'streets lined with galleries full of modern pictures' – Matisse, Derain, Picasso, Braque – and had already tasted something of the excitement of the modern movement. How different it all was from London and from shivering Oxford! Most memorable event of all, he had just met Modigliani. The invaluable Nina Hamnett had provided him with a note of introduction and he had met the painter in his studio. Here, if anywhere, was a genius who required a patron. The hovel of a studio was crammed with paintings, but the artist was sick and in despair, his mistress starving. Sacheverell was offered the contents of the studio for one hundred pounds. Osbert agreed that when they returned to Paris they would most certainly see what could be done about it all.

In the meantime there were other painters to attend to. One in particular was El Greco. After a ceremony in which Osbert symbolically pushed his uniform in a hamper into the Bay of Biscay the two brothers journeyed south towards Toledo searching out every church and gallery where they could see his paintings. They also discovered Spain itself, Spain which had seemed to them throughout the war an unattainable island of peace, with its idiosyncracy and austere beauty. It was a new, exciting place for both of them, an unknown, unexpected land which even more than Italy seemed to be offering the sort of extravagance and fantasy which their imaginations craved.

They left for Paris in mid-April 1919, but it was a sombre journey back. Osbert was contracting jaundice, and neither he nor Sacheverell could raise the hundred pounds required for the contents of Modigliani's studio.

The first effect of that Easter holiday was to settle any lingering desire of Sacheverell's to stay at Oxford. He had made what brief use of it he could and, as Walton says, 'there was absolutely no point for him in staying. He was so well read that he knew far too much about absolutely everything.' Besides, he had work to do in London.

For once, Sir George appears to have had no say in his son's decision, and Sacheverell moved in with Osbert at Swan Walk. William Walton followed some months later. One of the most enterprising households in artistic London had finally been formed.

It was a hectic summer. When Osbert recovered from his jaundice he launched into the first of what Arnold Bennett called his 'seven professions' – as a contributor to the newly founded organ of the Labour Party, the *Daily Herald*.

Although it was edited by the veteran socialist George Lansbury, the *Herald* at this time was at least partly written by young intellectuals who had become radicals as a result of the war, and who, as A. J. P. Taylor has pointed out, 'contributed a gay, self-confident contempt for the doings of the governing class within its pages'. This, rather than doctrinaire socialism, was Osbert's own approach to politics. Sassoon had been appointed the paper's literary editor, and on 19 July he published the first of Osbert's contributions. It was called 'Corpse-Day' and must rank as one of the bitterest poems ever written on the war. It begins with Christ looking down from heaven at the victory celebrations and being suddenly disturbed at the cries of mourning from the widows and the children for their men who had been slaughtered. But the cities of the earth were still rejoicing:

Old, fat men leant out to cheer
From bone-built palaces.
Gold flowed like blood
Through the streets;
Crowds became drunk
On liquor distilled from corpses.
And peering down
The Son of Man looked into the world;
He saw
That within the churches and the temples
His image had been set up;
But, from time to time
Through twenty centuries,
The priests had touched up the countenance
So as to make war more easy
Or intimidate the people –
Until now the face
Had become the face of Moloch!
But the people did not notice
The change
. . . But Jesus wept.

Then, a week later, Osbert contributed the first of three poems which attracted so much notice that the *Herald* published them as editorial leaders. In them he pilloried the one man who appeared to represent all the worst vices of the warmonger, the unrepentant warrior of 'Corpse-Day', Winston Churchill.

Osbert's dislike of Churchill started in childhood, when Sir George had rented the London house of the great statesman's aunt, Mrs Morton Frewen, and the Sitwell children had to endure a wall full of photographs of the young hero in the nursery. Even then he must have represented just those qualities that Osbert hated – skill at games, aggressiveness, unquestioning patriotism – the attitude incarnate of the Golden Horde. Nothing he had done since then had made Osbert moderate this view of him. He had emerged as Britain's man of blood, the man responsible for the slaughter at Gallipoli, and now that the war in France was over he seemed still hell-bent on continuing the carnage – by leading the allied military intervention against the Sitwells' friends, the Russian Bolsheviks.

The three short satires Osbert wrote against him – under the cautious title of 'A Certain Statesman' – contain some of the most effective lines he ever wrote. He caricatures the great warlord much as he had learned to caricature poor Ginger, showing him petulantly threatening to burn the *Daily Herald* because it had been horrid about his nice new war, and as he continues his imaginary Churchillian monologue he succeeds in mimicking the chilling crudity of Churchill's outbursts on the subject of war.

As I said in a great speech
After the last great war,
I begin to fear
That the nation's heroic mood
Is over.
Only three years ago
I was allowed to waste
A million lives in Gallipoli,
But now
They object
To my gambling
With a few thousand men
In Russia!
It does seem a shame.
I shall burn my *Daily Herald*.

Behind the buffoonery lay all the bitterness of 'Corpse-Day'. Some of the Churchillian monologue was closer to the statesman's actual words than most people can have realised. Not long before Osbert wrote the satires, his friend and editor, Sassoon, had been interviewed by Churchill at his ministry in London. Sassoon has described the whole improbable event in his book *Siegfried's Journey*. The talk had turned to militarism, and

Sassoon quotes Churchill's very words on the subject: 'War is the normal occupation of man.' Had Churchill been entirely serious when he said it? Sassoon was not completely sure, but he must have mentioned it to Osbert, who remembered the remark and venomously twisted it to the purpose of his poem.

I consider
That getting killed
Should be
The normal occupation
Of other people.
I enjoyed
Doing my bit in France
Immensely.
And am only sorry
That the war stopped
Before I could go out again.

These poems in the *Herald* represent the furthest left that Osbert's pacifism and satiric gifts would carry him. The poems were invaluable propaganda for the Bolsheviks – and the British anti-interventionists – and were hurriedly published in a rough brown cover by the revolutionary Mr Henderson of The Bomb Shop at 66 Charing Cross Road, under the title of *The Winstonburg Line*. It was Osbert's first book of his own – and was distributed at a 'Keep-hands-off-Russia' Rally at the Royal Albert Hall, priced sixpence.

Osbert's pro-Bolshevism and his contributions to the *Daily Herald* seem not to have disturbed his high-flying socialising in the least. Rather the reverse. There is no sign that Mrs Ronnie Greville ever worried over Osbert's publishers – if she even noticed them. Snobbery has its uses – not least among them is the way that everything can be forgiven – or ignored, or simply overlooked – to anyone whose social credentials are entirely in order. Osbert's were, and this would be something he would always make the most of. By accent, birth, profession and appearance he was an aristocrat, and he wisely made no bones about the fact. It was a useful passport to a multiplicity of worlds.

One of these, which he and particularly Sacheverell were now intent on entering, was the world of modern art. Here Roger Fry had paved the way with his two important post-impressionist exhibitions before the war, and some of Bloomsbury's enthusiasm had spread even to its fringes. Those rooms of Monty Shearman's at the Adelphi where they had welcomed in the Armistice, for instance, housed a magnificent collection of the post-impressionists. But it was not Bloomsbury which was to have the honour now of bringing the latest offerings of the school of Paris across the

English Channel. (Nor, for that matter, was it a London dealer.) It was the Sitwells.

With that same mixture of effrontery and flair with which they adopted William Walton, they now went overboard for modern art, and, as with Osbert's Liberalism, they may have been a bit vague over the principles, but they were certainly staunch. Modigliani's dealer, the melancholy Pole Zborowski, collaborated with them, and thanks to him they managed to assemble an exhibition at Heal's Mansard Gallery in the Tottenham Court Road, which Roger Fry himself would hail as 'the most representative show of modern French art seen in London for many years'.

As well as Modigliani there were paintings by Derain, Vlaminck, Soutine, Matisse, Utrillo, Picasso and Dufy, and sculptures by Zadkine and Archipenko. The two Sitwells may have been doing little more than offering the French dealer a shop-window in the heart of London for his wares, but it must have seemed as if Sacheverell was rapidly achieving his ambitions as a high-powered impresario of the arts. More important still, it suddenly established Osbert and Sacheverell as the protagonists and sponsors of the Modern Movement. They, after all, had made the arrangements with the gallery, personally hung the pictures, and Osbert had seen to the publicity with his customary flair, even persuading 'Uncle Arnold' Bennett to compose a short preface to the catalogue. It was this preface which publicly proclaimed the Sitwell brothers as the fashionable exponents of modern French art in Britain. It hardly mattered that they had scarcely been aware of several of their painters until Sacheverell's brief trip to Paris six months earlier.

'This collection has been brought together by Osbert and Sacheverell Sitwell', stated Uncle Arnold's preface, and from then on they were held responsible for them.

It is hard to avoid detecting a touch of masochism in the way Osbert reacted to the public's reception of the show. There was the usual crop of public idiocy and ill-informed reaction in the press that has always greeted any artistic innovation. But for Osbert this was something far more sinister. This 'public rage', this 'fury of the Philistine' was part of that same persecution which he had suffered from the Golden Horde, the yellow press, the military mind and burgess orders of Scarborough. It was the same mentality which crucified his hero, Oscar Wilde, and it compelled him to adopt that role he loved and had pledged himself to follow on the night of *Firebird* – the 'champion' of the artist.

In fact the real problem was not the rage and fury of the philistine – but his indifference. Few of the paintings sold – nor did the exquisite Modigliani drawings which were offered for a shilling each – and the whole exhibition represents one of the greatest bargains at which the buying public ever turned up its unadventurous nose. Shrewd Uncle Arnold bought a superb Modigliani – for which he would live to thank the Sitwells – and Osbert and Sacheverell bought another Modigliani for themselves, his 'Peasant Girl', now in the Tate. They paid four pounds for it.

But when Sacheverell and Osbert approached Sir George to put up the few hundred pounds Zborowski wanted for the remainder of his stock, Sir George refused, and several million pounds' worth of prime French paintings (judged by today's inflated prices) returned to Paris.

Edith had played no part in her brothers' exhibition. Her interest in painting was more or less confined to portraits of herself, and it was now that portraiture began to play its part in the creation of her public myth. That constant and uncomfortable awareness of her appearance which had been forced on her in childhood persisted. Sargent had been the first painter who had taught the thirteen-year-old girl at Renishaw how those extraordinary looks of hers, which caused her such acute embarrassment, could be transformed by art into a source of unexpected beauty. It was an alchemy she understood.

In her own poetry she was learning to transmute the misery of childhood into fantasy and myth. Painters, it seemed, could do the same with her appearance. The first to attempt to do so now was Roger Fry; not that that kindly, dedicated follower of Cézanne was the man to make the most of Edith's looks. But he must have realised Edith's strange potential as a model, and when he started painting her in January 1918 she clearly enjoyed the process.

Acutely sensitive though she was about her whole appearance, she wrote to Robert Nichols that her sittings were her only relaxation: and though the painting turned out far from flattering, she was apparently quite happy to look 'like a new version of a symbol out of the "Sacre du Printemps", secretive and crude'. She often said she longed to be an actress, and had something of a great actress's delight in the conscious creation of an external personality: one can see how this personality was forming in the next portrait after Roger Fry's.

Towards the end of the war Edith had met the Chilean painter, boxer and bisexual, Alvaro Guevara, and for a period had formed another of her deep, non-physical attachments with him. He was an attractive figure, tall, good-looking and very much a part of that 'Higher Bohemia' which swirled around the wild and gilded presence of Nancy Cunard. Like Robert Nichols and, later, Aldous Huxley, he was unrequitedly in love with her, and, as with Nichols, in his unhappiness seems to have made a confidante and friend of Edith.

This was a role which was starting to become a habit with poor Edith, but it was one that suited her. No one could understand an artist better, no one was more enthusiastic in success or more consoling in adversity. She was a useful friend as well. She had no jealousy about her friendships, and generously introduced people she was fond of to each other – and to her invaluable brothers. She did this for Guevara, even inviting him to Scarborough on one occasion. But though she was undoubtedly attached to 'Chile' and he to her, there were clearly limits to their affections.

She would always be attracted to romantic but essentially impossible men with whom she could conduct a passionate but totally non-physical relationship. Hers really was a love 'begotten by despair upon impossibility' – the more impossible the better; for this protected her from the insults and rejections with which Sir George had scarred her adolescence, and yet permitted her a sort of love which, like her poetry, could rise above mere bestial reality.

For characters like Nichols and Guevara – both of them miserably in love with someone else – there must have been something restful and reassuring about Edith's unattainability. She could be part nun, part sister, loyal, witty, dedicated to her art, and yet a being set apart. One feels this very strongly in the portrait which Guevara painted of her now.

There she sits, eyes lowered, long fingers folded on her lap, like some secular madonna gracefully enthroned upon her hard-backed chair in Moscow Road. It is already a commanding presence with its serenity and bloodless elegance. The thirteen-year-old girl of Sargent's painting has emerged as this extraordinary woman. And just as the Sargent portrait was triumph of a sort for Edith over her family, so did Guevara's painting proclaim her position among the poets of her day. Her appearance was becoming one of her greatest assets; and Guevara's portrait, entitled simply 'Edith Sitwell, Editor of Wheels', caused a sensation in the autumn of 1919 when it was shown at the Grosvenor Gallery.

One of the most touching things about Edith was her lack of jealousy towards her brothers as they pursued a way of life so much more comfortable, exciting, elegant than hers. The total loyalty which had always linked the 'closed corporation' was as strong as ever. For her they could really do no wrong, and this solidarity became a source of growing strength for all of them. In some ways they were like a Mafia family, with Osbert at the head and all pledged to assist each other to the death. They shared their friends, their interests, their ambitions, yet since their essential talents were so varied they achieved an exemplary division of labour.

Osbert was the front man, publicist and social figure, champion of the arts and resident satirist of the *Daily Herald*, Liberal politician, editor of

Art and Letters, and 'paladin' of modern art. Edith duplicated some of Osbert's roles. She was every bit as aggressive, witty and anxious for publicity, but she was also the priestess of the trio, the dedicated poetess – and, of course, the editor of *Wheels*. But it was Sacheverell whose role was possibly the most intriguing of them all. In all traditional Mafia families there is a shy, sensitive, much-loved younger brother who is protected and encouraged by his elders. Sacheverell was this, but he was something more. His learning, taste, and far-reaching ambitions as an artistic impresario endowed the trio with a range and brilliance they would otherwise have lacked.

During that summer and early autumn this combination of the Sitwell talents went from success to fresh success.

Thanks to Osbert, Edith had secured seven of Wilfred Owen's finest poems for *Wheels* and was already making plans to edit the full collected edition of his work. Sacheverell had met Picasso – at some now forgotten studio near Charing Cross, where he was painting the historic backcloth for Diaghilev's London production of *Tricorne* – and was already planning how he could use *his* talents.

Meanwhile, the extraordinary artistic–social life of the Sitwells was rolling on as vitally as ever despite the fact that the lease on the Swan Walk house had fallen through. Osbert had rapidly discovered a far better house – at 2 Carlyle Square on the corner of the King's Road. It was a spacious house with a fine first-floor drawing-room, and enough room for Osbert, Sacheverell and William Walton to set up home here permanently with their housekeeper, Mrs Powell. Here they could play host to the most brilliant gatherings in literary and artistic London.

> Dined at Osbert Sitwell's [wrote Arnold Bennett in his journal]. Good dinner. Fish before soup. Present W. H. Davies, Lytton Strachey, Woolf, Nichols, S. Sassoon, Aldous Huxley, Atkin (a very young caricaturist), W. J. Turner and Herbert Read (a very young poet).... A house with much better pictures and bric-à-brac than furniture.... But lots of very modern pictures of which I liked a number. Bright walls and bright cloths and bright glass everywhere. A fine Rowlandson drawing. Osbert is young. He is already a very good host. I enjoyed this evening....

In the autumn of 1919 there was even talk of Osbert and Sacheverell lecturing in America. British war poets were still in some demand on the American lecture circuits. Nichols and Sassoon had both responded, but in the Sitwells' case the idea fell through because of Osbert's health. That autumn Edith was writing that he was suffering 'a dilated heart', and the brothers finally went off to Italy instead. For some of the time they were in Venice with Sir George, who was going through one of his more amenable periods; and it was now that Sacheverell had a brain-wave which will always rank among the might-have-beens of modern art. Picasso's Russian wife was pregnant, and the painter was anxious to take her away from Paris until the child was born. Why not engage him to fresco the great hall at Montegufoni? For a while it seemed as if Sir George might just agree, and the arrangements with Picasso got to the point of serious discussion. Picasso apparently had the idea of reinterpreting the Gozzoli frescoes in the chapel of the Palazzo Medici in Florence; and the result, around the walls of that enormous room, would have been quite staggering. A price was even fixed – £1,000.

But then, as with the idea of purchasing Modigliani's paintings, Sir George played his usual game which Osbert called 'they may think I shall but I shan't'. The offer to Picasso had to be withdrawn, and Montegufoni lost its chance of entering the history of twentieth-century painting.

CHAPTER EIGHT

Man's Natural Occupation

1919–1920

IF CHURCHILL was wrong about war being the natural occupation of man it was certainly the main occupation of Osbert Sitwell during the early twenties. That reservoir of anger which had been filled to overflowing by the Great War and his conflict with his father would still take many years before subsiding to a comfortable level. In the meantime it was liable to spill over on to anyone he disapproved of, and these early months of European peace saw the beginning of a number of Sitwellian wars.

Osbert ascribed them to ancestral temperament. 'My heredity, coming as I do on all sides of stock that for centuries have had their own way, and have not been enured to suffer insolence passively, made it hard for me, and for my brother and sister, not to fight back.' This was emphatically true of Edith too. Sacheverell says that one of her recurring nightmares was that someone had been impertinent to her, but she was unable to reply. With her, however, as with Osbert, one might surmise that it was more the irritation and frustration of her way of life than some warlike configuration of the Sitwell genes that really accounted for her state of chronic touchiness, particularly as Sacheverell, with the same heredity, had nothing like their love of battle.

Indeed, throughout this period he did his best to keep his head down during most of their controversies. 'I was never terribly interested in them, and Edith and Osbert were so much better at that sort of thing than I was. They were much wittier – particularly Osbert – and then Edith finally became entangled in all her controversies which seemed a dreadful waste of time.'

In Edith's later life this was undeniable, but during these early months of peace Edith and Osbert's sheer aggressiveness helped to establish them as figures of influence and power, and as the leading artistic 'rebels' of their day.

It was a period which suited them. One of the dominant ideas persisting from the war was that of 'The Enemy'. Paul Fussell convincingly ascribes this to the experience of prolonged trench warfare, with its fears and obsessions over 'what "the other side" is up to'. This was something Osbert had learned at first hand out in Flanders, and in his satires he had already set his sights on some of his peacetime enemies. It was a warlike period, with little quarter asked or given in the controversies that started with the peace, and as both art and politics increasingly divided into 'them' and 'us' Osbert, the angry captain from the Grenadiers, could go into action.

He had a lot of friends around to egg him on. Wyndham Lewis had identified 'The Enemy' in *Blast* as the bourgeois 'philistine', reincarnating Matthew Arnold's celebrated enemy of 'culture', and advocated the strenuous use of modern art to root him out. Sassoon had railed against the smug middle classes and the hard-faced government. Sacheverell's hero, Marinetti, had preached the aggressive gospel of the artist as a man of action, and even the idealistic Herbert Read was vigorously opposed to anyone who failed to respond to 'creativeness' in art. But Osbert's staunchest comrade in the battle was undoubtedly his sister, Edith. They, after all, had fought together for so many years in most of the causes that were now in vogue – for the young against their parents, for 'art' against the Golden Horde, for 'modern' poetry against Victorian tradition and for freedom and fulfilment against tired convention.

There was an extended line-up for the battle. On one side, under the banner of tradition, stood a host of disparate but more or less united allies – the popular press, the middle classes, provincial Little Englanders, admirers of academic art and Georgian poetry, puritans in general and the old in heart. Against them stood the forces of what Herbert Read still optimistically termed 'the future'. They were less numerous but more vocal than their enemy. They included left-wing politicians, 'intellectuals', 'dandy-aesthetes' with Wilde's green carnation still in the button-hole, admirers of experimental poetry and the latest painting from the school of Paris, and the whole of Bloomsbury.

The groupings were, of course, amorphous, and could change: so could the line of battle, with groups and individuals switching sides or even fighting on both sides at once. But the Sitwells never wavered. Lewis entitled them 'The Phalanx', Arnold Bennett pictured them with battle 'in the curve of their nostrils', and as the twenties started they plunged into the thickest of the fray.

This fighting really started with Osbert's slightly shocked reaction to the opposition which the French art exhibition called down on their heads. He has described how he and Sacheverell were staying at Polesden Lacey during the exhibition, and how the stuffier of Mrs Ronnie's other guests ganged up on them, as they had not done over the Winston Churchill satires, or his pacifism, or his writing for the *Daily Herald*. It was the same with the outraged reaction of the popular press. For some reason it appeared as if 'The Enemy' was far more worried about modern art than modern politics. This confirmed what all three Sitwells had long suspected from their own experience: that for some reason it was the artist who would always be the victim and sworn enemy of entrenched tradition. Sir George enraged at Edith's poetry, the apoplectic letters in the *Spectator* claiming that Modigliani's paintings were a direct encouragement of prostitution, were merely symptoms of a particularly unpleasant English disease called hatred of the artist. It was a disease the Sitwells were determined they would kill or cure.

In the mopping-up campaign that followed the French art exhibition, Bloomsbury was firmly on their side. Indeed, it was Bloomsbury's established publicist for modern art, Clive Bell, who was the most effective marksman in the press, picking off various angry correspondents including one who signed himself quite simply 'Philistine'. But the next encounter turned out to be one which the Sitwells made peculiarly their own, and in so doing took on the role which Bloomsbury might have otherwise espoused as champions of the *avant-garde*. This was the battle with the group who called themselves, with the simple man's good humour, the 'Squirearchy'.

John Collings Squire was in some ways an unlikely leader of 'The Enemy'. He was a popular, good-hearted fellow, a poet of sorts and a professional countryman, a hard-walking, deep-drinking, tweed-clad man known to his many friends as 'Jack'. As a very young man he had translated Baudelaire, but with the onset of the war and middle-age he had discovered, as Richard Aldington put it, 'that Britishism and moral disapprobation of "the Gallic" went better with the multitude'. This discovery led him to become a passionate and an increasingly embattled low-brow, a mere 'writing chap' as he preferred to call himself by now, but, since he was also the literary editor of the *New Statesman* and critic-in-chief of the *Observer*, an extremely potent one.

Just as it might have seemed as if the Sitwells had been custom-built to champion the *avant-garde*, so Jack Squire was by nature and by nurture the obvious protagonist of that symbolic creature whom the Sitwells hated – the English common man. Squire had been strenuously defending him for years, but it had been the war that had given him his greatest chance to adumbrate his basic verities – good patriotic love of the English countryside, particularly Sussex, the sound wisdom of simple folk, cricket, the hatred of the Hun and a devotion to the poems of Rupert Brooke. He was not above suggesting that there was something even treasonable about high-brow writers. Reviewing D. H. Lawrence's *The Rainbow* in 1915, he had remarked that Lawrence might be under the spell of 'German psychologists'. No fate could possibly be worse.

Squire was personally a kindly, sentimental man, and if one ventured even mild disapproval of one of his innumerable friends the answer invariably was, 'Oh, but he's such a nice fellow.' But for the Sitwells 'niceness' was irrelevant – and faintly suspect: so too was kindliness when more important matters were at stake.

By 1919, when Squire founded his magazine, *The London Mercury*, Bloomsbury had already come to loathe him and all he stood for. For Lytton Strachey he was 'that little worm, Jack Squire', and for Virginia Woolf he was 'more repulsive than words can express, and malignant into the bargain'. It took the Sitwells a little while to reach the same conclusion. As we have seen, Edith was enraged by the *New Statesman*'s rude remarks about her tiny talents in the Christmas 1918 review of the third *Wheels* – and held Squire responsible. But she calmed down, and in November 1919 the two were still on speaking terms. They met at a party given by Edmund Gosse, and, as she recounted in a letter that she wrote to Robert Nichols, she had 'such fun with him'.

It was the sort of situation which the malicious side of Edith never could resist: 'My whole conversation with him consisted of lamentations that people would not be "more natural". He never smiled once – either because he is too good-mannered or because he never realised his leg was being pulled. I can't help liking him personally, though I know the feeling is not returned.'

But Jack Squire for once was *not* being good-natured – and Edith should have known better than to pull the tweed-clad legs of editors in public, as she learned a few weeks later when Squire reviewed the fourth *Wheels* in the *London Mercury*. Having dismissed *Wheels* in general as 'full of wayward pot-shots', he then wrote off the whole of Edith's verse as well in one infuriating, condescending sentence: 'Miss Edith Sitwell's verses, though incomprehensible, contain a good deal of vivid detail, pleasant because it reminds us of bright pictures.' The most that he could bring himself to say of Wilfred Owen was that 'Strange Meeting' had a 'powerful, sombre beginning', and the one poet he actually praised was Aldous Huxley, on the unlikely grounds that he reminded him of the youthful Rupert Brooke.

As one reads Squire's words one can almost hear Edith's 'Damn them, – oh damn them!' echoing again across the page. From that moment hostilities broke out between the Sitwells and the *London Mercury*.

It was to be a complicated war, sputtering and thundering on for several years, with sallies and diversions as well as one considerable set-piece battle. But its importance for the Sitwells was that it gave them something they were really good at and in the process set them up in rivalry, if not in actual opposition, to their former friends in Bloomsbury. For where Bloomsbury was feline, they were bellicose; where Bloomsbury tended to look down its intellectual nose at the distasteful work of giving battle to the 'Squirearchy', the Sitwells clearly revelled in it, loving the publicity, the stunts, the chances of discomfiting their conventional and all too vulnerable opponents.

Inevitably this gave them a reputation for frivolity, although their war aims were profoundly serious – not that a little fashionable frivolity came amiss in the early twenties, and the Sitwells were undoubtedly becoming 'fun'. But the real value of the battle was that it gave them a heaven-sent cause round which to rally their dependents, allies and still grander figures in the establishment.

Squire too was clearly spoiling for a fight. By now he had taken over from Sir Edward Marsh as generalissimo of the traditionalist Georgian poets, and was all set to use his *London Mercury* as 'an influential platform from which he could damn the dangerous literary bolsheviks of the period'. Besides the Sitwells, these were to include Ezra Pound, the ghost of Wilfred Owen, and T. S. Eliot. The printing of *The Waste Land*, Squire would soon lament, was hardly worthy of the Hogarth Press.

After the *London Mercury* review of *Wheels* and an attack on Osbert's *Argonaut and Juggernaut* – 'he is revealed as an ordinary immature writer of verses' – the next attack upon the Sitwells came not from Squire himself but from a keen and blustering admirer of his poetry, an otherwise obscure Belfast journalist and poetaster called Louis McQuilland. In July 1920 he used Osbert's permanent *bête noire*, the popular press, in an attempt to smother them with ridicule. Writing in Lord Beaverbrook's *Daily Express*, he attacked what he chose to describe as 'The Asylum School' of poetry, complaining bitterly that 'quite an appreciable number of Londoners do read the absurdities of Osbert, Sacheverell and Edith Sitwell (that curious Chelsea trinity) . . . the clever gibberish of Aldous Huxley, and the clotted nonsense of lesser members of a semi-fashionable coterie, whose work has scanty rhyme and no reason'.

The article, in fact, bears all the marks of having been inspired by

Beaverbrook himself. He, after all, was a close friend and admirer of Winston Churchill, and the Sitwells were exactly the sort of targets he enjoyed selecting for his journalistic feuds. As his biographer remarks, his lordship's feuds 'were as likely to be conducted for purposes of entertainment as for any serious reason', and the mischievous press-lord knew the publicity value of controversy. So did Osbert – despite his extreme annoyance at the enemy's 'impertinence' – something which he would not forget. Huxley, too, expressed his irritation at the 'attentions of the *Daily Express* which has elected to make me and several other youthful poets its Giant Gooseberry for this year's Silly Season'. Two days later a most composed Osbert was replying to McQuilland in the *Express* itself, under the cheerful headline: 'Poet's Defence of Asylum School – Are they or their critics mad?'

It was a clever piece of journalism, certainly far smarter than the unfortunate McQuilland's lumbering attempt to send them up. Having paid tribute to McQuilland's 'nimble Ulster wit' in calling the Sitwells 'the Asylum School', Osbert advanced the theory that life needed to be expressed in a way that was typical of its age 'and that the characteristics of the modern school of poets are the same as those manifested in the best modern painting and music'. To illustrate his point he proceeded to quote at length from two of Edith's poems and one of Sacheverell's.

It was, of course, a considerable stroke of luck for the Sitwells to be published at length in the *Express* and to find themselves proclaimed as the leaders of the modern school of English poetry. Whether they were or were not, Osbert had no intention of allowing the impression to be lost if he could help it. Sir William Walton recalls him now as 'willing to do absolutely anything for publicity', and when the *Express* controversy died down Osbert launched a most astute attack upon 'The Enemy'.

One of the great strengths of the Squirearchy lay in its power within the literary establishment, a power which Squire himself assiduously employed to reward, encourage and keep loyal the poets he approved of. In this respect the Sitwells – with only *Wheels* and *Art and Letters* – were no match for the traditionalists who had the *London Mercury*, the *New Statesman* and the *Observer*, and most of the popular press behind them. Another area where the Squirearchy had power was in the awarding of the best-known prizes for literature and poetry, which had a curious habit of going to devoted members of the Squirearchy. In 1919 the Hawthornden Prize was won by Edward Shanks, and in 1920 by another eager nature poet and contributor to *Georgian Poetry*, John Freeman.

Using the pseudonym 'The Major', and posing as the betting man's adviser on poetic 'form', Osbert set out a plain man's guide to the next

Hawthornden Stakes in an article which he entitled 'Jolly Old Squire and Shanks's Mare's Nest.'

Mr Freeman writes for the *London Mercury*.
Mr Shanks writes for the *London Mercury*.
Mr J. C. Squire is editor of the *London Mercury*.
Mr J. C. Squire is chief literary critic of the *Observer*.
Mr Iolo (or I.O.U.) Williams, a poet, writes for the *London Mercury* of which Mr J. C. Squire is editor, and reviewed Mr J. C. Squire's book of poems for the *Observer* of which Mr J. C. Squire is chief literary critic. Mr Squire has now written a preface to an anthology edited by Mr Iolo Williams.
The Major, wiring from Newmarket, advises sportsmen to back Mr Iolo Williams as a future Hawthornden Prize Winner.

It was a shrewd tip – and a neat exposure of the literary log-rolling that was going on. On several occasions in the future Osbert would resuscitate 'The Major' to embarrass literary committees when he felt a poet he disapproved of had become the odds-on favourite for a prize. The Sitwells also staged their own rival presentations, which were adroitly planned to steal the headlines. On one of these all three of them and Aldous Huxley trooped ceremoniously into the Savoy Hotel to present a wreath of bay and myrtle to the great singer Madame Tetrazzini – 'as a tribute from the young writers of England', as the assembled journalists and press photographers were carefully informed. An even better stunt they thought up later was to present an annual literary prize of their own. It was a stuffed owl, which they awarded for the dullest literary work of the year. (Recipients included J. C. Squire, Edward Shanks and Harold Nicolson.)

Stunts such as these did much to liven up the literary scene – and kept the Sitwells' name before the public. They had enough experience of 'Ginger-baiting' to know exactly how to deal with Squire. But behind such entertainments there was still a profoundly serious sense of purpose and affront – just how serious Osbert would reveal when he published a manifesto against the Squirearchy entitled *Who Killed Cock-Robin?*

This is a fascinating little book – part literary squib, part satire and part private affirmation – in which he explains what poetry is – and is not.

Its form was clearly influenced by Cocteau's celebrated tract on art and music, *Le Coq et l'Arlequin* – the influential testament which Cocteau had published in 1918, with illustrations by Picasso, giving his artistic philosophy in a series of dogmatic, gnomic and prophetic propositions.

Cocteau stands out in many ways as the great original whose style and work and influence the Sitwells were adapting for the English market. How much of this was conscious imitation, how much a similar response to the same modernistic spirit of the age is hard to tell. Sacheverell had met Cocteau briefly during his trip to Paris at the Easter of 1919, and seems to have absorbed a great deal of the authentic Cocteau magic in the process; and both brothers must have picked up still more via Diaghilev. Cocteau stands out as the great prototype for almost everything that they were doing. By now he was the supreme example of the poet as man of influence and fashion – dandy, self-publicist, and impresario, friend and inspirer of painters and musicians, high priest and founding father of the artistic twenties. The staging of his ballet *Parade* by Diaghilev in 1917 was one of the historic moments in the modern movement, and Cocteau, as his biographer Francis Steegmuller wrote, had now established himself 'as one of the most prominent bridgers of the gap between Left and Right, in revealing to the fashionable Right Bank world the artists who came to be known as the School of Paris'.

This, as with so much else that Cocteau did, seems to have been a source of inspiration to the Sitwells – particularly Sacheverell – and Osbert instinctively adopted Cocteau's style and language as he composed *Who Killed Cock-Robin?* – so much so that one brief, uncanny paragraph actually prefigures one of the most striking episodes in Cocteau's film *Orphée* by a good seven years. '*Poetry*', writes Osbert, 'is the conversation of Gods through the medium of man. Poets transmit, like wireless apparatus, what the Gods say.'

But most of *Who Killed Cock-Robin?* is concerned with developing the fighting theme which Osbert had first expounded in his second preface, 'In Bad Taste', for *Wheels*, when he complained of poets writing of the larks that were singing in the trenches.

'*Poetry*', he insists, 'is *not* the monopoly of lark-lovers, or of those who laud the Nightjar, any more than it belongs to the elephant or the macaw.'

For Osbert the poetic skylark was symbolic of the great sell-out by the Georgian poets of the Squirearchy. During the war they unforgivably attempted to prettify the slaughter. Now with the peace they were attempting much the same unforgivable exercise by turning their attention from what was vital, crucial, often ugly in the world, to sentimental nature poetry and outdated country ecstasies. Since they appeared so smug and favoured by the philistine, Osbert coined a special word for them. He called them 'Mammon Poets'.

Mammon poets, he insisted, were essentially puritans (and the Sitwells hated puritans); they were also academic poets (and the Sitwells hated academics even more) who were trying to place 'daisy chains and cowslip-balls' on the tomb of late Victorian English poetry in the hope that the corpse would then miraculously revive.

But instead of scrawling 'sentimental observations . . . about that very cruel, matter-of-fact old hag, Nature', poets should turn to their true role as 'magicians' – 'Poetry', he wrote, 'is more like a crystal globe, with Truth imprisoned in it, like a fly in amber.'

This was a dictum which Edith and Sacheverell were following in the poetry they were writing now. Both were creating worlds of crystal fantasy. Edith's were still formed around the nursery-rhyme figures which she was re-creating from her childhood memories, like 'Queen Circe, the farmer's wife at the Fair', 'Clown Argheb, the honey-bee' and Sir Rotherham Redde, who rode 'on a rocking-horse home to bed'. But she was using these jaunty figures rather like characters in a ballet, so that they move within a brightly coloured world of fantasy which Derain or Picasso could have created for Diaghilev himself:

> Miss Nettybun, beneath the tree,
> Perceives that it is time for tea
> And takes the child, a muslined moon,
> Through the lustrous leaves of afternoon.

She seems to have forgotten her intentions of writing that 'genuinely feminine poetry' which she had once expressed to Robert Nichols. Instead she is developing an imagery all her own, an imagery which with its clarity and colour clearly complements the painters whom her brothers now admired – 'blue-leaved fig-trees', 'fruit-ripe heat', 'swan-bright fountains'.

The other interesting development now is the emergence of her actual tone of voice within the poetry she writes. Not for nothing was she a would-be actress, not for nothing did she so enjoy intoning her poetry aloud in that extraordinary voice so like the cello which Sir George had always wanted her to play. From now on the texture and the actual music of her words become an absolute component of her poetry.

Sacheverell's poetry set out to make a different sort of 'crystal globe' from Edith's – larger, more grandiose, and baroque, drawing upon a whole range of landscape, situation and character which he created for himself from the abstruse corners of his reading, from the *commedia dell' arte*, and from his love of the theatre. In his poem 'Laughing Lions will Come', published in *Wheels, Fifth Cycle* for 1920, he achieves an uncanny sense of drama and hallucination with an armoured Zarathustra (sounding like one of the faceless, threatening figures in a painting by the Sitwells' own friend, Wyndham Lewis) striding through a carnival of tight-rope dancers, cripples, jumping dwarfs and bird-like men on stilts playing mandolines.

As for Osbert, his most memorable poetry was still his satirical verse. He was no 'magician' but was very funny, and at his best when at his bitterest, as when attacking Lady Colefax in his two short poems, 'At the House of Mrs Kinfoot' and 'Malgré Soi'. It was uncharitable of him to pick on her, especially as she seems to have done nothing to offend him since the night she brought her inoffensive husband round to dinner with the Sitwells at Swan Walk. But Osbert was not particularly concerned with charity. He found her tiresome and pretentious, and the manic nature of her lion-hunting irritated him. So did her appearance. That was enough.

Osbert was never one to sacrifice a good joke for a friend and 'Malgré Soi' is one of the wittiest short satires that he ever wrote, with worthy Mrs Kinfoot going up to heaven all set to give a dinner 'to meet the Judge upon the Judgement day!' God is accordingly impressed, forgives her her few sins and packs her off, along with Mr Kinfoot, into everlasting life. But then comes the catch. As she departs for bliss she sees that all the great names she has lionised, 'the Counsellors and Kings/And brilliant bearers of a famous name', are being sent by God to Hell: the thought of the distinguished company she is missing is too much for her, and she cries out that she too is a sinner. 'I swear I beat my children! . . . I will tell/How I escaped a Ducal Co-respondent/Last year – my God – I must insist on – Hell.' But God is not deceived, and Mrs Kinfoot joins the heavenly choir. Yet even here her lionising tendencies torment her:

> Night never comes, – so when she tries to flee
> To that perpetual party down below,
> The angels catch her, shouting out with glee
> 'Dear Mrs. Kinfoot – you are good! – We know!'

But even when mocking Lady Colefax – he used to call the scene of her parties, Argyle House, 'Lion's Corner House' – Osbert was expressing something far more serious – his profound rejection of the respectable British upper middle classes which he believed she represented. There was a good deal of conscious snobbery in this. The heir to Renishaw was putting the *parvenu* 'Coalboxes' in their social places in a way he would never have dreamt of doing to the equally ridiculous but much grander Mrs Ronnie Greville. But snobbery apart, there was also his uneasy anger with the safe, smug, wartime middle classes which Osbert and his friend Sassoon had brought back from the trenches. The middle classes had been 'The Enemy' then – with their uncomprehending eagerness for the fighting to continue – and they were still the enemy.

Just as the Squirearchy seemed to represent the artistic complacency of the middle-class mind, so did the Mrs Kinfoots of this world reflect middle-class social complacency:

> Listen, then, to the gospel of Mrs Kinfoot:
> 'The world was made for the British Bourgeoisie,
> They are its Swiss Family Robinson;
> The world is not what it was.
> We cannot understand all this unrest! . . .

The War was splendid, wasn't it?
Oh yes, splendid, splendid.'

Mrs Kinfoot is a dear,
And so artistic.

In this poem Osbert expresses the fullness of his contempt for the 'burgess orders':

The British Bourgeoisie
Is not born,
And does not die,
But, if it is ill,
It has a frightened look in its eyes.

Apart from showing up the Squires and Kinfoots, all three Sitwells seemed determined to give the world a clear example of just how life *should* be lived; and since they spent such efforts on avoiding boredom, they were rarely boring. Indeed, a certain brilliance seems to surround them now as they set out to 'conquer' London, rather as Cocteau had already conquered Paris.

Here their success is understandable. Osbert and Sacheverell in Carlyle Square could put on an irresistible double act when they wanted to. They complemented one another, and their charm, their sense of fun, their Sitwell courtesy made them impressive hosts. Even the prickly Richard Aldington was hardly proof against them. 'Osbert Sitwell attracted me by his wit and honesty, and his brother by his passion for beautiful things and a sensitive taste which amounted almost to genius. . . . I had a slightly romantic feeling about them, because they seemed to be carrying on the tradition of the cultivated English aristocrat.'

A little of this sort of snobbishness did them no harm, particularly when so many writers were so desperate to clamber up the social ladder – nor did the impression of their house. Just as those rather chilly and refined eighteenth-century Bloomsbury residences expressed something about their intellectual owners, so 2 Carlyle Square expressed a great deal about the brothers Sitwell. Most of the taste – and pictures – now were Osbert's. He had a magpie instinct for Victoriana – in itself almost a half-century ahead of its time – which made him collect whatever caught his fancy; ships of spun glass, pictures made of shells, wax flower arrangements under elaborate glass domes. Alan Pryce-Jones described it all a few years later:

There was a fine jumble: from sketches of décor for Diaghilev to the Neapolitan conch-shaped silver chairs round the dining-room table. Small pretty things abounded – the house was a temple of a now-forgotten style called the 'amusing'. And there would be music and delicious cocktails, very dry, preceded by a ritual in which the wet rims were dipped in icing sugar. It was most dashing.

No one could accuse the Bloomsberries of that; and whereas Bloomsbury, as Gerald Brenan said, resisted all new arrivals who did not bear their trademark of literary culture, the Sitwells were quite catholic in their choice of friends. W. H. Davies was nominally a Georgian poet, but he dined frequently in Carlyle Square; so, for a while, did W. J. Turner, who was a founder member of the Squirearchy. Most of the livelier Bloomsberries were at least partially seduced by Carlyle Square. Aldous Huxley had migrated almost totally from Garsington into the Sitwell camp by now. Clive Bell and Roger Fry were frequent visitors, and Michael Holroyd gives an intriguing picture of Lytton Strachey, in the very early twenties, having to choose between 'the Sitwells and the Bloomsberries, and various luncheon parties with Lady Colefax and Princess Bibesco' when he came up to town.

In Mecklenburgh Square the conversation might be more penetrating, but with the Sitwells there was more variety, possibly more wit – and unquestionably the food was better.

It would be wrong to picture Edith living a deprived and waif-like life in Moscow Road while both her brothers, never out of evening dress, basked in a life of gilded, overfed gregariousness in Chelsea. To a considerable extent, her life overlapped with Osbert and Sacheverell's – she was more often than not among their dinner guests – and despite the scruffiness of Pembridge Mansions, and the absence of the elaborate Carlyle Square décor, she had adopted much the same pattern of social, intellectual and poetic life as theirs. She shared their friends, and their beliefs – and most of their ambitions. Her main difference from her brothers – apart from lack of money – lay in her lack of interest in the visual arts. Otherwise she was duplicating most of their social successes, and despite her comparatively slender means – she was now existing on around four hundred pounds a year – her flat was rapidly becoming one of the most important intellectual drawing-rooms in London. As Geoffrey Gorer says,

> those tea-parties of hers really *were* one of the most extraordinary literary affairs of the twenties when you think of them. For there she was, all but penniless, in a dingy little flat in an unfashionable part of London. All she could offer was strong tea and buns. Yet because of who she was she attracted to that flat almost every major literary figure of the twenties.

Writing in old age, Seán O'Faoláin, of all people, remarked how much he envied the Sitwells 'who had travelled in leisure through Italy . . . in the good old days when you could do it freely on a few pounds, all in their gilded youth while I was wasting my years on Irish politics'.

Not that Edith did much travelling in the early twenties, lacking as she did even the few pounds that were necessary. But there *is* something enviable about the way that Osbert and Sacheverell now made the most of the unspoilt bargain Europe offered anyone with time to spare and sterling in their pockets. In 1920 they were back in Spain – for Holy Week in Toledo. Even here, though, there was no escape from Bloomsbury. As they stood watching the Easter procession file up to the Cathedral, they saw a strange sight on the far side of the street,

the lean, elongated form of Lytton Strachey, hieratic, a pagod as plainly belonging as did the effigies to a creation of its own. Well muffled, as usual, against the wind, and accompanied by his faithful friend and companion Carrington . . . who, with fair hair and plump, pale face, added a more practical, but still indubitably English-esthetic note to the scene. . . .

Easter in Spain would be something of a ritual for the Sitwells for several years to come, but it was Italy that really claimed them, and in the autumn of 1920 Osbert and Sacheverell were comfortably aboard the train for Naples. Osbert would always say that he and his brother virtually had to flee the country to escape Sir George. Sir George, he would lament, was always interfering, always dogging them with good advice, and even working to kidnap them and put an end to their independent lives. And certainly the old gentleman, apparently rejuvenated with the ending of the war, must have become extremely irritating to three dedicated writers now that he was insisting that it was dangerous for his children to lose contact with him for a single day – 'you never know', he used to say, 'when you may not need the benefit of my experience and advice'.

This makes it all the odder that the place which Osbert and Sacheverell chose to get away from him was in exactly that part of Italy which he loved – Amalfi and the Bay of Naples. They even picked on one of his favourite hotels – the Capuccini, once the haunt of Wagner, Longfellow and Samuel Morse, converted from an eleventh-century monastery, whose terraced gardens offered what the hotel without undue modesty claimed as the most beautiful view in the world.

In the winter it was comparatively cheap. The weather was delectable and the cloistered calm of its ancient rooms – hardly changed at all from the days when they were still the cells of the Capuchin monks – made it appear the perfect spot for any writer who wanted respite from the noise and the distractions of the London social round. There and then the Sitwells both decided they would be back. But, in the meantime, London called and there was still one strange and significant adventure for them both before they returned to Carlyle Square.

It was at the end of November 1920, on the shores of the Neapolitan Bay, that the idea of Fiume first laid hold on us. My brother suddenly remarked to me, as we stood on a terrace overlooking the sea, mountains and islands:

'We never saw Lenin seize power in Petrograd: let us go to Fiume to see D'Annunzio. It may be the beginning of something else.'

Alas, Sacheverell's memory of the beginning of their historic visit to Fiume does not support the languid sense of history with which Osbert credits him in his autobiography. According to Sacheverell's account, the true reason why they went was not political or historical, but purely artistic, and very much in line with his ambitions to continue as a catalyst and patron of the arts. He was still anxious to employ Picasso. Sir George may have vetoed his enormous fresco in the hall at Montegufoni, so that they finally made do with Cocteau's friend and neighbour, the Italian painter from Cortona, Severini, in his place. On their instructions – and Sir George's bank account – the Italian painter was at that moment decorating one of the rooms at Montegufoni with a succession of *commedia dell' arte* figures which were to form the Sitwells' most important monument as patrons of the visual arts. But Sacheverell had plans for publishing a *de luxe* edition of one of his favourite books – Urquhart's translation of the works of Rabelais. Picasso, he hoped, would do the illustrations, and who better to compose the introduction than the greatest living poet in Italy, Gabriele d'Annunzio? The fact that at the time d'Annunzio was in open rebellion against the Italian government and desperately engaged in the survival of his revolutionary 'Regency of Carnaro' at Fiume was neither here nor there. D'Annunzio was an artist, a great war hero, a survivor from the nineties that they loved, a man whom Osbert would compare with Byron in the life he led and the influence he had upon his times. The two Sitwells never doubted for a moment that 'Il Commandante' would receive them – or fail to understand the reason for their visit.

There is something splendidly assured about their journey to this gangster principality, crammed as it was with desperadoes and adventurers from all over Italy, simply to get an introduction to an edition of Rabelais. They even took their own translator with them, Pino Orioli, the diminutive Florentine bookseller and friend and publisher of Norman Douglas and D. H. Lawrence. Relying on their British passports, and the welcome which d'Annunzio was promising to 'fellow-artists', they calmly took the train from Venice, sat up all night with several warlike volunteers joining the revolution, and finally reached Fiume.

Writing about it later – in an article commissioned by H. W. Massingham for the *Nation* – Osbert could not resist describing some of the crazier elements about Fiume: the bald-headed little demagogue with his glass eye and threadbare beard retiring to his study for eighteen hours at a stretch when inspiration seized him; the battles his supporters waged with live grenades to stave off boredom; and the d'Annunzian harangues from the balcony of the town hall, which d'Annunzio assured him had

had no equivalent in European history since the great days of ancient Athens.

In fact, d'Annunzio's adventure was fast breaking up when the Sitwells arrived in Fiume. The 'Regency' was doomed, d'Annunzio himself was bored and in decline, and Osbert and Sacheverell had to wait several days, living apparently on prawns and cherry brandy, before the great man received them.

His first words seem to have redeemed the situation. 'Well,' he began, 'what new poets are there in England?' There was something wildly improbable about the interview that followed, as d'Annunzio discoursed upon Shelley and on English greyhounds 'running wild over the moors of Devonshire'. He was noncommittal about Rabelais, but lyrical about his beloved 'legionaries' and described how four thousand of them had staged a real-life battle to welcome a visiting orchestra from Trieste. Over a hundred of them had been injured in the fighting – including five members of the orchestra. Osbert, the apostle of non-violence, was strangely undisturbed, although he did add that in d'Annunzio's discourse 'there was not a little to northern ears of absurdity'.

And yet, absurd or not, d'Annunzio undoubtedly impressed him, and for both the brothers this brief, uncomfortable visit to Fiume was of extreme importance. They genuinely believed, as Osbert wrote a few weeks later, that there was a chance that 'the small dominion' of 'this frail little genius' could yet 'develop into an ideal land where the arts would flourish once more on Italian soil'. More important still, Osbert felt that the experiment of Fiume, if it survived, could even offer 'an escape from the Scylla and Charybdis of modern life, Bolshevism or American Capitalism'.

Someone who felt this still more strongly was a recusant-socialist newspaper editor from the Romagna, Benito Mussolini. From d'Annunzio and Fiume he was already copying many of the techniques and a great deal of the style of his emerging Fascism, which would triumph in his 'March on Rome' in two years' time – the Nietzschean myth of the heroic superman, the stage-managing of revolt, the circus uniforms, rhetorical harangues and operatic bombast which inspired the early Fascist movement. Fiume also, of course, showed Mussolini the extraordinary weakness of the democratic liberal regime of Giolitti before a determined military coup.

This all-important link between d'Annunzio and Mussolini helps to explain a great deal of Osbert's own relationship with Fascism, which he supported – somewhat erratically, it is true – from the seizure of power in Italy until Mussolini's attacks on Britain in 1936 made such support impossible. For Osbert's real sympathy was not so much with Mussolini as with the original d'Annunzian myth as he had seen it in the 'hypnotic' personality of d'Annunzio himself in his office sanctuary in Fiume in 1920.

For despite all the d'Annunzian absurdities and terrible vulgarities, despite his warlike utterance and grotesque ambitions, the poet struck an instant chord in Osbert Sitwell. The fact that Osbert could never really read or understand Italian left him in grateful ignorance of d'Annunzio's political demands, enabling him to see him as a Byronic figure, a quaint but powerful Italian version of his own rhetorical ideals in all their romantic, mythic, irresistible extravagance – the poet as the man of action, the Wildean aesthete turned fearless man of war, the disillusioned hero pledged to revive the arts, destroy the money-grubbing burgess orders and restore the lost purity of Italy in all its idealistic, and nostalgic, fervour.

The fact that by the time the Sitwells saw him d'Annunzio had become a played-out mountebank hardly mattered. The d'Annunzian poetry and myth endured, and for Osbert, a would-be man of action and a creature of nostalgia and myth himself, this was what counted.

CHAPTER NINE

Work in Progress

1921–1922

SO FAR, the great achievement of the Sitwells had been to build their power and growing reputation on a minimum of published work. Edith was still known primarily as the editor of *Wheels*, Osbert for his *Daily Herald* satires, his social life, controversies and stunts, Sacheverell for his erudition and his air of youthful genius. These activities – and the distractions that surrounded them – might well have kept the Sitwells from the productive work that they had set themselves. But a strange series of coincidences checked them now and launched them on a period of intense creation which was a fresh surprise from this perpetually surprising group.

The period started badly. Osbert and Sacheverell returned from Italy to face another family Christmas at Scarborough, and to find Edith far from well. Her back was troubling her. (She always blamed Sir George and his 'Bastille' for this weakness which pursued her all her life.) Her 'love affair' with the ebullient Guevara had proved unsatisfactory, as all these passionate relationships of hers were bound to. And she had not got over her profound annoyance at the publication of the collected poems of Wilfred Owen.

Since she alone had been responsible for the first appearance of Owen's poetry in *Wheels*, she felt, not unnaturally, that he was 'hers', and believed in him with all the passion of her childless nature. 'Each time I write to you,' she had ended one note to Mrs Owen, 'I write with the kind of reverence and humility with which I should have written to Dante's mother. You have given the world the greatest poet of our time.' Thanks to Mrs Owen, she had had charge of the poet's manuscripts and, as she told her, had 'worked heart and soul at getting his work ready for publication' by editing and 'disentangling' the various versions of the poems he had left behind.

Then came enormous disappointment. All too trustingly, she had written to Sassoon asking for advice.

He came to see me [she told Mrs Owen]; and told me it would have been your son's wish that he (Captain Sassoon) should see to the publication of the poems, because they were such friends. In those circumstances, I could do nothing but hand them over to him; though it has cost me more to relinquish them than I can tell you.

It had indeed, and now that the book had finally appeared and Wilfred Owen was being hailed as the great poet she had always said he was, Sassoon was getting all the credit for discovering him. She did her best to tell herself that all that mattered was that justice was being done to a dead and splendid poet, but it was hardly in her nature to suppress her anger at 'Captain Sassoon's' behaviour, or the feeling that justice had not been done to *her*.

Edith suffered badly from suppressed rage all her life, and Christmas and 'the Gingers' seemed to finish her. She had what she would later call a 'breakdown' – although she angrily resisted her doctor when he tried to make her spend the next two months in bed.

It must have been an even touchier family Christmastime than usual. Osbert spent these days at Scarborough trying to bait his father. According to Peter Quennell, who visited Renishaw shortly after this, Osbert would tease Sir George whenever they met. Osbert 'had recently learned how to imitate a popping champagne bottle with a finger in the corner of his mouth; and while Sir George at the dinner table discussed Italian garden planning . . . Osbert would emit a succession of loud pops, which delighted his friends and Lady Ida, but entirely failed to disturb or interrupt his father's nicely balanced periods'. Lady Diana Cooper says that Osbert often carried a false ginger beard around to imitate his father. Hardly surprisingly, in the circumstances, Sir George began to take more and more of his meals alone – since people distracted him and their company prevented the gastric juices from following their normal course, as Osbert put it tactfully. Since he also went to bed at an extremely early hour, he would never meet some of his children's guests at Renishaw or even realise that they were there. Instead his mind was on the fresh improvements he was still planning avidly for Renishaw. 'Am afraid Sir G.'s plans for the garden are too exciting to be ever carried out,' Sir George's agent, Hollingworth, had written to Osbert a few months earlier.

To complete the merry picture, Sacheverell fell ill as well, as if infected by the tensions in the air. Immediately he arrived from Fiume, he had begun to write a book on Naples and the baroque. He had filled seven pages of a thick blue exercise book when he mysteriously collapsed. A few days earlier he had fallen on the rocks and grazed himself badly: he believes now that his illness was partly blood-poisoning, partly 'a sort of nervous breakdown, caused by nervous exhaustion and over-stimulus'.

He was to spend the next three months in bed, and one effect of the illness was to put him into an appalling state of nerves. 'At one point I wondered how I'd survive, and I made a vow that if I did recover I would go to Lecce and the south of Italy.'

He did recover, slowly, and by April, true to his pleasant vow, was well enough to set off with Osbert for the south. The journey was part pilgrimage, part celebration, and the two brothers duly took their time. At Rapallo they descended from the Rome express to take tea at the Villa Chiaro with one of the great survivors of the nineties, that immaculate and most peacefully resigned of exiles, Robert Ross's friend Max Beerbohm. Beerbohm's biographer assures us that Max enjoyed receiving visitors but made few close friendships now, having quietly withdrawn into his gentle, self-sufficient world of almost ritualistic idleness, with his beautifully pressed linen suits, his summer boater and his carefully selected buttonhole. But the old dandy's pale-blue eye responded to the vision of the Sitwell brothers, and the sketches that he made of them 'place' them for ever in their period. He drew them now and on several subsequent occasions, and each time emphasised the resemblance to the Edwardian dandies he had known, as they stood, elegantly poised in their perpetual double act, Osbert's convexity shown off by Sacheverell's concavity, Sacheverell's wispy nonchalance the perfect foil for his brother's vehement solidity.

After Rapallo it was not far by train to Florence and Sir George's castle, but they had other reasons now for stopping for a few days in the city they remembered from before the war. With the declining value of the lira, Florence was attracting new members to that 'queer collection; a sort of decayed provincial intelligentsia' as Aldous Huxley had described the English colony. During the early twenties Florence would become a Mecca for artistic exiles, and an important city for the Sitwells on account of them. There were the permanent inhabitants like Vernon Lee and Berenson and Geoffrey Scott, whose book, *The Architecture of Humanism*, published in 1914, had heralded reviving interest in the architecture of the baroque. There were rich aesthetes and collectors such as Henry 'Bogey' Harris, Arthur Acton with his rich American wife, who had refurbished an enormous *quattrocento* villa, 'La Pietra', out on the Via Bolognese; while Mrs Keppel had invested some of her profits from Canadian railroads in a fine villa overlooking Florence from the hill of Bellosguardo. There were survivors from the world of Oscar Wilde, like Reggie Turner – considered by his friend Max Beerbohm to be the wittiest man in the world – and Ada Leverson, Wilde's faithful 'Sphinx', who welcomed Wilde from gaol and was now living in stone-deaf old age in the fly-blown Hotel Porta Rossa. Firbank was often to be seen in Florence too; so was D. H. Lawrence, who in 1921 was finishing his *Aaron's Rod* with his hero turbulently in love beside the Arno. Norman Douglas, who enjoyed the Tuscan food and Tuscan boys, would soon settle in a flat on the Lung' Arno too, with Osbert's bookseller friend, Orioli.

But in the spring of 1921 the most important visitor to Florence, as far as the Sitwells were concerned, turned out to be Aldous Huxley: the few days that they spent with him would have a bitter and far-reaching aftermath.

By now they had all become great friends. Huxley had come a long way since the days when he thought Edith 'insufferably boring' and called the three of them 'the Shufflebottoms'. The Sitwells, too, were obviously delighted to have been able to attract this gangling and all-too-promising young genius into their growing circle. Two years earlier there had even been talk of Huxley and his newly married wife Maria spending a month or two at Montegufoni while he wrote a book. Osbert presumably suggested this in an expansive moment as a solution to the Huxleys' chronic money problems. Luckily for everyone, particularly Maria who would soon be pregnant, the idea fell through, but Huxley must have been impressed by the idea of Florence as an ideal, inexpensive place of refuge and escape from England. For in the spring of 1921 he finally did bring Maria and their year-old baby, Matthew, out to Florence while he wrote a novel, recently commissioned by a London publisher. The family had been settled for barely three weeks in their two-pounds-a-week apartment at Montici, on the edge of Florence, when the brothers Sitwell called.

During the next few days the Huxleys and the Sitwells saw a lot of one another. Huxley was run-down after his London winter, and during these few spring days, before he actually began his novel, Maria was encouraging him to relax and see the countryside. He and Osbert went off together to visit Lucca.

It must have been a pleasant trip. Some forty miles to the west of Florence, the old Roman city with its narrow streets and fine Renaissance walls offers a very different world from Florence. In the 1920s it was off the tourist beat, and still had something of the atmosphere of the independent, rather sleepy city-state that it had been when Napoleon's sister, Eliza Bonaparte, ruled it as her private principality. In 1921 the journey there was still an expedition from Florence. One changed trains twice, so the visitors naturally dined and then spent the night at the impressive old hotel in the main piazza, with its chandeliers and eighteenth-century dining-room.

Here they were not the only English visitors. Staying at the same hotel was a good-looking, silver-haired old gentleman whom Osbert recognised. He had met him two years earlier at Monte Carlo with Mrs Ronnie Greville – Harry Melvill, known as 'The Talking Harry Melvill', the once legendary raconteur of Edwardian London.

In his heyday people had given parties and inscribed on the invitation the brief legend: 'To hear Mr Harry Melvill Talk', and at Monte Carlo Osbert had still been intrigued to meet this garrulous old relic of the nineties. But the 'great conversational geyser', as Connolly called him, had outlived his time, and even Osbert finally found his anecdotes about Wilde and Pater something of a bore.

Osbert never said what he and Huxley talked about to Melvill that evening at the hotel, but as it happened they were not to leave Lucca without hearing 'Talking Harry Melvill' performing just once more. They had retired to bed when Huxley knocked on Osbert's door. He called him to his room, then pressed his ear against the communicating door to the room beyond.

> 'Who can it be? Who is it? Who is it?' he whispered. And then, quite clearly, each word taking on a greater significance . . . I heard . . . 'As I was saying only a few days ago to a man, a great friend of mine, who has, I think, really one of the most amusing and interesting personalities – a man who, I know, would delight you, with your knowledge and genuine appreciation of modern art. . . .'

It was Harry Melvill, neglected and alone in the room beyond, practising one of his once-famous and interminable monologues. Huxley was fascinated, and Osbert did his best to explain the pathos of the situation, as this friend of Whistler and Pater desperately tried to keep in practice with his waning talents. The incident, unlike poor Harry Melvill, would not be forgotten.

After Florence it was time for Naples and the heel of Italy, and suddenly Osbert and Sacheverell found themselves in the unlikely role of explorers, just as they had in Spain at the end of the war. Despite the books of Geoffrey Scott and Norman Douglas – and Sir George Sitwell, for that matter – few visitors would suffer the discomfort and the risks of venturing far beyond the Greek ruins of Paestum. But thanks to his reading – and to what Sir George had taught him – Sacheverell was the most informed of travellers. Both brothers were already devotees of the neglected art of seventeenth- and eighteenth-century Europe, if only as a way of shrugging off the heavy shades of Ruskin and the inheritance of Victorian aesthetics. What they discovered now at first hand was something of the power and splendour of the baroque architecture of southern Italy.

Even today there is something hallucinatory in one's first impressions of the town of Lecce, once described by Gregorovius as the Florence of rococo art. Nothing can quite prepare one for the sheer improbability of finding so much architectural lushness and extravagance in a small, forgotten town in the heel of Italy. When the two Sitwells came here it was more remote – and more neglected – so that there seemed something quite miraculous about the place, but the most exciting experience of all that summer came when they arrived at the vast, all-but-empty Carthusian monastery of Padola. It was the Sitwells' great discovery.

According to Osbert, 'in the last eighty years probably not more than ten Englishmen have visited it altogether', the guide-books had ignored it, yet here on its 'distant, unvisited, poverty-stricken plain' barely seventy-five miles to the south-east of Naples stood this building which Osbert himself described as 'one of the greatest baroque monuments in Italy', with its two enormous courtyards, its empty chapel and a great double staircase guarding its entrance.

The Sitwells had had to make their way to the monastery in a hired car over broken and neglected roads, and it struck them as 'a building as romantic and remote as any that could be encountered in the central forest of Brazil'. That word 'romantic' is important, for the two brothers were now discovering the baroque not as a mere academic chapter in the history of taste but as a personal and vivid revelation. It was not unlike the 'revelations' they had experienced from the Russian Ballet or the paintings of the School of Paris, and this idea of the baroque adapted perfectly to the whole Sitwell attitude to life. They who were in revolt against the whole heavy intellectual legacy of conformist nineteenth-century art, who abjured 'democracy' and all the horrors they believed it stood for, had found a style that instantly appealed to them. It was amusing and extravagant, aristocratic and nostalgically unencumbered by the dreariness of mass society. More than their beloved nineties even, the baroque expressed something which they felt inherent in themselves – their wit, their taste in music and in poetry, their gentle hedonism, and their escapist unconcern with the dour realities of modern politics. As they saw it, they and the modern movement in the arts were in fact restating much of the tradition of the baroque, which would be henceforth a style, a cause, and part of their cherished way of life.

While Osbert and Sacheverell were in Italy, Edith had recovered from *her* breakdown and was busily attempting to enhance her public role as a poetic personality and leader of the arts. Beneath her paralysing shyness lay all her old assertiveness and personal ambition. She was beginning to develop a unique public presence. Cocteau's example had not been lost on her either; and like Osbert she had a determined sense of mission, a passion for publicity, and great eagerness to do 'something' for the poetry and poets she believed in. All this resulted in the sad but instructive saga of the Anglo-French Poetry Society.

The society began in 1920 as something of an answer to the problems Arnold Bennett had been having with his young wife, Marguerite. Marguerite was French and bored and socially ambitious. Like Edith, she believed that she could act, and with her husband to encourage her she gave a number of successful readings of French poetry among their friends. Edith and Helen Rootham were invited, and out of this the society was born, with Arnold Bennett as the president.

To start with it was informal and successful. After one early session Osbert wrote effusively to Mrs Bennett to congratulate her on her 'superb Baudelaire. I have never heard even you recite better – and that is the highest praise that one could give anyone.' (The fact that Osbert's French would always be extremely shaky did nothing to diminish his enthusiasm.)

But the society looked rather different to Lytton Strachey. That lively cynic's sense of the ridiculous has left us a malicious but heartfelt account of the whole scene in a letter that he wrote to Carrington.

> Last night the Sitwell dinner was dreadfully dull, and they took me off afterwards to an incredibly fearful function in Arnold Bennett's establishment. *He* was not there, but *she* was – oh my eye, what a woman! It was apparently some sort of poetry society. There was an address (very poor) on Rimbaud etc. by an imbecile Frog; then Edith Sitwell appeared, her nose longer than an ant-eater's, and read some of her absurd stuff; then Eliot – very sad and seedy – it made one weep; finally Mrs Arnold Bennett recited, with waving arms and chanting voice, Baudelaire and Verlaine till everyone was ready to vomit. As a study in half-witted horror the whole thing was most interesting.

One can be sure that this account soon made the rounds of Bloomsbury – and lost little in the telling.

But by the spring of 1921 the danger signs began to loom around this harmless-seeming small society, as Marguerite Bennett's social ambitions started to clash with Edith's poetic ones. The first indications of an Edith takeover came with the impressive headed paper that was printed early that spring: Edith was now the official secretary, and the address of the society was firmly 22 Pembridge Mansions. Stranger still is the list of new members Edith proposed to add to the society. They included Bernard Shaw, H. G. Wells, Granville-Barker, Sassoon, W. H. Davies, Holst,

Honegger, Ravel, John Galsworthy, Walter de la Mare, Hugh Walpole and even Mrs Kinfoot's *alter ego*, Lady Colefax.

These, as Edith hastened to tell Marguerite, were only her *preliminary* suggestions. More would follow, and the society was now all set to hold 'meetings which will combine recitals of poetry (both English and French) and causeries, which will be varied by music of the period. There will be occasional readings at which poets will read their own verses.'

There are strong echoes here of the way that Edith once took over *Wheels*, and one detects an obvious ambition to make herself the centre of a high-powered artistic group, rather as Osbert had begun to do at Carlyle Square.

The response was disappointing, but Edith was undeterred. In February 1921 she was writing to Marguerite informing her that she was addressing the 'Lend a Hand Guild' at Lady Baring's house in Cadogan Square.

> I said I proposed to lecture on modern English poetry and the influence on it of modern French poetry, and that I was going to ask *you* if you would *most* kindly recite some French poems by way of illustration of the lecture. She was absolutely *thrilled*. *Please* do recite for me. . . . They are just the sort of people we want. They adore the idea of recitations, and it may be another drawing room to make use of.

So far, so good. A few days later Edith was thanking Marguerite – 'Your recitation even penetrated the stupidity of the people who were listening. Even Lady Baring, (a nice sleeper with the best intentions) realised she had never heard anything like it before.'

Edith's sense of her poetic dignity was simmering. There had already been trouble over a Mlle Galbain who had read some poetry and had had the temerity to make her own selection of what she was to read. An outraged Edith wrote immediately to Marguerite. It was agreed, she wrote, that

> It was *I*, as the poet on the committee, who had the casting vote. Consequently, when Mlle Galbain, instead of merely stating that she did not feel equal to doing the poems chosen, wrote a long letter criticising the merits of the poems in question (of which it is impossible that *she* should be such a good judge as I am) – and in an extremely impertinent way set her extremely amateur opinion against my expert one – I was naturally extremely angry and she was very properly put in her place.

Shades of Sir George who always happened to know best – and a foretaste too of a succession of Sitwellian slaps which would go echoing down the corridors of literature whenever Edith felt that her poetic dignity was threatened. Edith's friendships would always be at risk to the embattled poetess within her. From now on it was clear that the days of the Anglo-French Poetry Society were numbered.

It just survived throughout the rest of 1921 – chiefly because Marguerite spent most of this time in Paris. There was one meeting in December, which she attended, and Edith sent her an enthusiastic letter.

My Dear Marguerite,
... I am longing to see you again and have a talk. You recited *divinely* (there is no other word for it) the Baudelaire and Samain.

Je vous embrasse,
Yours affectionately.
EDITH.

Then, four weeks later, came the inevitable break.

Dear Marguerite,
I don't know how you thought you dared write such a letter about Helen to me. Your spiteful impertinence merely throws a most unpleasant light upon yourself. Conceit about any form of art is really the last thing Helen can be accused of! As for her art – I do not choose to discuss Helen's art with you.

Yours,
EDITH SITWELL.

And the Anglo-French Poetry Society came to an abrupt end.

Apart from the comic aspects of this whole affair, the episode is revealing for the important weaknesses it shows in Edith's character – her prickliness, her desperate anxiety to be accepted as an authoritative poetess, and her lack of humour, strange in so humorous a woman, where her artistic reputation was at stake.

Several of her friends maintain that during this period Helen Rootham used to aggravate these tendencies. This is a theory that the details of the final break with Marguerite Bennett seemingly support. But one can also see the fatal flaw which would always tend to trip up Edith's passionate ambitions during the years ahead – that dreadful sense of insecurity which could produce hysteria and rage and finally prostrate her when her artistic soul felt threatened.

Osbert was only slightly tougher than his sister, but he was more adroit at dealing with his enemies, and he could generally conceal his vulnerability behind the bright veneer of the accomplished social figure he had now become. This makes it interesting to compare the way that he conducted an important feud which cropped up now with Edith's behaviour to Marguerite.

He and Sacheverell were back in Italy again that summer – and once again met Aldous Huxley. Huxley was in the best of spirits. His time in Italy had been a great success, and he had finished his novel, which he decided he would call *Crome Yellow*. In August 1921 the two Sitwells took him for the day to Montegufoni. It impressed him greatly. Just a few days later he was reporting to his brother that he was all set now to write a big Peacockian novel against an Italian background set in some

> incredibly large castle – like the Sitwells' at Monte Gufone [*sic*], the most amazing place I have ever seen in my life – divided up, as Monte Gufone was until recently, into scores of separate habitations, which will be occupied, for the purposes of my story, by the most improbable people of every species and nationality.

This was the origin of Huxley's book *Those Barren Leaves*, although by then the *castello* had been thoroughly disguised and shifted a hundred miles north to the fringes of the great Carrara Alps. There was a reason for this change, for by the time he wrote it Huxley's relations with the Sitwells were not as easy as they had been that spring and early summer in Tuscany. Indeed, after this summer things would never be the same again between Osbert and his great friend Aldous.

The first offence occurred that autumn with the publication of *Crome Yellow*. The book had considerable success – and caused something of a scandal in the small literary world of London because of its unmistakable portrayal of many of his friends. Like some replacement surgeon, he had built up his characters from recognisable bits and pieces of half Bloomsbury and, as Sybille Bedford noted in her biography, 'he took little trouble to cover up his traces'.

The couple who were most offended were his erstwhile hosts and benefactors from Garsington, the Morrells, who saw themselves cruelly caricatured in the main characters of Henry Wimbush and his wife, the eccentric host and hostess of the country house called 'Crome' which was clearly derived from Garsington.

But as a novelist Huxley was a *pasticheur*, and the offending Wimbushes were not constructed solely from the Morrells. The Sitwells could not fail to notice that, without asking them, Huxley had made use of some of their liveliest stories about Sir George and Lady Ida – and had not been over-tactful in the process. The outrageous 'Old Priscilla' Wimbush is in outline and appearance very like Lady Ottoline, but she was also shown to have some disturbingly un-Ottoline-like habits. To start with, she had once gone nearly bankrupt – through extravagance and gambling. 'The number of thousands varied in different legends, but all put it high.' Worse followed when the Sitwells read of how her husband Henry had had to sell some pictures to redeem her debts and then asserted himself forcefully against his erring wife, just as Sir George had done. Indeed, there was even more of George in Henry Wimbush than of Ida in his wife – his dotty archaeological passions, the book he was writing on the history of his family, and even his collection of expensive antique beds.

All this, of course, was really unforgivable. It was an unwritten rule among the trio that they alone could mock 'the Gingers': others did so at their peril. Any other blundering offender would have been dealt with summarily; and that would have been the end of it – and probably of him as well. But Aldous was slightly different. In the first place, Edith undoubtedly had a soft spot for him, and was less inclined to be upset by slighting references to her parents than her brothers were. More important still, Huxley was having an extraordinary success now with his wretched novel. How could the Sitwells, who adored success, break with the most promising young novelist in London, particularly if in doing so they showed themselves to be as thin-skinned and ridiculous as Lady Ottoline?

So, for a while, the appearances of friendship were maintained. Aldous continued to be one of the brightest ornaments around the marble dining-table at Carlyle Square – and on Thursdays now the Sitwells often dined, less opulently, with the Huxleys at their house in Westbourne Terrace, and afterwards all went to the cinema together.

But Osbert was wary – and obviously watching for a recurrence of those magpie-habits of his friend. In 1922 it duly came. Despite the critical success of *Crome Yellow*, Huxley was still obliged to struggle hard to support himself and his young family. Under great pressure he was writing short stories for the monthly magazines. One of the best of them was called 'The Tillotson Banquet', but before publishing it even Huxley might perhaps have paused a moment to consider its effect upon his friend in Carlyle Square.

Tillotson was an aged and forgotten painter, suddenly rediscovered in his nineties by an ambitious critic who persuaded a rich art collector called Lord Badgery to hold an impressive banquet in his honour – with disastrous results. Badgery is obviously drawn from Osbert, and so closely drawn that it would have taken the most thick-skinned of friends not to be mildly offended. Osbert was not thick-skinned, and the irreverent, faintly mocking tone with which Lord Badgery was treated clearly went far beyond what could be permitted even the most talented young novelist in London; particularly as what he wrote was so uncomfortably near the truth. Look at the way that he described the Badgerys: 'In the eighteenth century, when life had become relatively secure, the Badgerys began to venture forth into civilised society. From boorish squires they blossomed into *grands seigneurs*, patrons of the arts, virtuosi.' Boorish squires indeed! And look, too, at the way he pictured Osbert. 'Behind the heavy waxen mask of his face, ambushed behind the Hanoverian nose, the little lustreless pig's eyes, the pale thick lips, there lurked a small devil of happy malice that rocked with laughter.' Perhaps it was affectionately meant – perhaps – but it *is* a little hard when somebody you dine with every Thursday night describes your eyes as lustreless and pig-like and your lips as thick and pale. Even Osbert's erratic mode of conversation was described in detail, and by the end of the story one has a clear impression of this Osbert-like Lord Badgery as a rich, physically repellent, callously malicious, aristocratic ass.

As if all this were not enough, there was the character of the ancient Tillotson. The whole idea of this rediscovered idol of the nineties was plainly taken from the incident at Lucca with old Harry Melvill – and Harry Melvill had been Osbert's discovery in the first place.

Clearly the publication of 'The Tillotson Banquet' had to result in the expulsion of the Huxleys from the Sitwell circle, but what else could Osbert do to gain at least nominal revenge? Had it been Edith there would have been a resounding verbal 'slap', and a scene that echoed all the way from Westbourne Terrace up to Bloomsbury. Osbert's revenge was different, and more practical. In the autumn of 1922 he and Sacheverell were once more at the Cappuccini in Amalfi, and as he pondered his revenge on the offending Huxley he decided he would retaliate in kind. He had never written prose before, but now was perhaps the time to start. Slowly and lovingly he constructed the first short story that he ever wrote. It was entitled 'The Machine Breaks Down', and it could almost have been written by Aldous Huxley.

It described in detail the whole incident of the visit to Lucca and the encounter there with Harry Melvill, and in the course of it Osbert allowed himself to get in various retaliatory digs against Huxley. After that insolent portrait of Lord Badgery, how satisfying it must have been to write about Huxley now as 'tall and thin as a young giraffe, with the small head of some extinct animal . . . that subsisted on the nibbled tops of young palm-trees in the oases'! How pleasant too to mock the maddening way in which he had already become 'more Italian than the Italians' after four and a half weeks in their country, 'even talking the language with such an exquisite *bocca Romana* that the Romans were unable to grasp his meaning'. But most important of all, Osbert was calmly telling Huxley off for his ungentlemanly lack of taste in using his friendship and his 'Olympian', languid air as a cover for his search for journalistic copy. 'And how often, when I saw silly little jokes of mine appearing under the guise of musical or scientific articles in the weekly papers, did I wish that his character had been true to his appearance. . . .'

It was all very elegantly done. Lord Badgery was reading his presumptuous young friend a lesson in good manners – and in such a way that there was no possible reply. But in the process Osbert was doing something more important: he was discovering his talent as a writer of short stories. 'The Machine Breaks Down' is a most polished and successful story in its own right, and the side-swipe at Huxley – he is called 'Erasmus' in the text – is almost incidental to the picture Osbert gives of sad, forgotten Harry Melvill, the lovingly evoked account of Monte Carlo, and the description of the journey in the spring through Tuscany. The one person no one apparently considered in all this was Melvill himself. He was cut to the quick when he discovered he was the inspiration of the two stories.

But Huxley's behaviour would rankle with Osbert for many years to come. When he was writing his autobiography, twenty years after the event, he still referred to it with some asperity and took pains to conceal Huxley's identity, although acknowledging the debt he owed him.

> Indeed, it was only because a friend of mine, a respected contemporary novelist, stole my stories, and because I resented the manner in which he twisted and spoiled them, and because I felt I could write them better myself, that I took to prose.

The collapse of the Anglo-French Poetry Society seems to have done little to deflect Edith's missionary zeal to become the poetic leader of her time, but it was uphill work. Her published poems were not yet sufficient on their own to have established her as 'Prufrock' had established Eliot as the most promising younger poet of his generation. *The Wooden Pegasus* published in 1920 was a collection of her poems, almost all of which had previously appeared in *Wheels* or *Art and Letters*, and the other book she published in that year, her delightful *Children's Tales from the Russian Ballet*, while acknowledging the debt she owed Diaghilev, did nothing to advance her standing as a poet.

Nor had her friendships with the poets she admired and liked done much to further her ambitions. She was still 'Edith Sitwell, Editor of *Wheels*', but even *Wheels* was now in trouble. Outside the poems of her family, its only recent real poetic coup had been the publishing of Wilfred Owen – and that was in 1919. None of the other important poets who had called for tea at Moscow Road – Graves, Eliot, Sassoon and Ezra Pound, or even Robert Nichols – had been persuaded to appear in it. Pound was now in Rapallo, and 'really quite impossible' as she explained in a letter that she wrote to Robert Nichols. Eliot, though friendly (particularly with Osbert) and quite willing to be roped in for an occasional verse-reading if Edith insisted hard enough, now had his own *Criterion* magazine, and would rely on Ezra Pound for advice about *The Waste Land* when he finished it. She was friendly, for a while, with Robert Graves, and even stayed with him and his first wife, Nancy, for a few days in his cottage on Boar's Hill, near Oxford. Although, as he put it in a revealing phrase, he 'always felt uncomfortably rustic' in the presence of Osbert and Sacheverell, he got on well with Edith: 'It was a surprise, after reading her poems, to find her gentle, domesticated, and even devout. When she came to stay with us she spent her time sitting on the sofa and hemming handkerchiefs.' But when he left his wife and eloped with the poetess Laura Riding, Edith could not forgive him.

But the desire to lead, to influence, was very strong, and it was now that Edith first experienced the perilous but heady possibilities of becoming both a leader and a source of inspiration for the poetic young.

It was in many ways a tempting prospect, especially at a time when the Modern Movement had been developing its own sweet cult of youth. Just as the Squirearchy looked backwards to a nostalgic past and seemed to enshrine the safe, paternal values of the old, so the Sitwells and their allies were emphatically, doctrinally the party of the young. It was not by chance that the stars of the Modern Movement were a bright constellation of 'young geniuses' – Rimbaud, the father of them all, whose poetic life was over by the age of twenty-two, Cocteau himself, Picasso who started painting at the age of twelve, the young cyclist, Georges Braque, and that most extraordinary young genius of all, the schoolboy Raymond Radiguet, who wrote two classic novels in his teens and then, in the words of his lover, Jean Cocteau, 'died without knowing it, on December 12th, 1923, after a miraculous life'.

The best hope for the artistic rebirth which the Sitwells stood for seemed to lie with the unsullied young, and although Osbert had adroitly cultivated so many useful members of the literary gerontocracy in his time, youth was what really counted now. Huxley was getting dangerously old when he published his *Crome Yellow* at the age of twenty-seven. That young genius, Sacheverell, had 'discovered' a still younger genius, William Walton, and was looking out for more: in 1921 he got to know a promising composer, sixteen-year-old Constant Lambert. (It was this tendency of the Sitwells to attract the young which the age-conscious Wyndham Lewis mocked when he christened them 'a sort of middle-aged *youth-movement*'.)

In these circumstances it would have been strange if Edith too had not 'discovered' a young poet of her own, and in 1921 she did, a sixteen-year-old Anglo-American Etonian called Brian Howard. Howard was precocious, snobbish, cosmopolitan and passionately addicted to all that was new and dashing in the arts. Early that year he was writing in his diary: 'Have my first boxing lesson. . . . Buy *Cock and Harlequin* by Cocteau. Polish off some Dadaïste poems and send them to Edith Sitwell.' And on 14 February Edith had replied with a four-page letter full of intoxicating appreciation and support.

Dear Mr Howard,

I am going to do a thing that I have never done before, – I am going to write you detailed criticism and advice about your poems. I get a very great many manuscripts sent to me, and I invariably return them with a short note of regret. But in your case it is different. You have quite obviously very real gifts, and I hope to publish some work of yours in *Wheels* – perhaps this year, perhaps next. It depends on you.

Howard replied instantly from Eton with yet more poems, and four days later Edith was writing to him again. Any qualms she may have had about his work were over.

There can be not the slightest doubt that your gifts and promise are exceedingly remarkable. You are undoubtedly what is known as a 'born writer'.

Magisterial advice ensued. 'One must feel, you know, as well as see. But never *never*, pretend to yourself that you are *feeling* something when you are not.' She advised revision. But the hortatory note in Edith's message now was loud and clear. She had made up her mind to make Howard a poet as she had once done with Sacheverell. 'I am perfectly determined that you are to become an extremely distinguished writer, and I must not spoil your gifts.'

Edith was an erratic judge of other people's poetry – particularly where her unprofessional emotions were involved. One thinks of her effusive praise for the gimcrack poetry of handsome Robert Nichols, her eagerness to print the now forgotten verse of Sherard Vines and Arnold James, her preference for the early poems of Iris Tree against the early poems of Ezra Pound. But even so it is strange to see the apparent eagerness with which she let herself be taken in by the pretentious juvenilia of Brian Howard. He was in fact a gifted schoolboy mimic who had read his Cocteau and Huysmans and his Edith Sitwell, and who could reproduce passable pastiches of her verse. Instead of this she saw him as a disciple she could form and shape into an important poet, as she had once shaped the poetic talents of Sacheverell.

In your *Musical Marvels* [she wrote], nobody but a writer of real talent could have written . . . *Exquisite little chopped up canary wings. Falling down in jerks and bursts and jangles out of a blue-gilt sky*. That is simply *admirable*.

She also thought 'admirable' Howard's extraordinarily silly poem 'Barouches Noires' which she insisted on publishing in the sixth and last *Wheels*.

Howard's great achievement as a schoolboy was *The Eton Candle*, a magazine which he co-edited with his then friend Harold Acton. It was dedicated to Swinburne's memory, and 'the print was to emulate that of Max Beerbohm's early works'. Osbert and Sacheverell, as the rather special sort of Old Etonians the editors admired, inevitably contributed.

But Howard remained Edith's protégé, and since the relationship formed something of a pattern for the numerous young poets she took up and fostered for the remainder of her life, it is worth following a little further. For Howard the Sitwells were a useful bandwagon he could join, a chic and flatteringly accessible part of that whole Modern Movement he was in love with. As he wrote to his mother, who was concerned that his friendship with the Sitwells would be getting him a bad reputation when he went to Oxford:

I can't sit around reading Carlyle and James Russell Lowell... I've *got* to have my cubist posters, my dances, my visit to the Sitwells.... I don't SEE why you should think I'm ruining my literary sense and being petty and ephemeral, and mixing myself with worthless people. Let me smell green carnations while I am still of an appropriate age.

Nevertheless, to reassure his mother he had his poem in *Wheels* published under the ninetyish nom de plume of 'Charles Orange'.

As for Edith, whose thwarted maternal urges would always make her sadly vulnerable to would-be young male poets on the make, she was soon asking him to visit her – in terms of almost girlish eagerness.

Do you live in London when you are not at Eton? Or if you do not live in London, do you ever pass through? If so, let me know and come and spend the afternoon with me. We could work on your poems together before tea, and after tea we would talk about books, pictures, music.

Howard was far too Etonian and smart for this sort of thing, and he expressed his disillusion at actually meeting Edith in a letter that he wrote to Harold Acton.

Well, I went to see Edie (Sitwell) and was very disappointed indeed. I arrived at Moscow Road – an uninviting Bayswater slum – and toiled up flights and flights of bare Victorian stairs (very narrow stairs). From outside Pembridge Mansions looked like an inexpensive and dirty hospital. I arrived at a nasty green door, which was opened by Edie herself. She has a very long thin expressive face with a pale but good complexion. She looks rather like a refined Dutch medieval madonna. She had on an enlarged *poilu* hat of grey fur, an apple green sheep's wool jacket, and an uninteresting 'old gold' brocade dress. Her hair is thin, but of a pleasant pale gold colour, bobbed at the sides and bunned at the hind. I do not care for the people she usually has around her – to Saturday teas, for example, at all. They are common little nobodies. Also I don't like her teas – *as teas* at all. Tell William I got *one penny bun and three-quarters of a cup of rancid tea in a dirty cottage mug*. Also I don't like her apartment, or, rather, room. It is small, dark, and I suspect, dirty. The only interesting things in the room are an etching by (Augustus) John, and her library, which is most entertaining. The remainder seems to consist of one lustre ball and a quantity of bad draperies. Miss Helen Rootham, whom she lives with, is one of those terrifyingly forceful women – she proffered me a picture by Kandinsky – an incoherent smudge of colour and murmured with great vim into my ear, 'Isn't that *pure* beauty?' I replied in the affirmative. I always do that when I meet muscle.

Horribly snobbish and unkind, and yet there is a certain ring of truth to

Howard's letter, and it helps explain who so many of her anxious friendships foundered, and so many of her grand poetic gestures ended in fiasco. Behind that desperate, would-be dominating presence lay a naïve and very vulnerable spinster in her middle thirties. Howard saw this and it put him off. Once he had met her, Edith became for him a curious mixture of 'genius' and '*grotesquerie de cauchemar*', as he uncharitably called her, and one still senses in another letter that he wrote to Harold Acton his embarrassment at the visit that she made to Eton.

Edie came down here looking like one of her own poems. . . . She whispered her speech into a table, and as a result it was horribly difficult to hear anything at all. Ostensibly lecturing on Modern Poetry, she went on for hours about Stravinsky. What people *could* hear of her speech astounded them so much that they sat there smiling, with their mouths open, and tittered once in a while – genteelly, of course . . . at the end . . . the beastly old Vice-Provost, McNaghten – got up and advised Edie to *speak louder in the future!* Eton is a queer place, isn't it?

1922 was the year which Cyril Connolly described as 'the *annus mirabilis* of the modern movement, with *Ulysses* and *The Waste Land* just round the corner, and the piping days of Huxley, Firbank, Waley and the Sitwells, of the Diaghilev ballet and Stulik in his pride at the Eiffel Tower'.

Certainly for all three Sitwells it was a year of real achievement and enjoyment, a year in which, despite the manifold attractions of their way of life, they all began to *work* as they had never worked before. The chief distraction still seems to have been Sir George, though it is a little hard to tell how much of a menace he really was, and how much an excuse for all the gallivanting off abroad, which Osbert claimed was vital to escape his palsied presence. During that year's statutory August spent at Renishaw, Sir George seems to have been more tiresome than usual and his children certainly continued with their teasing. His energies increased with every year that passed: also his sense of always knowing best, so that he was a constant source of interruption and gratuitous advice. Whether Sir George, as Osbert claimed, was actually 'determined to prevent' his children's work 'as far as lay within his power' is another matter. Osbert pretended to believe he was, and it was this belief which Osbert used to justify that autumn's flight, along with Sacheverell and William Walton, to the cloistered refuge of Amalfi.

Typically, they made their flight the occasion for the sort of carefully worked-out conspiracy that Osbert loved. The old paternal bogey-man was never far from Osbert's father-haunted thoughts. Even in southern Italy Sir George could easily pop up at any moment, advice and green umbrella at the ready. To put him firmly off the trail, Osbert devised the happy fiction of the Steam Yacht *Rover*. As they left Renishaw early that

September they told Sir George that they were cruising around Europe with a rich friend in a yacht. Paper was printed specially with '*S.Y. Rover*' at the head, and for the next few months Osbert and Sacheverell wrote Sir George a baffling succession of letters which they got friends to post for them from different ports of call around the Continent.

They took Walton with them to their beloved Venice. With their friend Richard Wyndham they had taken Victor Cunard's palazzo on the Grand Canal for a fortnight. Wyndham Lewis was there as well, and has described the scene:

> Osbert and Sacheverell Sitwell were both there, the pleasant corpulence of the former vibrating to the impact of his own and Hugo [Rumbold]'s pleasantries; Sacheverell with the look of sedate alarm which at that period was characteristic of him. We would meet in the café every day. Eventually Bob MacAlmon turned up too. An impromptu combat was arranged in the café one night, when an Italian painter challenged me to a match of draughtsmanship. Egged on by the Sitwells, I took him on. William Walton . . . was the object chosen for the exercise of our skill. Walton sat with his head on the plush back of the seat and we drew him. I won this match hands down.

Lewis, alas, does not record whether Hugo Rumbold's 'pleasantries' that autumn included one eccentricity which fascinated Osbert. According to a note in Osbert's posthumous papers, 'the most singular feature of Hugo's life and character was that he could disappear from circulation for weeks at a time and live and dress as a woman'. Rumbold, a distinguished stage designer and a brother of Sir Horace Rumbold, British ambassador to Berlin, was, Osbert adds, no ordinary transvestite. He was apparently heterosexual and liked nothing more than acting out the part of lady's maid or companion to any woman with whom he was in love. 'Hugo', writes Osbert, 'made an excessively neat woman.'

After Venetian high jinks it was time for work, and even William Walton, who should have known them by now, was surprised by the energy and discipline with which Osbert and Sacheverell pursued their day once they had reached the Cappuccini. Earlier that July Sacheverell's full-length collection of his poems, *The Hundred and One Harlequins*, had appeared, dedicated to Ada Leverson. It was a revelation of the range and the extraordinary assurance of his gifts, with poems as varied as three 'variations' in the style of the seventeenth-century poet Peele, his lyrical 'Serenades' and his full-scale, grand-operatic meditation, 'Parade Virtues for a Dying Gladiator'. But he was not writing poetry now. Instead he was continuing the book he had begun before his illness and recreating his own imaginative version of the lost but passionately glimpsed baroque world of seventeenth-century Italy and Spain – the singing of the great *castrato* Farinelli, the serenades and fireworks that greeted King Ferdinand

I of Naples in the gardens of Caserta, the work of the forgotten sculptor Cosimo Fansaga for the Certosa di San Martino above Naples. At the time he was not, he says, drawing a very precise line between poetry and prose, and possibly for this very reason he found the book 'marvellously easy to write'. It was, he says, written that autumn at Amalfi in a state of 'exaltation'.

According to William Walton, Osbert too was 'slogging away' from breakfast until lunch-time. The quiet of the hotel and the example of his brother's dedication carried him on once he had finished polishing 'The Machine Breaks Down' to his satisfaction. The experience of writing prose appealed to him, for at the end of that November he was writing cheerfully to his publisher, Grant Richards, announcing, 'I don't propose to return to England for two or three months . . . The weather', he went on, 'is divine – bright warm sunshine, and I am busily engaged on completing a book of travel sketches and in writing some short stories.'

It is interesting that the second of these short stories which he wrote (and eventually published under the collective title *Triple Fugue*) was once again inspired by literary revenge – in this case on Lord Beaverbrook's poetic hatchet-man and the discoverer of 'The Asylum School of Poetry', Louis McQuilland. It was a satisfying exercise to sit in that whitewashed cell above the azure waters of Amalfi quietly composing a short story which would bring retribution to an enemy a thousand miles away. In McQuilland's case, Osbert based his story on a bizarre conversation he had actually overheard with the pugnacious Ulster journalist. As he told Grant Richards, 'When I last heard him (I don't know him of course) he was promising two elevated ladies at a luncheon party in Fleet Street that he wouldn't, no, he couldn't really, commit suicide.'

This was to be the basis of 'Friendship's Due', the unimproving tale of Ferdinand H. McCulloch, a lark-loving, Squire-following poetaster, whose name 'was one to conjure with in the more serious salons of West Hampstead', and how he promised his romantically inclined admirers that he would kill himself in the cause of poetry, only to disappoint his audience when his nerve failed him.

Revenge in a warm climate, while your enemies are already shivering in the London winter – small wonder Osbert was beginning to enjoy Amalfi!

CHAPTER TEN

Façade

1922–1923

WHILE OSBERT and Sacheverell could spend the winter in Amalfi, Edith was still enduring the discomforts and the stairs of Pembridge Mansions, so that there seems a certain justice in the fact that it was the poetry which she was hatching now in darkest Bayswater which was to bring the Sitwell name its greatest fame – and greatest notoriety: the verse and music entertainment known as *Façade*.

Façade would ultimately become so much a part of the legend of the Sitwells that it is hard to disentangle the precise facts of its genesis from the fictions that surround them. But what is indisputable is the degree to which all three Sitwells were involved in it. The verse was Edith's and the music William Walton's, but the whole inspiration and development of *Façade* into a key-work of the twenties was a combined achievement. And as so often happened when all three Sitwells worked together, the whole affair got curiously out of hand.

Edith had recently become increasingly involved in experimental poetry as she had moved away from the heavy realism of her earlier verse, and began evolving a style which was akin to abstract art. She made a lot of what she called her personal poetic 'technique', and was absorbed by such things as the 'texture' of particular words, with assonances, dissonances, rhythms and repetitions to produce particular effects. She had begun conscientiously performing what she would term her poetic 'exercises' – teaching herself to use words much as a composer uses notes and phrases in his music. Music had always meant a lot to her, and according to Osbert she was experimenting to obtain 'through the medium of words the rhythm of dance measures such as waltzes, polkas, foxtrots'.

This naturally interested William Walton, and according to what Edith told her secretary, Elizabeth Salter, 'Willie gave me certain rhythms and said, "There you are, Edith, see what you can do with that." So I went away and did it. I wanted to prove that I could.' She added, not unimportantly, that she first of all wrote the poems in *Façade* 'for fun'.

Sir William Walton's memory of what occurred is slightly different, although it confirms the distinctly casual way *Façade* began. He has no recollection of providing Edith with her 'rhythms'. 'It was simply that Edith had written a number of poems already which were calling out for music. When I began writing it it was just one big experiment.' And, certainly, Edith had already published more than half the poems which were included in the first production of *Façade* before the idea of writing any music to accompany them was even mooted.

Not that it matters greatly how these 'experiments' began. What was important was the way that they developed into the full-scale entertainment of *Façade*. Here there is no dispute about the nature of the driving force. Again, according to Sir William:

> Osbert and Sachie were both very much excited and involved with it all once I had started, and they were the ones who were really keen on making me continue with the music. I remember thinking it was not a very good idea, but when I said so, they simply told me that they'd get Constant to do it if I wouldn't – and of course I couldn't possibly let that occur.

As usual, Osbert and Sacheverell were very much aware of what they were up to. Ever since Schoenberg composed music to accompany Albert Giraud's *Pierrot Lunaire* ten years earlier, the whole idea of having verse declaimed to music had become something of a cliché in the Modern Movement. The great precedent had been Cocteau's *Parade*, with décor by Picasso and music by Erik Satie, which had caused such notable furore when performed in Paris in 1917. The idea of creating a similar artistic triumph – and issuing an artistic challenge to the diehards and the Philistines in the midst of London – must have been irresistible.

Walton, by now an ancient and self-confident nineteen, had fitted in surprisingly well with the whole *ménage* at Carlyle Square. However prickly the Sitwells may have been with those outside their circle, they were extraordinarily generous and easy-going with those inside it. In their desire to find some way of being 'of use to him, and of advancing his chances and genius', the Sitwells had accepted him quite naturally as a supernumerary member of the family. According to the pianist Angus Morrison, who lived in Oakley Street, Chelsea, and was a close friend of Walton's at the time, all three Sitwells treated him quite casually, 'and so were the ideal patrons for a young musician like Willie. They were completely undemanding, never seemed in the slightest bit concerned to get their pound of flesh, and simply left him to get on with his own work in whatever way he wanted.'

Walton confirms this. He says that he regarded Osbert as 'part elder brother, part father-figure', and according to Constant Lambert's biographer, Richard Shead, 'friends visiting the Sitwells noticed that from time to time Osbert would say, "Oh, Willie, do fetch that book for me," or something of the sort, and so they sometimes teasingly referred to Walton as "Osbert's fag".'

But none of this appeared to worry Walton, who was by all accounts extraordinarily self-possessed. 'Had I been a writer, things might have been uncomfortable, but as a musician I just had my own room with my piano at the top of the house, and I was left in peace to get on with my work.'

Mrs Powell mothered him and darned his socks, his keep was free, and when the Sitwells went to Italy or Renishaw he would travel with them.

As an updated exercise in artistic patronage, the Sitwells' behaviour to Walton is remarkable, and far more reminiscent of the traditions of eighteenth-century European aristocratic patronage than of the usual fate in store for young composers. In Walton's case the greatest benefits were freedom and sophistication. As he says, 'for a raw young composer who knew nothing about anything it was an extraordinary education simply to be with them, and especially to meet their friends who came to Carlyle Square'.

And now, as *Façade* progressed, there was a sense of the Sitwells' patronage beginning to pay off. Osbert has described the long sessions Edith and William Walton spent together going over her poems again and yet again, 'while he marked and accented them for his own guidance, to show where the precise stress and emphasis fell'. The aim was to achieve a fusion of poetry and music and to get through for once to 'that unattainable land which, in the finest songs, always lies looming mysteriously beyond, a land full of meanings and of nuances, analogies and images'. The essence of *Façade's* success would also lie in the uncanny way the music would complement the wit and rhythm of the poetry. The sense of 'fun' was wonderfully preserved – and something of the Sitwells' attitude to life as well. One feels that only a composer who was virtually a member of the family could have managed it.

In composing the music for *Façade*, the only modern work Sir William now admits to having influenced him is Stravinsky's *The Soldier's Tale*. This had been performed first in Switzerland in 1918 and, working directly with his librettist, Ramuz, Stravinsky had composed an extraordinarily complex score which included a waltz in Viennese style, a ragtime serenade, an Argentinian tango and a Bach-like chorale.

It was a difficult model for an inexperienced young composer to attempt to follow, and it inevitably took time – and quite a number of rehearsals, fiascos and performances – before *Façade* could hope to be complete. But Walton had the determined Osbert and Sacheverell to spur him on. It was Sacheverell who solved the problem of how the poetry should be declaimed by discovering the Sengerphone – a large papier mâché megaphone with an elaborate mouthpiece that fitted round the speaker's face. It had been invented by one Herr Senger, a Swiss opera singer, to enable him to sing the part of Fafner the dragon with great clarity and resonance at Bayreuth. It was, Sir William says, 'a lovely instrument', with a mouthpiece covering the lower part of the face, and it allowed the poetry to be spoken clearly and impersonally; it was Sacheverell's idea that the Sengerphone, the speaker and the orchestra should all be carefully concealed from the audience behind a painted curtain.

The resemblance of the whole production to that of Cocteau's *Parade* was growing: Cocteau had even had the idea of using a hidden megaphone for declaiming words above the ballet, and Picasso would have been ideal to paint the dividing curtain for *Façade* just as he painted Cocteau's scenery. But the days were over when the Sitwells had been able to commission him. Instead Osbert picked Frank Dobson who had been working on a Brancusi-like portrait bust of him in brass.* His studio was round the corner in Manresa Road, and together they devised a large drop curtain with a huge mask in the centre with an open mouth – through which the Sengerphone projected.

Finally there was the question of the title. Various versions of its origins have grown up across the years – as things have a habit of doing with the Sitwells – but Sir William Walton firmly insists that it came from Edith's charlady at Pembridge Mansions. 'All this carry-on is just one big façade,' she is supposed to have grumbled, a remark which Edith thoroughly enjoyed.

There were a number of experimental performances of *Façade*, and with each the production and poetry and music were improved. The first was in private in January 1922 – in the cramped, ice-cold, L-shaped, first-floor drawing-room at Carlyle Square, before an invited audience of friends, acquaintances and 'fellow-artists' of the Sitwells. With varying degrees of bafflement or pleasure they heard what the typewritten programme officially described as 'Miss Edith Sitwell on her Sengerphone, with accompaniments, overture and interlude by W. T. Walton.' The poems

* This bust was seen by T. E. Lawrence during his own sittings at the artist's studio, and so impressed him that he presented a copy to the Tate. It was, he told the gallery trustees, 'appropriate, authentic and magnificent in my eyes. I think it's his finest piece of portraiture and in addition it's as loud as the massed bands of the Guards.'

that she read included 'Madame Mouse Trots', 'Said King Pompey' and 'Jumbo's Lullaby'. Walton conducted his somewhat mutinous sextet behind the Dobson curtain, and the evening ended with placatory offerings of hot rum punch to all the audience.

There was a further performance again in private at Mrs Mathias's house on 7 February 1922. During the next fifteen months the Sitwells, and especially Walton, rallied round 'to smooth out the imperfections of the entertainment' for a full public performance of *Façade*. This memorable event took place on the afternoon of 12 June 1923 at the Aeolian Hall.

The Aeolian Hall (or the A-E-I-O-Ulian Hall, as Noël Coward christened it) is in Bond Street, and the performance that took place there was to become part of the artistic history of the twenties, and central to the whole mythology of the Sitwells. According to the 'official' Sitwell view, as expressed by both Edith and Osbert in their memoirs, it was the nearest that they ever came to actual confrontation with the embattled philistine. A large part of the audience was there to indulge their favourite English sport of 'artist-baiting' and were so incensed at their contact with original artists in the flesh that they became infuriated as *Façade* continued and 'manifested their contempt and rage', first by hissing, and finally by threatening to attack poor Edith, who had to wait until the hostile crowd dispersed. 'She was', she used to claim, 'nearly lynched by the inevitable old woman, symbol of the enemy, who had waited, umbrella raised, to smite her.'

Next day the popular papers took up the battle and continued to insult the Sitwells in their sacred role of artists. All Osbert's wartime hatred of the yellow press revived when he read one item in a gossip column under the headline 'Drivel They Paid to Hear', and in another, an interview with the fireman at the Aeolian Hall who was quoted as saying, quite reasonably in the circumstances, 'that never in twenty years' experience of recitals at that hall had he known anything like it'. 'For several weeks subsequently,' wrote Osbert, 'we were obliged to go about London feeling as if we had committed a murder' – and from then on the battle of the Aeolian Hall appeared to epitomise all that was obscurantist, vulgar and ill-natured in the way society at large reacted to the artist. But was it really?

Angus Morrison was there and remembers something very different from the scenes described by Osbert and by Edith.

Everyone was perfectly good-mannered and no one objected violently at all. There were certainly no boos or catcalls. On the other hand there wasn't much enthusiasm either. As a performance, the Aeolian Hall *Façade* simply wasn't very good, and then a hot summer afternoon in a stuffy London hall isn't the place to listen to a new production like *Façade*. The hall wasn't very full, and the whole thing fell distinctly flat.

Harold Acton and Evelyn Waugh were both there too; and neither of them made mention of any demonstration of 'contempt and rage' from the audience. Nor did Virginia Woolf who was also there and who wrote afterwards, 'I understood so little that I could not judge.' As for Sir William Walton, he admits quite freely that 'the performance at the Aeolian Hall was a shambles. It was badly performed and the music wasn't right. By the time of the next performance at the Chenil Galleries, we got it right, and it was all a great success, but at the Aeolian Hall it was disastrous.'

He adds that he was 'terribly upset, far more so than Edith and Osbert, who really loved a fight, and saw this as a chance to weigh in against the opposition'.

This seems to have been the simple truth of what occurred: faced with a setback, Osbert and Edith's immediate reaction was always to discover an opponent they could blame it on. Even a failure had to be interpreted in terms of the pattern of betrayal and aggression life had taught them. Given their confidence and taste for battle, this was a source of strength, and much of their wit, their energy, their social life, even their work, was fuelled by this constant need to fight back at a dimly comprehended enemy. And if, as now, there really was no enemy, one had to be invented. Even a gossip columnist would do.

But there was something more than this to the myth that rapidly grew up around the philistine 'furore' at the Aeolian Hall. To understand it one must remember what Osbert and Sacheverell had been hoping to achieve with their production of *Façade*. At the end of his account of the affair in his autobiography Osbert congratulates himself on the fact that 'we had created a first-class scandal in literature and music', a somewhat strange remark until one recalls again the all-important precedent of Cocteau's *Parade*. For one of the great achievements of *Parade*, when originally performed in Paris, had been the controversy it roused. It had produced *un scandale* – which in French means much more than that distinctly querulous and narrow English word 'scandal'. *Un scandale* in the French theatre means jeers, catcalls, critics up in arms, and a full-scale battle in the stalls between the impassioned enemies of the work and its supporters. It is the traditional French way of shocking the bourgeoisie and the reactionaries, and showing where the lines of battle lie. All this was triumphantly achieved by the *scandale* of *Parade*, and instantly established Cocteau's name at the centre of the Modern Movement.

But *Façade* at the Aeolian Hall was simply not *un scandale*. Osbert did his best to make it one but, unlike the French, the secret weapon of the

English philistine is not aggression but indifference. And the whole 'scandal' at the Aeolian Hall would have been quietly forgotten but for the smooth-haired young man Angus Morrison says he saw walking out 'a little more ostentatiously than perhaps he need have done' just before the end of the performance.

Osbert had got to know Noël Coward a few months before that afternoon. Although penurious and more or less unknown, the twenty-three-year-old author of *The Young Idea* was, as his biographer Cole Lesley says, by now always in demand in the houses of the rich and famous, thanks to his conversation and the songs he sang at the piano; and Osbert had met him at the house of Mrs Beatrice Guinness. Early that summer he was writing a revue for Gertrude Lawrence and Maisie Gay, and a few days before the business at the Aeolian Hall Osbert had met him lunching at the Ivy. According to Coward's recollection, Osbert remarked, 'I hear you're doing a review. What *fun*!' He then invited him to come and see *Façade* with what would prove to be the ominous words, 'It might give you some ideas.' Accordingly Coward went, and Cole Lesley says, 'Noël always told me that he did *not* walk out of *Façade*. "I wouldn't have missed a minute of it for anything in the world," were, I am afraid, the words he used.'

Whether he did or not hardly matters, for Coward saw enough to give him what he needed. He had been looking for material for Maisie Gay, and when *London Calling!* opened at the Duke of York's Theatre at the beginning of September 1923 one of the great hits of the show was her performance as the poetess Hernia Whittlebot in a sketch entitled, 'The Swiss Family Whittlebot', reciting her 'poems' with her two brothers, Gob and Sago Whittlebot.

If Osbert really had been longing for publicity, this was it. The trio were unmistakably the Sitwells and there was a compliment of sorts in being singled out as *the* representatives of modern English poetry. But Maisie Gay was just a little too successful. Her round face with its comedienne's pop eyes could hardly have been less like Edith's, but it was not difficult to send up the way that Edith spoke *Façade* and, as Coward said, Miss Gay was soon out-Sitwelling the Sitwells with 'poems' like 'Peruvian Love Song' and 'Sonata for Harpsichord'.

In point of fact, the 'poems' which Coward wrote for her were not particularly Sitwellian – nor particularly witty. One of the most memorable of them, 'Poor Shakespeare', began

> Blow, blow thou winter wind,
> Rough and rude like a goat's behind . . .

And there was another lyric which went simply,

> Your mouth is my mouth,
> And our mouth is their mouth
> And their mouth is Bournemouth.

But everything depended on the way they were delivered, and even William Walton, who made a point of visiting the show, had to admit that the Swiss Family Whittlebot was 'really not unfunny'.

Then came trouble. As Coward himself related,

> During the first two weeks of the run I received, to my intense surprise, a cross letter from Osbert Sitwell; in fact, so angry was it, that at first I imagined it to be a joke. However, it was far from being a joke, and shortly afterwards another letter arrived, even crosser than the first. To this day I am a little puzzled as to why that light-hearted burlesque should have aroused him, his brother and his sister to such paroxysms of fury.

Coward was being more than a little disingenuous when he wrote this, for as he must have known there were several reasons for the Sitwells' anger. Possibly the worst of them was sheer frustration. Osbert had been all set to face the artistic scandal of *Façade* and to perform his usual act of ridiculing his reactionary opponents, as he had ridiculed the Squirearchy, the generals, Lady Colefax and Sir George. But the reverse had happened. At the Duke of York's the Sitwells and modern poetry were being ridiculed themselves, and the philistines, instead of being laughed at, were now laughing in their turn. An actor from the London suburbs and the lower middle classes had taken on *their* role of mockery – at their expense. Worst of all, the Sitwells were in the situation Edith always dreaded – of having someone being most impertinent and not being able to retaliate.

Osbert was all his life a great believer in the prompt despatch of lawyers' letters, but in this case legal action was impossible. For apart from the cost of a full-scale libel action, one can imagine what a field-day Coward and the irrepressible Miss Gay would have had in the witness box. In the circumstances, the advice that Beverley Nichols says he would have given them was probably the best – 'Yes, Noël should be stopped; but darlings, if you ask for it in *la Vie de Bohème* you get it, and you *have* asked for it . . . and you should all rise above it.'

But this was the one thing Edith and Osbert could never bear to do, and in the months that followed their repressed indignation steadily increased, as *London Calling!* went on to become the hit show of the season.

Edith never saw it. Perhaps it would have been better if she had, although, as Sir William Walton says, she would have made such a scene in the theatre that anything could have happened then. As it was, she obviously received wildly exaggerated reports of Maisie Gay's antics, until she genuinely believed that she was depicted in the sketch with extraordinary obscenity.

One must remember here her abnormal sensitivity over her appearance, and also a vein of curious naïveté and credulousness in her make-up. She became convinced that Maisie Gay depicted her suffering from some unspecified loathsome disease. Edith never brought herself to say what the disease could be, but she was convinced that Maisie Gay had also insinuated that she was a lesbian. For somebody like Edith, whose extraordinary dignity and angry sense of pride covered the insecurity and childhood scars left by Sir George's snubs, this must have been a harrowing experience. That autumn she fell ill again, first with her usual nervous trouble, then with an acute attack of jaundice which left her more or less prostrated for the remainder of the year.

The long-term effect of the affair on her was even worse. Edith had always been insecure and touchy, but it was now that the idea that she was always liable to become assailed by what she called 'gross public insult' started to obsess her. It would do so, off and on, for the rest of her life, and as one sees from the desperate letter that she wrote to John Lehmann about Noël Coward as late as 1947 (see page 391) her most pathetic fears would always come back to this hateful incident when, as she saw it, she was being guyed in public in what she called a sketch 'of the utmost indecency'.

Edith's unhappiness and illness after *London Calling!* made Osbert's own reaction to the strange affair more understandable, as the row went reverberating on in the years ahead, often with the most surprising results. With ever-growing bitterness Osbert blamed Coward for its effect upon his sister, and there were shades here of the Lady Ida scandal and the way the family drew together to defend themselves in the face of public insult.

Coward replied to Osbert's angry letters with a well-simulated air of puzzled apology, and the remark that he had merely taken the idea of the sketch from the performance of *Façade* – as Osbert had in fact suggested. Osbert replied with scarcely veiled fury.

We are delighted to have been the means of suggesting an idea to you. It is always as well to have one, even if it isn't your own; it must be a novel experience to you. All you want, now, is a little self-confidence – and, of course, to use your voice more.

Insulting my sister is a fine beginning for you. We look forward to other triumphs. Have you tried cheating at cards?

Sacheverell was, as usual now, a pacifying influence, and did his best to calm things down. Like William Walton, he believed that these violent controversies and feuds were self-defeating and upsetting, and should at all costs be avoided. But Osbert disagreed, and the vendetta was bitterly maintained for years.

One can gather something of the venom that he felt for Coward at the time from a short – and wisely unpublished – poem that he wrote and called 'The Missing Link':

> In one smug person, Coward sums
> Up both the suburbs and the slums;
> Before both, nightly boasts his race
> By spitting in a lady's face.

Coward's private view of the Sitwells was perhaps more telling. He dismissed them as 'Two wiseacres and a cow.'

As with Edith, the Coward feud brought out the irrational, aggressive element in Osbert's nature and led to a series of related feuds that would bedevil and distract him for several years to come. And yet, ironically, the fuss stirred up by *London Calling!* was the nearest that he got to the artistic *scandale* he had been counting on after that June's production of *Façade*. With the spiralling success of *London Calling!*, publicity for the Whittlebots became immense. Coward repeated the offence, with Joyce Carey impersonating Hernia Whittlebot at a theatrical garden party; and he even issued some slim volumes of her poetry. Osbert's anger became public knowledge, and the whole incident aroused general awareness of the Sitwells as nothing ever had before.

Earlier that year a correspondent writing in the London *Star* had announced, 'Plain people are now writing to "The Star" asking, "Who are the Sitwells?"' Answering their question, the *Star* replied in words that must have gladdened Osbert's heart, despite himself.

'They are all poets. But they are more than that. They are a cult.'

CHAPTER ELEVEN

Break-up

1924–1926

THE BEST description of the Sitwells in their prime as leaders of the 'cult of Sitwellism' is Cyril Connolly's account of how he first saw them at the Washington Irving Hotel in Granada at Easter 1925. He was twenty-one, on vacation from Balliol with twenty pounds in his pocket from his unloved, shell-collecting father, Major Connolly, and he had spent four 'heavenly' days walking across the mountains from Algeciras. Spain was still 'totally unreal' for him, 'the complete earthly paradise', and there to complete it all, in the hotel in Granada, he came upon the Sitwells and their followers 'in force'. It was a sight that he was never to forget.

They were really quite alarming – alarming rather than forbidding. All of them were wearing black capes and black Andalusian hats and looked magnificent. Dick Wyndham was with them and Constant Lambert and Willie Walton. I'd never met them before, but they *seemed* like the Sitwells. I heard them joking about Scarborough and got into conversation with them, and we immediately became great friends. I was totally bowled over by them. Edith had written 'The Sleeping Beauty', which I thought one of the loveliest poems I'd ever read, and Osbert told endless funny stories about Sir George and Scarborough. I particularly remember telling Edith how much I admired *The Waste Land*, and how she replied: 'I'll tell Mr Eliot next time I see him. He'll be delighted.'

Such boyish adoration could not last. Back at Oxford, Connolly received the inevitable summons from Edith for one of her 'Saturdays' at Moscow Road. He had no money for the journey and, embarrassed, failed to answer. 'A little later, Harold Acton, who did go, told me that Edith had remarked "What an extremely rude young man Connolly appeared to be," and that, for the time being, was that.'

In fact, this was only the beginning of the ambivalent relationship – part worshipping, part mocking, part secretly resenting – which Cyril

Connolly, like so many intellectuals of his period, managed to maintain with the Sitwells for the remainder of his life.

The worship was quite genuine. Like Brian Howard and a host of other talented, socially ambitious, middle-class members of his immediate Oxford generation, Connolly saw the Sitwells as social and artistic paragons, embodying an attitude, a total way of life which, more than with any other artists of the time, appealed to him.

Martin Green is pitching their claim a little high when he describes the Sitwells as 'the major preceptors of the postwar generation', but they were certainly precursors, particularly over taste, and for the sort of influential, mainly Oxford-educated young that he calls 'dandy-aesthetes' – people like Harold Acton, Evelyn Waugh, Peter Quennell and Kenneth Clark – the Sitwells possessed an irresistible allure.

It was extraordinary how acute the Sitwells' cultural antennae had been to what would fascinate this postwar generation: their passion for the figures of the nineties, their spiritual affinities with Cocteau and Diaghilev, and their rejection of Cambridge-oriented Bloomsbury intellectualism, their contempt for the plodding, homespun virtues of the Squirearchy, and their attacks upon the conventional paternal wisdom of the old.

'For all their faults,' said Connolly, a few months before he died, 'the Sitwells were a dazzling monument to the English scene. They absolutely enhanced life for us during the twenties, and had they not been there a whole area of art and life would have been missing.'

Desmond MacCarthy, who was no great admirer of the Sitwells (Edith used to call him 'Dismal Desmond', although a less dismal man it would be hard to find), compared Osbert with Comte Robert de Montesquiou, the great Parisian dandy, and Proust's model for the Baron de Charlus; and Harold Acton, following and then surpassing the Osbertian dandy's style and manner, rapidly became the cynosure of Oxford. His efforts to follow in Edith's poetic footsteps proved less successful, but he caused many a hearty Oxford eyebrow to be raised by reciting her poems through a megaphone from a college window. *Façade* had made megaphones intellectually fashionable (a contemporary caricature of Acton by his friend Evelyn Waugh even shows him holding one) and 'Sitwellism' had suddenly become a banner to be waved against the tedious, the puritanical, the insular, the old, so that the young Oxford dandy-aesthetes seemed to be fighting the old Sitwell battles on their own account.

More than this, the Sitwells – Sacheverell in particular – were becoming what Connolly called 'trail-blazers in the arts', and setting new fashions with the whole range of their enthusiasms. For the young aesthete Kenneth Clark, 'it was wonderful to find people so liberated from accepted thought and values – particularly from those of Bloomsbury, and the domination of Roger Fry and all that muddy-coloured, pseudo-classicism'.

The Sitwells had something else that Bloomsbury lacked – straightforward glamour. Lœlia Ponsonby, who later became Duchess of Westminster, met them now,

> and believe me, it really *was* something to meet the Sitwells for the first time in those days. They were so utterly unlike anybody else, and held a position in the arts that no one aspires to today. The nearest to it, I suppose, would be some very elevated, cultural pop-star, but they excited far more awe than any pop-star would. They were so extraordinarily clever and funny and there were three of them which made them still more disconcerting.

But the most telling testimonial from this period of the effect the Sitwells could produce on the vulnerable young comes not from Oxford but from Cambridge – in the youthful diaries of the ambitious but still naïve young Cecil Beaton. During his early twenties he describes himself as having been snobbish, with artistic aspirations, in revolt against his dominating father, and secretly concerned at what to do with his life. When he asked a worldly friend for advice, the letter he received from him was short and to the point. 'I wouldn't bother too much *being* anything in particular. Just become a friend of the Sitwells and see what happens.'

He did – and in Osbert and Sacheverell found what he was seeking.

> The Sitwell brothers . . . had established a mode of existence that completely satisfied my own taste. No detail of their way of life was ugly or humdrum. They managed to give a patina of glamour to a visit to an oculist, a bootshop, or a concert. Each catalogue they received from a wine merchant or a book seller in their hands became a rare volume. With their aristocratic looks, dignified manner, and air of lofty disdain, they seemed to be above criticism.
>
> A whole new world of sensibility opened to me while sitting in candlelight around the marble dining table in Osbert's house in Chelsea.

Here in Beaton's slightly breathless adoration of these aristocrats on whom the aspiring bourgeois aesthete saw a chance to mould himself, we have the measure of the influence the Sitwells were exerting over the young dandy-aesthetes of the twenties. Class was of extreme importance to them all. Almost without exception, they were evading or escaping from their bourgeois backgrounds by posturing as aristocrats – in style, in speech, in manner, dress and attitude to life. It was not all that easy for them. Acton, the most successful, was half American and had the great advantage of an Italian background and the enormous Acton villa on the edge of Florence, but in the painful snobberies of an Evelyn Waugh one sees how desperately important sheer social aspiration had become.

For such aspiring intellectuals the Sitwells were unique and heaven-sent. Not merely were they aristocrats who actually wished to meet young artists and young intellectuals – this in itself was rare enough – but they had also made themselves epitomise the myth of a legendary romantic and artistic aristocracy, and seemed quite willing to impart its secrets to their followers.

The middle-class philistines could giggle at Maisie Gay if they wanted to, but for Cecil Beaton, and the favoured group of other socially ambitious young men like him, a 'whole new world of sensibility' really did appear to be on offer from the Sitwells.

By mirroring the Sitwells' 'aristocratic looks, dignified manner, and air of lofty disdain', as most of them attempted to, they too could do their best to rise above mere vulgar criticism. They might manage, as Beaton certainly did in his photography, to give something of the Sitwells' own 'patina of glamour' to the humdrum details of an ordinary life.

Most important of all, the Sitwells were providing by their example, and the hospitality and interest that they always gave their talented admirers, something of an *entrée* to the aristo-artistic world these young men craved.

Cyril Connolly admitted frankly that it was 'Social success . . . which most appealed to writers up to the Slump, for social success, besides gratifying the snobbery which is inherent in romantic natures, also provided them with delightful conditions, freedom and protection in large country houses.'

All this the Sitwells – part aristocrats, part artists as they were – were supremely fitted to dispense. Carlyle Square was always ready to receive the young 'geniuses' they currently approved of. So was Moscow Road, and so, if the genius behaved himself, was Renishaw. There he could enjoy that feeling of supreme and ultimate success that meant so much, share in the jokes about the 'Gingers', gossip about the other writers that he knew, enjoy the food and comfort of a splendid country house, and even feel a tenuous if temporary equality with these courteous, amusing and perceptive members of the aristocracy.

But these protégés of the Sitwells were never quite permitted to forget their place. That easy Sitwell friendship could be as easily withdrawn – and often was. A solecism, a suspected insult or disloyalty would give offence. When this occurred the victim would be warned – or suddenly discarded.

This seemed to happen to almost all the Sitwell followers and friends at some stage in their lives, and it is interesting that both Connolly and Beaton picked on the same word to describe the Sitwells during this period – 'alarming'. As Beaton says, 'One just never knew when one would be dragged over the coals for something one had not expected.' This occurred with him when he had the temerity to refer to Sir George as 'Ginger' after a weekend at Renishaw. 'All the others did, so I failed to see why I should not.' A grave misapprehension. Offence was taken, 'and

Osbert gave me such a dressing down that to this day I've not forgotten it'.

While publicity and snobbery undoubtedly played their part in the growing reputation of the Sitwells, an equally important factor was their sheer hard work. For all three of them, the two years following that first disastrous public performance of *Façade* in 1923 turned out to be the most productive period of their lives so far.

Those 'experiments' which Edith had conducted during the writing of *Façade* had led on to a whole new vein of poetry and two impressive books – *The Sleeping Beauty*, which had so delighted Cyril Connolly, and in 1925 *Troy Park*, a set of poems built around her childhood memories of 'haunted summers' in the 'austere and elegant' world of Renishaw. *Troy Park* included her magical autobiographical poem 'Colonel Fantock' (dedicated to Osbert and Sacheverell), which, more than anything that any of the Sitwells wrote, proclaimed their mythical poetic status as romantic aristocrats. Here in the 'sweet and ancient gardens' of their ancestral castle she pictured herself, Osbert and Sacheverell as a group of fantastic, other-worldly beings with their 'long eyelids', stately beauty, and immaculately royal descent, moving with gazelle-like grace through the splendid tapestry of their legendary habitat.

The same year also saw Edith in a new role – as the embattled theorist and defender of what she called 'modernist poetry' in a short essay, *Poetry and Criticism*, published by Bloomsbury's own Hogarth Press. Edith's didactic nature was never far below the surface. She was not one to cast her poetry upon the waters and calmly wait to see what happened. Poetry, once written, had to be defended, publicised, explained, and *Poetry and Criticism* is particularly interesting for the eager way that she insists upon the kinship now between her poetry and the Modern Movement in the other arts. Lofty moral themes, she says, are as irrelevant in modern poetry as in modern painting. Technique is all important – 'Poetry is primarily an art and not a dumping-ground for the emotions' – and she explains her poetry by comparing the way that she has stylised the abstract patterns in her words with the stylised abstractions of 'Douanier Rousseau, Picasso, Matisse, Derain, Modigliani, Stravinsky and Debussy'.

Osbert too was now becoming more ambitious in his work. His second book of poems, *Out of the Flame*, had come out in 1923 and had shown him developing beyond the strident manner of his earlier verse and hinted at his range as a social satirist, an anecdotal versifier, and a descriptive poet of the Italian scene. But his most interesting development was not really as a poet but as a story-teller and a travel-writer. He had published his first collection of short stories, which he entitled *Triple Fugue*, and which included 'The Machine Breaks Down' and 'Friendship's Due'. Like them, the title story 'Triple Fugue' was based on another of Osbert's old vendettas – this time with the one-time editor of *Georgian Poetry*, the secretary and friend of Winston Churchill, Edward Marsh – and he also used it for a fresh attack upon the log-rolling that he was convinced went on behind the scenes in the award of literary prizes. The malicious portrait of Marsh as Matthew Dean, 'or poor Mattie as he is to his friends', was a telling one, particularly the remark that 'He is of the type that, like certain Orientals, enjoys obscure power, loves power without the public appearance of it'. All the more reason, possibly, for Osbert not to have also exposed Marsh's weaknesses in such cruel detail – his priggishness in art, his social snobbery and, most unforgivably of all, his physical appearance, 'with the nervous wriggling waggle . . . as if a ventriloquist's dummy is speaking – so thin is the wooden stalk of the neck moving in the too-wide socket of the collar bone'.

It makes amusing reading. So, for that matter, do the first few pages of the account of the award of the annual Pecksniff Prize for English Literature. But the story as a whole shows Osbert's dangerous tendency towards long-winded, self-indulgent, mandarin prose; and a still more dangerous tendency to pile up unnecessary enemies against himself. It was becoming quite a habit. He had no need to savage Edward Marsh so cruelly, particularly as they had been on perfectly amicable terms until *Triple Fugue* was published. Osbert already had enemies enough in London and, understandably, the influential Marsh never forgave him for the caricature of 'Triple Fugue'.

A safer – and, in fact, a more effective – line of literary development for Osbert was contained in another of the stories that he published in *Triple Fugue*. This was called 'Low Tide', and describes the touching and pathetic lives of two spinsters, the Misses Cantrell-Cooksey, in the nostalgic, pre-war world of Scarborough which he knew so well, and which he would make peculiarly his own in the most successful novel he would write. But where he really found his *métier* as a prose writer was in the travel sketches he was writing, and which were published in April 1925 under the all-embracing title of *Discursions on Travel, Art and Life*. Here, in these leisurely, Beckfordian pieces on Catania and Lecce, his travels in Sicily and his researches on the Bourbon kings of Naples, he managed to convey a tone of voice, an easiness of style and learning that guarantee their place in the travel literature of Italy.

But although both Edith and Osbert were producing with such signal energy, the most extraordinary and certainly the most original Sitwell book of all was Sacheverell's – his *Southern Baroque Art* which was published by Grant Richards in 1924. Despite its considerable *succès d'estime*, it was never a best-seller. (The Sitwells' essential audience would remain a resolute minority until the publication of Osbert's autobiography and Edith's poems in the 1940s.) But *Southern Baroque Art* was a remarkably influential book from the moment it appeared. Along with Edith's poems, which in a strange way it complemented, it soon became a sort of bible for Sitwellians, and put the seal upon the trio as fashion-setters in the arts. As Kenneth Clark puts it, 'the Sitwells were using the baroque to express something of their own slightly frivolous, un-English attitude to life'.

On reading *Southern Baroque Art* one soon sees that in southern Italy and Spain Sacheverell had suddenly discovered a fantastic and forgotten land which he could appropriate through his poet's eye and fervent imagination. It was a land to which his father introduced him first, but during his travels with his brother he had enlarged its frontiers so that it formed a zone of magic inhabited by creatures of bright fable and elegant excess – Farinelli, the legendary *castrato*, the Bourbon King Ferdinand who built the last great palace of Caserta with slave labour, and even the lands of far-off Mexico, which he had never visited except in his endless reading and imagination.

This world of the baroque provided an important landscape which the Sitwells could believe in and employ as a background to their lives and to their work, much as they had already used the world of the *commedia dell' arte* from the ballets of Diaghilev. For the idea of the baroque appealed to them on several levels. Politically it offered an imaginative escape from the dead-end world of 1920s politics back to the world of art and artists, aristocrats and men of taste, untroubled by the common man, the middle classes, democracy or the homespun insularity of a Squire or Edward Shanks.

To believe in the baroque was to believe in a world before the Fall – the Fall in the Sitwells' case being the multitudinous disasters of the nineteenth century – industrialisation, the Reform Acts, total war and the decay of taste and manners in the arts. And, artistically as well, the baroque naturally attracted them. It linked up with the world of Alexander Pope which had already influenced so much of Edith's poetry, and it was elegant, fantastic and theatrical, creating an imaginative illusion which could supersede the dull, common-sense realities of life. Originally it seemed that they had found this in the affinity they felt for the aestheticism of the 1890s, but the world of the baroque had more to offer now than the world of Swinburne, Wilde and Walter Pater. It was an elastic concept which could enfold their poetry, their taste in architecture and their love of southern Italy, while for their followers and friends a feeling for the baroque became the touchstone for a certain attitude to life, a certain species of intelligence, distinguishing the elect from the puritanical, the commonsensical, the insular, the old.

But with Osbert and Sacheverell this growing interest in the baroque also led them to be serious connoisseurs of the art of seventeenth-century Italy. Their eyes had been opened by the big 1923 exhibition of seventeenth-century painting in Florence at the Pitti Palace, and as an expression of their enthusiasm they formed their own society in England the following year to encourage its appreciation. Provocatively, they named their new society after one of the most neglected Italian painters of the seventeenth century – the Genoese Lissandrino Magnasco.

Osbert remarked, quite truthfully, that he discovered classical art by way of modern art, and with both Osbert and Sacheverell it is interesting to see how these admirers of Picasso and Matisse and the School of Paris now became increasingly concerned with connoisseurship of the paintings of the past. It was at this point that they really abdicated any leadership they ever had as amateurs of modern painting as their interest turned to the spendthrift century of the baroque.

Its lavishness, its doomed, nostalgic splendour appealed to them far more than the vigorous assertiveness of Renaissance art. Magnasco apart, the painters they enjoyed were Guido Reni and Carlo Dolci. Osbert found their paintings 'bathed in an autumnal glory scarcely less beautiful than the outburst of spring', and loved to see the greatest artists of the baroque 'squandering the great accumulations of technique and feeling just as the Italian dynasties, the Medicis, the D'Estes, the Gonzagas and Farneses of the same period squandered and lavished in every direction the accumulated treasures of their houses'.

He and Sacheverell joined forces with one of the acknowledged experts on seventeenth-century art, the Finnish art historian and declared enemy of Berenson, Tancred Borenius. Like many dealers and art experts, Borenius was quite ready to cash in on the new market in seventeenth-century paintings which the Sitwells were incidentally helping to create; and with his help the Magnasco Society became an influential publicity and pressure-group promoting the great and frequently neglected painters of the baroque.

The society was conducted with true Sitwellian panache, the high point of its activities being a splendid annual dinner at the Savoy Hotel, surrounded by magnificent seventeenth-century paintings borrowed for the occasion, and with a future Duke of Wellington as chairman for the evening. In 1924 the main speaker was Edith's old friend Walter Sickert. Under the patronage of the Sitwells, the art of the baroque was in energetic hands.

But although the Sitwells were so fashionable and powerful, they were by no means as happy and united as they seemed as they sat there in Granada in their big black hats at Easter 1925. Paradoxically, this was a time of crisis for them all, and the whole future of the trio had never been in graver doubt.

Edith was feeling very sorry for herself. Her health was far from good: indeed, it was because of this that she was in Spain at all. At the beginning of the year she had had to have an operation on her back. According to the note she wrote John Freeman, the poet, it had been 'horrid' and she had been 'quite helpless for some time before and after and in great pain'. When the pain left her, she was still suffering what she called 'the inertia of delayed convalescence', and to speed her on her recovery Osbert had invited her to Spain and paid her fare. But Edith had no taste for travel, and little more for Spain. 'I was put in the train like a penny in a slot,' she complained irritably to Freeman, 'and after three days of nervous agony, as I can't speak a word of Spanish, I found myself here with Osbert and Sachie.'

She was even gloomier about her work. *Troy Park* had just been published, but she now seemed dissatisfied with it – 'one always hopes to do so much better' – and the reviews had deeply disappointed her. In the *Morning Post* there was a 'positively imbecile one, comparing my poetry with almost everything I most dislike', and the inevitable 'impertinent' one had been published in the *Express*.

This was particularly galling after the success of Eliot's *The Waste Land*. Edith herself admired it enormously. Only that January she had been in angry conflict with the novelist and critic L. A. G. Strong because of what she termed his 'gross attack' upon the poem. 'I confess I was absolutely furious,' she told John Freeman, 'and charged Mr Strong, head down, like a bull. But he is going to behave well, and, I hope, apologise.'

But no one was charging, head down, to defend *Troy Park*, and her dissatisfaction with the poem came at least partly from contrasting it with Eliot's masterpiece.

Osbert had even greater problems and dissatisfactions. That aristocratic and Olympian calm, those anecdotes delivered in that 'wonderfully fruity, rumbling voice of his', indeed, the whole impression that he gave of being a *grand seigneur* of literature in his cloak and wide-brimmed Andalusian hat, covered what had become by now a most precarious stability.

In the first place he was very short of money. True, he received just under £2,000 a year from the Sitwell estate, but out of this he had to run Carlyle Square, pay Mrs Powell, and at least partially support William Walton and his brother; nor were his own tastes exactly cheap. Not surprisingly, he looked to his writing as a source of additional income, but none of his books had paid for the labour he put into them (his total American earnings from *Triple Fugue* came to £18) and that spring saw him penning what would soon become a crescendo of angry letters to his financially erratic publisher, Grant Richards, demanding payment of the advance of £200 promised on publication of his *Discursions*. As he explained to Richards, the trip to Granada had proved expensive: he had had to pay for Edith, and he was 'in financial straits at the moment, a thing for which (though I always expect it) I am never prepared'.

These financial worries aggravated the strange contradictions of his character: now that he had the fame and notoriety he craved, he was faced with the price of his uncomfortable success. His problem was an odd one. His congenital aggressiveness seemed to have a key ingredient in his creative processes as a writer; and, as he showed in his attack on Edward Marsh, once embarked on battle he simply could not relinquish it, whatever the consequences might be. But at the same time he was unusually vulnerable himself. As he put it, 'any artist [and that emphatically included him] must have as part of his necessary equipment nerves more sensitive and susceptible than those of other people', and it was this almost masochistic sensitivity that plagued him now. Those artistic nerves of his were playing up that year, and the more frustrated and aggressive he became, the more his suspiciousness seemed to verge on paranoia.

One sees how bad things were in the letter he had sent Grant Richards as early as 29 August 1924, complaining about Michael Arlen who had just published his phenomenally successful novel, *The Green Hat* – a letter all the more extraordinary in the light of the warm friendship that would later on develop between them.

I am sending you Michael Arlen's new book, 'The Green Hat'. Please see pp 141–152. The horrible little Armenian has 'lifted' passages wholesale from the 'Machine Breaks Down' and put them in his rotten book. If you consider anything can be done will you do it?

I met the beastly little creature in the street in February and he stopped to congratulate me upon the story in question, pretending that he had only just seen it in 'The Best Stories of 1923'; but it is probable that the nasty little creature had seen it a year before in the 'English Review' when it was first published, and, not knowing that I was going to publish short stories, thought that he could get away with the goods. I think he has spoilt that whole thing, covered it with slime, and missed any point there may have been in the original. But he certainly wants a lesson.

As Richards must have realised, even the most prejudiced of readings of the relevant passages of *The Green Hat* shows nothing of the kind. It simply was not true that Arlen had plagiarised 'The Machine Breaks Down' to create his own runaway best-seller, and the fact that Osbert believed he had and wanted 'the beastly little creature' taught a lesson is a disturbing symptom of the crisis that was building up. By the spring of 1925 the crisis had grown worse.

By then Osbert had learned about a fresh betrayal, a play just published by a one-time friend and visitor to Carlyle Square, who had recently deserted to the Squirearchy, the poet W. J. Turner. It was entitled *Smaragda's Lover – a Dramatic Phantasmagoria*. It was a patent take-off of the Sitwells, but a take-off so pathetically inept that one can only wonder that a successful man of the world like Osbert should have given it a moment's thought. But the mockery of the Whittlebot affair must still have rankled. Turner had been 'disloyal', and Edith was upset again. 'He has been cad enough to make out that I write verses which are disgusting and common,' she wailed, and Osbert in turn complained to Richards of the fact that Turner had been 'offensive and vulgar about us. . . . The weather', he went on, 'is heavenly, but I am so angry with Turner that my day has been spoiled for me. . . . Aldous, Arlen, Coward and now Turner!!'

But in the midst of so much anger and self-pity Osbert could not help remembering that even Turner's rudeness could be turned to excellent publicity, and at the end of this letter to his publisher he slipped in the neat suggestion that he might advertise his latest book in *The Times* on the lines that the Sitwells 'seem to afford almost as many ideas to other authors as we do to ourselves'. (In another letter to Grant Richards he had suggested a formal announcement in *The Times* to the effect that *Triple Fugue* was *not* awarded the Hawthornden Prize that year.)

By what must have appeared a maddening coincidence, Coward, Huxley and Arlen were all receiving fame and riches by their writings: all that the Sitwells seemed to get was personal notoriety and literary indifference. But Osbert was determined now that this would change. He and Sacheverell were finishing a play. With so much public interest in the Sitwells it could hardly fail.

They had begun the play in the winter of 1924–5 at Rapallo. Since their first visit to Beerbohm in 1922, the prim little seaside town on the main line down to Rome had continued to attract them both. It was convenient and comfortable; they could meet Beerbohm there from time to time and there was now another friend who lived there – Ezra Pound, 'covered with buttons in the form of triangles of red cornelian', as Osbert rather strangely described him in another letter to Grant Richards. Pound was, he added, 'very romantic and rather disillusioned'.

As for Max, they saw him occasionally and there was always the chance of picking up the sort of reminiscence of the nineties which Osbert treasured. But true to their claim to justify these winter holidays abroad, they had been working hard and their play was all but finished when they reached Granada. It was, Osbert told his publisher, 'a sort of *Charley's Aunt*'. It seemed to him to be 'very amusing' and he hoped anxiously some manager might take it. 'Then we might all be rich!'

When most people write a play they wait until production before the publicity begins – not so the Sitwells. Early that July Osbert, impatient as ever, decided he would hold a small press conference at the Savoy Hotel to announce the play's completion and presumably to drum up interest from any impresarios with money to invest. On 18 July the *Evening Standard* dutifully announced: 'A PLAY BY THE SITWELLS – Burlesque comedy from the two brothers' – and spent a column on the interview at the Savoy. Osbert as usual did the talking – 'Sacheverell Sitwell sat opposite and nodded approval' – and the whole interview shows Osbert's skill at handling the press. The play, 'a pleasant play, and that in itself is a distinction', was set in an Atlantic liner. Its title was as yet 'a dead secret', and as for Edith, she 'has had nothing to do with it, for she contends that any time not spent writing poetry is a waste of time. We do not agree with that.'

So far so good, but Osbert never could resist a battling controversy, particularly in the presence of a journalist. It started innocently enough. Why, the reporter asked, had they written a burlesque? 'For several reasons,' replied Osbert owlishly, 'the main one being that the English stage at the moment is so rich in actors and actresses who excel in farce. Most of the other actors allow their talents to be destroyed by the insane mania to play the part of gentlemen on the stage as well as off. If they would only devote their attention more to being actors on the stage and gentlemen off, it might be better for their art. Golf', he concluded, in what he knew quite well would be an infuriating *non sequitur* to certain *Evening Standard* readers, 'is surely one of the great curses of the acting profession at the moment.'

That simple sentence was to become the opening volley in what would soon be known as the 'Actors' War'. It undoubtedly had its origins in the resentment Osbert was still feeling against Noël Coward, but it soon widened out to include any successful actors who, like Coward, had the temerity to cut a dash within polite society. With all the cool assurance of his class, Osbert was intent on putting these pretentious *parvenus* quite firmly in their places.

Inevitably this caused offence – and produced a great deal of publicity, especially in the following year when the 'war' became more complicated and bitter. But it is strange that a man of Osbert's stature and essential seriousness should have involved himself in such a very silly fight; especially as, beneath it all, he was extremely edgy, and had a lot to occupy his mind during that summer of 1925.

Renishaw had suddenly become his own responsibility, for, as he put it in his autobiography, 'the object for which I had striven so long was near to attainment. I had persuaded my father to emigrate – to Italy.' It must have seemed a great achievement – an end at last to Sir George's interference in his affairs, an end to that maddening paternal presence, and the opportunity at thirty-three to enter into his inheritance as the *de facto* owner of the ancestral home.

How much part Osbert really played in finally persuading his parents to emigrate is problematical; for Sir George was very much a man who pleased himself, and he had long been at his happiest in Italy. His preferences apart, the most pressing reason for the baronet's departure now was clearly fiscal. At sixty-five the sturdy hypochondriac was more convinced than ever that his health was failing, and that the best way to preserve the Sitwell fortune would be to shift his capital abroad, entail Renishaw, and beat what he saw as the looming threat of death duties by taking up residence officially in Italy.

It was now nearly twenty years since Sir George's car had broken down on the Via Volterrana, but the restoration of the castle remained far from finished. The last of the two hundred squatters had been ejected, some of the accretions of the centuries were cleared, but the great challenge of that monstrous house remained. There were the terraces and gardens to be laid out (he obviously knew better than the Italians how these things were done) and there were yet more rooms to be re-plastered, more antique beds to be collected, more vineyards to be put to rights, and more builders to be driven mad with his pernickety instructions. It was a tempting prospect.

'I shall be known as the Italian Sir George', he remarked to Osbert, and that September, after firing off two final letters – to the Archbishop of Canterbury and the Chancellor of the Exchequer – complaining of the iniquity of a poor old man being driven from his ancient home by the burden of taxation, 'the Lady Ida and I set out,' as he described later, 'with a fair wind, for France, and thence to Italy, to make a new home for ourselves'.

The Lady Ida, as with most things now, took the change philosophically; the only fear which she expressed was whether her English newspapers would arrive on time. But although their departure seemed to have freed Osbert from the menace of Sir George, it brought him little else. Renishaw remained so much infected with Sir George's presence that he found it gloomy and distasteful and had certainly no plans to live there on his own except somewhat briefly in the summer. Nor had the acquisition of the house done much to benefit his rickety finances. For even in that self-denying act of vesting all his property in England, Sir

George was still playing his financial power-games with Osbert. He still distrusted him. The house and the lands were carefully tied up and put in the control of incorruptible trustees. A special trust was even formed to protect the Renishaw commode and several other tempting heirlooms which were in the house. Sir George had gone – but his influence remained. And Osbert was as short of ready cash as ever, at just the moment when he needed it.

In the spring of 1924, at a dance at Arnold Bennett's (since Uncle Arnold had left Marguerite for Dorothy Cheston Bennett, the Sitwells were back on good terms with him again), Osbert had met a startlingly beautiful Canadian brunette called Georgia Doble. Her father was a banker and a friend of the arch-fiend, Max, Lord Beaverbrook; but despite this, young Miss Doble was, she tells us, 'very, very high-brow in those days', and preferred writers to press lords. She danced with Osbert. He invited her to tea at Carlyle Square, and there, in the famous dining-room, she met Sacheverell, who fell in love with her immediately. She, who, like all good young high-brows, had read her *Southern Baroque Art*, was suitably impressed, and a romance began. It had continued, somewhat stormily, ever since, with Miss Doble fearing that 'I had bitten off more than I could chew', and the Sitwell family doing their concerted best to convince her that indeed she had. Sir George and Lady Ida were for once agreed, and felt it highly inadvisable for Sacheverell to get married, much as they liked Georgia as a person. Edith considered it unwise for a dedicated poet to marry anyone at all. And Osbert was horribly upset at the idea of the break-up of the Carlyle Square *ménage*, and the loss of his devoted brother, having taken it for granted that they would stay contentedly together now until they died.

But, despite the clear impression that one gets of the shy young poet overshadowed by the monolithic egos of his family, Sacheverell showed resilience at this latest crisis in his life. Behind the apparent harmony in Carlyle Square, he had, it seems, been growing restive for quite some time. 'It had become obvious to me that my relationship with Osbert could not go on like that much longer. For some while I had been longing to escape, and once I met Georgia I was determined I would marry her.'

By the late spring of 1925 Osbert was already starting to put his best face on what was looking like a *fait accompli* with his brother and Miss Doble. He made point of dedicating his *Discursions* to Georgia, and a press report quoted him as saying manfully that his brother's marriage, far from breaking up the Sitwell trio, would actually strengthen it, with the bridal pair settling after marriage in the house next door. He even told

Rebecca West that he was glad Sacheverell was getting married now: it would ensure that the Sitwell line continued, a remark which surprised her, as she had never heard Osbert previously express the faintest interest in the family.

Then, finally, that October Sacheverell, with much romantic resolution, carried his fiancée off to France; and on the twelfth they married 'quietly' – as the *Continental Daily Mail* described it – at the Anglican Church of St George in Paris. The Doble parents were present, the Sitwell parents not. Edith wore her most impressive Cossack hat, and in the wedding photographs one can still see the look of stately misery on Osbert's face.

Doubts and prophesies of doom notwithstanding, it was to prove an exceptionally happy marriage from the start. Sacheverell was the most uxorious of husbands and Georgia the perfect wife for a supposedly impecunious, shy, twenty-seven-year-old poet. Edith's fears proved groundless. Her brother was the sort of artist marriage suited perfectly, and he felt nothing but relief at being out of all the drama and the battles that membership of the trio had involved. Through Georgia he had escaped into a happy, normal private life at last.

But this new-found happiness inevitably made him feel concerned and guilty about Osbert, and he knew quite well how miserable he was. 'I felt dreadful about leaving him and wondered how on earth he'd manage. But, of course, he finally *did* manage perfectly, although his whole life was to change totally in the process.' It was this need for drastic readjustment that was behind the turmoil that assailed Osbert during the months ahead.

By the time Sacheverell married, Osbert had made it quite clear that he never would. He had had many close friendships with women in his time, and Sacheverell believes that, for a period at least, he had actually been in love with that flightiest of literary *femmes fatales*, Nancy Cunard. He had known her from just before the war, and for a time was certainly extremely fond of her, but her free-wheeling love-life and determined indiscretions quite ruled out any deep involvement, had he wanted it – which is doubtful. Ever since he had had to dodge Violet Trefusis's attempt to get engaged to him, he had kept clear of such entanglements with all the tidy skill of the professional bachelor. Extremely masculine in his appearance, solicitous, amusing, and with his courtly manners, he was attractive to a lot of women, but ever since Eton, as he confessed, he had been used to wearing a 'disguise' to hide his real feelings. According to Peter Quennell, 'quite a number of society ladies were in love with Osbert, who in return was in love with their fine houses and good food and pictures, but he would *never* go too far with any of them. He always drew the line.'

This seems to have summed him up. One form of Sir George's curious, incongruous puritanism that Osbert seems to have inherited was over sex: many who knew him well have commented on the unexpected vein of almost Victorian prudishness about him. Harold Acton, for instance, makes a revealing remark about Osbert's reactions to his own uninhibited conversation when he was with him in Peking in 1932. 'He flinched visibly from the Rabelaisian jest. I suspect I used to shock him with my absence of disguise: his expression at such moments delighted me, for the twinkling eyes and twitching mouth struggled with his resolve to remain impassive.'

It is a telling picture, and throughout the agonies of 1925 one can see Osbert struggling to maintain that same resolute disguise of manly impassivity. Whatever his true sexual tastes, he had always managed to present an adequate façade of the confirmed and perfectly contented bachelor. He had his social life, his feuds, and those seven separate professions which Arnold Bennett had ascribed to him to absorb his driving energies. His frustrations could be vented on the philistine. His almost oriental powers of discretion could deflect unwelcome questions, and the establishment at Carlyle Square had made a wife unnecessary. Mrs Powell looked after him, saw to his clothes and fed him splendidly. More important still, Sacheverell had always been his constant ally, friend, collaborator, confidant and antidote to boredom. Now he had gone, and the carefully constructed balance of his life broke down.

It took some while for the collapse to show, and for the first few months it almost seemed that nothing had really changed because of Sacheverell's departure from the house in Carlyle Square. Georgia, 'slight, exquisite, young, and warmly and brightly coloured as a flower', had rapidly become, wrote Osbert, 'not only my sister-in-law but one of my most intimate friends'. What more natural than for Georgia and Sacheverell to spend their honeymoon at Amalfi and to stay there as usual with Osbert through the early winter with the old pre-marital regime of literary hard labour to keep everybody busy.

Walton was still with them, and they invited Peter Quennell – sent down from Oxford for what was then termed 'fornication' – to join them. There was a sixth member of the party too – the golden-haired young art historian and critic, Adrian Stokes. According to Richard Wollheim, Stokes in those days looked 'rather like a great blond hawk', and Quennell describes Osbert's attitude towards this very beautiful young man as 'romantically protective'. He had known his wealthy mother – she was a Montefiore – years before, and had persuaded Stokes's father to provide him with an income which would enable him to develop the strong literary gifts that he enthusiastically believed in.

Not surprisingly, the attempt that autumn to act as if nothing had been altered by Sacheverell's marriage was doomed to failure. The old routines of work were rigorously maintained, with everybody slaving through the morning in their cells and meeting in the evening over their cocktails. Walton was composing on the hotel piano, Osbert writing a full-length novel set in pre-war Scarborough, and Sacheverell was working on a book of poetic reminiscences. But for once, in Amalfi, the weather failed them. Storms swept southern Italy, and before long Quennell – himself struggling on a historical novel on the romantic life of Alexander the Great – was detecting hints of 'tension, of malaise and ennui' in their daily round.

Osbert's romantic attitude to Stokes appears to have been the cause of much of the tension. Quennell disliked him, finding him pretentious. Osbert became increasingly defensive of his protégé, and there were frequent rows at dinner. Quennell admits that he was tactless – one afternoon when Osbert seemed beside himself with worry because Stokes had not returned from swimming, he said airily, 'Oh, I expect he's drowned!' Osbert was furious, and finally the atmosphere was such that he felt obliged to leave. It would be several years before Osbert spoke to him again.

But plainly Quennell himself was not the only cause that year of the 'malaise' at Amalfi, and long before the winter ended everybody must have realised that, however anxious they might be to maintain the former Sitwell way of life, there was no way now of combining Sacheverell's marriage with Osbert's protective longings for the handsome Adrian. After that year joint working winter holidays at the Cappuccini were not repeated.

But in one way at least the time in Italy had been a great success. By March, when they all returned to London, Osbert had virtually finished the Scarborough novel, and was calling it *Before the Bombardment* (the bombardment being that celebrated shelling of Scarborough by the German battle fleet which spelled the opening of hostilities for Sir George); Sacheverell had the completed manuscript of his autobiographical *All Summer in a Day*; and William Walton had the score for a much restructured and revitalised *Façade*.

After the dolours and distractions of the previous performance, it says a great deal for the collective courage of the Sitwells that they were willing once again to put themselves at risk with a public rendering of *Façade*. But courage – or masochism – was one virtue they had never lacked, and on 27 April 1926 *Façade* received its second public performance at the Chenil Galleries in Chelsea. 'And this time', as Sir William Walton says, 'we finally seemed to have got it right, and I can remember Ernest Newman banging his umbrella on the ground with excitement.' *Façade* had suddenly become a glittering success, and a few days later in the *Sunday Times* Newman, the greatest musical panjandrum in the country, confirmed the judgement of his umbrella by comparing the musical technique of *Façade* with that of *Pelléas* and *Tristan* and hailing the whole performance as 'the jolliest entertainment of the season'.

For the Sitwells there was even more to it than that. Unlike the Aeolian Hall disaster three years earlier, this was a performance on their own home ground, and it became a demonstration of the support and interest which their name could now attract among the socially artistic cream of London. Diaghilev was in the audience, and Cecil Beaton found the hall so crowded that he could not get a seat and had to stand 'along with masses of other thrilled and expectant people. Half the audience seemed nicely arty and the other half merely revolting arty.'

The nicely arty and the revolting arty were for once united in their enthusiasm. There were repeated encores, and afterwards the Sitwells and their allies dined in triumph at the Eiffel Tower.

It was a sweet revenge for *London Calling!* but for Osbert there was soon more serious business to attend to. Within a few days of this production of *Façade* he was to enter what he later called 'in some ways the most exciting period of my life'. This was his one attempt to dabble in high politics during the strange episode of the Wimbornes and the General Strike.

He had been fortunate to have been away from the threats of double-dealing that had gone on that winter and which had led the miners and the T.U.C. into the tragic gesture of the historic nine-day General Strike, which began on 3 May. But when the strike began he was back in London and took a typically dramatic view of it, seeing the situation as the probable beginning of a civil war. As a pacifist he was against a conflict. As a pessimist he feared the worst. As a frequent visitor to Italy, before and after Mussolini's march on Rome, he was more aware of the extremities of civil violence than his compatriots who stayed at home. And as a lifelong enemy of Winston Churchill he was inevitably opposed to the policy of warlike confrontation with the workers which the great man stood for.

All this made Osbert feel that something emphatically must be done, an emotion which on 7 May he expounded with considerable force to Lady Wimborne, after a stiff Martini over lunch. The result was more than even he had bargained for. It is hard to know if Alice Wimborne falls into that category of Peter Quennell's very grand society ladies who were in love with Osbert. She was certainly extremely grand and rich and rather beautiful. Her husband, Ivor Wimborne, was heir to the Guest steel fortune and one of the richest industrialists in Britain. Their London house in Arlington Street, just next to the Ritz Hotel, was one of the splendid London houses Osbert loved. And Alice Wimborne was a free and a romantic spirit living her own life and inclined to fall in love with men of talent. She had admired *Triple Fugue* enormously. She had admired Osbert too, and a close friendship had developed.

During their lunch together Osbert suggested that as an influential political hostess she should make a bid to help to end the strike by bringing a group of leading politicians and trade unionists to lunch at Wimborne House. She did. On 8 May at Wimborne House Lords Londonderry and Gainford, for the major coal-owners, met the chairman of the T.U.C., J. H. Thomas, along with Lord and Lady Wimborne, Lord Reading, Philip Snowden, J. A. Spender of the *Westminster Gazette*, and Osbert Sitwell.

It seems to have been a most placatory affair. The endearing Mr Thomas – a great crony of King George V – was an unlikely man to lead a revolution, and was already actively engaged in finding a formula to wind up the strike without leaving the country's miners too cynically in the lurch. In the country, public apathy – and Stanley Baldwin – were making civil war, to say the least, unlikely. And everyone who lunched that day at Wimborne House was decently convinced of the importance of a return to the *status quo* as soon as possible – which duly happened four days later, when the whole General Strike quietly and gratefully collapsed.

But during those four extremely busy days Osbert, as the historian of the General Strike records, 'clearly relished the atmosphere of high drama in which he was now immersed'. He telephoned and talked and was in constant contact with the Wimbornes, but it is hard to see that his initiative finally accomplished very much. The lunch at Wimborne House may have convinced poor worried Jimmie Thomas that the upper classes were behind him. Equally, Lord Reading's intervention probably did something to offset the warlike influence of Birkenhead and Churchill on the government. But for Osbert this was undoubtedly his finest hour, the one time in his life when the frustrated politician which was in him held the stage, and as an episode in the history of Osbert Sitwell these few days of the General Strike are most revealing. He clearly loved the whole idea of power, and years later was somewhat ruefully to describe the way he had envisaged himself now like one of the heroes from a novel by d'Annunzio 'who within their monumental palaces of travertine stone awaited the signal'.

This was very much in line with his romantic, anti-democratic view of politics, and embodied his belief that if only the old upper classes and the workers got together – without the beastly burgess orders to upset things as they always did – most of the country's ills could soon be amicably settled. Now that his call had come, he felt the time had come to play his part as the *éminence grise* of Carlyle Square, using his friendships with the great to set affairs in motion, chiding the complacent, offsetting the efforts of the wicked, and using the drawing-room and dining-room to make his presence felt.

It was essentially an eighteenth-century, Whig's-eye-view of politics, and one can see how admirably Osbert would have fitted into an unreformed House of Commons. (One of the many grudges that he bore against Sir George was that he had not bought a peerage from Lloyd George when he had the chance.) And it is interesting to see that, despite his aggressiveness and admiration for the d'Annunzian man of action, he was essentially a man of moderation, decency and peace in a real-life political situation. One of his deepest reasons for disliking Churchill now was the Churchillian rhetoric of 'No surrender' to the strikers.

But one must not forget where Osbert's own real interests lay, for he never did himself. The Sitwell fortunes had been built on coal, and although he had frequently expressed his sympathy and feeling for the local miners – he admired their cleanliness and neatness and their affection for their whippets – he showed no great concern for their inconvenient demands to earn a living wage. Once the danger of a General Strike was over, and the miners' own beleaguered strike dragged on through the months ahead, Osbert, like most of his fellow citizens, soon found himself with other things to think about.

Then came a fresh involvement in the 'Actors' War' – this time with the unsuspecting Mr Bobbie Hale (father of Binnie and Sonnie), the comedian. It brought a lot of press publicity, and Sir William Walton remembers it as 'one of the most successful stunts that Osbert ever did'. It also shows him at his most ingeniously malicious, and since this helps explain the sort of wariness and the dislike which he could almost wilfully promote among the so-called 'enemies' and 'philistines', it calls for just a brief examination.

As great protagonists of the Modern Movement in the arts, the Sitwells were quite naturally attracted by the latest media: first there was their film and then that summer they became involved in broadcasting. All three of them agreed to take part in a reading of *Façade* on a programme for the B.B.C. entitled 'The Wheel of Time', and Osbert gave an interview about it to the *Evening Standard*. Clearly the news that three poets would be reciting on the radio was of no overwhelming public interest: Osbert soon gave it some. Why, the reporter asked, were the Sitwells doing it?

'The stage', replied Osbert, savouring the moment, 'is in so deplorable a condition that so far as I can see the importance of broadcasting cannot be too strongly emphasised. Actors and actresses', he added wickedly, 'are so busy trying to be ladies and gentlemen and golfers that they have no time left to pay attention to their jobs.'

Uproar followed, as he knew quite well it would, but he had also set a trap, and two days later Mr Archibald de Beer, the emotional and very vocal general manager of the Vaudeville Theatre, blundered nose-first into it. For, as Osbert knew, there were a number of actors from the London stage already booked to take part in the same programme as the Sitwells, and two of them were under contract at the Vaudeville Theatre. As a direct result of Osbert Sitwell's 'impertinent and insolent attack upon the theatrical profession', spluttered Archibald de Beer, he was instantly withdrawing his permission for his actors to appear with the B.B.C. in any programme with the Sitwells. This was front-page news. But Osbert had not stopped there. A reporter from the *Evening Standard* had interviewed Bobbie Hale who had already withdrawn from the programme. A few days earlier he had had lunch with Osbert and Sacheverell and a producer from the B.B.C. to discuss his taking part in the programme called 'The Wheel of Time'. It had been most extraordinary, for throughout the meal he had found it difficult to understand a word that they were saying.

It was such high-brow stuff that I, who am not a poet but only an actor, was very often altogether at sea.

They talked about poetry and they quoted poetry, and some of it seemed to me peculiar stuff. This sort of thing, you know:–

Evelyn Laye, Evelyn Laye
Evelyn Laye, Evelyn Laye,
Why do you go on a tram,
Evelyn Laye, Evelyn Laye?

As a comedian himself, Bobbie Hale might have realised when he was having his leg firmly pulled in the gentlemanly pursuit of pure publicity. For this too made another front-page story, and Osbert's little stunt continued through the silly season, by the end of it the B.B.C. had had to re-engage fresh actors for the programme, and the Sitwell name, as Osbert himself admitted, was good for a conventional laugh on any music-hall stage in Britain.

But there was more to this than mere publicity-mongering. Sacheverell seems to have disliked the row intensely. 'This is most distressing,' he replied when the man from the *Evening Standard* asked him for his views on the whole affair. 'There is some sort of conspiracy against the family. Many people have said much harder things about the stage than Osbert did, and no notice has been taken of it, but when a Sitwell says them it is looked upon as a very dreadful affair.' Which, of course, was very true, for such was the nature of the sort of fame which Osbert had been consciously attracting to himself. So why, one can only wonder, did he do it?

A year later he would offer something of an answer when he published a brief confessional account of these few months under the unassuming title 'A Few Days in An Author's Life'. This was the strangest, most uncomfortably revealing piece that Osbert ever wrote about himself, the one public occasion when the impassive, bland façade of the successful author was allowed to drop, and one can suddenly appreciate the pain and anger and frustration of the man behind it all. It was, complained the pontifical Cecil Scott in an unsigned review in the *New Statesman and Nation*, 'a long-drawn-out whine, utterly unnecessary, and utterly unworthy either of the writer or of his brother and sister – a tedious and tortuous tale of trivial complaint'. And so at first sight it still appears. For this is Osbert at his touchiest and most irascible, listing and documenting his complaints against the world at large.

He recapitulates the battles of the previous three years – *Façade*, the 'Actors' War', the insults thrown at Edith, and the attentions of 'the Noble Company of Judas', the social gossip writers who 'chatter' on in print about their social and their intellectual betters. But more revealing are the three vitriolic chapters which he directs against 'the Mysterious Mr X', a writer he describes as 'our most virulent enemy' who was responsible during 1926 for 'one or two incidents which would have even laid an ox low with nervous prostration'. The offence of Mr X, as far as one can gather, was that he had continued to attack and vilify the Sitwells – 'in a queer bastard Catholic-Socialist-ultra-Conservative paper' – mocking their looks, their antecedents, and finally their parents. All this had come to an insufferable climax during that miserable summer which Osbert describes as follows:

> At that particula. instant our literary fortunes were at their lowest ebb: we could find no publisher to accept our work, and a new plan was being experimented with by our literary enemies of pretending we were not there, and of hoping then to find that by dint of suggestion such a pretence had carried itself into actuality. This ingenious scheme did not succeed; but there was in addition, behind this silent façade of resolutely unseeing ox-eyes, quite an amount of spite and plotting in progress.

Osbert recounts how, shortly after this, he had met the offending Mr X in a London theatre. X had offered him his hand. Osbert refused it. 'The poor creature then asked me why I would not shake hands with him. And I was forced to tell him, "Because I have for a long time disliked you, and because you have been impertinent."' There was more to come, for

Osbert was intent on drawing blood from this anonymous opponent. No quarter given, he describes the man's defects and offences – his 'snappy, northern surliness', his 'absolute lack of interest or originality', the way that he had 'desecrated the tombs of my ancestors' with his 'grisly articles', and finally the 'ghoulish process' by which this drab and offensive literary hack was trying to inhabit 'the dead body of a great mind' by trying to translate a great French author.

There is a certain irony in this attack, for the wretched Mr X was none other than the translator C. K. Scott Moncrieff, and his ghoulish attempts at inhabiting the dead body of a great mind – the translation of Proust's *À la Recherche du Temps Perdu* which he had just begun – would become one of the major influences on Osbert's own most ambitious work, and one of the favourite consolations of his declining years.

Yet as with all Osbert's other enemies around this time – Arlen and Noël Coward and Aldous Huxley – it takes considerable time and energy to discover his exact offence. Certainly the articles that Scott Moncrieff contributed to that mysterious 'Catholic-Socialist-ultra-Conservative paper' – in fact it was the *New Witness* – are scarcely even critical, much less impertinent, about the Sitwells. Scott Moncrieff himself was undoubtedly a prickly character. Orioli said of him that 'nobody was more prone to take offence. He liked quarrelling on every pretext and with every friend of his, though the quarrels never lasted for long . . .'. He had been badly wounded in the war. He was a dedicated spiritualist, he was extremely short of money and, as Evelyn Waugh discovered some years later, when he considered working as his secretary, he was a homosexual. Scott Moncrieff had settled in Florence, and the virulence of Osbert's hatred of the man must have begun with some affront which he sustained from him in Italy.

But beyond all this the real importance of the squabble lies in the light it throws on Osbert's state of mind during these troubled months of 1926. Short of money, 'deserted' by Sacheverell, no longer with Grant Richards and with his literary fortunes at their lowest ebb, he was wretchedly unhappy. He has described how at an earlier period in his life he had overcome attacks of rage by smashing cheap china dinner-plates on his doorstep. (He had for some years kept a supply of crockery for just this purpose.) The savaging of Scott Moncrieff was somewhat similar: so were the other battles he engaged in at this time, most of which show a level of undoubted paranoia.

It was now, for instance, that he became involved in an unnecessary squabble with Cyril Connolly, who described it all as follows.

I was in Paris that autumn and one evening happened to meet William Walton, Constant Lambert and Philip Heseltine. We all got very drunk, and one of them started to insult me, saying I was Desmond MacCarthy's bum-boy. I shouted back that I was no more Desmond's bum-boy than he was Osbert Sitwell's. As I say, it was all very drunken and extremely silly and that, I thought, was that. But word of what happened was duly reported back to Osbert, as always seemed to happen with the Sitwells. They had their hatchet-men who loved to stir up trouble, and shortly afterwards, when I was staying with Harold Nicolson in Berlin, I received a letter from Osbert's solicitor saying that I must instantly withdraw the slanderous accusation I had made or he would sue. Harold advised me how to deal with it, but the affair produced a total coldness between us that lasted for a long, long time.

In contrast with Osbert's angry isolation, Sacheverell's life had never been happier than now and, far from marriage having ruined his talent, his career was flourishing. In 1926 he published two books of poetry – *Exalt the Eglantine* and *Doctor Donne and Gargantua, Canto the Third*. He seemed to have accepted his role as the most private and *recherché* of all 'high-brow' poets and made a virtue of the smallness of his audience. A note on the title page of *Doctor Donne* proclaims that of the sixty-five copies printed 'Fifty are for sale'. But there was no such self-denying feeling to his other work. *All Summer in a Day*, when published in October 1926, was enthusiastically received and proved a popular success. The publisher, Thomas Balston at Duckworth's – himself an enthusiastic Sitwellian – had taken over from Grant Richards as the trio's publisher, and had commissioned Sacheverell to write on the art of the German baroque in succession to *Southern Baroque Art*. And most satisfying of all for a passionate balletomane like Sacheverell, the great Diaghilev had finally commissioned him to write the book for a ballet of his own.

This had originated with Diaghilev's visit to *Façade* that spring. The Sitwells knew that he was anxious to include new British work in his repertoire – if only to encourage Lord Rothermere, his most hopeful backer for his London season. Eager as ever to promote the work of William Walton, the Sitwells rapidly arranged an audition with Diaghilev, but after listening to some of Walton's latest compositions the impresario smiled and said with splendid tact, 'Young man, you'll write me a much better ballet in a few years' time – so wait.' (Sir William Walton adds, 'I waited, and of course I never wrote it.')

But during this same period Diaghilev was taken with an idea of Sacheverell's to create a balletic 'English Pantomime', based on the characters of the Victorian 'tuppence coloured' theatrical prints. Sacheverell even took Diaghilev off to Hoxton to look at the prints in Pollock's famous toy museum – and the result, the Russian Ballet's *Triumphs of Neptune*, with music by Lord Berners, had an extraordinary success when produced at the Lyceum in December 1926. (Sacheverell took a curtain call at the première, but when Diaghilev tried to persuade the composer to follow him Berners refused with the remark that his aunt had threatened to disinherit him if ever he went on the stage.)

Sacheverell and Georgia did their best to include Osbert in their lives, but this was clearly no solution to his problems now. Georgia was soon pregnant, Sacheverell had his own life and individual career to follow, and he had no desire at all to get involved in Osbert's latest battles. Most of that summer Edith was abroad, staying with Helen Rootham at her sister's flat in Paris, and Adrian Stokes was in America. Not that the handsome Adrian could have done much for Osbert's loneliness had he been in England. However 'romantically protective' Osbert may have felt towards him – and however much he played up to Osbert's interest – he could not fill the gap left by Sacheverell's departure from the house in Carlyle Square.

During the bleak autumn of 1926 Osbert had one considerable success to sustain him. The publication of his first full-length novel, *Before the Bombardment*, brought him such critical acclaim that at thirty-four a new life seemed to beckon – as a novelist. A new country beckoned him as well. The Doran Company was bringing out the novel in New York and was anxious that he should promote the book in person. Stokes was still in America, and Osbert suddenly decided that the time had come to make a break.

It was all done with the style his friends expected of him now, including a large farewell party which he threw at the Savoy. Wells, Gosse and Bennett were the guests of honour. The press was there in force, and to give them something they could write about the meal included a salad of chrysanthemums. Next morning, when he boarded the *Majestic* at Southampton, he was consciously concluding, he said later, an entire stage in his life. It was a rough, five-day voyage, and those troughs of despair, those terrible valleys of the Atlantic, seemed to reflect the desolation that he felt within him. His old life was over, the 'closed corporation' had been broken. What lay ahead now was America – and the lonely life of a solitary writer.

CHAPTER TWELVE

Osbert, Edith and the Chatterleys

1926–1927

WHATEVER OSBERT may have hoped for in New York, the S.S. *Majestic* had no sooner docked than he was discovering that the New World was no answer to the troubles that pursued him. It was a perfect day as the great liner sailed up past 'the formidable, star-spangled harridan that acclaims American liberty and guards the harbour of New York', and he was pleasantly reminded of San Gimignano by 'the slender towers that form the skyline of New York City, chanting hosannahs to a summer sky'.

Thanks to the publication of his book, and to the sheaf of high-powered introductions from (among others) Siegfried Sassoon and Somerset Maugham, he was soon treated as a visiting minor-league celebrity, and inevitably met a number of people he enjoyed. Marianne Moore was one. He found her 'charming and very unpretentious'. Alexander Woollcott was another. They met at a party given in Osbert's honour by the legendary theatrical hostess, Bessie Marbury, thanks to an obliging note from Maugham, and he found the great humorist 'so funny in conversation' that when they shared a cab even the driver was soon overcome with mirth, and had to stop the cab until he had recovered his equanimity.

He warmed to Harlem (which he saw with Carl Van Vechten), danced the Charleston, and was entranced by the stately treasures of the Frick Collection. The union of American plutocratic wealth and high European art that the Frick personified was one small segment of New York that he could unreservedly applaud. Not so the rest of it. The monstrous Vanderbilt 'château' on Fifth Avenue struck him as unforgivably vulgar: so did his hostess, Mrs Cornelius, 'with her invariable bandeau and her grey hair', although he was too polite, for once, to say so. Even her request that he convey her fondest love 'to the dear boys', the Prince of Wales and the remainder of the royal dukes, failed to impress him.

Osbert was seeing New York in the final throes of Boom and Prohibition. Neither was to his taste. Exuberant naïveté and trusting unsophistication never appealed to him and, as he wrote, he found New York now 'as innocent as Adam and Eve before the Fall'. At one party given in his honour he ended up beneath the supper table with Muriel Draper, simply to escape the din of the other guests and a troop of celebrating Cossacks. It seemed to be, he said, the one place in the room where one could hope to carry on a civilised conversation. It is a picture that somehow sums up his whole American experience. Osbert's New York stay might have been happier had Adrian Stokes been there to keep him company, but the young man left for London. World-weary, disillusioned, sick of London though he may have been, Osbert was not ready for America on his own.

Just before Christmas he was gratefully aboard an Italian liner bound, first class, for Naples. Soon he was safely at Amalfi. As he wrote later with undisguised relief, 'no one can know the sheer joy of being in Europe or being a European until he has visited America'.

When he returned to England, early in 1927, it was to find Sacheverell on the edge of fatherhood, and curiously delighted at the prospect. All fears and warnings to the contrary, he was devotedly in love with Georgia, and the birth of a son and heir in April now served to emphasise, if emphasis were needed, that the whole dedicated shared life of the trio was gone beyond recall. The child was christened Sacheverell Reresby – names which echoed the ancestral battle-honours of the Sitwells – just as his birth inevitably roused the genetic passions of Sir George. From Italy the baronet expressed his grandfatherly joy, and all seemed more or less forgiven.

As if to emphasise the way that he was now developing his independent life with Georgia and Reresby, Sacheverell had moved his family from 4 Carlyle Square to another, but more suitable, rented house in Tite Street, Chelsea. And as if to disprove the gloomy prophecies about the effect of marriage on his writing, he was also going from success to fresh success. He had a car, and – since he could not drive and had no aptitude for learning – a chauffeur. (The car was, in fact, a wedding present from the Dobles, and it was mainly thanks to the largesse of Sir George that they were able to afford the rent, and a nurserymaid and cook.) He had the *réclame* of the continuing success of the *Triumph of Neptune* with Diaghilev. And his book on *German Baroque Art* was published to appreciative reviews that November. As Osbert must have realised, his younger brother had now proved conclusively that he could do without him.

Edith's star was rising too. Sacheverell's marriage had not upset her life in the way it had Osbert's. She had her own experience firmly arranged around her life at Moscow Road with Helen Rootham, and 1927 proved a satisfactory year both for her work and for her private life. *Rustic Elegies*, her latest book of poetry, came out that spring to more appreciative reviews than any she had had before – tempting her to think that she was beginning to attract the audience she wanted for her work. At the same time she was starting to enjoy at last the conscious role of a celebrity, as 'Edith Sitwell the modernist poet' became increasingly accepted as the resident English high priestess of modern poetry.

She was undoubtedly encouraged in this hieratic role by close contact now with an even more determined and established priestess of the modern movement in the arts, Miss Gertrude Stein. Given Edith's habit of regarding other living female poets as a personal affront, it might have been predicted that the only contact possible between these two embattled ladies would have been a state of war – and in the end it was. But to begin with Edith was remarkably impressed by Gertrude Stein – and Gertrude Stein was equally impressed with Edith.

Contact had first been made as early as the summer of 1923 when Edith had reviewed Miss Stein's *Geography and Plays* in the *Nation*. Edith must have recognised a powerful potential rival from the start, and treated her with unusual circumspection. It had taken her, she said, several weeks before she could even clarify her feelings about what Miss Stein was up to. Finally she gave her the benefit of the doubt about being 'perfectly, relentlessly and bravely sincere' in her attempts to give 'fresh life and new significance to language', and even condescended far enough to praise her for her 'valuable pioneer work'. But her verdict on the book remained equivocal – 'I find in it an almost insuperable amount of silliness', but also 'great bravery, a certain real originality, and a few lines of exquisite beauty'.

A year later, writing in *Vogue* in October 1924, Edith advanced her estimation of Miss Stein, putting her above Katherine Mansfield and Dorothy Richardson as a literary 'pioneer', and soon afterwards, during a stay in Paris, despite that painful trouble with her back, Edith was able to call upon Miss Stein and Miss Toklas at their house in the Rue Fleurus. The meeting was a great success, despite Miss Toklas's first impression that Edith looked exactly like a *gendarme* in her large, grey, double-breasted coat with its big brass buttons. More flatteringly, Gertrude Stein wrote afterwards that she thought Edith's nose 'one of the wonders of creation', and praised the 'delicacy and completeness' of her grasp of modern poetry. Edith, not to be outdone, compared Miss Stein with 'an Easter Island idol', found her 'immensely good humoured' and said, highest praise of all, that she was 'verbally very interesting'. From that moment Edith had found what she felt to be a new and influential ally in her battle to resuscitate the corpse of English poetry. In October 1925, writing again in *Vogue*, she fulsomely proclaimed that 'in the future it is evident that no history of the English literature of our time could be of any worth without a complete survey of the work Gertrude Stein is doing for our language'. It was now, too, that she assured Virginia Woolf that Gertrude Stein was 'gigantic'.

The following year, true to her convictions, Edith continued to campaign on Miss Stein's behalf, and became adamant that she must come to lecture at an English university as a way of furthering the great 'cause' of modern poetry. 'I do feel your actual presence in England would help the cause,' she wrote. 'It is quite undoubted that a personality does help to convince half-intelligent people.' Miss Stein was not convinced, and peremptorily turned down an invitation from the Cambridge University Literary Society. Edith was 'bitterly disappointed' as she told her, and used all her power of flattery and exhortation to make her change her mind: 'a great writer like yourself is absolutely bound to win through. . . . Do if you can reconsider this question of coming over here. . . . There are *many* fresh admirers.'

Harold Acton, now in his last year at Oxford, and a friend of Miss Stein's on his own account, was one of them and put in his own plea that she speak at Oxford as well as at 'the rival university' – and finally the great lady gracefully agreed. When she arrived in London at the end of May all three Sitwells came to welcome her, to calm her stage-fright at her lectures, and to make her stay a happy and successful one. They virtually took charge of her. She found Osbert particularly helpful and agreeable, and this was when she wrote: 'He had that pleasant kindly irresponsible agitated calm that an uncle of an english king always must have' – and Miss Stein was properly grateful for the way the trio showed their solidarity by taking her to Cambridge and to Oxford and then sitting on the platform through the lectures.

The Oxford reading was a particular success, with a large, enthusiastic audience of those 'half-intelligent' young creatures Edith was now so anxious to attract towards the 'cause', and, according to Acton's own account of the affair, Edith did her best to appear unembarrassed as Gertrude Stein recited a 'word portrait' of her in her honour; while Sacheverell 'looked as if he were swallowing a plum' and Osbert 'shifted on his insufficient chair with a vague nervousness in the eyes'.

After what Gertrude Stein herself termed 'the glory and excitement' of her trip to England, Edith inevitably became a still more welcome guest at the Rue Fleurus. Whenever she found herself in Paris now with Helen

Rootham, she would make a respectful visit to Miss Stein, and the influence of the Caesar of the Rue Fleurus began to show in Edith's manner and ambitions. There was a most decided ring of the Steinian presence and authority to the way that Edith was beginning to conduct herself.

In Moscow Road the Thursday and Saturday gatherings continued over the strong brown tea – but with a difference. Edith at forty was beginning to project the personality with which she had always faced the world into something more and more formidable. Her public presence had become a role she played with a total, all-absorbing dedication. Poetry was her religion, being a poet was her life, and she performed her role with the high seriousness of a great actress mastering her craft.

She had an actress's awareness of her appearance, and had created her inimitable style much as Miss Stein had done. 'I am', she was to write, 'as highly stylised as it is possible to be – as stylised as the music of Debussy or Ravel. . . . I have, if I may say so, my own particular elegance.' She had indeed, although she would not have been Edith Sitwell if she had not been simultaneously tormented by the thought that people were still managing to laugh behind her back. She even wrote about this in an attempt to exorcise her fears by a sort of brazen jollity at her own appearance.

> From time to time, a loud cry arises from certain portions of the Press: 'There she goes! *She's* sighted! Heave something at her!'
>
> Then will follow, either a sympathetic description (under the heading: 'Plain Jane makes good') of my sufferings as one not gifted by Providence with anything causing the eye to dwell willingly upon me.
>
> When I am not being called 'Plain Jane', a description is given of me that seems to have been culled from a medieval description of sea-monsters.

Behind the extraordinary public self that she projected now lay an attempt to outface her 'enemies' as she had once outfaced Sir George, to dare them to be impertinent so that she could slap them down.

Even the clothes that she concocted, the long, Renaissance dresses, the enormous rings and outlandish hats, the great jewelled cross around her neck, had become the standard costume for the self-assertive role she felt she had to play. It was a role with which she took unceasing trouble, for she was hardly ever off the stage. During her trips to Paris she had already started taking lessons in speech and declamation from a retired actress of the Comédie Française, and just as she practised her 'technical exercises' to get in training to compose her poetry, so she was continually practising her appearance, manner, voice and the whole apparatus of her public presence.

One sees how effective this could be in the description of her by Allanah Harper, society beauty and editor of the influential poetry magazine

Échanges, describing the first time she heard Edith read her poetry in public.

She was already a legendary figure who seemed to have moved out of the tapestry of *La Dame et la Licorne* – her flat, fair hair like that of a naiad, her hands as white as alabaster, huge rings of topaz and turquoise on her long gothic fingers, on her wrists, coral and jet bracelets. She began to recite and a window opened on to an enchanted world. Each vowel and consonant flowed and she seemed to weave her poetry in the air. The world became heightened and transformed until I could see a whole landscape there behind her eyes.

There speaks, of course, a true disciple, totally won over by the Sitwell magic. Another was Cecil Beaton; and in his case one can still see the immediate effect in the photographs he took of her.

Edith first visited him for lunch in December 1926. Although he already knew her brothers, and had worked on a Sitwell film, this was the first time they had been alone together. He had been hoping to photograph her there and then and his luck was in. His diffidence and obvious enthusiasm must have been reassuring. Edith allowed her guard to drop, and Beaton's account of what occurred, as recorded in his youthful diary, is interesting, showing as it does the two quite separate Ediths. The first was the public one as she came in through the door, 'a tall, graceful scarecrow with the white hands of a mediaeval saint'. Then slowly the private Edith was permitted to appear. 'Gradually, I found her formidable aspect less striking than her sympathetic girlishness. In spite of her cadaverous appearance, her complexion is as fresh as a convolvulus.'

Edith began to make him laugh, and as she 'fired comic broadsides at Drinkwater and Squire' and shook 'a little pepper on Ethel M. Dell, Hannen Swaffer and Tallulah Bankhead', Edith began to laugh herself. Edith was hungry.

Lunch was a success. Edith ate heartily of a piping hot fish *soufflé*, which was a triumph. . . . During dessert she recited a bit of Gertrude Stein's 'Portrait of Tom Eliot': 'Silk and wool, silken wool, woollen silk'. How precisely and richly she spoke. She could make any rubbish sound like poetry.

But there was work to do, the serious work of posing for her photographs, and here again the true professional in Edith took over.

She posed instinctively. No matter how many positions I had already taken, I felt loth to call a halt. Surely this was an unique opportunity. I must perpetuate the image in front of me, of a young faun-like creature sitting against my leaping-fawn design, looking surprisingly Victorian in her crudely-cut Pre-Raphaelite dress, with her matador's jet hat, and necklace, her long mediaeval fingers covered with enormous rings. When the hat was discarded, she became a Brontë heroine, and her pale silken hair fell in rats' tail wisps about her face, while the big teapot handle bun made the nape of her neck appear even more impossibly slender.

As the afternoon wore on, I suggested more exotic poses. I even persuaded her to asphyxiate under the glass dome. She became quite hysterical kneeling on the floor, her knees and joints popping and cracking. A Chinese torture she called it, but loved it all the same.

These were the first of countless pictures Beaton took of Edith – and her brothers – and they were to play their part, just as the portraits by Guevara, Dobson and Wyndham Lewis had already done, in fleshing out the public Sitwell myth. But the Beaton photographs did something more than that. Here was a young photographer in love with the Sitwell legend, and using his camera to depict them in a world of fantasy which somehow seemed to complement their own. He had all three of them dancing ring-a-roses in the middle of Sussex Gardens, and the pictures that he took of them appear like the sequence of a dream and not at all ridiculous. His pictures of Sacheverell lying like some early saint upon his tomb could illustrate one of his own verses. But it was Edith who remained the favourite and irresistible objective for his camera, Edith whom he thought entirely beautiful and a most wonderful aesthetic object, and Edith who could create a vision of herself to match the vision of her poetry. 'With her eyebrows like tapering mouse-tails,' he wrote, 'with the little pale blue veins at the temples of her noble tissue-paper forehead, with her wrists the most delicate stems, she possesses the mad moon-struck ethereality of a ghost.'

Only a very few admirers were actually permitted to enjoy the private, unpretentious side of Edith as Beaton did. Someone else who did so was Geoffrey Gorer, later to become a distinguished anthropologist, but in those days just down from Cambridge where he had run the Poetry Society and declared himself a fervent and uncritical Sitwellian. He saw Edith frequently at Moscow Road and remembers her 'for the way she giggled when she got going'. He also tells of how much she enjoyed steak and onions and a glass of beer during an off-duty evening with him at the music hall, watching her favourite, Nellie Wallace.

But the less favoured visitors to Moscow Road felt, as they did with Osbert, that in Edith's presence laughter, although encouraged, had to be carefully directed and confined. Ian Laing, for instance (later in life the Foreign Editor of the *Sunday Times*, but in those days a would-be poet), was at first flattered and amused to be told about some of Sir George's latest follies, which seemed to form the staple background of amusing small-talk at the Thursday gatherings.

No aberration of Ginger's was too trivial to be worked up into an elaborately grotesque anecdote – how he had shamed Sacheverell by taking him to lunch at a fashionable hotel and then refusing to pay in full the *prix fixe* on the menu:

or how, when persuaded to risk a journey on a London bus, he had then started haggling with the conductor over the fare.

But funny as these stories were, before long, says Laing, he started finding them a shade embarrassing. 'I could not escape the feeling that one's laughter had to be discreetly modulated, that any comment on Sir George's peculiarities would have been resented, and that it would have been unforgivable for anyone but a Sitwell to refer to him as Ginger.'

Edith, the *grande dame*, was strongly present at these weekly gatherings of hers. Consummate actress that she was, she was engaged in acting out her most important part – that of her own authoritative self. The visitors at Moscow Road, the listeners at her readings, all had a role to play on which her own depended – that of the assenting and applauding audience to the extraordinary performance which was now her life.

Another of her keen admirers now was Virginia Woolf, who seems to have felt simultaneously sorry for her, intrigued by her, and attracted by her unexpected beauty. After she saw her at one Bloomsbury party she wrote of 'poor Edith Sitwell in her brocade dress', looking 'so remote and washed up on a rock', and she repeated the same wave-swept metaphor when she described her at a party Edith gave at Pembridge Mansions in honour of Gertrude Stein: 'all Edith's furniture is derelict, to make up for which she is stuck about with jewels like a drowned mermaiden'.

According to Edith, Virginia Woolf hinted that the Hogarth Press might be interested in publishing her books and would increase her literary earnings (they had published a pamphlet of hers in 1925), but Edith was evidently wary of getting too involved with Bloomsbury. 'I was not an unfriendly young woman, but I was shy,' and 'I do not think I should have "fitted into" the closely serried company of Bloomsbury,' Edith wrote later.

But this did not alter their genuine affection for each other. (It was much later that Edith tartly dismissed Virginia Woolf the novelist as 'a beautiful little knitter'.) Virginia thought she was a 'character', that her hands were the loveliest in London, and she genuinely enjoyed her poetry. 'Of course you are a good poet: but I can't think why. The reason may strike me in Sicily,' she wrote. Above all, it was Edith's ethereal appearance that attracted her. 'She is', she told Vita Sackville-West, 'like a clean hare's bone that one finds on a moor with emeralds stuck about it.'

Allanah Harper describes one of the occasions when she arrived at Moscow Road to discover that Virginia Woolf had come for tea as well. She found the two impressive ladies sitting together on the sofa, 'like two praying mantises putting out delicate antennae towards each other', as they began discussing Miss Sackville-West's long, long poem, *The Land*, which had just been awarded the Sitwells' favourite booby-prize, the

Hawthornden. Edith, who, Gertrude Stein apart, disapproved on principle of any female poet after Sappho, was considerably put out. 'It is not poetry,' she insisted, 'although it might conceivably be of use to farmers as a help to counting ticks on sheep.'

Perilous words, considering Virginia Woolf's devotion to Miss Sackville-West – but Edith appears to have got away with it. According to Allanah Harper, all that Virginia Woolf could manage was the meekest of rebukes. 'Edith,' she said archly, 'must one *always* tell the truth?'

If Edith could have this effect on the reigning queen of Bloomsbury, one can imagine how she could dominate the lesser members of her own small court. Some refused such fealty, and left. Others managed to evolve their own way of preserving a degree of independence in her regal presence. One of the most interesting of these was the mysterious young man whom Iain Laing met one Thursday sitting alone on the periphery of the tea-time circle.

His good looks alone would have marked him out, but what was equally unusual was that he did not react to anything that Edith said, nor for that matter did he speak a word until the time came when he bade her a brief good-bye.

'Who', I asked, 'is that extraordinarily beautiful youth?'

'Oh, didn't I introduce you?' said Miss Sitwell. 'His name is Constant Lambert – only eighteen years old but already a very gifted musician. Willie Walton thinks he has a brilliant future.'

'He's not very talkative, is he?'

'Well, no, it's sad,' replied Miss Sitwell gravely, 'but the poor boy's almost stone deaf. . . . Like Beethoven, you know.'

Six or seven years passed before I saw Lambert again. I'd started going to the Fitzroy Tavern, and it was there one night that Lambert's first wife, Florence, said to me, 'Why do you shout at Connie so?'

'Well,' I replied, 'he's deaf, isn't he?'

'Deaf? He's no more deaf than you are, Iain.'

'And it was then,' says Laing, 'that I realised that Lambert's pretended deafness those years before at Pembridge Mansions had been *his* way of avoiding getting too involved in Sitwellism.'

It seems more likely that the whole idea of Constant Lambert's deafness was just Edith's little joke with Iain Laing, who was the sort of somewhat serious young Scot she would have enjoyed taunting.

When in extreme old age Edith came to distil the remembered bitternesses of her life into her memoirs, which she entitled *Taken Care Of*, she made it clear that there were still two people that she really hated, and singled them out for a double dose of the old Sitwellian venom. 'Provoked beyond endurance by their insults,' she would write, 'I have given Mr Percy Wyndham Lewis and Mr D. H. Lawrence some sharp slaps.' As indeed she had. It is not hard to see why Wyndham Lewis qualified for such rough treatment. He after all had turned from being a close companion of the Sitwells into their strongest, most vociferous critic. But why D. H. Lawrence? How had he ever come to provoke poor Edith 'beyond endurance', and what were the 'insults' he had hurled? One looks for them in vain in Lawrence's letters and his published works, while Edith, sibylline as ever in old age, gave no hint of what his great offence had been. In *Taken Care Of* there is an uncomplimentary description of the novelist – 'Mr Lawrence looked like a plaster gnome on a stone toadstool in some suburban garden . . . he bore some resemblance to a bad self-portrait by Van Gogh. He had a rather matted, dank appearance'; and an unexplained and furious attack on *Lady Chatterley's Lover* – 'a very dirty and completely worthless book, of no literary importance'. Mellors the gamekeeper should have been soundly 'thrashed' for the 'unutterably filthy, cruel and smelly speech' to the unfortunate, castrated Sir Clifford Chatterley on the general subject of his sexual inadequacies. Edith dismisses Constance, Lady Chatterley, as 'that nasty little nymphomaniac'.

Again one wonders why this one book among so many other scandalous productions should have so incensed her. Clearly the 'insults' that provoked Edith so sorely somehow involved Lady Chatterley and centred in particular on the two loathsome characters of its hero and heroine. But why should they have given rise to this hysterical attack?

Reading between the lines of *Taken Care Of* there is a single clue to what it was that so infuriated Edith. This is her powerful, indeed, her passionate, defence of the character of Sir Clifford Chatterley, whom she describes as rousing Mellors's jealousy because he was a baronet, and 'a famous writer', who had 'fought like a tiger in the First World War, instead of remaining safely at home, fornicating and squealing, shrilly, about the oppressions from which he had suffered'.

Interestingly, there is no reference by Lawrence in the book to Sir Clifford's fighting 'like a tiger'. That is an Edithian gloss, but a revealing one. Tigers and lions were the approving metaphors she often used for members of her family, and the simple and extraordinary fact of the whole matter is that Edith believed that D. H. Lawrence had based the character of Chatterley on that of her own beloved Osbert – who, like Sir Clifford, was a baronet, a famous writer and a soldier from the war. Worse still, she was convinced that Lawrence had intentionally written to attack Osbert on two forbidden grounds – of class and, more distressing still, of sex. This was the monstrous yet unspeakable 'insult' which haunted Edith to the day she died. This was what had provoked her 'beyond endurance' and finally accounted for that bitter cry of hatred from her deathbed. But was it true? And was she being fair to Lawrence? The answer is more involved than it might seem and lies in the somewhat complicated movements of the Lawrences and of Osbert and Edith around that spring of 1927.

That May Osbert and Edith were in Italy together. He had been rushing wildly around Sicily in March and April on his own (at Syracuse he had bumped into Leonard and Virginia Woolf driving from the station in a cab), and had persuaded Edith to join him in Florence. Now that Sacheverell was married and a father, Edith to some extent had stepped into his place as travelling companion. She was devoted to her 'dear old boy', as she called Osbert, and since their brother's marriage they were closer than they had been since childhood. She was concerned about him now, and they could back each other up during the days they had to spend visiting Sir George and Lady Ida at Montegufoni. Osbert also had his usual round of Florence friends to call on – that 'old scamp', Norman Douglas, Orioli the bookseller from whom he was now collecting big Venetian books, and scatterbrained old Ada Leverson, who was still as deaf and still as hopelessly in love with him as ever. This time, however, he was also hoping for a chance to call on the D. H. Lawrences who had rented the Villa Mirenda at Scandicci five miles south of Florence.

They had been trying to effect a meeting for quite some time but without success. Edith knew Lawrence, having met him, several years before his travels to Australia and Mexico, at London parties in the early twenties with their mutual friend 'Chile' Guevara, but Osbert had never met him – apart from that brief, unlikely glimpse he had of both the Lawrences at Monty Shearman's Bloomsbury-encrusted Armistice night party in the Adelphi.

Edith admired Lawrence's poetry and earlier novels, whilst *Aaron's Rod*, with its unsympathetic portraits of a host of Sitwell enemies – or non-friends – like Ottoline Morrell, Bertrand Russell and Michael Arlen, had raised him to the status of an instant ally. When Lawrence returned from New Mexico at the end of 1925 and published *The Plumed Serpent*, Edith expressed continued admiration, and in the spring of 1926 there were tentative arrangements for a meeting at Montegufoni. However, there was now a muddle over dates, and at the end of June Osbert was still in England when he received a letter from his mother.

A Mr D. H. Lawrence came over the other day, a funny little petit-maître of a man with flat features and a beard. He says he is a writer, and seems to know all of you. His wife is a large German. She went round the house with your father, and when he showed her anything, would look at him, lean against one of the gilded beds, and breathe heavily.

Osbert has described the change which life in the *castello* (and possibly, although Osbert does not say so, freedom from the disturbing presence of his children) had wrought on Sir George's personality. Here he was in his element at last. He loved the enormous house and 'He began, too, almost to *like* people in the excitement he derived from showing them the Castello, and what he had done to it.' The Lawrences were asked to stay to lunch, and then, according to Sir George's version of what happened, Frieda Lawrence 'jumped on all the beds after luncheon – to see if the mattresses were soft'. Before leaving, she signed the visitors' book 'Frieda Lawrence, *geborene* Richthofen' (the baronial splendours of Montegufoni evidently bringing out the baronial splendours of Mrs D. H. Lawrence). Lawrence may not have quite matched up to the eminence of that great writer 'Mitchell' – the mythical novelist Sir George pretended to admire as a way of teasing any of his children who mentioned some new literary discovery of theirs – but the visit was clearly far from the fiasco that it might have been. Lawrence seems to have liked the Sitwell parents, and he was now more keen than ever to meet Osbert and Edith.

At the end of that July he was at Baden-Baden, *en route* to England, where he would make a brief, depressing visit to his native Midlands, and we find him writing sadly to his friend, Dorothy Brett. 'And what will England be like? I shan't see Murray, he is too much, or too little, for me. Think I shall see the Sitwells and Rolf Gardiner.'

This in itself is interesting. J. Middleton Murry, now editing the *Adelphi* magazine, was a long-standing friend and correspondent of Lawrence, and Gardiner ranks as perhaps his most dedicated disciple. For Lawrence to be preferring the company of the Sitwells to that of Murry, and to be placing them on a par with the devoted Gardiner, shows the extent to which they had aroused his interest.

But once again the visit failed. Lawrence came to Renishaw, but missed them. The house was closed, and Lawrence, on what proved his final visit to his native Midlands, was left with an impression of Renishaw at its most unwelcoming. He would not forget it.

But he had other things now to be concerned with. Early that October he was back with Frieda at the Villa Mirenda starting a new novel which he would finally, and *faute de mieux*, entitle *Lady Chatterley's Lover*. It was a difficult, unsatisfactory book to write. His health, which had been so much better all that summer, started to decline. All thoughts of any meeting with the Sitwells had to be shelved until that spring of 1927 when Osbert and Edith arrived in Tuscany.

Some five years later Osbert would describe how he and Edith drove out one afternoon at the end of May to visit the 'square, blue-painted farmhouse' among the 'gentle hills' of Scandicci, how, when they knocked,

Lawrence himself opened the front door to them, and how he realised for the first time 'what a fragile and goatish little saint he was: a Pan and a Messiah . . . a curious but happy mingling of satyr and ascetic. . . . Unlike the faces of most genuises,' he added, 'it was the face of genius.'

He says that he and Edith spent two hours with Lawrence. There is no mention of what they talked about – nor, for that matter, is there any mention of the abundant Frieda. 'He was extremely courteous, I remember, and prepared the tea himself, doing all the work: which grieved one for he looked so ill.' Osbert's only criticism was of the Lawrence paintings. 'I thought and still think [them] crudely hideous and without any merit save that he painted them and in so doing may have rid himself of various complexes, which might otherwise have become yet more firmly rooted in his books.'

Edith's version of the visit is, as one expects, more scathing, for by the time she wrote it she had become convinced of Lawrence's unmentionable offence. After the lapse of nearly forty years, she remembered the Lawrence villa as 'their tall, pink house, that looked as if it were perched upon a hen's legs', and Lawrence was soon 'trying to make us uncomfortable by references to the contrast between his childhood and ours'. 'Poor Mr Lawrence', she added, 'had a very bad chip on his shoulder. He hated men who were magnificent to look at. He hated men who were "gentlemen".'

There is certainly no sign of such an attitude to Osbert at the time in anything that Lawrence, or, for that matter, Frieda, wrote, and if Lawrence really had behaved so rudely Osbert would hardly have referred to him as having been 'extremely courteous'. In point of fact Lawrence, in the few references that he made about the visit, expresses nothing but quite genuine affection, and a sort of puzzled concern for Osbert which, coming from a man in Lawrence's then state of health, is both moving and intriguing. This is how he describes the visit in a letter to his friend, the Ukrainian translator, S. S. Koteliansky:

> Osbert and Edith Sitwell came to tea the other day. They were really very nice, not a bit affected or bouncing: only absorbed in themselves and their parents. I never in my life saw such a strong, strange family complex: as if they were marooned on a desert island, and nobody in the world but their lost selves. Queer!

Meanwhile, to Richard Aldington Lawrence wrote that, while he had liked Osbert so much better than he had thought he would, Osbert all the same 'makes me feel sort of upset and worried. Of him the same I want to ask: But what ails thee, then? Tha's got nowt amiss as much as a' that!'

One can only wonder at the reply that Osbert might have given if he had. As it was, it is clear that Edith and Osbert – especially Osbert – were in such a nervous state that May that they actually managed to upset the Lawrences. As Frieda wrote,

They moved us strangely. They seemed so oversensitive, as if something had hurt them too much, as if they had to keep up a brave front to the world, to pretend they didn't care and yet they only cared too much. When they left we went for a long walk, disturbed by them.

During those two hours of the visit Osbert gave D. H. Lawrence much to ponder on.

So what of Edith's angry certainty that Lawrence distilled the bitterness that Osbert supposedly had caused him by his visit and used it as the basis for the character of cuckolded, castrated, poor Sir Clifford Chatterley? In the first place, all the evidence about that meeting – Edith's own angry, aged testimony apart – is entirely against any such reaction by Lawrence towards Osbert on this one occasion when they met. And in the second place – and far more conclusively, it seems – Lawrence had definitely fixed the outline of his plot and the character of Clifford Chatterley at least eight months before. According to Richard Aldington, who was seeing Lawrence off and on throughout this period, the first version of *Lady Chatterley's Lover* was begun at the Villa Mirenda in October 1926, immediately after Lawrence's unsatisfactory trip to England, and according to Aldington 'the starting point of *Lady Chatterley's Lover* is not from the characters of "Constance", "Clifford" and "Mellors", but from Lawrence's own recoil from the ugliness of "his" Midlands', which he had just visited again. Aldington says the book was finished in February 1927, but it was not published by Frieda Lawrence until 1944 under the title *The First Lady Chatterley*.

This on the face of it disposes of the whole idea of Clifford Chatterley originally having had the faintest possible connection with Osbert. Certainly it frees Lawrence from all accusation that he might have been trying to use the novel to attack him on the grounds of class or childhood privilege, sex or lack of it. On all these grounds at least Edith was quite wrong, her hatred of Lawrence totally misplaced.

Osbert's nephew, Reresby Sitwell, has a theory which almost certainly explains the true origins of the character of Clifford Chatterley. Half way between Renishaw and Eastwood, where Lawrence was born, stands the house of Sutton Scarsdale, seat of the Arkwright family, whose fortunes were founded on the eighteenth-century Sir Richard Arkwright's invention of the famous 'spinning jenny'. Some time before the First World War the heir to the family, young Joseph Arkwright, fell from his horse and, breaking his spine, was paralysed. Lawrence would certainly have known about this local tragedy – and also would have heard the local gossip when, in the early twenties, the heir of Sutton Scarsdale, who, like Clifford Chatterley, was 'not more than half a man', married and brought his young wife back to the house of his ancestors.

One will never know for certain whether this was the origin of Clifford Chatterley, but one does know for certain that it was not Osbert. There were, however, in *The First Lady Chatterley* two important references which concerned the Sitwells as a whole. The first was to Renishaw. That day when Lawrence called and found the great house closed had left its mark, and Wragby Hall, home of the Chatterleys, bears an obvious resemblance to Renishaw:

> a low, long old house, rather dismal, in a very fine park, in the midst of newly developed colliery districts. You could hear the chuff of winding engines, and the rabble of the sifting screens, and you could smell the sulphur of burning pit-hills when the wind blew in a certain direction over the park.

Lawrence, like his friend Aldous Huxley, clearly was a novelist who wasted nothing, and his visit to Montegufoni also appears to have yielded material for *The First Lady Chatterley*. In this version Clifford has an aunt called Lady Eva, who is unmistakably based on Lady Ida.

> She had the remains of the *grande dame* about her, being daughter of one of the really big families. But she had rather got into disrepute, what with gambling and brandy. . . . Lady Eva was somehow like a ghost, in her black clothes and her curious naïveté, which was almost girlish and winning. Yet the odd, straightforward girlishness rested upon a very hard determination never to yield an inch, essentially, to anyone on earth. A hard imperviousness and isolation, like the Matterhorn, was at the bleak centre of Lady Eva.

It is a memorable portrait from a single sitting, but for some reason Lawrence deleted it from the second version of his book which he began almost immediately the first was finished, and which was published, this time in 1972, under the lively title *John Thomas and Lady Jane*. It is just possible that Aldous Huxley, whom Lawrence had been visiting at Viareggio about this time, had mentioned something of the offence that he had caused the Sitwells when he incorporated his observations upon Lady Ida into the rakish character of Priscilla Wimbush in *Crome Yellow*. Certainly this second version of the novel which Lawrence finished some time that spring before Osbert and Edith's visit, has even less connection with the Sitwells than the first.

In it, however, Lawrence makes it clear that he is particularly obsessed now by the character of Clifford Chatterley, which has developed quite considerably since *The First Lady Chatterley*. This is not so surprising when one remembers that Lawrence himself had the unenviable role of cuckold at the time – Frieda had recently begun her passionate affair with the dashing Capitano Angelo Ravagli, who was to leave his family and marry her after Lawrence's death. In such a situation he must have known the fears of impotence as well. Certainly this second version of the baronet is far less sympathetic than the first – self-pitying, unpopular, irritable, and giving vent to his frustration by game-shooting in the woods from his wheelchair. Sick, irritable D. H. Lawrence must have put a great deal of himself, and of his relations with the full-blooded Frieda, into the second Clifford Chatterley. He gives only one small hint of the way his mind was working – and of the extraordinary changes he would make to Clifford's character when he came to compose the third and final version of the book.

He [he writes] had always been too high-brow, too flippant, thought himself too clever and behaved in too democratic a fashion, to please the county. They knew he made a mock of them all. He had even pretended to be a sort of socialist, besides having written those pacifist poems and newspaper articles.

John Thomas and Lady Jane had been finished – and discarded by the still unsatisfied Lawrence – by the time of the Sitwell visit to Scandicci, and, by a strange coincidence, Lawrence was just embarking on his third and final version of the book around the time that they arrived. If one wants proof of the impression Osbert and Edith could produce during a single tea-time, one has only to compare the final and completed version of *Lady Chatterley's Lover* with its two predecessors. For on to the two existing layers of the character of Clifford Chatterley Lawrence has now spread a third. The main outline still persists – Chatterley's sexual impotence and social status, his role as the lord of Wragby Hall, and his relationship with his unhappy wife and the priapic keeper. But the additional details, almost without exception, have obviously been drawn from Osbert. When Osbert and Edith drove off back to Montegufoni, Lawrence quite clearly went on asking that insistent question he would have liked so much to put to Osbert – 'But what ails thee, then?' The answers Lawrence gives are there in his final portrait of Sir Clifford, and they provide the clearest, most perceptive summary of Osbert Sitwell during the crisis of his middle thirties that we possess.

In this third version Lawrence makes Clifford Chatterley, for the first time, one of three children: 'he was aristocracy. Not the big sort, but still *it*. His father was a baronet, and his mother had been a viscount's daughter.' Then, in describing this aristocratic trio of Chatterley children, Lawrence expands on that first impression he expressed to Koteliansky, of Osbert and Edith being so alone, 'and nobody in the world but their lost selves'. The three young Chatterleys:

had lived curiously isolated, shut in with one another at Wragby, in spite of all their connections. A sense of isolation intensified the family tie, a sense of the weakness of their position, a sense of defencelessness, in spite of, or because of the title and the land. They were cut off from those industrial Midlands in which they passed their lives. And they were cut off from their own class by the brooding, obstinate, shut-up nature of Sir Geoffrey, their father, whom they ridiculed, but whom they were so sensitive about.

For Lawrence, this cruel isolation of Osbert and of Edith epitomised the alienation which he saw between the Midlands aristocracy and the land and the sources of their wealth. Since this was one of his obsessions – and one of the main themes of his book – it is not surprising that he should have seized on it. What is more surprising is the way he then goes on to put his finger on the causes of Osbert's own profound unease – both with himself and with the world around him. It is an extraordinary example of Lawrence's own uncanny sensitivity, and an acute diagnosis of Osbert's troubles, that Lawrence should have picked on fear as the emotion lying behind his rage and vulnerability.

He was at his ease in the narrow 'great world', that is, landed aristocracy society, but he was shy and nervous of all that other big world which consists of the vast hordes of the middle and lower classes, and foreigners. If the truth must be told, he was just a little bit frightened of middle and lower class humanity, and of foreigners not of his own class. He was, in some paralysing way, conscious of his own defencelessness, though he had all the defence of privilege. Which is curious, but a phenomenon of our day.

Lawrence goes on to describe how fear had led this Osbert/Clifford to assert himself against 'that outer world of chaos' that he could never really master. This led him to become a rebel:

rebelling even against his class. Or perhaps rebel is too strong a word; far too strong. He was only caught in the general, popular recoil of the young against convention and against any sort of real authority. Fathers were ridiculous: his own obstinate one supremely so. And governments were ridiculous: our own wait-and-see sort especially so. And armies were ridiculous, and old buffers of generals altogether, the red-faced Kitchener supremely. Even the war was ridiculous, though it did kill rather a lot of people.

Touching briefly on the subject of the way the two other Chatterley children had reacted to Clifford's marriage, Lawrence gives a hint of something that Edith or Osbert must have said that afternoon about Sacheverell's marriage.

'The three had said they would all live together always,' but Emma, the sister, who like Edith is exactly ten years older than her brother, 'felt his marrying would be a desertion and a betrayal of what the young ones of the family had stood for'.

There are countless similar small points throughout the book which Lawrence clearly picked up from the Sitwells. He even has Sir Clifford's father, the old baronet, talking about his early attempts at writing exactly as he must have heard Sir George do about Osbert's – 'As for Clifford's writing, it's smart, but there's nothing in it. It won't last!'

But it was Osbert who really interested him, and he achieved a memorable portrait after that one brief session over the teapot at Scandicci. Lawrence even gets Osbert's manner to perfection: 'His very quiet, hesitating voice, and his eyes, at the same time bold and frightened, assured and uncertain, revealed his nature. His manner was often offensively supercilious, and then again modest and self-effacing, almost tremulous.' Lawrence also hit exactly on his sense of isolation.

He was not in touch. He was not in actual touch with anybody, save, traditionally, with Wragby, and, through the close bond of family defence, with Emma. Connie felt that she herself didn't really, not really touch him; perhaps there was nothing to get at ultimately; just a negation of human contact.

Lawrence had evidently read Osbert's stories, and must have discussed them with him that afternoon, for the paragraph that he devotes to Clifford Chatterley's attempts at writing is in effect Lawrence's own critique of *Triple Fugue* and *Before the Bombardment*, and an account of Osbert's own reaction to his work.

Still he was ambitious. He had taken to writing stories; curious, very personal, stories about people he had known. Clever, rather spiteful, and yet, in some mysterious way, meaningless. The observation was extraordinary and peculiar. But there was no touch, no actual contact. It was as if the whole thing took place in a vacuum. And since the field of life is largely an artificially-lighted stage to-day, the stories were curiously true to modern life, to the modern psychology, that is.

Clifford was almost morbidly sensitive about these stories. He wanted everyone to think them good, of the best, *ne plus ultra.* They appeared in the most modern magazines, and were praised and blamed as usual. But to Clifford the blame was torture, like knives goading him. It was as if the whole of his being were in his stories.

One of the strangest features of this whole strange incident is Osbert's reaction to it all. The third and final version of *Lady Chatterley's Lover* was printed and published in Florence at the end of 1928 by Osbert's friend and favourite bookseller, Orioli, so that he must, without a shadow of a doubt, have known of the book and read it. And yet this most vulnerable of men who could erupt so violently at Arlen's imaginary plagiarism, or Turner's play or Scott Moncrieff's impertinence, gave not the slightest sign of wrath. It was not until some three years later – in an article he wrote for the *Week-End Review* – that he pronounced on Lawrence and then, unlike Edith, it was to hail him as a genius.

There would seem to be two possible explanations for this odd and apparently untypical reaction. The first is that Osbert considered it all a private matter, which would only become public if he drew attention to it. The second is that he, unlike Edith, must have known the truth about the composition of the book so that he could appreciate the parts of Clifford Chatterley's character which had been drawn from him, and those with which he had no connection. Given the circumstances, it is not entirely unflattering to read a portrait of oneself, however candid, from the hand of a novelist you honestly believe to be a genius, and Osbert's silence on the subject of the book can perhaps be taken as an acknowledgement, however tacit, that Lawrence had not been all that wide of the mark in what he wrote of him.

CHAPTER THIRTEEN

Pastures New

1927

WHEN THE Russian Ballet had given the French première of Sacheverell's ballet *The Triumphs of Neptune* in Paris at the beginning of January 1927, Edith had been there, for the performance had happily coincided with her New Year visit with Helen Rootham to Madame Wiel's flat in the Rue Saint-Dominique. It was a proud occasion to be witnessing this latest triumph of Sacheverell, proof of that genius which she had been the first to divine, the most devoted in believing in. It was not to be Sacheverell's success alone that made this an evening she remembered all her life, but the presence of a 'tall, desperately thin, desperately anxious-looking young man', who for some reason circled around her without speaking throughout the interval, 'staring at me as if he had seen a ghost'.

Three days later, at the Rue Fleurus, she met him once again. She had been sitting with Miss Toklas and Miss Stein when the young man was announced. 'If I present Pavlik to you,' said Miss Stein, 'it's your responsibility because his character is not my affair.' Edith presumably absolved Miss Stein from all the consequences, and it was with these irresistible words that the great love affair of Edith's life began.

As Gertrude Stein presumably foresaw, it was to bring her untold pain, frustration and despair. Whether it also brought her compensating happiness is more difficult to tell. It certainly would change the whole course of her life and work, and helped to complete the break-up of the Sitwell trio which had begun when Sacheverell married. Many years later Edith would describe it as 'exactly' like the relationship 'between Vittoria Colonna and Michelangelo', a typical remark that both overstates and oversimplifies the roles of the two protagonists in this curious affair.

The 'Michelangelo' of Edith's life was Pavel ('Pavlik') Tchelitchew, a twenty-six-year-old *émigré* White Russian painter and stage designer who was living at the time in poverty in Montparnasse with his consumptive sister, Choura. A draughtsman of genius, he was a leader of the profoundly disunited, so-called Neo-Romantic Movement, which, with painters like Cocteau's friend Christian 'Bébé' Bérard, Eugène Berman and the young Dutchman Kristians Tonny, was reacting violently against the by now accepted masters of modern art, Picasso and Matisse, and also against their fashionable successors, the abstract expressionists.

Miss Stein, as ever on the look-out for some fresh young genius, had for a period enthusiastically pursued him – and his pictures. The story goes that at her first visit to his studio he was away, and the imperious Miss Stein had battered down the door to the room where he stored his canvases and had appropriated several on the spot, paying for them later, and hanging them in the Rue Fleurus in place of her Picassos which, like their creator, were beginning to annoy her.

By now it seems that Tchelitchew was about to do the same. Possibly Edith was as well. Miss Stein bored easily, and she later told the composer Virgil Thomson that she had introduced Pavlik to Edith because she had 'had Miss Sitwell on her hands in Paris for ten days, and introduced them as a way of helping to ease the burden of both'. For Tchelitchew was by no means the easiest of geniuses for a lady like Miss Stein to patronise. He was very Russian and excitable, proud, and abnormally suspicious even for an *émigré* White Russian, and more than something of a monster. For Miss Stein his paintings were beginning to approach the dangerous stage where, as she put it in a useful phrase, they were beginning to fade into the wall. Pavlik's manners irked her too. Miss Toklas said she always found 'something very frank about Pavlik's *méchanceté*'. The less tolerant Miss Stein told Edith grimly that there was something seriously at fault with his 'procedure with *grande-dames*'. Frankly, she said, 'He has no respect for them.'

Too, true, alas, as Edith would discover in her turn, but Edith at forty, with her pale, gothic face and her pale hair starting to go a little thin, was in no mood to resist the battering display of charm, exuberance and talent with which 'The Boyar', as she called him, started to pursue her. One will never know whether his strange behaviour at the ballet had been quite as spontaneous as it appeared. From what one hears of the Russian and his way with potential patrons it would seem unlikely. But one must give him credit for the imaginative way in which he made the most of his opportunities. As he explained to Edith, when he first saw her he was sure she must be Russian because she possessed so striking a resemblance to the original of Father Zossima, the saint in *The Brothers Karamazov*. 'He was my father's confessor,' he explained in the sort of throw-away remark that Edith rather specialised in herself, 'and one of my father's greatest friends.'

In fact, Dostoevsky may well have modelled Father Zossima on a certain Father Amrovsky, who really had been the Tchelitchews' confessor in the far-off days before the Revolution, when the family had owned a large estate at Dubrovka outside Moscow. Edith was plainly flattered by the idea of such an unlikely literary resemblance, while Tchelitchew himself, the ruined but inordinately proud Russian aristocrat, must have appealed to the Plantagenet in Edith.

Indeed, he had almost all the qualities that attracted Edith in a man – striking good looks, distinguished pedigree, a spirit she could never hope to dominate, and an artistic 'genius' she could mother, champion and fight for. He also embodied what she required in all her protégés – a cause. Even the fact that Choura, and Pavlik's current lover, an expatriate American pianist called Allen Tanner, were involved in it as well failed to discourage her. Generous in battle, Edith saw them all as comrades in the cause of Pavlik's art. Had Tchelitchew been heterosexual the situation might have become more complicated, with wives, mistresses, jealousy, and even the unthinkable prospect of sexual involvement, betrayal and rebuff. Fortunately he was not, and the whole situation was reasonably straightforward – at the start, at any rate. This is not to say that Edith was necessarily aware of the exuberant and unconventional nature of the Boyar's sex life. He was as obsessional over sex as over almost everything in life, but Edith had a great capacity for seeing only what she wished to see. As a Victorian she had been taught that privacy was privacy, while as Osbert's sister she naturally assumed that men could be close friends with men without thinking the unthinkable.

As a self-appointed priestess of the arts, she had a very Pauline view of sex. For artists sex was best avoided or ignored, if for no other reason than that art was so much more important. If, as with Sacheverell, it became unavoidable, then, like St Paul, she felt that it was 'better to marry than to burn' – but, all the same, it was a pity.

Edith's determined innocence was an important element in her relationship with Tchelitchew, for it allowed her to respond to his outrageous charm without immediately becoming hurt – or frightened. (Both would come later.) It also meant that she could become great friends with the gentle, somewhat ineffectual Allen Tanner. Above all, it allowed her to believe that she was simply encouraging a great artist and his work. The fact that she had little taste for painting, and that Tchelitchew might possibly be trying to exploit her, was neither here nor there. Before she left Paris Edith had decided she would support her latest genius with all the energy and passion of her devoted nature.

As for Tchelitchew, the unstated but quite clearly understood message at the introduction by Miss Stein had been that Edith would somehow 'take him over', help to support him, and in time introduce him to an influential clientele in England through her brothers and her friends. True, Edith had no money. (Tchelitchew would have been worried had he known quite how poor she really was.) But to Edith the role of patroness was almost second nature now. If Gertrude Stein could patronise her painters, so could she, and there is something touching in the driving optimism with which she started drumming up support for him. This was Edith, the old campaigner of the Modern Movement, at her energetic, slightly gushing best, as she wrote off to friends like Allanah Harper, early that May of 1927, proudly announcing that beloved Pavlik's *vernissage* was being held in Paris on 2 June at the Galerie Vignon and that, since Choura was too shy, the great man had actually asked her to 'receive' the guests.

He has asked me to absolutely *beg* you go to over there *as soon as possible*, and either to send him lists of people to whom to send cards, – or, – better still, to send them yourself. And I do *beg* you to do so, too, as I want the exhibition to be a huge success. . . .
[P.S.] Let us have a real campaign.

Edith was able to attend Tchelitchew's exhibition on her way back to England after Tuscany and the Lawrences, but despite her efforts it was not that 'huge success' that she had hoped for. Paris was not ready yet for Tchelitchew but, far from being discouraged, Edith's warlike nature was profoundly challenged, and she began to plan the next stage of her Tchelitchew 'campaign' in London. Osbert was naturally enlisted in the cause. Neo-Romanticism was not obviously to his taste, but he could be counted on to respond loyally to Edith's bold artistic bugle. So could Sacheverell, and both would play their parts in the Tchelitchewian battles in the months ahead.

In the meantime, there were indications that Osbert's aggressive phase was drawing to a close. Perhaps the handsome Adrian Stokes was more amenable that summer and acting as a more acceptable substitute for Sacheverell. They had been to Spain together in the spring. (Osbert was planning to make him the hero of a novel, and they were making plans to spend part of the winter together once again.) Perhaps too the flattering support of his new publisher, Thomas Balston, and the success that he was having with *Before the Bombardment* had convinced him that the ox-eyed conspiracy of silence by his enemies was over, or, at any rate, was ineffectual. There were even hopes now that the play might be produced. The management of the Arts Theatre Club were interested in something by the Sitwells; and this interest inevitably revived one of Osbert's secret dreams and most passionate ambitions – a success on the West End stage.

The 'Actors' War' was mercifully drawing to a close with Osbert clearly trying to cut down his enmities. Just a few months before, Edith had written a brief note from Pembridge Mansions.

Dear Mr Coward,
  I accept your apology. Yours sincerely, Edith Sitwell.

This marked a grudging armistice in the Coward conflict. Although resentment still festered, Osbert observed it too. It was now that he also turned from loathing to a genuine friendship with that 'beastly little creature' and arch-offender of the Sitwells, the Armenian Michael Arlen. (These abrupt swings from hatred to approval, and sometimes back again, were typical of both Osbert and Edith all their lives.) But the most spectacular result of Osbert's peace offensive was the restoring of relations with one of the most important former allies who recently had slipped from grace – T. S. Eliot.

The ups and downs of the Sitwells' relations with the great man form an intriguing saga, the details of which were in Osbert's unpublished papers at his death. For the Sitwells were close spectators of the drama of his unhappy marriage, and their relationship with Eliot was intimately involved in it. After the early post-war days when the Sitwells were such close friends and allies of Eliot and his wife Vivienne, dining together every Thursday at a restaurant in Piccadilly, and eating their tea and crumpets in the tea-shop close to Marble Arch, relations had become distinctly strained. Osbert blamed Vivienne, and certainly that sad, neurotic lady – 'delicate and totally unsuited to him in character and outlook', as Osbert wrote of her – can have done little to endear her struggling but already famous husband to his friends. She was suspicious, jealous, and a little mad. Rebecca West remembers dining one night at Carlyle Square when the Eliots were there, and how Vivienne went off, leaving her shoes behind under the table. More seriously, according to Osbert, 'she had developed a capacity, derived equally from observation and a flair founded on instinct, for wounding and angering her husband, and in this showed a certain malign comprehension of his nature'.

The Sitwells had other sources of suspicion and mistrust apart from Vivienne. Eliot himself was too proud – and too aware of his unique abilities – to have accepted any artistic *diktat* of the Sitwells. In time he had acquired his own magazine, the *Criterion*, and his own fairly generous backer, Lady Rothermere. His closest friendship as a poet was with Ezra Pound, whose advice he took over the cutting and editing of *The Waste Land*, and there could be no question of his entering the Sitwell camp. On the other hand, a certain core of mutual friendship and respect seems to have survived. Edith was a great admirer and defender of *The Waste Land* when it was published, and Eliot remained an honoured if occasional visitor to Moscow Road throughout this period. For a while it is more than possible that she was in love with him – certainly in old age she implied as much. Poetically she regarded him, if not as an outright ally, at least as a valued co-belligerent in the great cause of the Modern Movement; while personally she seems to have retained a slightly piqued affection for him, which was always liable to be soured by those bouts of inevitable jealousy and mistrust which extreme success in others could arouse in her.

With Osbert, Eliot's relationships were more straightforward. They genuinely liked each other, and Osbert was one of the few people to whom Eliot, most stoical and buttoned-up of men, finally chose to unburden himself on the depressing subject of his marriage.

> From the first [writes Osbert], I had observed how solicitous he was for Vivienne, how patient with her, behaving to her, even during her most irresponsible moments in public, with a rather formal courtesy. Indeed he seemed anxious to carry the whole weight of the marriage on his shoulders.

For a long time Osbert puzzled over this. Then finally, in 1924, Osbert plucked up courage as an old friend to ask Eliot why he put up with her.

> Though by that time their existence together had reached a state of despair and hopelessness, even then he was most careful and kind in the way he referred to her: albeit in one respect, when he revealed a fact of considerable importance, he spoke with some bitterness. During their engagement, he said, she had never mentioned to him the appallingly bad health from which she had suffered even as a small child. She had been afflicted by tuberculosis of the bones, or a form of it, and had undergone so many operations before she was seven, that she was able to recall nothing until she reached that age. Tom considered that since he possessed no means of his own at the time, and since it was plain that he would have to support them both by what he could earn as a bank-clerk or in some other similar capacity, she should have told him, and thus have prepared him in advance for her later illnesses, which were to impose so crushing an additional burden on him during many years.

Osbert goes on to say that in the past he had always found Vivienne kind and attentive and good company, but after this conversation he saw the way she was behaving and the effect that it was having on her long-suffering husband, and began to change his mind.

He had yet to realise quite how sick she was, but he and Sacheverell soon saw how she could make trouble between her husband and his friends. Eliot had appealed to them for help to escape the drudgery of the bank and devote himself full-time to writing.

Though the present position was, of course, farcical, it was not easy to think out an alternative, and no sooner had my brother and I begun to consider the practical side of the problem, than Vivienne assailed us in a tremendous letter of denunciation, in which she rated us for wishing Tom to stay in the bank (whereas for years we had been engaged in every effort made to get him out of it), and had declared that she could never see us again. She so cunningly contrived her rigmarole that it even for a week or two sounded like the truth to Tom himself, until finally we were able to see him and convince him that it was a lie. . . . Subsequently we found out from Virginia Woolf and from Lady Ottoline Morrell that each of them had undergone separately a more or less identical experience when engaged in a similar endeavour to find for the poet a more congenial occupation.

Although Osbert's sympathies so obviously lay with Eliot against his wife, it is clear that Eliot was very strange and difficult to live with. Osbert admits that during this period Eliot himself was acting strangely, so that, as Osbert says, the marriage tended to be lived out in 'an ambiance permeated with tragedy, tinged with comedy, and exhaling at times an air of mystification'. As an example he describes how he and Sacheverell went to dinner with the poet in the small top-floor flat in Charing Cross Road which Eliot had rented so that he could work alone, while Vivienne remained in their larger flat at Clarence Gate Gardens. According to Osbert, the atmosphere surrounding the flat in Charing Cross Road was bizarre. At Clarence Gate Eliot was known quite simply as 'Mr Eliot', but here for some unexplained reason he was 'Captain Eliot',

and visitors on arrival had to inquire at the porter's lodge for 'The Captain', which somehow invested the whole establishment with a nautical – for I cannot say why, I took the title to be naval rather than military – a gay, a gallant feeling. . . . The room in which we dined was high up, at the back of the block, and looked down on St Martin's Lane, being almost on a level with the revolving glass-ball lantern of the Coliseum music-hall. I sat next to Tom on one side, Sacheverell on the other. Noticing how tired my host looked, I regarded him more closely, and was amazed to notice on his cheeks a dusting of green powder – pale but distinctly green, the colour of a forced lily-of-the-valley. I was all the more amazed at this discovery, because any deliberate dramatisation of his appearance was so plainly out of keeping with his character, and with his desire never to call attention to himself, that I was hardly willing, any more than if I had seen a ghost, to credit the evidence of my senses. Indeed, I should probably have come myself to disbelieve in what I had seen, had I not gone to tea with Virginia Woolf a few days later. She asked me, rather pointedly, if I had seen Tom lately, and when I said 'Yes' asked me – because she too was anxious for someone to confirm or rebut what she thought she had seen – whether I had observed the green powder on his face – so there was corroboration! She had been equally astounded and though we discussed it at considerable length neither of us could find any way of explaining this extraordinary and fantastical pretence; except on the one basis that the great poet wished to stress his look of strain and that this must express a craving for sympathy in his unhappiness.

Virginia Woolf's diary confirms the face powder, if not its colour. On 12 March 1922 she noted, 'Clive . . . says he [Eliot] uses violet face powder to make him look cadaverous', and on 27 September she added, 'I am not sure he does not paint his lips.' But Osbert never did discover why T. S. Eliot called himself 'The Captain' and wore make-up.

By the beginning of 1926 the Eliots' marriage was entering a state of crisis, and it was then, in the midst of Osbert's own angry, miserable summer, that the second and most serious estrangement between him and Eliot suddenly flared up. Once again Osbert lays the blame squarely on unhappy Vivienne.

In 1926, on the first or second day of the General Strike, my sister in her flat in Moscow Road, Bayswater, and I in my small house in Chelsea, both of us received long and incoherent letters from Vivienne, couched in almost identical terms. She wrote from Rome, where she had gone to stay, and declared that we should inevitably have heard of the scandal to which she was referring and in which she was involved. We should be aware, therefore, that if she returned to Tom, it would inevitably bring disgrace upon him, of which he would never be able to rid himself. She appealed to us, in consequence as old friends of his, to tell her what to do. Unless, she proceeded, she heard from us by cable or letter, advising her to go back to him, she would continue to remain abroad. We were on no account to let him find out that she had written to us.

We were cognisant of no rumours – indeed there were none – but the letters were those of an unhappy, distracted, almost demented person, and we felt ourselves bound to give considerable attention to them. Indeed they caused us much worry as we did not know how to act. For one thing, here was Vivienne, demanding an immediate answer, yet we could not judge what would be Tom's feelings in the matter: he was the chief person to be considered, and yet we could not talk about it to him without admitting that his wife had written to us. Eventually, as it seemed to us that he very probably might not want her to go back to him, we decided not to answer her – in any case, from a practical point of view, owing to the General Strike no letter or telegram from us could reach her. . . . Vivienne, who of course, returned on her own initiative as soon as it was possible to travel, subsequently, it transpired much later, told Tom that when, finding herself in an agonising personal quandary, she had appealed to us for counsel, neither of us had bothered to reply.

Eliot took umbrage. So did Osbert, and a long frost started.

Untypically, it was Osbert, inspired by the mood of reconciliation which started to affect him during 1927, who made the first move to establish peace, by telling Eliot exactly what had happened. Eliot, it seems, accepted his version of events, and as a gesture of the renewal of their friendship Osbert dined at Clarence Gate Gardens just a few days later – this time with Vivienne as well.

Vivienne had just finished a course of treatment in a home, and we were to celebrate her recovery. This in itself imparted a certain awkwardness to the festive occasion that had been arranged.

When I was shown into the drawing-room, I found that for once I was not the first arrival, and Mr Faber, Mr James Joyce and Mrs Joyce were already talking to our host. A few moments later, Vivienne entered in a flustered manner, on her lips her rather twisted smile, which was opposed by the consternation and suspicion in her greenish eyes. Not quite certain what tone was suitable to adopt, I opened, by remarking to her with heartiness,

'It *is* splendid to see you again, Vivienne.'

Looking me straight in the eye, she replied slowly,

'I don't know about *splendid*: but it is strange, very *strange*.'

Thereupon we went in to dinner. The table seemed set for a gala: for by each place was set a bouquet or a large buttonhole, of sweet-peas, struggling through a white misty rash of gypsophila. The food was good, and accompanied by excellent hock. . . .

As for our hostess, she was in high spirits, but not in a good mood. She showed an inclination to 'pick on' Tom across the table, challenging his every statement, and at the same time insinuating that the argumentativeness was his; that he was trying to create a scene. All this he parried with his calm and caution, remaining patient and precise, with on his face an expression of wary good nature. Indeed, it was impossible, except for those who knew them both well, to tell when he was put out with her. Her name was usually pronounced *Vivien*: but if he were irritated by her, we would notice that he would call her *Vivienne*. However, on this occasion he allowed no faintest sign of strain to escape him. And, as dinner went on, things became a little better: and she began to leave him alone, and instead to talk to me about him. He was being very *difficult*, she averred; only one human being seemed now to interest him, an ex-policeman of about seventy years of age, who acted as odd-job man, and was an habitual drunkard.

When, after dinner, the men came in from the dining-room, in an effort to win our sympathy, our hostess alleged that Tom would never give his guests enough to drink, and that she was always having to complain about it. Tom stayed on in the drawing-room, to talk to Mrs Joyce and Mrs Faber, but looked rather as if he were keeping an eye open for a squall, while Vivienne led us into the dining-room, found another bottle of hock for us, and left us at the table in order to join her other guests. The door was open onto the passage, and, while Joyce was talking to us of Italian opera, which he so greatly loved, and was even singing passages to us of his favourite works, a door in the further wall of the passage suddenly swung back, and out stepped the figure of an elderly man in a dark suit with white hair and moustache, blinking as if he had suddenly emerged from darkness into a strong light, and – rather singularly inside a house – crowned with a bowler hat.

My attention had been focused on him from the moment he appeared, as it were out of his trap-door, since I had at once identified this rather tortoise-like individual as the ex-policeman of whom Vivienne had spoken. Silence now fell on the company. The newcomer stopped in the doorway opposite us for a moment, and made to each of the three of us – Joyce, Faber and me – a sweeping bow with his hat, saying as he did so, 'Goo' night, Mr Eliot!', 'Goo' night, Mr Eliot!', 'Goo' night, Mr Eliot!', and then, while the last syllable was still on his lips, and without giving himself time to discover the failure of his ingenious method of insurance by address in triplicate against the possible charge of inebriation, he turned and went on his way humming to himself.

The incident possessed atmosphere, and I was delighted by it, as was Vivienne when we returned to the drawing-room and told her about it. For once Tom refused to see the joke, looking rather as I imagine Dante would have looked had someone ventured to make a stupid pun in his presence.

Soon the moment came for the party to break up. Mrs Faber rose . . . and observed, 'It's been lovely, Vivienne.'

Vivienne looked at her mournfully, and replied, 'Well, it may have been lovely for you, but it's been dreadful for me.'

Mrs Faber, rather at a loss, rapped out at her,

'Nonsense, Vivienne, you know it's been a triumph.'

Vivienne repeated in desolate tones, 'A triumph! . . . Look at Tom's face!'

This dinner with the Eliots took place in the early summer of 1927, and one detects a curious hiatus in the lives of all three Sitwells now. Osbert has written enigmatically of the year 1927 as the time when a new era in his life began, although he was too discreet to say publicly what it was. But with hindsight one can see the beginnings of this new era forming for them all. Before long they would begin their individual lives, and the united trio they had been would soon become only part of the false image which the public had of them. The bonds were weakening with every month that passed. Sacheverell and Georgia had their own private life in London with their small son and their separate friends, and that autumn they did not repeat the experiment of joining Osbert at Amalfi. Edith, increasingly involved with Tchelitchew, was also on the edge of a dramatic new phase in her life and in her poetry. Soon the determined effort they had all three made to maintain the trio after Sacheverell's marriage would be quietly forgotten.

They had one final public swan-song as a trio – the performance that Christmas of the play which Osbert and Sacheverell had written together two years earlier during their visit to Max Beerbohm at Rapallo. They had finally decided on a title, *All at Sea*, and Osbert himself described it as 'a satire on current silliness so near to that silliness itself that the silly would feel at home with it'.

As a verdict, this was all too true. Cecil Beaton did the scenery, and all three Sitwells appeared on stage. But not even this could save the play which seems to have been a total and deserved disaster after its three-night run at the Arts Theatre Club. For a trio who made such a point of wit, it was extraordinarily unwitty. It was also very dull. After all the bold publicity that Osbert had drummed up, after the vast hopes he had nourished for a theatrical success, it proved an infantile and boring piece of nonsense on the stage. Small wonder that the critics panned it, and that there was speculation that the Sitwells were in decline.

CHAPTER FOURTEEN

Separate Lives

1928–1930

THE FAILURE of *All at Sea*, followed by the scathing reviews that greeted 'A Few Days in an Author's Life' when it was published with the play in November 1927, had a painful effect on Osbert. It was as if he were attempting to court disaster, as if in some masochistic way he was anxious to present himself as the artist-victim of the fates, the philistines and all the enemies he was convinced were waiting to destroy him as an author.

A longing for rejection and betrayal had always formed a dangerous part of his psychology. All had been well as long as he was able to project his inner fears on to the world outside and then strike back at them, in the shape of Sir George or the generals or the Squirearchy or Noël Coward. All had been well as long as Sacheverell had been there to support him and to keep him company.

But now his aggressiveness was turning inwards. Osbert, who needed flattery so much, who was so desperate to succeed, was courting failure. It was a time of terrible depression, but there seemed no escape. Destruction was changing into self-destruction. Even his choice of a companion merely added to his misery. Adrian Stokes was no longer the carefree, golden boy Osbert had known originally at Rapallo and Amalfi, but was beginning the long, painful breakdown which ultimately landed him in Freudian analysis with Melanie Klein. (As a result of this, Stokes's whole personality would be dramatically restructured, his interest switch abruptly from art to psychology, and he would marry twice and father several children.) Despite Stokes's own depressions, Osbert decided to escape with him from London at the height of winter, taking a house together at the village of San Bartolomeo, some miles inland from Alassio on the Italian Riviera. The adventure ended just a few weeks later with

Osbert at a hotel in Rapallo in the middle of a breakdown of his own.

Unable to work, bereft and in despair, Osbert appealed to the one person who could help him now – his brother. 'It must be *so* horrible for you, and you must *long* for Sachie's arrival,' wrote one of the very few friends who knew what was going on, the painter and illustrator Rex Whistler. Then, at the beginning of February, Sacheverell arrived in Italy; and Osbert gratefully joined him at San Remo.

Osbert had loved San Remo as a boy. It still appealed to him as a lovely tropical place, full of old ladies and English churches. The weather suddenly broke into a premature Riviera springtime, and Sacheverell's reassuring presence helped to restore his equanimity. The doctor had forbidden him to work. Instead he gambled at the casino, lost moderately, and within ten days was sufficiently recovered to be off with Sacheverell again, this time to Marrakesh, for it was change and movement that he needed. Here he began to work at last. Like Edith he possessed great powers of recovery after apparently prostrating illness, and during these few days in North Africa Osbert was able to complete the slightest but also one of the most delightful books he ever wrote – *The People's Album of London Statues*, illustrated by his old friend, the irrepressible Nina Hamnett – a personal guide-book and critique of some of the best and least-known London statues, those trebly distilled spectres of the once great and famous, haunting the streets and squares of London. As for Sacheverell, this glimpse of what he later called 'the most southerly town of civilised history' was his introduction to a continent which would increasingly attract him, both as a traveller and poet, although it would be eleven years before he finally distilled his knowledge of North Africa into a guide-book which is also something of a classic – his *Mauretania*.

But Marrakesh was not enough for Osbert. He was not anxious to return to London. His old housekeeper, Mrs Powell, was seriously ill in hospital (she died in 1929). Osbert distrusted illness and without her Carlyle Square was not the same. Rather than trail back to the empty house at the tail-end of a London winter, he and Sacheverell took another fortnight on a short cruise around the Mediterranean. It solved nothing, but Osbert was enjoying life again, travelling with Sacheverell as they had travelled in the past. It was extravagant, of course, but for a writer extravagance can always be excused, if only as a source of new material, and this was very much how Osbert saw the voyage, particularly when he found that a London hostess – Violet Trefusis's friend, Violet Hammersley – was aboard the ship. The cruise finally restored his health, although almost everybody else aboard succumbed to a mysterious malady, and it was this that gave him the idea for the short story he entitled 'That Flesh is Heir to'. It was about a cruise, and the way a certain 'Mrs Chitty' carried a fatal germ which spread disease and death to everyone she met.

It amused him to work out this quietly malicious fable, and thanks to Sacheverell, thanks to Marrakesh and the gentle tonic of the Mediterranean, Osbert returned to London just before Easter, ready for new adventures.

Osbert had first met David Horner at a London party some time in 1921, and even then had been impressed by his extraordinary good looks. He was just twenty-one and still an undergraduate at Trinity Hall, Cambridge. 'Beauteous Adonis' was how a future admirer (a permanent head of the Civil Service, no less) used to address him, and he lived up to the description, with his small head of tight blond curls, his matinée-idol's profile, his willowy elegance and pale, rather melancholy features. He was immensely chic and stylish, and rather out of place in Trinity Hall (an uncomfortably straightforward college, famous for rowing Blues and lawyers). Osbert used the word 'orchidaceous' to describe him, and there was something of the hothouse flower about him in those days – a certain air of luxury and *fin de siècle* decadence. He had a personality to match – a studied wit, a cosmopolitan sort of worldliness and a faintly mocking, knowing attitude to life.

His father, John Stuart Horner, was a civil engineer with his own private company who spent much of his working life in France, so that as a boy David had lived in Paris and in his parents' villa at Biarritz. His holidays were spent in England at the ancestral home of the Horners, Mells Park in Somerset. The Horners had owned it since the Reformation, when, according to later legend, his most famous forebear, 'Little Jack Horner' of the nursery rhyme, had first put his thumb into the rich Christmas pie of the local monastic lands. The 'plum' he pulled out was Mells Abbey – and Horners had lived there ever since.

During David's boyhood it was the home of his uncle, Sir John Fortescue Horner, and his famous wife Frances, who was 'beautiful as well as scholarly'. She was a famous hostess and a 'Soul' – a member of that high-thinking and influential late Victorian coterie that included Balfour, Curzon, George Wyndham, the Duchess of Sutherland and the future Margot Asquith. But Frances Horner's most important friendship was with that rising politician, Herbert Asquith. When he became prime minister she remained his closest confidante and friend, and in 1907 Asquith's son Raymond married the Horner's daughter Katherine, thus linking the Horners by marriage with the whole Liberal aristocracy of Charterises, Wyndhams, Tennants and Bonham Carters.

This was very much the world that Osbert himself had known just before the war. It was Margot Asquith who had invited him to parties at

Downing Street, and his friend 'Bimbo' Tennant's poems which were published in the first number of *Wheels*: David's cousins, Edward Horner and Raymond Asquith, had both been friends of Osbert until they were killed in France. So when Osbert first saw David Horner they were not exactly strangers.

David himself appears to have possessed a commendably relaxed attitude to the problems which assail most undergraduates. As the younger son of a younger son, he had little money of his own, but this did not stop him from deciding he would live an interesting life, unencumbered by ambition or by the vulgar need of earning himself a living. His looks were literally his fortune. He dressed superbly, had an amusing line of gossip about all the best people, which he recounted in an engagingly *basso profundo* voice, and after leaving Cambridge was soon floating, as unattached, good-looking, upper-class young Englishmen could float in those more gentle, far-off days, through a rarely failing world of dinner-parties, long weekends and holidays abroad. He was the perfect guest, the ideal ornament for any party, charming to women and agreeable to men, better connected and far better read than the usual run of gilded, social butterflies, and equally at home in the best society in Paris or in London.

Indeed, it was Paris that soon became his home, thanks to his friendship with an elderly aristocrat and art dealer, Vicomte Bernard d'Hendecourt, whose house in the Rue Bazard provided just the sort of Parisian *pied-à-terre* that David needed. D'Hendecourt possessed both wealth and taste – two things that David valued. He had another virtue. He was not possessive, and made no attempt to interfere with David's far-flung social life. So Osbert and David Horner had somehow kept in touch throughout the busy early twenties. This was the period when Osbert was inclined to be romantically protective to good-looking, well-brought-up young men, and David was not immune to the Sitwell legend. The one reference Osbert makes to him in the entire five volumes of his autobiography is to the few days of the General Strike in 1926 when he describes how he was living on his own in Carlyle Square. Apart from Sacheverell and Edith, and William Walton who was 'in undisturbed communion with his piano' in a room upstairs, he saw no one there,

> except for a very dear friend of mine, whom I had known for some few years, David Horner; who was my main link during these ten days with the more useful members of society; for he had volunteered as a lorry-driver, and he would now come to see me, between his spells of duty, with a cut over one eye which he had earned for his patriotic effort, and would describe to me the people he had met on his journeys and the feeling that was in the air.

The strike settled, it was time for David to return to Paris – and explain his cut eye to the Vicomte – while Osbert's attention during the following two years would be engaged by Stokes. But no sooner had Osbert parted from Adrian at San Bartolomeo than he was writing off to David with the news; and once he was back in London more letters followed saying how anxious and indeed determined he was to see him again. Could he come down to Montegufoni some time in May when Osbert was planning to be there? He couldn't. Could Osbert then come across to Paris at the end of June? That proved impossible as well, since he had to read some verse and make a speech at Liverpool. Not until July did they actually meet – thanks to the Earl of Ellesmere, whose ball at Bridgewater House they both attended during Newmarket Week. Osbert had written previously to invite David to stay at Carlyle Square while in London, 'should you be short of a lodging, and if you don't mind having Willie's bedroom with a piano (grand) in it. He, I need hardly say, will be away.'

The friendship flourished. Early that September they were together once again at Evian-les-Bains – where David had already picked up the habit of taking the waters at least once a year. No sooner had they parted – David for Paris, Osbert for the Grand Canal Hotel, Venice – than Osbert was writing anxiously to thank David for 'my charming stay with you', and to 'remind you how devoted I am to you'. The letter had a postscript whose importance David cannot have overlooked. 'I haven't seen Adrian yet, and don't know whether I'm going to,' Osbert wrote.

Whether by coincidence or not, Edith's uneasy love for Tchelitchew was also flourishing that summer. They were in constant contact, and had seen one another both in London and in Paris. For a few days in July they had even stayed together in a country cottage romantically set in an orchard outside Paris. This cottage, which belonged to Stella Bowen, friend of both Tchelitchew and Gertrude Stein, and mistress of Osbert's old sparring partner, the novelist Ford Madox Ford – or 'Freud Madox Fraud' as Osbert used to call him – was at Guermantes, and during the whole period of Edith's visit the proprieties were carefully, indeed, painfully, observed. For with 'The Boyar' was the inevitable Allen Tanner, and Tchelitchew's sister, the consumptive Choura, was also there to chaperon Edith from the improbable attentions of the two males in the house.

Despite – or, possibly, because of – this, it seems to have been a most idyllic time for Edith, one of the few rare periods of unalloyed and total happiness she ever had. The fragile balance between the curious quartet was magically maintained. They were artists and they understood and loved each other. Pavlik began to paint her portrait – the first of many that he did of her – and she responded to his sincere enthusiasm over her appearance. She was his 'Sitvouka', the virgin sibyl of his personal mythology, symbol, so Tchelitchew's biographer assures us, of 'the purely passive, female sensibility that lives forever' like a 'glass flower under glass'. 'One can speculate', he adds, 'how immensely successful must be his masculine preenings, his glances, his random endearments, his cosiness, especially in moments withdrawn from witnesses.'

But with Tanner and Choura there, neither was in the slightest danger of overstepping the proprieties – nor did the outside world as yet intrude. In years to come her holidays in the cottage at Guermantes would always be something she looked back to with a terrible nostalgia, as if they had offered her a taste of perfect happiness of which malign fate cheated her.

By mid-September the Sitwells were together once again, this time at Montegufoni, bringing with them all the allies and supporters they could muster. However much the children were inclined to grumble about the house, and about Sir George and Lady Ida, they were never slow to take advantage of them all when the need arose – as it did now. The Chigi family of Siena had invited them to present *Façade* as part of that year's annual music festival and Osbert was all set to make the most of it – despite the fact that relations between him and William Walton had been worsening of late. (Walton was beginning to become a shade too independent, and his flirtations with young ladies upset Osbert.) One visitor was Richard Aldington, who arrived with his mistress, Brigit Patmore. In her diaries she left a curious picture of Sir George – 'a courtly little gentleman* with bright ginger hair. At lunch . . . he was both charming and intelligent, not at all the ogre I had imagined.' The only shadow on the festival – *Façade* was received enthusiastically by the Italians – was that David Horner was not there. D'Hendecourt was sick. David's place was suddenly in Paris by his side, and on 24 September Osbert wrote to tell him what he had missed in Italy.

Siena was great fun, and Willie a success beyond his merits (as a person). He was referred to as 'Il world-famous Wa*n*ton' in one paper. *Façade* went extremely well, and was really a lovely performance. Christabel [Aberconway] is here and Siegfried and Willie and Constant. Edith left yesterday. Yes, please do come to Amalfi (where I go next Monday).

Osbert reached Amalfi to discover the little town 'enjoying a heavenly false spring, with all the trees flowering', and he was able to go bathing every day. Walton was with him, and relations seem to have been restored by now, with the young composer busy on an important new work. The B.B.C. had been anxious to commission Walton to compose an oratorio, 'on a suitable religious subject that everybody knows about, apart from the Crucifixion'. He had been stuck for an idea, and it was Osbert – prompted perhaps by the memory of the great regimental pre-war feasts

* He was in fact six foot one.

in the Guards which were called 'Belshazzars' – who had suggested the subject of Belshazzar's Feast, and had composed the words, based on the Book of Daniel, for him.

And so the bright November days went by with the calm of the Cappuccini, broken only by the tinkling of a cheap piano from the cloisters as William Walton hammered out his oratorio. It must have been distracting for Osbert, who had started a fresh novel of his own. He based it on an idea he had discussed with Stokes – about a young novelist who meets himself grown old and famous. It might have made a very short short-story, but the only excuse for *The Man Who Lost Himself* as a full-length novel is that it was written when its author's mind was constantly on other things. For apart from the noise of *Belshazzar's Feast* there was also the uncertainty of when David would arrive.

He duly came aboard the train to Naples, and reached Amalfi by 20 November. But no sooner had he settled into the Cappuccini than he was summoned back by telegram to Paris. The Vicomte was ill again. The next thing Osbert heard was that the Vicomte had died. It was deeply shocking – and an opportunity for Osbert to play the experienced, understanding friend to David.

What can I say except that I am glad it is over and thankful that at the end it was painless. . . . Remember how many devoted friends you have and let us thank God that there was not to be another two years of agony for both of you. . . . Meanwhile, don't think about anything more than you need to, till you can get away – And then come as soon as you can, for I'm sure that quiet is very necessary for a while for anyone who has gone through what you have gone through.

Soon after Christmas David returned: the friendship deepened with a David shaken by D'Hendecourt's death and an Osbert in the role of the protective older man. But Osbert was already the more dependent of the two, and when at the end of February David went off for a brief visit to some friends on Capri, Osbert wrote to him the very night he left.

My Dearest David,
You would never believe how much I miss you, even though it's only for a day or two. I am miserable and pining: you'll find me pale and thin (and distinguished-looking) when you return: which, please, at once.

Return he did – and stayed until the end of February as Osbert ploughed on with his novel. And finally, when David departed once again – for a few days in Paris arranging his effects – Osbert was disconsolate. Walton had left for London, and the only guests remaining in the whole hotel were Godfrey Winn and a balloon-like German lady who did Swedish exercises every morning on the beach. They were no substitute for David.

Not that Osbert had long to wait. From Amalfi he had already planned to go on to Athens where he was meeting Georgia and Sacheverell for a short holiday around the monuments. David must obviously come as well, and no sooner was Osbert safely installed in the Hotel Grande Bretagne than he was writing anxiously again: 'Athens is exactly your town. And ever so much nicer than one could imagine.'

The loved one did not take too much persuading, and in the second week of April arrived by train, first class, from Paris. Everyone was pleased to see him. He was so civilised, so full of life and so amusing. Sacheverell was particularly glad to have someone as company for Osbert. Osbert, whose temper up to now had been uncertain, brightened instantly.

What the others could not know was how desperate Osbert had become, and how unreservedly he was prepared to throw himself on David's mercy. David seemed suddenly less anxious to commit himself. Coolness followed. In a panic, Osbert scribbled a pencilled note and left it in his room.

I have been so unhappy today and yesterday. Perhaps writing to you will ease me more than it will bore you.

I love you so much that if you were to stop caring for me, I should be nothing but an empty, unfortunate, large shell walking about with nothing inside it in a world with which it had no connection.

You must know this yourself: and please, that I think of you much more than anyone else. You are never away from my thoughts for an instant, and, this being so, for heaven's sake, sometimes put up with me if sometimes I'm irritating from an excess of affection.

But it was not that easy. Osbert was painfully possessive. David found other friends in the hotel and insisted on his right to see them if he wished. Osbert exploded in a storm of jealousy – and then repented miserably. A second note was left in David's room for his return.

This time it's my turn to be quite blind and write you a letter. If ever you think I don't love you, you are a goose of the first water (which you aren't): if you stop loving me, the same, if I may say so. But being in love is a very painful business: I want to see you always, and hoped I would tonight. And was furiously jealous at your preferring those, to quote their own vocabulary, 'drabs' to me.

Surely, to be disappointed, and in consequence odious, at not seeing you so regularly for so long a time, is an indirect, if tedious at the moment, compliment? You will never know, or probably you do all the time – the agony that such things can be.

And so the courtship, stormy as it was beneath the necessary layers of discretion, rose to a new intensity during these days of Athenian spring.

During another visit to Athens the presence of that genial joker, Gerald Berners, was a source of light relief – even if he did snore so thunderously at night that everyone was soon obliged to shift their rooms to get some sleep.

One of his jokes back-fired. Since his arrival he had insisted that the hotel manager, a vulpine, rather toothy Greek with an ingratiating smile, was in fact a werewolf. Late one night Berners caught sight of him outside the hotel and began baying like a wolf. It was a fearsome sound, and he had evidently thought that it would send the werewolf scuttling away. Instead, the manager was most concerned and with great solicitude inquired, 'Is anything wrong, my lord? Perhaps I should fetch a doctor?'

Ugly, eccentric Gerald Berners often maddened Osbert with his irreverent compulsion to prick what he felt was Osbert's pomposity. It was Berners who supplied the best riposte to Osbert's perpetual mania for publicity. After seeing Osbert's famous bowlful of newspaper cuttings on the Sitwells always so prominently displayed at Carlyle Square, he placed an even larger bowl in the entrance hall of his house in Halkin Street when Osbert was invited for dinner. In it was one lonely press cutting on Lord Berners – a two-line announcement from *The Times* that his lordship had returned to London from abroad.

This sort of thing was irritating, but Osbert also envied many of the features of Gerald Berners's way of life. There was the title – Berners was the fourteenth baron, and the title went back to 1455. There was his jewel of a country house at Faringdon near Oxford – the house which so enchanted Nancy Mitford that she made it the abode of the character she called 'Lord Merlin' in *The Pursuit of Love*: and Berners was himself a sort of Merlin figure, making the most of all his talents, painting and writing and composing, and then flitting like some wealthy wizard from Faringdon to Paris to his house in Rome, then on to wherever fancy took him in a custom-built Rolls-Royce fitted with a harpsichord. Above all, it seemed that Gerald Berners had the crucial thing that Osbert lacked – absolute freedom, with all the money he desired, great luxury and no Sir George to worry him.

Now that there was a chance of sharing his life with David, such things mattered, and one can see how Osbert's mind was working in a short story based on the character of Gerald Berners. He called it 'The Love-Bird', and the hero, Robert Mainwroth, is a rich, eccentric aristocrat. Osbert describes how sensibly Mainwroth had 'adapted his situation to the day' by refusing to be weighed down with vast inherited possessions and converting them into ready cash.

> It was pointless and hopeless, he felt, in these days, to own vast draughty, machicolated mansions, ugly in their conglomerated selves, even though full of beautiful objects. . . . The modern world dictated its terms to the rich, and the

moneyed nomad, with a few tents pitched ready for him in various parts of the world, in let us say Paris, London, New York, Seville, Budapest [in the manuscript Osbert also wrote Rome, then crossed it out] and with very easy means both of reaching them and leaving them, was the fortunate man of today. His heirs, as much as himself, ought to be gratified at the firmness and foresight he had evinced, for many would have been intimidated by the sheer weight of these possessions into keeping them. Now they, too, would be equipped as modern men out of the increased income into which all these things had been transmuted, and would have no desire, and certainly no room, in their small houses and large motor-cars, either for monumental pictures and pieces of furniture, or for loads of clustering, clattering things.

So much for Sir George, with his passion for his heirs and for keeping the Sitwell inheritance intact at any cost. So much for 'draughty, machicolated' Renishaw itself, crammed with its lumbering, lovely things, but really such a bore and responsibility for penurious Osbert. But for Sir George he could have sold the place, invested the proceeds, and then become a 'moneyed nomad' with a Rolls-Royce like fortunate, enviable Gerald Berners.

All he could do was to try once more to get money from his father, as he explained in a letter which he wrote to David at the end of April 1929:

At present I'm engaged in a frantic long-distance row with Ginger. Rather fun. But the real object of this dreary letter is to ask you if it is possible for you to come with me to Renishaw for Whitsun. I've got to go up there on business for a day or two. And it would change the melancholy of the visit to golden happiness if you could come too. I insist, of course, if you can come, on doing your railway fare as it is 'an 'ell of a long way off', to quote Mrs Harper. The prospect is this . . . good plain food, quite good really. Masses of drink of whatever sort you like imported by me. Empty house, living in a corner. And then we'd motor to Chatsworth and other places. It might be heavenly? But perhaps you'll be engaged.

David was not engaged, and he arrived to see Renishaw at its best with the willows breaking into pale green leaf around the lake and the hills now clear of mist in the bright spring weather. David loved the house – but there could still be no question of a full-time life in such a place, and the money problem was increasingly pressing and absurd. For here was Osbert, nominally the owner of this great house in Derbyshire with all its lands, and heir to an even larger house in Italy, and yet he was in debt and seemed to have no prospect of relieving this chronic shortage of ready money for as long as his father lived. But without more money now Osbert could hardly hope to maintain the house in Carlyle Square, let alone continue to indulge in the sort of life he had to be able to offer David if he hoped to stay with him. It was a tantalising situation, surrounded as he was by so much Sitwell wealth, and yet unable to obtain an

D. H. Lawrence and Frieda at the Villa Mirenda

Below: Rex Whistler's letter-heading for Renishaw Hall, 'A long, low, old house' (enlarged twice original size)

Above: Sacheverell and Georgia at Weston Hall

Below: Edith and Georgia

David Horner

Below: Osbert and David with their house-servants, Peking, 1934

Edith by Tchelitchew

Tchelitchew, self-portrait

House parties at Renishaw Hall, 1930

Above: Rex Whistler, Raymond Mortimer, Reresby and Georgia

Below: Tom Balston, Lady Ida, Georgia, Bryan Guinness, Edith, Diana Guinness, Tom Driberg

Dylan and Caitlin Thomas

Osbert, Georgia, Edith, Sacheverell at the *Reynolds News* libel action

The Royal Poetry Reading
Above: Edith reading, Osbert, Vita Sackville-West, Walter de la Mare, Arthur Waley, on the platform
Below: The distinguished audience: Arthur Waley, Princess Elizabeth, Osbert, The Queen, Princess Margaret, Walter de la Mare

Osbert at Renishaw

extra penny from Sir George when his whole happiness depended on it. Never before had money meant so much – or been so difficult to find. Never before had his father's avarice been quite so maddening – or threatened everything that he had set his heart on.

For Edith 1929 saw the publication of her most important poem to date – the long and passionate tirade against London and the remorseless drift of mass philistine society which she entitled *Gold Coast Customs*. It seemed a radically new departure for her poetry, both in style and content. For nothing could be further from the wistful and nostalgic elegance of *Troy Park* and *The Sleeping Beauty* than this impassioned cry of bleak revulsion at the world around her. Nor, for that matter, could there be a stronger contradiction of her earlier disdain for 'lofty moral themes' in poetry. For here she is in the territory of Eliot's *Waste Land*, crying out against the cruelty and filth and spiritual bankruptcy of her time. The title – and a great deal of the imagery of the poem – is taken from the so-called 'Customs' of Ashanti, which once supposedly prescribed a period of ritual cannibalism following the death of a tribal chief. And in fact there is a line of continuity between the poem and some of Edith's earlier work – those melodramatic and uncomfortably intense effusions of her later twenties like 'The Mother' and 'Thaïs'. She has returned now to the same well-springs of her private misery and personal affront against the world around her. But the technique is more sophisticated and controlled, and she extends the poem to achieve what she claimed as 'not so much the record of a world as the wounded and suffering soul of that world, its living evocation, not its history, seen through the eyes of a protagonist whose personal tragedy is echoed in that vaster tragedy'.

Its chief importance lies in the emergence of a new poetic role for Edith – as the impassioned prophetess of doom – a role she would take up again in the war poems of her middle fifties.

> . . . Do we smell and see
> The sick thick smoke from London burning . . .

Edith would later claim that just as the poem itself described the 'state that led up to the second World War', so lines like these were intended as 'a definite prophecy of what would arise from such a state – what *has* arisen'.

Yeats hailed the poem, saying that it restored something that departed from English poetry with the death of Swift – a strange remark – while other critics – in particular the Marxist, Jack Lindsay – have seen the poem as prefiguring the social protest of the left-wing poets of the thirties, citing especially the rhetorical finale of the poem:

Yet the time will come
To the heart's dark slum
When the rich man's gold and the rich man's wheat
Will grow in the street, that the starved may eat, –
And the sea of the rich will give up its dead –
And the last blood and fire from my side will be shed.
For the fires of God go marching on.

Edith herself would always refuse to be drawn over exactly what she *did* mean by these powerful and revolutionary lines; but for her personally *Gold Coast Customs* would always be a poem of particular significance. For her it always marked the point when, as she wrote, 'My time of experiments was done.' And although she did not know it at the time, it would also be the last important poem she would write for more than a decade.

Although Osbert had set his heart on a permanent relationship with David Horner, it was not so easy to achieve. One problem was David himself. Osbert was the lover, David the beloved; and whereas Osbert was intensely, single-mindedly in love, David was not. He was extremely fond of Osbert, flattered by his friendship, charmed and diverted and amused by him. Whether he was ready to commit himself to living with him was another matter; and throughout the summer and the autumn he made it very plain that he had other friends and another life apart from Osbert.

For Osbert another problem was still the profoundly boring business of finance, and throughout that autumn of 1929 he continued his campaign to make Sir George see sense and release some capital. But Sir George had had long practice at evading such requests, and then in November the baronet decamped from Montegufoni with Henry Moat and Lady Ida on a tour of Germany. It must have been a doubly frustrating time for Osbert. He was in Amalfi as usual, and no sooner had he managed to persuade David to join him there than he fell sick with influenza. He had it badly – 'pains in the eyelids, stiffness and pain in the spine and terrific swelling in the nose' – but worst of all it meant that he was in isolation in his own room on doctor's orders and totally cut off from David. All he could do was write: 'Life is almost too tantalising, with the sound of your brazen voice in the passage and I unable to reply. What is fate doing with me, keeping me in bed all the time you are here, snatching me away like that, and rendering me incapable of thought or speech?' He nobly insisted David kept his distance. 'You must be careful not to catch this bloody thing as with your bronchial tubes it might be serious, apart from the frightful boredom anyway.' But as he also tried to make quite clear: 'Influenza, though perhaps not romantic in itself, makes one sentimental, and I lie pillowed on your letters . . . you can never know how much I miss you and fret.' Scarcely was Osbert well again than it was time for David to be off, but as some consolation he had agreed on a winter holiday with Osbert shortly after Christmas. They would meet in Venice on Christmas Eve, then make their way as comfortably as possible down to the thrice-blest sacred zone of southern Italy.

Apart from the pleasure of reunion with David in his favourite city, Osbert had a more serious reason for going to Venice now. On his way back from Germany Sir George had arranged to spend a few days before Christmas at the Grand Hotel. It was a useful opportunity to see him and to try to make him see sense on the vexed question of his allowance. As Osbert knew, Sir George was always at his ease in Venice. If there was any chance of getting what he wanted, it was there.

Osbert described his mid-December journey to the city at the beginning of the happiest book he ever wrote, his collection of travel sketches which he entitled, with good reason, *Winters of Content*. Even the wheels of his train seemed to spin out 'an insistent waltz lilt' as it carried him across the plain of Lombardy, past 'the golden wreckage of the fields' and the pruned vines and leafless mulberry trees and on towards 'the proudest, most challenging city that man has attempted'.

On 17 December Osbert was writing David his first letter from a table at Florian's

Arrived here safely this evening. Ginger in tearing form, and goes to bed at 9.30 every night now – such a blessing. So I have a very quiet booze here, by myself, with the prohibited cigar, and write, my angel, to you.

He had proposed to leave here for Florence on the 24th. But I've stopped that. I've told him you were coming to join me and wished to consult him on your inheritance. He was delighted as he loves interfering.

Ostensibly you and I go to Brindisi on the 26th. It will be heavenly.

The journey here was heavenly. All Italy spun past in a blue wave of sunshine.

The perfect weather, Venice, and the prospect of David in just a few days' time kept Osbert in a state of permanent high spirits. Not even Sir George, it seems, could dent them and there is no greater contrast than between Osbert now and as he was when he composed his self-pitying 'A Few Days in an Author's Life'. Left to himself in the all but empty city, he found that contrary to the common persuasion, Venice was of all cities the most pleasant to stroll about in. He visited the Veneto. He called on his friend Bertie Landsberg at the Villa Malcontenta. And all the time he was quietly preparing the first chapters of his *Winters of Content*.

The only cloud on the horizon was Sir George. His German trip had been a great success. As Henry Moat told Osbert, Sir George had taken him to see the countless castles, palaces and museums that interested him.

We have become well known in Germany, Ginger visiting the above places over and over again and giving the attendants a hell of a time that when we enter a door and they see him they scatter like scalded cats some through doors, some through windows and others up the chimneys. One fat old woman wanted to take his umbrella from him and then commenced a vigorous tug-of-war result the fragments of the umbrella has been set to the castle to be put in the armoury.

Perhaps because of these exertions, Sir George was not at his best in Venice. 'Ginger today in bed with a rupture,' Osbert reported to David on 19 December, '(does it, poor sweet, come from being disagreeable?) and awaits the doctor.'

Osbert had evidently raised the vexed question of finance – unsuccessfully.

Yesterday was a bloody day. 'Ginger' bullied and stormed. He wants me to become an Italian subject and give up living in London. I've refused. But I'm depressed as I realise every day how dependent I am on you and lost without you, only half a rather depressed vacuum of a man when you're not with me. And I'm afraid that these long, wearing bothers and, as Burton very rightly says, 'suchlike' will decoy you away from me. But they won't really. And even if it's to be a garret, a very lovely garret we'll find. And above all, I hope you're a little lost without me, too?

In the same letter Osbert gave a brief schedule of his day – and an intriguing imitation of Sir George at dinner.

My life is: 'Bekkast' at 8 o'c. T-T lunch at 12.15. Dinner at 7 with 'littla laita, whaite, waina', (like a spinster). No brandy, and lights out at 10. 'Ginger' all his life has made a speciality of making other people feel dissipated if they stay up after 9.15. However, they've some excellent old brandy at Florian's and we'll look into the matter when you come here.

Sir George's strange suggestion that Osbert should become an Italian must have been part of some complicated plan to make him give up Renishaw, as well as Carlyle Square, save on his income tax and cheaper Continental living, and give Sir George himself a chance to keep an eye on him. ('Such a mistake to be too far away from me. You never can tell when you'll be in need of my advice.')

Osbert was hardly likely to agree. Much as he loved Italy, he was determined to enjoy the best of every world available – and so was David. If Sir George refused to help, all was by no means lost. Osbert was hardly likely to end up in a garret, however lovely, nor had he the least intention now of losing David or abjuring the gentle pleasures of the rich. He had his own plans now – and if they enraged Sir George, so much the worse for him.

All that really mattered was that David had arrived, and it was Christmas Eve. Venice, as Osbert said, was the perfect place 'to bury Christmas' – 'a terrible festival', he called it, 'and a time of unavoidable sadness' – but here in Venice all was happiness, and since Sir George was finally despatched to Tuscany on Christmas Day, Osbert and David were able to enjoy their brandy at Florian's in peace. On Boxing Day they took a slow train to Bologna, missed their connection (one of the very few occasions Osbert ever did), walked around the wintry city for two hours (source of another chapter in *Winters of Content*) and finally picked up a night train down to Naples. They were at Lecce just in time to toast the 1920s out and the 1930s in.

CHAPTER FIFTEEN

Partial Eclipse

1930

JUST AS the twenties were pre-eminently the Sitwells' own decade in Britain, so the thirties would provide their season of partial eclipse. Almost overnight, it seemed, the times had turned against them. They would survive and go on working, but for this grey decade the scandal and excitement and suspense exhaling from their earlier work would go. As poets they would almost seem to peter out, as personalities no longer rouse the daring adoration of the young. No second Harold Acton would declaim 'The Sleeping Beauty' through a megaphone along the High, nor would another Cyril Connolly find glamour in their presence. Only their most persistent audience, the philistine, would remain faithful to them now.

All this was partly age. Edith was forty-three in 1930, Osbert thirty-eight, Sacheverell thirty-three, and a new generation of younger, grimmer, more committed poets was waiting in the wings. The Sitwells' novelty was over, and middle-age is rarely glamorous.

Also, of course, the world was changing – to their disadvantage. With the Slump on its way, Osbert's highly romantic, highly personal political reaction to events would soon appear irrelevant, the attitudes he struck a little frivolous. Fashion, which once abetted them, would now reject them; their sort of cosmopolitan, bright culture slowly fade, the shores of the higher Bohemia recede: so would the once important twenties salon-life of London which had been such a potent source of Sitwell influence and power.

But when all this is said it still only partially explains the way the trio seemed to wander from the limelight as the thirties started. To understand why this occurred one must pursue the rocky and diverging pathways of their private lives, as the new decade brought quite dramatic changes to all three of them. And since, as so often with the Sitwells, it was money and Sir George which proved such decisive factors in their lives, it is at Montegufoni in mid-May of 1930 that we take up our story.

The gardens and the countryside were wonderful that year: the season was unusually advanced, and Tuscany in early summer seemed an earthly paradise. New vineyards planted by Sir George promised a splendid vintage for the autumn, the enormous blood-red oleander in the Cardinal's Garden was coming into bloom, and banks of misty blue plumbago tumbled along the now replanted lower terraces. It was a time of long hot days when nothing stirred across the valley, and silent nights with fire-flies and lightning from the early summer storms.

Normally Osbert would have enjoyed these few weeks at the castle, enduring his father's disapproval and his mother's luncheon parties, and making the most of Florence. True, some of his former circle of acquaintances had gone. Lawrence was dead, Huxley was out of favour, young Harold Acton had decamped from the great house on the Via Bolognese for the gentler pleasures of Peking. But Norman Douglas was still around and still willing to be bought dinner in Fusi's Restaurant, where Osbert and Sacheverell had introduced him years before to the most devoted of his admirers, Nancy Cunard. (Osbert was always mildly aghast at the blatant nature of Douglas's taste in boys. Only recently he had brought a child in a sailor suit along to dinner, sitting throughout the meal with him on his knee.) His great friend Orioli still had his bookshop on Lung' Arno, and Ada Leverson, 'the Sphinx', was still alive, still very deaf, and still in love with him. (She promised to leave everything to Osbert, but when she died three years later it would prove a nonexistent legacy.)

Some of the English colony were worried by the growth of the *fascisti*. Florence was a Fascist stronghold from the start and that very May was staging a fervent welcome for the Duce. Osbert felt no enthusiasm for military-style parades and uniforms and so kept clear, but he admired Mussolini if only as the protagonist of the d'Annunzian ideal of the man of destiny. Thanks to him, Italy was not bedevilled by the insidious nonsense of the 'little man' and middle-class democracy, large castles owned by Englishmen were safe, and the trains ran on time – an important point for Osbert, who was a great traveller by rail.

But Osbert had other things to think about that May than Mussolini and the Fascists. David had been staying at the castle earlier in the month, but had departed in the second week for London. Anxious and already more than a little jealous, Osbert was writing every day:

> Here I am racing to write to you before the light goes, and having just received your letter am in consequence full of the perturbation I always feel when parted from you. . . . Do understand what you mean to me please. And you don't say if you miss me. Do you? A future without you would be something I could never face. . . .

Osbert's anxiety grew worse, and just three days later he was writing:

> Darling, what a miserable man I am, so far away from you and with streams of jealousy and anxiety. No romps or cocktails please,
>
> Your saddened little friend,
>
> O.

But even as he wrote Osbert already knew that his ordeal would soon be over. True, David was often unpredictable and gave him cause for jealousy. Given the one-sided nature of their love, this was perhaps inevitable. But Osbert also knew by now that David relied on him as well, was grateful for his kindness as an older man, and basked in the security he offered, the influential friends, the holidays abroad, the interesting life. They also had a lot in common – their taste in literature and art, their brand of wit, their choice of friends. As well as this, David had provisionally agreed to move into Carlyle Square once the holiday was over. But before this could occur there were still things to settle with Sir George. Sacheverell and Georgia and infant Reresby were staying at Montegufoni and much would depend on what the old baronet decided.

For none of those angry questions of finance which he and Osbert had discussed at Venice had been settled. Now the whole weary business was reopened: 'Life here is trying:' Osbert wrote to David on 19 May. 'I'm not well – pining I suppose. And Sachie and Juggins [Osbert] are having "financial" talks – not to mention screams and tears – with Ginger.' Money – and family possessions – were still emotive subjects for Sir George. As he witnessed what he saw as Osbert's unredeemed rake's progress, he was increasingly repentant of the settlement entrusting Renishaw to him. Nor was Sir George all that much more disposed to trust Sacheverell, who was always loyal to Osbert in these matters. As might have been expected, Sir George's policy and passion now were to attempt to keep the whole patrimony intact for Reresby and unborn generations of future Sitwells to inherit. This was the creed which he had lived by and which came near to offering a sort of *raison d'être* to his whole existence.

But while Sir George was adamant against what he saw as robbing unborn Sitwells to pay Osbert's debts, Osbert was adamant as well. He was not leaving Renishaw, and he was certainly not becoming an Italian. 'Screams and tears' notwithstanding, he must insist that he had not the least intention of compromising over the life he wanted. And it was then that Osbert dropped what proved his most effective and most devastating bombshell. He was arranging to take out a mortgage for £10,000 as prospective heir to Renishaw. He regretted it, but Sir George left him no alternative. Sir George made it clear that he regretted it still more, but since Renishaw was so entailed there was nothing he could do, except to show his fine paternal rage in his iciest paternal manner.

Osbert was not particularly concerned, for by then he had heard from David. The news was even better than he dared to hope. Elatedly he wrote back:

Darling,

I am so delighted with your letter to hear you are coming to Paris. I'm writing off to the Lancaster for a room for you today: and have booked a 'Golden Arrow' ticket for you from Paris with me when we leave.

I am so delighted that you're staying at Carlyle Square.

Willie returns to London on Monday. I've told him to find out from you about the decoration of Mrs Powell's room . . . whether you have your own decorator or will employ ours? Also to have it fumigated as soon as possible.

The idea of seeing you is heaven. I long for the day.

The last days at the castle dragged appallingly. Since the big row Sir George was not appearing much at meal-times, but Lady Ida, with her passion for arranging titanic meals as some answer to her perpetual boredom, more than made up for the lack of company. '16 awful people to lunch, 22 to tea, and Sphinx to din-din', Osbert reported gloomily. But the worst was over; on 30 May he stepped aboard the Rome–Paris express. David was waiting for him at the Gare de Lyon, and when in the second week of June the two of them returned to London and David moved into poor Mrs Powell's now fumigated room, a whole new era in Osbert's life had started.

Indirectly, all this affected Sacheverell and Georgia as well. Until now Sir George had been maddeningly indecisive over the question of their future. They had to have some help from the estate: that much was obvious. Sacheverell had only a small income of his own (£1,000 a year) and even smaller hope than Osbert of living off his writing. Sir George had enjoyed the great good fortune of living his whole life on the Sitwell money he had inherited. Now in his old age one would imagine he would have wanted to endow his favourite son and father of the ultimate heir to Renishaw with something that would let him live a life of travel, scholarship and leisure, as he had done himself. Sir George apparently felt otherwise.

One of the several properties he possessed in England was Weston Hall, a gothicised seventeenth-century manor-house in one of the 'lost' villages of Northamptonshire. It was an enchanting house, and had been lived in by a succession of old ladies. One of these had been the wife of Colonel Henry Hely-Hutchinson, the gallant and romantic survivor of the Peninsula and Waterloo, and Sir George's most unlikely grandfather. The house had been descended through their daughter, Sir George's redoubtable Aunt Harriet, Lady Hanmer, and she in turn had left it to Sir George for the intended benefit of Sacheverell.

Sir George was not particularly interested in the house. 'I don't intend to do much here; just a sheet of water and a line of statues,' he had remarked to Georgia when showing her the garden. But, all the same, he had no intention of giving it to Sacheverell outright. Like Renishaw, it would be administered by trustees. It was an ideal house for a second son with a wife and family, and he and Georgia would have the use of it for life.

Sacheverell and Georgia both loved London. Most of their friends were there, and the museums and libraries Sacheverell needed for his work. He had been hoping, not unnaturally, to be able to go on living there. But Sir George's experience with Osbert had made him less inclined to help Sacheverell. Why should he help him live a life of extravagance in London which had already proved disastrous for Osbert? The 'screams and tears' during the great financial rows at Montegufoni during May must have settled things for good. Sacheverell had his £1,000 a year, and he had Weston Hall. That was as much as he was getting – until Sir George decided otherwise.

This was a crucial moment for Sacheverell – and for his work. In some ways, Weston suited him and Georgia to perfection. The house was a treasure-chest of eighteenth- and early nineteenth-century objects, carefully hoarded by those generations of long-dead magpie ladies in its dusty rooms – pictures, books, letters, eighteenth-century clothes. It had an old walled garden, fine lawns and ancient trees, and was the perfect place for a scholar–poet to pursue a life undistracted by the noise and cares of the great world outside. Georgia loved it and began the hard work of restoring and refurbishing the house with taste and iron dedication. For her it would be a life's work, and today Weston must be one of the most perfect lived-in small country houses in the country. Osbert described it as 'full of ineffaceable shadows of the past'. All this remains, and thanks to Georgia there is also comfort, elegance, and all the style one would expect in a Sitwell residence.

But none of this could change the fact that Weston was, for all its beauty and tranquillity, a place of exile for the couple who were living there. Sacheverell had always lived a far more private life than Edith or Osbert. Marriage to Georgia had given him a separate existence from his siblings. Weston helped complete the process of withdrawal, and from now on they would have less and less influence on each other. They would still meet, of course, still passionately defend each other's work, still gather every August up at Renishaw, and still present a more or less united front against their parents or against the world when the times required it. Sacheverell and Osbert were sufficiently united still to form a front politically – by their support for the d'Annunzian superman, Sir Oswald Mosley, in the early days of the New Party. Sacheverell and Georgia had been fairly frequent visitors to the Mosleys' house at Denham, which was quite a centre for 'artistic' intellectuals at the time; Oliver Messel and Cecil Beaton were among the Mosleys' other frequent guests. And in August 1931 Osbert would open Renishaw for the big rally which launched the New Party, and during these early days Sacheverell would join with such unlikely bedfellows as John Strachey, C. E. M. Joad, Harold Nicolson and Peter Howard in campaigning for the new messiah of the Right.

But for Osbert and Sacheverell the years of real intimacy were over. Like it or not, the centre of Sacheverell's life henceforth was Weston, just as for Osbert it was Carlyle Square with David Horner. As for Edith, she was now on the verge of the greatest break of all with past and family and friends.

In March 1930 Edith's first full-length prose book – her biography, *Alexander Pope* – was published by Faber & Faber (of which T. S. Eliot was now a full director) to mixed reviews. It is a strange biography. It is a popular book, written quite clearly to make money, which she badly needed at the time, and staking no claims to original material or research. But while it gives nothing that is very new on Alexander Pope, it tells a great deal about Edith Sitwell. The essayist Robert Lynd, in a reasonably sympathetic review in the *Westminster Gazette*, pointed out the underlying bias of the book.

> Miss Sitwell seems to have begun her book on Pope under a curious misapprehension. The temper of her opening chapters suggests that she regards him as a man whom a base world tortured while he was alive and a still baser world belittles to-day.
>
> The truth is, of course, that the world – the English world, at any rate – has never treated any poet so well as it treated Pope, with the possible exception of Tennyson.

Certainly there is no mistaking the vehemence and passion with which Edith wields her literary machete on behalf of 'the small, unhappy, tortured creature, who is one of the greatest of our poets'. And as Lynd pointed out, all this posthumous mayhem was more than a shade uncalled-for and beside the point, at least on Pope's behalf. But Edith was not so much concerned with Pope as with her own emotions at the prospect of his life. 'I am moved almost to tears', she admitted, 'when I see Richardson's portrait of Pope in the National [Portrait] Gallery'.

And she laments that 'Only his small misshapen body, and the fits of fury to which the sufferings and illnesses of that body made him liable, have remained to us.' Change that word 'small' for 'large', and the sentence could apply, of course, to Edith. And when one looks at Richardson's portrait of the poet, one sees how her description of the painting also applied to herself – the 'visionary eyes that have a look of almost childish anguish and loneliness, the wide and beautiful brow, the worn cheek-bones, and the sensitive, pain-stricken mouth'. This was how Edith saw herself, and throughout her book this self-identification with Pope continues – the attacks of Grub Street, the hatred of his enemies, the insults he endured and the real nature of his 'genius'. '. . . being a man of genius, he is naturally able to deliver harder blows when attacked, than the majority of his aggressors are able to administer to him.' And so could Edith. Even in the introduction to this book on Pope she does not do too badly, using the occasion to attack a whole range of her favourite targets and Aunt Sallies, from Matthew Arnold's 'chilblained, mittened musings' and the effusions of the Squirearchy, to the 'shrill moronic cacklings of the Sur-Realists', and what she calls 'the Jaeger School of Poetry', rejoicing in the discovery of sex and 'advocating health at all costs'.

For her these variegated miscreants were all part and parcel of what she felt to be the unforgivable betrayal of the great tradition of English poetry. as symbolised by Pope and continued in herself, a poetry grounded, as she put it, in 'rhetoric and formalism' and not concerned with preaching messages, providing political panaceas, or to 'comfort the dying world', as the young poets of the thirties were already doing.

But being Edith, she could not be satisfied with simply defending what she felt to be the unsullied stream of English poetry. She also needed to defend herself under the guise of artist–genius rather as Osbert did in his 'A Few Days in An Author's Life'. She even managed to include one of his favourite phrases – 'artist-baiting is one of the national sports' – and it is this uncomfortable, slightly hysterical subjective note throughout the book that makes it so revealing. She was intent on battle – and it duly came. Rather surprisingly, the spearhead of the attack was launched in a review of *Alexander Pope* in the Sitwells' local newspaper, the *Yorkshire Post*.

Miss Sitwell asks for it. By her arrogance, her self-satisfaction, her superiority to mere critics and scholars, she exasperates the reviewer into forgetting his rules and his politenesses and retaliating with her own weapons. . . . The damning fault of this book is *suppressio veri*. This alone makes it negligible, though if there were space and cause it would be easy to pillory the manifold inaccuracies of Miss Sitwell, her ignorance of psychology, her occasional ineptitude in analysing even the technical qualities of Pope's poetry, her parrot repetitions, her tricks of fine writing which cover absence of original research and shallowness of thought.

And so on. It was not merely a damning and an extremely personal review, but also something of a portent. The tide was on the turn against the Sitwells. They who had always been so firmly on the side of youth were now becoming vulnerable to the young. The review was signed 'G.G.' These were the initials of a twenty-five-year-old poet, Geoffrey Grigson, and before long he would figure in her private demonology, only a slap below the unspeakable D. H. Lawrence and the unforgivably unforgiven Wyndham Lewis.

It was an aggressive summer for Edith as she began to vent her wrath at what she felt to be the conspiracy of the reviewers ranged against her – and Alexander Pope. To make things worse, Lord Beaverbrook was keeping up his old vendetta. It amused him now to bait the Sitwells, as he did through his music critic on the *Daily Express* when *Façade* was broadcast from a performance at the Central Hall, Westminster.

'"Sitwellian Orgy." Broadcast of Fantastic Inanities,' trumpeted the *Express*. Lord Copper was simply up to his old tricks again, but for Edith this was one more insult from the enemy, and in April she struck back – against the inoffensive, somewhat gushing popular lady novelist, Ethel Mannin.

Quite why she picked on Ethel Mannin is difficult to know – apart from a somewhat patronising article the lady wrote claiming to have known Miss Sitwell. In Edith's current state of mind this was gross *lèse-majesté*: all the power of her monumental rage went into the resounding slap she swung against the heedless lady. 'I do not want Miss Mannin's feelings to be hurt by the fact that I have never heard of her. . . . At the moment I am debarred from the pleasure of putting her in her place by the fact that she has not got one.' A good phrase – and so wasted on Miss Mannin that Edith would use it several times in years to come against far tougher targets. But all this rage and eloquent frustration seemed to be coinciding – as it usually did with Edith – with one of her mysterious bouts of lowered health. In July 1930 Edith was writing to Virginia Woolf, explaining somewhat vaguely that she had 'been ill, on and off, for weeks with all my glands poisoned. A horrible disease, believe me.' And in August Osbert was describing an even more alarming set of symptoms in a letter to David Horner:

It now transpires that poor Edith, who has been very odd in her health for months, has been suffering from concussion. She had a bad fall in March in her bedroom, and the idiot woman doctor she went to never treated her for it. She has a distinct small dent in her skull (forehead) and all the symptoms of sickness and loss of sense of balance are typical concussion symptoms. Then when she saw her Doctor Wilson, she never told him she had had a fall, as she did not want to own that she had seen another doctor. So like a woman! But she is better now.

It needed more than mere concussion to interrupt her social life – or stop her from delivering a slap to anyone she felt had been impertinent, however eminent, as Evelyn Waugh discovered late that June when he sat next to her at dinner at the Duchess of Marlborough's.

The dining-room was full of ghastly frescoes by G. H. Watts. Edith said that she thought they were by Lady Lavery. She talked mostly about Ethel Mannin's book and what she said and was going to say about what Mrs Mannin had said she had said. That evening [Dame Nellie] Melba said, 'I read your books, Miss Sitwell.' 'If it comes to that, Dame Melba, I have heard you sing.'

One might have cynically predicted – as several did – that Osbert's infatuation would not survive David's installation at Carlyle Square. The reverse was true. The more they were together, the happier they seemed – Osbert, at any rate. Two weeks after their return from Paris, Osbert could not resist writing a brief note from the St James's to David threatening that 'if I'm not wanted I shall decamp (if I may use that expression) and disappear'. There was no need for that. The experiment was working out. David may not have been as desperately in love as Osbert was with him, but he was enjoying life at Carlyle Square. He found Osbert endlessly kind and thoughtful and amusing and was becoming more and more attached to him. As for Osbert's feelings, they were clearly more than mere infatuation. Passion and love apart, David could do for him what no one had been able to since Sacheverell departed – amuse him and cheer up the day and help to banish the black rages and loneliness and paranoia of the past. In almost every way it was an ideal partnership – which was to last for more than thirty years.

But during these early days there were inevitable misunderstandings. These came partly from Osbert's possessiveness and fears of losing David, partly from David's equal fears of being overwhelmed by Osbert, and partly from the strain of keeping up appearances to the world outside.

At the end of June, for instance, there was a scene involving Emerald Cunard. Osbert was speaking to her on the telephone and David overheard him saying he was lonely. It was the sort of thing that cautious Osbert would say to someone as flamboyantly indiscreet as Lady Cunard, but David took offence. That same evening Osbert did his best to repair the damage, with an anxiously repentant note:

I am really deeply grieved that you should think me capable of ingratitude for your lovely and precious companionship; the chief joy I find in my life. You *must know* that when I said that I was lonely on the telephone, I did it *on purpose*; because I don't see why one should ever tell the truth – which is secret and a personal possession – to the Emeralds of this life. And I thought you knew me well enough – and you do – to know *why* I said it.

It's a sort of desecration to let people one doesn't want enter into the heart and its secrets. It has made me so unhappy this afternoon not to be able to explain properly, that's why I wrote. Bless you.

There were other problems too. High on the list of them loomed Renishaw and the family. As usual every August, the Sitwells would be gathering for their annual feudo-comic jamboree at Renishaw. Sir George and Lady Ida were already on their way from Tuscany (although for fiscal reasons Sir George would insist on sleeping at the Sitwell Arms). Guests were invited. Sacheverell and Georgia and Edith would be there to do their duty. It was unthinkable for Osbert as principal resident of Renishaw to duck it.

On the other hand, David had his own friends he was keen to visit in Kent and in the South of France. He insisted that he had to live his own life when he wanted to. Wisely, but with considerable effort, Osbert agreed. And so, at the end of that July, they parted – David for Tunbridge Wells, Osbert for Renishaw. David left first, and from the empty house in Carlyle Square Osbert scribbled him a note:

I never dreamt, until you came, that a house could be so uninhabited as mine is, when you go away. . . . I can never tell you how grateful I am to you for all that you make life for me. I love you more than you can ever imagine.

Renishaw that summer was *en fête* – and crammed with guests and visitors. Lady Ida was still a great inviter, and all her children seem to have decided that the best way to make the family ordeal bearable was to pad the place with friends. The turn-out was impressive. Young Evelyn Waugh turned up with his Oxford friend (and later bitter enemy), the traveller and writer Robert Byron. Waugh, who was firmly in the grip of desperate social aspirations, was predictably impressed by this sudden glimpse of *la vie de château*. According to Cyril Connolly, the Sitwells now became 'exactly the sort of aristocrats he [Waugh] always longed to be himself', and certainly his diaries for these few days offer the best account we have of Renishaw and the Sitwells in all their oddity and glory.

He was struck at once by the extraordinary contrasts of the house, the ugliness of the north front with its dark 'discoloured Derbyshire stone', the gloomy, pillared hall within, and the squalor and sheer ugliness of much of the surroundings, Sir George's ornamental lake 'black with coal dust' and his beloved gardens hemmed in by a prospect of pit-heads and slag-heaps. At the same time Waugh was impressed by the unexpected beauty of the individual rooms.

After all that he had heard about Sir George, his first contact with the man whom Edith was comparing now with Cesare Borgia must have been something of an anticlimax. Wearing a white tie and tail coat, he was, Waugh said, 'very gentle'. This was the occasion when Sir George startled Osbert and his guests by changing out of tails into a grey suit and going down to Eckington Church 'to observe the effects of the moonlight'. He was writing his book *Tales from My Native Village* at the time, and, meticulous as ever, wanted to make sure of one of his descriptions.

Waugh also found, like many other visitors to Renishaw, that 'The household was very full of plots. Almost everything was a secret and most of the conversations deliberately engineered in prosecution of some private joke.' Most of the jokes and plots seem to have come from Osbert and inevitably involved Sir George. It was Osbert who put the story round that one of the less humorous female guests was obsessed with polar exploration and ecclesiastical instruments: laboured and very puzzled dinner conversation followed between the lady and Sir George. It was Osbert too who led the conversation on to the subject of incest one night at dinner when his friend Lady Aberconway was sitting beside Sir George.

'And is there much incest in these parts, Sir George?' she asked wickedly; but Sir George was ready for that one.

'Quite an extraordinary amount, I'm told,' he said, without the flicker of an eye, and deftly changed the subject.

Waugh seems to have found Edith even more disconcerting here than she had been at dinner with the Marlboroughs. He described her as 'wholly ignorant' (perhaps 'unworldly' would have been a better word), and gave some strange examples of her conversation. 'We talked of slums. She said the poor streets of Scarborough are terrible, but that she did not think the fishermen took drugs very much. She also said that port was made with methylated spirit; she knew this for a fact because her charwoman told her.'

One feature of Renishaw that particularly impressed the bourgeois side of Evelyn Waugh was the level of what he called 'feudal familiarity' maintained between the Sitwells and their servants. 'A message was brought by footman to assembled family that her ladyship wanted to see Miss Edith upstairs. "I can't go. I've been with her all day. Osbert, you go." "Sachie you go." "Georgia you go" etc. Footman: "Well, come on. One of you's got to go." '

This sounds like Osbert's old soldier servant, the original, incorrigible Robins, who was now at Renishaw, to Osbert's considerable pleasure. Osbert enjoyed the outrageous sallies of those outspoken members of the lower orders, such as Robins and the enormous Henry Moat, and secretly encouraged them: they in their turn played up to the roles that he allotted them. It also sounds as if Lady Ida was having one of her frequent days in bed.

But the most intriguing of all Waugh's Renishaw entries is concerned with Osbert. He was, he writes, 'very shy' – but despite his shyness it seems clear that Osbert had already got Renishaw much as he wanted it. In later years he would always claim that it was a unique country house, tailored to meet his slightest whim as artist, hedonist and man of taste. That August of 1930 Renishaw already bore his imprint.

'Osbert's breakfast was large slices of pineapple and melon. No one else was allowed these. Osbert kept cigars and smoked them secretly [Waugh had to bring his own]. The recreations of the household were bathing, visiting houses, and Osbert's Walk.' This latter was a very lordly operation, reminiscent of Lord Beaverbrook's own patent method at the time of taking exercise. It consisted of 'driving in the car quarter of a mile to Eckington Woods, walking through them, about half an hour (with bracken), the car meeting him on the other side and taking him home.'

Osbert's theory all his life was that games and all forms of violent exercise were totally unnecessary, and that a grown man could keep healthy on one sharp twenty-minute walk a day.

Another oddity about the house that year was the almost total absence of alcohol – owing to Sir George's principles and Lady Ida's tendencies. There was no pre-lunch sherry up in Lady Ida's room as in the good old days; guests who desired a drink were forced to patronise the Sitwell golf club by the entrance to the estate in Renishaw village. This did nothing to deter the Sitwells' visitors, a variegated lot who came in droves. On 17 August Tom Driberg came – as Osbert reported dutifully to David in his daily letter. 'He's been very ill, and looking plain poor lamb. But longing to go down a coal-mine. For the same reason as you, I suppose. Very much on his best behaviour though.' And the following night there was the local vicar and his wife. 'She is just going to Madrid for the first time. Very nervous and behaving as if it was Sodom and Gomorrah, which, of course, it is.'

And the next day, up rolled Mrs Keppel. 'I think "Ginger" is deeply in love with her,' wrote Osbert hopefully. 'I wish they'd elope together.'

But from the letters Osbert wrote to David it is quite clear that Waugh was wrong about Osbert being shy. 'I miss you absolutely and infernally. You are all my life,' he wrote to David on 1 August. Not even pineapple and cigars were any consolation.

Everyone here [he wrote two days later], including the chef [according to Evelyn Waugh, imported especially from the Ritz] is at his worst and most tiresome. Meanwhile I grow thinner every day. Do tell me in your letter that you love me. I adored our last days. They were too heavenly. . . . Don't forget to gargle and clean your teeth as I told you.

As the day of David's departure for the South of France approached, Osbert's uneasiness increased. The one consolation was that, come autumn, they would be reunited to spend the winter months abroad. 'The idea of winter alone with you is heaven to me; what do you feel about it?' he asked anxiously.

Just before David left for France, Osbert had one brief chance to speak to him: 'It was lovely to hear your "basso profundo" voice on the telephone. I've been writing poems all day as a result. Such a relief after years of prose. Today, a typical Sunday. Church-bells and housemaids singing and Reresby screaming.'

There was one other serious distraction now to Osbert's love for David – his social life. It was becoming very grand indeed, and had to be kept completely separate from his life with David. Osbert had met the Duchess of York through the Sassoons, and a lifelong friendship had begun. He was the perfect unattached male guest to liven up a royal weekend, amusing and informed, discreet and yet totally at ease, with the perfect manners of an Edwardian courtier, learnt in his London life before the war.

It was a strange role for Osbert the poetic rebel, champion of the *avant-garde* and satirist of the establishment to play, and yet it was one that suited him and that he thoroughly enjoyed. He never had entirely reconciled his social ambitions with his artistic ones. During the early twenties they apparently went hand in hand, although even then there had been murmurings about the way he played the gentleman among the artists – and the artist among the gentlemen. But with the thirties Osbert's high-flying social life increased as steadily as his status as an artist and champion of the arts declined. His old romantic love of pomp and power, his taste for luxury and noble birth were powerful incentives now, and help to explain the change that started to affect him and his work.

The other influence was David Horner. He calmed Osbert down. The days of rage and broken plates were over. There was less and less need now to set off an avalanche of anger against a Noël Coward or a Scott Moncrieff, a Squirearchy, a philistine or a Golden Horde. And David enjoyed society himself. He was not possessive over Osbert's friendships with the great, and certainly his presence was no bar to them. Osbert went to Glamis alone and thoroughly enjoyed himself. Another house he went to on his own was Philip Sassoon's big house, Port Lympne, on the edge of Romney Marsh. Osbert had always liked Sassoon since the days he fagged for him at Eton. He also liked his opulent, extremely ugly house, its splendid gardens and the ultra-respectable *mélange* of cabinet ministers and society ladies which graced his guest-list. Osbert went there that autumn, straight from Renishaw, while David was still in France. But even at Port Lympne he missed him.

'I've been very nervous lately,' he told David on 2 September, 'and wake up every morning at dawn, sad and miserable.'

The lonely dawns would soon be over. Two weeks later Osbert was off to the Balearics for the beginning of the second of his 'winters of content' with David.

If any proof is needed of the extraordinary change which David wrought on Osbert's life and personality one need only look at his reaction now to publication of Wyndham Lewis's vast literary satire, *The Apes of God*. It appeared in June 1930, and contained not only the funniest and most insulting caricature of Osbert ever penned, but also the bitterest critique of all the Sitwells and their works. It was, as Sacheverell rightly said, a sort of literary time-bomb, set to explode in the face of literary London – and the Sitwells.

For some time Lewis had been moving steadily away from his old position as their friend and ally. He had quarrelled with Edith while she was sitting for her portrait – thus accounting for the fact that the picture, now in the Tate, is devoid of hands. (This annoyed Edith who had always felt her hands her finest feature.) He had also become steadily more paranoid from poverty and ultra-right-wing politics, and was increasingly embittered by the failure of rich friends and patrons – the Sitwells among them – to support him.

This bitterness helped to fuel *The Apes of God*, and all the Sitwells are recognisably included as the three enormous members of the Finnian Shaw family, with Osbert the most memorable of all, 'the famous Chelsea Star of "Gossip", Lord Osmund Finnian Shaw'. Lewis, like Huxley before him, makes the most of Osbert's appearance, cheerfully mocking his 'carefully-contained obesity', and then describing him with memorable malice.

In colour Lord Osmund was a pale coral, with flaxen hair brushed tightly back, his blond pencilled pap rising straight from his sloping forehead: galb-like wings to his nostrils – the goat-like profile of Edward the Peacemaker. The lips were curved. They were thickly profiled as though belonging to a moslem portrait of a stark-lipped sultan. His eyes, vacillating and easily discomfited, slanted down to the heavy curved nose. Eyes, nose, and lips contributed to one effect, so that they seemed one feature. It was the effect of the jouissant animal – the licking, eating, sniffing, fat-muzzled machine – dedicated to Wine, Womanry, and Free Verse-cum-soda-water.

Not content with this, Lewis proceeds to detail what he saw as the absurdities and grossness of his former friends. The essence of Lewis's complaint was that for Sitwells everything – art, friendship, even enmity – was all part of an enormous game,

> a sort of ill-acted Commedia dell' Arte.... Their theatre was always with them. Their enemies – Pantalones, comic servants, detestable opponents (whose perfidy disrepect malice or cabal they would signally frustrate – unmask them, knaves and coxcombs to a man!) always this shadowy cast was present.

It was a telling criticism which the opponents of the Sitwells would voice increasingly during the years ahead. So was his mockery of their former cult of youth.

> The Finnian Shaw family group I should describe as a sort of middle-aged *youth-movement*.... This they have become in their capacity of 'rebels' against authority. The danger of the War first must have driven them into that attitude. The idea of 'youth' supervened – afterwards. It coloured with a desirable advertisement-value their special brand of rich-man's gilded bolshevism.

This idea of the Sitwells as mere elegant poseurs posturing before their little court really started with *The Apes of God*. Others would take it up, but it was Lewis who was unquestionably the most damaging enemy they had. It is this that makes Osbert's reaction to the book so very strange. He who had been so horribly upset by Huxley, incensed by Noël Coward, outraged by Arlen, agonised by Bobbie Hale and appalled by Scott Moncrieff made no sign at first of having even read *The Apes of God*. He had better things to do that autumn; and when he finally did bestir himself to get his statutory revenge on Wyndham Lewis, it would be very different from his vendettas in the past. David would see to that.

CHAPTER SIXTEEN

A Sitwell in Paris

1930–1932

WERE THERE a French equivalent of Moscow Road and Bayswater during the 1930s it would have been the faded quarter sprawling each side of the Rue Saint-Dominique in the seventh arrondissement of Paris. Certainly the Rue Saint-Dominique was almost everything the tourist's view of that romantic city in those days was not – pinched, dingy, lower-middle-class, the world of Céline rather than the Sitwells – and No. 129 was no exception, a tall, anonymous grey house with a coal shop opposite.

Here, in a third floor flat, with brown walls, no internal sanitation and a view that ended in a leafless inner courtyard, Edith came to live. And it was here, while Osbert was seeking happiness with David and Sacheverell with Georgia, that Edith would find herself trapped in what she later called a life of 'unmitigated hell' until the eve of war in 1939.

And yet, ironically, she had come here believing that in Paris she would find the peace and simplicity her life had lacked in London. On 16 July 1931 the London *Evening News* was to publish an article by Edith which she entitled 'Life's Tyrannies – and my gospel of happiness'. It was an uplifting, rather sentimental piece, but there is no reason to believe that she was being anything but strictly honest with herself, as she wrote that,

> meditating on the subject of my few, but precious, possessions one day, I came to the belief that our two most precious possessions are the soul and the heart. Nor do I believe, with some people, that the last is a tyrant. It is more precious to love than to be loved. The soul if it is kept clear, the heart if it is kept warm, will preserve our youth long after physical old age has come upon us.

In the cuttings book she kept the last sentence of this article has been angrily scored out, but she has left the rest of what she wrote, including the following revealing passage.

Another of the worst tyrannies is the possession of that false and spurious society notoriety that *must* remain in the limelight. Oh! the boredom, oh! the fatigue, oh! the shackles of it! . . . And believe me, to be celebrated in one of the arts is the worst horror of all. The art without the celebrity is the ideal possession.

Coming from Edith, of all people, this sounds a little rich, but her behaviour around this time makes it quite clear that this was precisely what she thought. One side of Edith always would remain the shy, retiring country girl from Derbyshire, and, just as with Osbert, her over-sensitive nature could be easily appalled by the trouble her aggressiveness called down upon her head. And while, again like Osbert, she possessed the toper's grim addiction to publicity, there were times when she was forced to realise that it was bad for her, a waste of energy and time and talent which she could ill afford.

In London the old supportive life with Osbert and Sacheverell had gone and there seemed less and less point now in going on with tea and buns in Bayswater, when Tchelitchew was writing ardently from Paris that theirs was 'a friendship that has neither beginning nor end and which brings me enormous happiness and a limitless feeling of tenderness'.

'It is more precious to love than to be loved,' she had written, and Tchelitchew was echoing her thoughts. 'But nobody has ever understood you better, or come closer to you than I have and nobody ever will,' he told her tenderly.

She was needed by this man of genius. 'There is sometimes such hell in my soul', he wrote 'that it is frightful to touch it, and you know that there isn't a single person who can understand and console me more than you.'

Much as she loved both Osbert and Sacheverell, neither of them needed her like this, and all Edith's energetic nature rose up to help him. Just before Christmas 1930 she had already written peremptorily to Allen Tanner saying that he must never hesitate to tell her whenever he and Pavlik were in financial difficulties. 'It is a question of *is* one a real friend or is one *not*.' If there was any chance of a recurrence of their poverty of two years earlier, 'you must tell me *at once* so that it can be prevented'. She went on to explain the useful contacts she was lining up to help sell Pavlik's pictures. 'I'll introduce you to the Comtesse Anne Jules de Noailles, the daughter-in-law of the old girl who writes bad poetry.' She was also 'asking Eddie Marsh to tea to see some of Pavlik's pictures. Unfortunately Osbert, who has got a running fight with him because he is a fool about poetry, has just gone for him again – but he *does* buy pictures and is useful. I barely know him and don't like him, but that doesn't matter.' It was a potent sacrifice for Edith to ask anyone to tea whom she and Osbert both disliked – particularly as Sir Edward Marsh had evidently not forgotten Osbert's caricature in *Triple Fugue* – but it was one that she was glad to make for the beloved Boyar. She took considerable trouble to get her friend Cecil Harmsworth to 'see about' Tchelitchew's passport at the Foreign Office. And above all, she abjured Tanner, 'Don't let poor Pavlik be too wretchedly unhappy. We *shall* pull through, honestly we will. I *mean* us to.'

Brave words – and Edith was braver still now on the subject of the once admired Gertrude Stein, who had turned against Tchelitchew for good. 'As for Gertrude,' she told Tanner, 'what can she do? . . . If she starts any of her games with us, we shall just laugh. I refuse to be rattled in any way at all. Life is quite hard enough without allowing impotent mischief makers to make one sadder.'

True to her word, Edith was soon to stage an impressive demonstration of her own against the old 'mischief maker' of the Rue Fleurus to make quite clear her solidarity with Pavlik. The occasion was an important one for Edith, a poetry reading she had been asked to give at Sylvia Beach's influential bookshop at 12 Rue de l'Odéon, Shakespeare and Company. It was a full-scale gathering of Miss Beach's literary friends, including Aragon, James Joyce, and naturally the Misses Stein and Toklas, who had been attracted by the promise that Miss Sitwell would be reading an appropriate selection of Miss Stein's poetry.

Edith turned up, nervous but defiant, in a dress that Tchelitchew designed for her, and read from Shakespeare, the Elizabethans, and finally from her own poems. But, as everybody noticed, there was nothing by Miss Stein. One can imagine the strained demeanour of Miss Stein as the reading ended, and later, according to Allen Tanner, Edith received a note from the Rue Fleurus expressing 'in her own curiously indirect way' Miss Stein's resentment at the way she was neglected.

As far as Edith was concerned, it was a loyal gesture on Tchelitchew's behalf – although one doubts that it meant as much to him as it had to Edith. It would also prove a rash one. Miss Stein was an influential enemy, and a lot of doors in Paris would slam shut on Edith as a result of that evening at Miss Beach's bookshop.

Not that this seemed to matter at the time, for throughout most of 1931 Edith was still stuck fast in England, working on her second full-scale prose book – this time on Bath. She seems to have genuinely hated writing it, as she hated every book of prose she wrote, and undertook it solely to make money. She refused to visit Bath. That autumn she was telling Allanah Harper that she would soon have 'broken the back of my odious book, and not be in such a bewildered and hopeless state. I *do* want to write poetry again. After all, that is what I was born for.'

But born for poetry or not, she had to live. Her prose books made her money, her poems merely brought her fame, and now that she had found this source of income the poems stopped. In fact, this was a crucial and a dangerous moment for her – the moment when she was ceasing to be a poet and becoming a professional prose writer instead. Her life was increasingly expensive. Apart from her trips to Paris and the support that she gave Tchelitchew, Edith needed every penny she could earn to keep life going. Helen Rootham had been taken mysteriously ill and was plainly unable to work. They had a maid at Moscow Road who helped look after her; and a legacy that Edith had been counting on when her Aunt Florence died failed to materialise. Edith was not one to economise. Instead she grumbled at abandoning her poetry, and laboured on for seven hours a day to finish *Bath*.

At the same time she had another book to keep her busy, her earliest anthology, *The Pleasures of Poetry* in three volumes. In the introduction she explained how she began it in the spring of 1930: 'to enliven a tiresome illness of three months' duration, I longed to gather together some poems that I loved; and so I have picked them, with delight, as one picks wild flowers, and have made them into a country bunch.'

All this took time, and the anthology brought headaches of its own.

> There has been a hell of a row in the *Times Literary Supplement* [she told Allen Tanner]. I misquoted some Baudelaire in my 'Pleasures of Poetry'. As a matter of fact, as a poet, I'm convinced Baudelaire would have written that particular thing if he'd got it in his head, because the effect is very subtle and beautiful. I wrote,
>
> Mon ange, ma sœur
> Songe à la douceur
>
> instead of,
>
> Mon enfant, ma sœur
> Songe à la douceur.

The inevitable 'suburban spinster', as Edith described her, had noticed and had written a furious letter to the editor of the *T.L.S.* But Edith had her allies in high places, even then. One of them was T. S. Eliot's friend, the editor and bibliophile John Hayward, who was working in the *T.L.S.* office at the time. He seems to have intercepted the letter and took it straight to Eliot who, according to Edith, dashed down to the editor of the *T.L.S.*, Bruce Richmond, and said that he would not allow anyone to be rude to her – for it was obviously a mistake.

Despite the book on Bath, and the worries and continuing labours on the anthology, Edith still managed to share part of August 1931 with the Tchelitchew *ménage* in the country, and again in December she was back in Paris to see Christmas in with Tanner, Tchelitchew and Choura. Back in gloomy Moscow Road once more, she would describe this as the loneliest Christmas of her life. But much as she longed to be with Pavlik, and much as she adored her time in France, there still seemed too much tying her to England for Edith to be able to escape.

During the spring of 1932 Edith was constantly referring to the 'Pandemonium' that ruled her life. The 'Gingers' were troublesome again. Harold Monro (an old friend and founder of The Poetry Bookshop) had died; his widow was relying over-much on Edith for solace and support. Fabers were suggesting that she write yet another wretched book of prose – this time on a subject that she almost knew by heart, English eccentrics, and it was unlikely that she would resist the offer. There was even talk that she might write her memoirs.

Then, at the end of March, Edith was briefly and dramatically involved in one of the closing scenes of T. S. Eliot's collapsing marriage. She described what happened in a letter that she wrote to David Horner – with whom, incidentally, she was now on the best of terms.

> We had a very exciting time yesterday. A certain lady (Osbert will tell you who I mean) came to tea without her husband. As she entered, a strange smell as though four bottles of methylated spirits had been upset, entered also, followed, five minutes afterwards, by Georgia and her mother. Nellie, (the maid) who was once what is known as an Attendant, enquired if she might speak to me on the telephone, and, as soon as she got me outside, said (looking very frightened): 'If she starts anything, Miss, get her by the wrists, sit on her face, and don't let her bite you. Don't let her get near a looking-glass or near a window.' I said, 'What *do* you *mean*?' and she replied that what I thought was an accident with the M.S. was really the strongest drug given by attendants when nothing ordinary has any effect!! She concluded, gloomily: 'Often it has taken six of us to hold one down.' – You can imagine my feelings, and when I got back into the room, I found that Mrs Doble had offered the lady a cigarette, and had been told that the lady *never* accepted anything from strangers. It was too dangerous.
>
> Poor Mrs D. was terrified as she thought that the Patient was going to spring at her throat. Georgia was terrified too, and tea was undiluted hell.

Her letter forms a dramatic postscript to Osbert's account of T. S. Eliot and Vivienne. By now Vivienne had been in and out of several institutions, and the smell that Edith noticed was the ether to which she had become addicted. There was no need for the maid to have acted so impulsively. Vivienne was never violent, but she was now impossible to live with.

At the end of his unpublished notes on Eliot Osbert described how Edith met Vivienne one last time walking down Oxford Street that summer.

'Hullo, Vivienne!' she called to her.

Vivienne looked at her suspiciously and sadly, and replied, 'Who do you think you're addressing? I don't know you.'

'Don't be so silly, Vivienne: you know quite well who I am.'

Vivienne regarded her with profound melancholy for a moment, and then said, 'No, no: you don't know me. You have mistaken me *again* for that *terrible* woman who is so like me. . . . She is always getting me into trouble.'

Eliot would leave Vivienne by going to America – and by the spring of 1932 Edith had decided that the one solution to the turmoil of her life was to get away – which meant a lengthy period in Paris. In May she wrote to Allen Tanner announcing her decision. She would arrive at Madame Wiel's in the autumn; 'I shall be there for much longer than you know. As a matter of fact, I cannot stand this life much longer. I want to be where I can work quietly, and see the few real friends I possess.'

Her letter shows her disillusionment with London and the high hopes she had set on that golden life in Paris.

My parents are here! . . . My mother is more tiresome than ever, and the awful thing is, that not having seen her for eighteen months, and having been through a good bit since then, I haven't got the patience I used to have. All this life of gossip, malice and tiny 'interests' wears me out! I am hoping that in Paris, all our lives will be more peaceful.

She adds that, once in Paris, she will definitely 'arrange' things for Tanner and for Pavlik, as she could not do in London. This would include getting 'good paying engagements' with her friend, the Princesse de Polignac; meanwhile, in London she was already planning to have lunch with the great collector and Lady Aberconway's admirer, Samuel Courtauld, to get him to buy a picture or two.

This need to honour what she felt to be her obligations as Tchelitchew's guardian and unofficial patron was becoming very real, thanks to the world-wide economic slump of the early thirties. With the rich looking to their investments, buyers for paintings as unfashionable as Tchelitchew's were hard to come by. Edith had done her energetic best to peddle his wares in London, but for the most part those few who bought his work were writers and intellectuals like Geoffrey Gorer, Kenneth Clark and Rebecca West, and not the rich collectors. (The one shining exception to this was Osbert's friend Edward James, who was to build up the greatest collection of Tchelitchew's work in Britain at his house, West Dean, in Sussex.)

Edith's contact with men like Courtauld seems to have confirmed that growing, and very un-Sitwellian, dislike of wealthy society which she had first of all expressed in *Gold Coast Customs*. 'I *am* beginning to dislike the rich, heartless and completely selfish people,' she told Tanner after she had failed to convince a number of them to invest in Tchelitchew. But this reaction did not lead Edith on to the sort of socialistic protest that might have given her a link with the younger left-wing poets of the thirties. Her attitude towards the Slump and the disorders of her day was pained exasperation. 'I do so wish to heaven this nightmare would pass, and that the world would become sane again,' she wrote to Allen Tanner.

Fond hope – and cold economic winds were hitting her as well, and forcing her to change her plans. Helen's illness, still undiagnosed, was worse. She was so weak and in such pain from her back that there was obviously no question of her working. She was anxious now to stay with her sister, who could look after her, and also to consult the specialists in Paris.

That spring, when she had first decided to go off to Paris for a while, Edith had not dreamt of giving up her flat in Moscow Road. Now, without Helen's contribution to the rent, it was impossible to keep it on, but it was a dreadful wrench to give up her home where she had lived for eighteen years, and the letter that she wrote to Allen Tanner on 15 August was already full of gloom and real foreboding as she described how 'the horrors of turning out everything in the flat, storing most things and packing others, sorting manuscripts – and the sadness of leaving my home, became too much for me'. In this same letter she was already talking brightly of her return to England, and how she was hoping to replace her flat with a home of her own in the country, just as Sacheverell had done. Since her Aunt Florence had died, the old Tudor dower-house at Long Itchington in Warwickshire had stood empty. Like Weston it was being treated as part of the whole Sitwell patrimony and was controlled by Sir George, but she had stayed there for a few days after leaving Pembridge Mansions, and she added in a postscript to her letter, 'I may live in this house soonish and you will be dragged here to pay me long visits. What a lovely time we would have!'

The summer of 1932 witnessed a brief reunion in print of Osbert and Sacheverell. Sacheverell had had the temerity to publish a short, enthusiastic book on Mozart, hardly the most fashionable of composers at the time. More daring still, he had attacked Wagner for what he called his 'vulgar, tweed-clad tunes', only to call down upon his head the thunder of that great Wagnerian Ernest Newman, in the *Sunday Times*. It must have been unnerving for Sacheverell to have had to face such rolling charges of inaccuracy and imprecision for what was essentially a warm-hearted tribute by a keen Mozartian amateur; and Osbert leapt to his defence. A long-drawn-out and not particularly enlightening correspondence followed in the *Sunday Times*, and James Agate, the theatre critic, has left an interesting impression of lunch with Osbert at the height of the controversy:

Osbert tried to defend his brother, and I had not the courage of my true convictions, which is that both brothers are artists who enjoy pretending to be asses. Osbert talked all the time, and I never got a word in edgeways. I don't dislike him, though I can understand why people may. There is something self-satisfied and having to do with the Bourbons about him, which is annoying, although there is also something of the Crowned-head consciousness which is disarming. His sister can be the handsomest woman in London.

Edith would have been delighted with this unexpected compliment, and was even more delighted to be able to persuade her brother to join her in attacking a more deserving and more vulnerable target than the great panjandrum of the *Sunday Times* – the unspeakable Wyndham Lewis himself.

Although Osbert had been cured of the need for the flailing, self-destructive battles of the past, he was not entirely averse to a little well-directed malice when the time was right. A couple of years later he even set out, in a brief essay, what he called his 'Rules for Being Rude'. Written only partly tongue in cheek, Osbert's battle-manual taught that being rude should always be a conscious, carefully directed operation to inflict the greatest damage to the self-esteem and reputation of one's enemy. This required self-control – 'never allow yourself at the moment to show any emotion, least of all anger. Dislike should breed contempt, but never a rage.' It required detailed knowledge of one's enemy – you must 'make a careful study of his ways' and leave 'no opportunity for your quarry to escape'. You must keep him guessing, to 'add perplexity to insult'. But 'do not dash at it; remember you must play your opponent for weeks, as though he were a trout in some limpid stream'. Then when the time is right, strike, 'do not spare his feelings' and with luck he will go on turning your insult over in his mind; 'In this way the wound you have given him will fester, scar him for life'. All this was carefully followed in the campaign he and Edith waged against Wyndham Lewis.

Lewis was still suffering from persecution mania. True to the precepts of his essay, Osbert set out to make the most of it. At Renishaw he had found an old picture postcard of two anonymous actors mysteriously cloaked and hatted in the style of Lewis himself. Osbert had several dozen copies made – and for months to come would have them sent off to Lewis from all over Europe with disturbing messages. Edith joined in. Lewis had recently published his book on Hitler. Edith composed a telegram to Lewis which she had sent from Calais:

Percy Wyndham Lewis, 21 Percy Street etc. Achtung. Nicht hinauslehnen. Uniformed commissar man due. Stop. Better wireless help. Last night too late. Love. Ein Freund. Signed. Lewis Wyndham, 21 Percy Street.

More telegrams were sent in gibberish or German. Osbert despatched a broken tooth in a small box with Gerald du Maurier's card, and followed this with a letter to the stridently non-Jewish Wyndham Lewis purporting to come from the organiser of an exhibition of Jewish painters, inviting him to join. To Osbert's delight, Lewis became involved in a lengthy quarrel with the exhibition organiser, a man called Sieff, as a result – or so Edith said. And one must make allowances for her usual exaggeration for the sake of a good story.

As Osbert had realised at last, this form of quiet persecution was more amusing – not to say less wearing – than the passionate controversy he had relied on in the past. Perhaps it was unkind – but then, as Osbert would have said, Lewis 'had asked for it', and he was none too scrupulous a fighter himself. A more serious criticism is that Osbert should have known better than provoke a man like Lewis at the time when he thought it necessary to carry a revolver on the most ordinary social occasions. As Geoffrey Gorer says, 'almost anything could have happened'.

At all events, this method of controversy was one that ideally suited Osbert. It was a technique perfected over years of battle with Sir George, and from now on the big controversies Osbert had seriously indulged in gave way to this sort of carefully thought-out campaign.

This was not good enough for Edith. She enjoyed her stint of Lewis-baiting. 'And when I feel cross, which is often, I tease Wyndham Lewis', she told her cousin, Veronica Gilliat, that December. But her aggressiveness, her passionate addiction to controversy would not be satisfied with teasing. Lewis must be annihilated, or, failing this, held up to ridicule; and so, as she began those memoirs she had promised, she could not resist starting with Wyndham Lewis. This was to be as far as she would get, but she devised a good fighting title – 'The Ape of God – or Next to Godliness' – and this one chapter describes the short, stormy history of her own relations with the author of *The Apes of God*.

It begins with an account of how the 'involuntary recluse' of Adam and Eve Mews had started to paint her portrait at the beginning of 1921; his eye-patch, the air of mystery that surrounded him, his complexion, 'darker at some moments than at others'; and most surprisingly of all, she says that when he removed his famous eye-patch, 'and both eyes were uncovered, they wore a blinking look of yearning and reproachful affection, extremely disconcerting to the object of the gaze'.

Whether she really had been 'disconcerted' by Lewis's yearning for affection one will never know, but she certainly implied to several friends that the real reason why the sittings ceased – and her portrait remains handless to this day – is that Lewis, as they used to say, 'forgot himself'. And it is interesting that she ends 'The Ape of God' with the promise that in years to come, when he has long been dead, I mean more dead than usual, and I have passed my prime, I shall write the story of his life for those who may still remember him and in it I shall show how every little fault and every little mistake made by this fundamentally gentle and affectionate character is the result of the fear that engrosses him, the fear that he is not loved, the crushing impression that those on whom he has set his affections do not return them.

Not uninterestingly, she blames a lot of Lewis's own aggressiveness on the way he is constantly incited by his two young allies – 'noisy, frothing little Mr Roy Campbell (that typhoon in a beer-bottle) and little imperceptible Mr Cyril Connolly'. (The 'Osbert Sitwell's bum-boy' incident was over but emphatically not forgotten, and Connolly would be in the cold for several years to come.)

But all the time that Edith was so angrily composing her reply to *The Apes of God*, Osbert kept silent, and the one sign he gave that any of Lewis's insults had actually struck home concerned his weight. Lewis had mocked his 'carefully contained obesity': in the autumn, while David was off visiting friends in Switzerland, Osbert began to bant. On 17 October he wrote miserably: 'Today I'm lunching at Ranelagh, in my best plus-fours with Hazel [Lavery]: but never a drop to drink, nor a morsel to eat. What a life! And I'm coming out in spots from it.' But either from lack of food and alcohol – or lack of David – he lost nearly a stone in six unhappy days.

That autumn, as Osbert dieted, Edith cast off the hawser that had bound her life for eighteen years to London. There was no time for social life, or much enjoyment of the Paris scene, even if she could have afforded it. She was behindhand with her book, and as soon as she arrived at the Rue Saint-Dominique she plunged into a gruelling routine to finish it. Only a writer who has had to push everything aside to meet a deadline with a pot-boiler can know the sheer exasperation of this sort of chore. With Edith, and particularly with *The English Eccentrics*, one must admire the standards she maintained despite her lack of real enthusiasm; one must admire too the way that she succeeded in producing something so personal and so light-hearted in the end. But how she loathed writing it: 'I am getting on with that boring book about the Eccentrics,' she wrote to Allanah Harper, 'but oh, how glad I shall be when it is finished and I can get back to poetry.' And on 8 October she was telling Allen Tanner: 'I am very tired from working; but I am glad to be tired, as it prevents me from thinking about things that give me pain – leaving England, etc. . . . I am going to be an Un-Ornamental Hermit, as I have so much work to do. Also I have no clothes and can't afford any.'

It was a bad beginning for what should have been a time of happiness and freedom, and for once Edith was not exaggerating when she spoke about her hermit's life. Three weeks later Lady Aberconway was in Paris and, at Osbert's suggestion, tried to see her. 'Xtabel called on Edith, (she's not on the telephone)' he wrote to David on 28 October, 'but she was ill and couldn't see her. And instead she saw Helen Rootham, who was rather impertinent to her.' The truth was that Edith was not so much ill as suffering from a bout of acute nervous depression brought on by overwork, as she makes clear in a postscript that she scribbled to a note to Tanner on 31 October, apologising for not sending him Helen's book on Rimbaud: 'The reason the Rimbaud has not come yet is that I had been too nervous to brave the Post Office, where the staring, etc., terrifies me.' But there was more to it than that. No sooner had she arrived in Paris than everything conceivable had gone wrong. Life in the Rue Saint-Dominique had turned very sour. Worse still, as she must have realised, Edith was firmly trapped within it.

In the circumstances, there was some excuse for Helen to be 'rather impertinent' to Lady Aberconway that October. Not long before, the French doctors had finally discovered what was wrong with her. The lethargy and the mysterious pains were not gall-stones or lumbago, but the first stage of cancer of the spine. As it was diagnosed so early there was hope of a recovery, but she would need an operation very soon. This would cost money. Helen had little, Evelyn Wiel even less. Edith said that she would naturally help meet the bills.

And there was trouble too with Pavlik. In place of those idyllic days with Choura, Tanner and the Boyar which she had been looking forward to so much, there was dissension now *chez* Tchelitchew. Edith had probably not realised it, but the painter had been treating gentle Allen Tanner quite outrageously for months. He bullied him and was unfaithful to him. There were more scenes than usual when Edith called, and soon after she arrived Tanner had told her his decision. He was leaving Pavlik. Edith was up in arms at once. 'You *cannot* part company,' she wrote, 'you can't do it. He is absolutely devoted to you, and you are absolutely devoted to him.' They would be lost without each other, and for a while it seemed as if Edith's imprecations worked. Tanner and Tchelitchew stayed together, but it was an uneasy household now. The happiness had gone. Pavlik suspected her of siding with Allen Tanner and against him. Choura kept silent. As December came there seemed little prospect of a repetition of the blissfully happy Christmas Edith had spent with her Boyar only the year before. Edith was sometimes kept awake by Helen groaning in the night. Misery descended on the Rue Saint-Dominique.

CHAPTER SEVENTEEN

Angkor and After

1933–1934

'WHOEVER HAS the chance of seeing Angkor and doesn't is mad,' wrote Osbert. This was to become one of the great escapist phrases of the thirties – at a time when war was looming, millions were unemployed and the West seemed threatened with collapse as the Slump worsened into 1933. Ten years earlier Osbert would have been up in arms against the chaos and appalling prospect of another war, but his days of angry protest were behind him. He had become more realistic and more pessimistic. There was no point in fighting the inevitable any more. He would leave that to others. The nearest that he came to political involvement was when he lent Renishaw for a Mosleyite rally. He had his very private, privileged own life to lead, and he would make the most of it while it lasted, as perhaps a man of taste and culture should.

For the remainder of the thirties this would become his private creed, as he and David set out to enjoy their freedom and good fortune in an ailing world. He made no attempt to justify himself. Why should he? *If you can go to Angkor and you don't, you're mad.* It was as simple as that, and in November 1933, true to his beliefs, he took a French boat from Marseilles with David Horner: three weeks later they were in Saigon.

Sacheverell and Georgia also had their escape routes away from Weston. They had a fairly active social life in London, and 1932 began with them seeing the New Year in at a riotous party with their old friends, the Mosleys. But active politics were not Sacheverell's style (any more than they were Osbert's), and the more determinedly Mosley moved away from the New Party towards active Fascism, complete with uniforms and violence, the less Sacheverell would see of him. More to his taste – and

Georgia's – was the Duke of Westminster, with his enormous country house at Eaton Hall, and his yacht to match – a converted Royal Navy destroyer. It was an unlikely friendship for one of Sacheverell's temperament to strike up with a rich and cantankerous character like 'Bendor' Westminster. It was more unlikely still to have endured. The Duke drank brandy which was bad for him, and few of his friendships lasted. He was a bully and an aggressively jealous husband to his young wife Loelia. He was restless, bored, and ultimately beastly to everyone around him, but for some reason the Sacheverell Sitwells kept the old monster happy. Georgia was a great friend of the Duchess while Sacheverell, with his learning, wit and gentleness, was simply not a person to be bullied. At the same time the Westminsters, with their extraordinary wealth, could offer them a taste of the rich and racy world not to be found at Weston. In the spring of 1933 they sailed for Egypt with the Westminsters, stopping briefly at Alexandria, then went on up the Nile to the Sudan.

Edith alone of the trio had no escape – save back to England. During the first half of 1933 she made one short trip to London, where she helped Pavlik stage his London show at Tooth's Gallery. Osbert composed the introduction to the catalogue and, largely thanks to the support of the Sitwells and his friends, Tchelitchew enjoyed a success that he had never known in Paris. 'Sam Courtauld says Pavlik is absolutely made, and his only danger is that he may be too much of a fashionable success,' Edith wrote prophetically to Tanner. During the remainder of the year Edith stayed on in Paris, slaving at the *Eccentrics*, helping to look after Helen Rootham, and 'longing', as she told her cousin, Veronica Gilliat, 'for my exile to be over'. She bore her sufferings courageously, but not in silence. Helen Rootham, brave though she was, was getting on her nerves. By June it was clear that the cancer, far from being cured, was back again, and Helen had to stay in bed. The lumbering Evelyn Wiel seems to have exasperated Edith in the tiny flat, and on 27 June Edith was moaning to Allen Tanner that it was no fun 'to do all the housework and marketing, and then, at night, have the poor creature's nightmares'.

It is hard to think of Edith tackling the housework very seriously. A graver source of anguish now was Pavlik. He was neglecting her quite shamefully: worse still, as she complained to Tanner, he had just sent an anonymous young man round to see her to read a novel he had written. The final insult came when the good-looking youth admitted that he had read none of Edith's poetry.

Edith came back to Renishaw that August, but the annual gathering of the family seems to have been more than usually contentious. Once again problems of money – or the lack of it – hung over Renishaw like the coal-dust from the pits. This year there was no chef from the Ritz Hotel, but this did not seem to stop the usual flood of quite disparate guests from joining in the antics of the family.

Osbert had invited Richard Wyndham. 'A most unfortunate skin, Wyndham's,' remarked Sir George before slipping off to bed to avoid meeting it across the dining-table. And Lady Ida, not to be outdone, had asked a group of Greeks after one of them had found her purse in Piccadilly. According to Edith, one of them claimed to be aide-de-camp to the ex-king of Greece – 'a bald, watchful man of a most unexampled boringness. Osbert and Sacheverell won't talk to him.'

Osbert was, as usual, missing David, but dared not phone him 'as Ginger is always on the prowl and on about telephone bills'. He was, he added, 'worse than I've ever known him and has been horribly beastly – and without reason – to Georgia and Sachie. Next week, I can see, he plans to start on me. But my lawyer's letter (which I've seen) should arrive on Tuesday and give him a jump. It is very good.' The Slump had obviously brought most of the family's financial battles to a head. Osbert's cause of trouble with Sir George seems to have been that, when the original trusts were set up in 1925 and he took over Renishaw, he was guaranteed £2,000 a year from various investments. Now their value had declined to a mere £200 a year, and Sir George was refusing to make up the difference. 'I *am so* tired: absolutely tired of Ginger and Mama, and long for a little rest,' Osbert wrote wearily to David on 23 August. 'One great bore is that Ginger is now staying in England till October: it's supposed, of course, to be my fault for not giving way about that £2,000.'

As for Sacheverell and Georgia, their chief conflict with the baronet was, according to Edith, over his attempt to 'arrange the settlement so that Georgia would get nothing' in the event of widowhood (a truly Ginger-like concern with the remotest of eventualities).

As usual with the Sitwells, these testamentary battles were conducted in an atmosphere of almost oriental plotting and intrigue behind a screen of careful courtesy. Perhaps Sir George did 'jump' when he received the lawyer's letter from his eldest son; but he relished lawyers' letters, and undoubtedly took his time composing a suitable reply. Wills and settlements were of the essence of the power-games he played with all his children. What better way was there to register the disapproval that he felt for them, what surer way of having the last word than in a will? 'Osbert and Sachie have definitely been cut out of Father's will,' Edith told Geoffrey Gorer when the holiday was over, 'and the dear old Baronet has left, attached to his will, a long list of all our misdeeds.' With so much angry plotting in the air, one is not surprised to hear that Edith's attempt to finish a long poem – her first for nearly two years – had not been too successful. It was impossible to write poetry in Paris. Renishaw that year was not much better. 'I am writing what *should* be a satisfactory poem,' she told Allen Tanner, 'but am interrupted every other moment and on a trivial pretext so that it is impossible to see the poem whole.'

The poem, which she called 'Romance', appeared in November 1933 in *Five Variations on a Theme*, with two old poems and two new songs – the only poetry she published during her years in Paris.

For Edith and for Osbert there was an interlude of a few weeks before David returned from Le Pradet and Edith to her bed of nails in the Rue Saint-Dominique. When at last August was over, Osbert spent his customary few days at Port Lympne, a welcome change from Renishaw, though he was missing David more than ever: 'I do miss you. I suppose I am very worried, for I sleep badly, and dream about Ginger and last night dreamt that I howled in my sleep: feel rather better for it.' To his horror he discovered that Lord Berners was among the guests and sleeping (and snoring) in the room next door. 'I threw books and books at intervals throughout the night,' he told David, 'but without result.' He was more amused by an R.A.F. officer staying with Sassoon. He 'explained to me that a lot of things one didn't know about went on in London, and that the other evening, when he was driving down Piccadilly in his car (it was a wet night) he had observed two constables holding hands under their mackintoshes. He seemed surprised!'

After Port Lympne Osbert spent a few days visiting around the country with Edith for company. They called on the Harold Nicolsons 'at Sissy Hurst (or whatever it is called): rather lovely, but such an "author's home",' he wrote to David. 'Vita's secretary wears trousers,' he added.

They went to Lady Aberconway's house at Bodnant, then on to Badminton, where they stayed a few impressive days with their Beaufort cousins, thus giving Edith an occasion to air her regal aspirations in the grandest manner. 'Mary Beaufort, my younger cousin-by-marriage is delightful,' she told Allen Tanner, '(although she is the Queen's niece and dangerously like her.)' This was not written simply to impress the impressionable American, but Edith knew that any letter that she wrote him always found its way to Tchelitchew, and no one was snobbier about ancestry than faithless Pavlik.

More interestingly, the Beauforts had asked Edith to create a pageant for them at Badminton the following summer, and Edith was quite taken with the whole idea.

I do think it will be fun [she wrote to Geoffrey Gorer]. I'll try to turn out some sort of Ben Jonson-like masque. It's an incredibly lovely and romantic house and park, and just the date I happen to know something about, so we ought to have great fun. But I do hope they won't want me to write 'parts' for the hounds.

Then, as October ended, Edith returned to Paris. 'I hope nothing will be *quite* as unhappy as this last year has been,' she wrote anxiously to Tanner. 'I could not live through it again.'

But Edith had to, for there was now no escape. That summer's quarrels over money with Sir George had quite destroyed whatever hope she had of help from the family. The dream of settling into the house at Long Itchington was over. It was the Rue Saint-Dominique or nothing, for it was only there that she could live in suitable economy and write the books she and Helen Rootham depended on for their survival. It was not a situation to bring out the best in any writer, least of all Edith. When she arrived at the beginning of November, Helen was worse, and the illness built up for the remainder of the month until she had what Edith described as a 'terrible attack' a few weeks before Christmas. She was, Edith told her friend Charlotte Haldane, 'one of the bravest women I have ever known', but the suffering and tension in the flat had reached the level of a constant nightmare. Tchelitchew's behaviour only made it worse.

Edith's instincts had been right, for once, when she took umbrage at the young novelist Pavlik had sent to her to read his work. His name was Charles Henri Ford. He came of an old but impecunious Southern family. He was in his early twenties, with extraordinary sky-blue eyes which, according to his lover, Djuna Barnes, were so large that they 'go around the sides of his head like an animal's'. Originally he came to Paris in an attempt to interest Gertrude Stein in a novel he had written in conjunction with Tchelitchew's future biographer, Parker Tyler. It was entitled, more than a shade self-consciously, *The Young and Evil*: Tyler describes it as 'an obstreperously naughty book' filled with 'crime-tinged amorous intrigues'. It was hardly likely to appeal to Edith.

For a while it had seemed as if Miss Stein would help publish the book, and it was through her that Ford met Tchelitchew. Tchelitchew became infatuated with this beautiful American who reminded him, he said, of Huckleberry Finn, and almost immediately Ford took the place of Edith as the painter's favourite model. Worse followed. When Gertrude Stein refused to have any more to do with Ford after her own break with Tchelitchew, it was Tchelitchew himself who took up the cause of the young man's writing – and then naïvely hoped that Edith could help with publication of *The Young and Evil*. In the autumn of 1933 Ford decided to throw in his lot with Tchelitchew for good. As Parker Tyler puts it: 'With his practical vision, Charles is sure that he wants this greatly gifted painter, only without his sister and his friend Tanner. His campaign is therefore obvious: the domestic combination must be broken up – and soon.'

All this cut Edith to the quick. For nearly five years she had been in love not just with Tchelitchew but with the small, adoring group around him. She had found happiness there. In a strange way she had even managed to express her love for Tchelitchew by making his lover, Allen Tanner, her confidant and dearest friend. Now, by deserting him, Pavlik was deserting her, and she reacted as if she, not Tanner, were the injured party.

I have just received a letter from Pavlik which has hurt me terribly [she had told Tanner at the beginning of October]. I did not need this extra blow to convince me, to prove to me, that he has lost any affection he ever had for me – if indeed he ever had any. When he has been betrayed by the last of the cheap, worthless people by whom he is so dazzled, perhaps he will realise the kind of friendship that mine is, and with what loving care I have watched over him and tried to help him in his worries and comfort him. But who knows but that it will then be too late.

By November she was writing bitterly to Geoffrey Gorer about Pavlik:

I hardly ever see him now, though, as he is too occupied with that creature who wrote the immortal book we know. He is always making excuses not to see me: first the *chauffage* went wrong, then Choura had influenza, then he had a cold. He then wrote and told me that it was so sad our both being in the same town and only seeing each other once a week – and I found out that in spite of all the disasters I have mentioned, the genius had spent his whole time there, every day!! I am not unnaturally a little hurt. I am also cross at being supposed to be a fool.

Hardly surprisingly, Edith's jealousy, and all the suffering around her, brought out a crop of ailments. At Renishaw, according to Osbert, she had already suffered three abscesses on the gums. Since then, as she wrote to Charlotte Haldane at the beginning of December, her health had steadily deteriorated:

for about three weeks now I have been having the most fearful headaches which are getting worse and worse. Why don't I see a doctor? Because he would tell me to take exercise, ride on horseback, eat cabbage, drink (like one unfortunate friend of mine) lukewarm Vichy water instead of early morning tea, lead a healthy life, sleep with the window open, and most of all, *not think about anything*. Ah! I know them!

One effect of all this suffering in Edith's private life was to make her appreciate the sufferings of others. At the end of the same letter, Edith wrote:

I've been feeling terribly unhappy lately. . . . I've seen people behaving in such a dreadfully ugly way lately. And I never can get used to it. I mean in my personal life it has been very bad, oh very bad, and outside, these horrible lynchings in America and the departure last week of the French convict ship for Devil's Island have upset me terribly. Nineteen hundred and thirty three years after Christ, and the 'righteous' are still behaving like this.

Her sympathy – and anger – were likewise roused by what was happening in Germany (in contrast with Osbert, who had visited Germany earlier that year without noticeable disquiet).

Indeed you are right [she wrote to Charlotte Haldane now], in speaking about the primal horror of Germany at present. It makes my blood boil. I did not know that you are a Jewess, but with the exception of one person, all my dearest friends – (I have very few) are either Jewish or part Jewish. So apart from the rage and horror that I feel in the cruelty of the Nazis, I understand the criticism of it. . . . I know the anguish you must feel in contemplating Germany.

But Edith was not one for saintly resignation in the face of suffering, and relations with Helen and her sister were becoming difficult. Edith complained to Geoffrey Gorer that they were treating her as if she were 'slightly insane' – 'I *know* I look wild and odd when I've just been writing: but they'd look wild and odd too if they wrote my poetry.' True to her nature, she also vented her frustrations on several of her favourite human targets. This was the time when her persecution of the unfortunate Wyndham Lewis was at its height, and Edith was always ready with a little ridicule for that 'impotent mischief-maker', Gertrude Stein. As she gleefully told Charlotte Haldane:

A lecture was given at the American Women's Club here the other day on the subject of G. Stein, and I am *told*, though I cannot vouch for the truth of the rumour, that G. wore a white Grecian robe, with her various chests surrounded by a gold – cestus, I think is the word – anyhow you will know what I mean. It must have been an impressive spectacle, and I can't think how she managed about her lingerie – for she normally wears dark grey knitted plus-fours under everything.

But soon not even honing her wit on Gertrude Stein could alleviate her wretchedness, such was the point she had reached in her relationship with Tchelitchew. In the spring of 1934 Tchelitchew finally left Allen Tanner and began the permanent relationship with Charles Henri Ford which lasted until his death. Edith was shattered – and she wrote to Tanner like a sister gallantly sharing his bereavement.

My dearest, very dear Allen,

I cannot come because I have no smart clothes, but tomorrow I am going to buy some stuff for a dress and have it made by a cheap dressmaker, and *then* we will go out together.

As for the wretched, faithless Pavlik,

I realise now that on the rare occasions when he attempts to behave with decency, or even ordinary politeness to me, it is only that he realises that, with your one exception, I am the only person on whom he can rely – well, I will *never* let him down . . . but he has murdered my love for him. He need never fear that I will revenge myself on him in any base or mean way. But at present I cannot bring myself to see him. I shall not come for a sitting on Friday, and he can *wait* for me.

Poor Pavlik. He is a terrible fool, – a real fool! A very great artist indeed, – as great an artist in a different medium as he is, and someone who loves me, – (not one of my brothers) said to me, when I was in England – 'What a fool you are, Edith, to waste yourself on someone who does not know what you are.' And then he paid me the greatest compliment that I think I have ever had. He said, 'Any man who was an artist could do great work if you were beside him.'

This was how Edith saw herself in her love for Tchelitchew, selflessly and nobly tending his genius. It was a role that she would seek to play many times again with other geniuses throughout her life, but never with the passion and devotion – or the misery – with which she had tried to play handmaiden to the Boyar. (As for the tantalising identity of the 'very great artist', this must have been T. S. Eliot. She always saw him during her trips to London, and he was one of the few men in England who would have measured up to Tchelitchew in Edith's eyes. Whether he actually said this is another matter.)

She obviously hoped that, even now, the letters that she wrote to Tanner would find their way – as indeed they did – to Tchelitchew, but this was the nearest that she ever came to actually reproaching him. She did her best to deflect the hatred that was in her heart on to the one person she could blame for what had happened – Charles Henri Ford. When *The Young and Evil* was finally published in the spring of 1934, she wrote bitterly to Allen Tanner that the book was 'entirely without soul, like a dead fish stinking in hell', and suggested she should send the author over to England '*with that book*, and I'll see that he gets three months' board and lodging free from the moment that he lands! Do you understand me? Just one little word from me, and what a reception from the authorities.'

That winter, which was such a wretched one for Edith, proved one of the happiest of Osbert's life. The Orient entranced him. He was with David. Europe and its worries were behind him, and he was travelling, as he explained, for one thing only – to enjoy himself. He was convinced that little of what he saw would last, and in China he particularly wanted to see 'the wonderful beauty of the system of life it incorporated before it should perish'.

Angkor impressed him vastly. It was, he wrote, 'the chief wonder of the world today, one of the summits to which human genius has aspired in stone, infinitely more impressive, lovely and, as well, romantic, than anything that can be seen in China'. He also liked the monkeys that swarmed across the great deserted temple. They inspired him to write a poem for David which shows the Osbert Sitwell that was hidden behind the aesthete mandarin and angry man of taste, and still conveys a little of the fun of his pre-war journeys.

> Oh, tell me, Aunt Ermintrude
> Is that my Cousin Horner?
> I think he walks most gracefully
> And neatly takes each corner.
>
> How like he is to Aunt Loo, too
> And also Uncle Joe,
> They say his pet name is Boo-Boo
> (Tho' others say Bo-Bo).
>
> He has a look, do you not think
> Of Grannie and Aunt Millie
> Though they, I see, are much less pink,
> . . . And don't look half so silly.
>
> Oh look at him, the swanky beast
> The airs he's putting on!
> The clothes and shoes and all the rest
> And no blue sit-upon.
>
> And just because you climb the trees
> He will not look at you.
> And yet he has as many fleas,
> And is as hairy too.

After Angkor, he and David were in Peking to welcome 1934. Harold Acton was there – spectacularly changed from the flamboyant dandy-aesthete of his Oxford days into a Chinese man of letters, and more at home in China than in Italy. Such drastic transformations were not for Osbert, but largely thanks to Acton and to his friend, the eminent sinologist, Laurence Sickman, the two visitors were in a privileged position. For three months they rented a small house in the middle of the Tartar City, and it amused Osbert to spend each morning writing his latest book – on Brighton (in collaboration with Margaret Barton) – while the winds brought the sands from the Gobi Desert on to his work-table. Through Acton they had the entrée to a Peking which has vanished utterly, so that Osbert's description of these three months in the city makes his book *Escape with Me!* more than a mere essay in escapism. It is one of the last and most readable eye-witness accounts of the Peking of the *ancien régime* and, like the best of Osbert's travel books, is instinct with enjoyment. He relished his good fortune while he had it, and it is this that almost justifies

Hugh Walpole's hailing of the book as one of the half-dozen best books of travel in English of the previous half-century.

When May came, and the peonies were over, it was time to leave. One of the last visits Osbert made before departing was to the ancient college of the imperial eunuchs. Osbert spent some time talking to the oldest of the eunuchs, a wrinkled, hairless man with a piping voice and an inquisitive manner. 'Tell me, young man,' the old castrato said, 'do you have no group of people like us where you come from?'

Osbert thought a while, then answered gravely, 'Yes, indeed we have. We call it Bloomsbury.'

Sacheverell's life was changing now, as he began to turn his phenomenal industry and powers of assimilation towards biography. Early in 1934 he published his monumental life of Liszt. Like his *Mozart*, this is the book of an enthusiast, but it is also a much more solid production and was not vulnerable to the criticisms that had assailed his earlier book. Today it still remains the best popular introduction to that extraordinary composer but, like Edith's book on Bath, it seems something of a waste of a unique talent. One could wish that he too had visited Angkor (as he did some twenty-five years later) and had been able to write another *Southern Baroque Art* on the Orient – something more substantial than the six sketches entitled *Touching the Orient*. But he was more or less tied to Weston and professional authorship. After his book of verse, *Canons of Giant Art*, was published at the end of 1933, he would write no further poetry for publication for the next forty years. Instead, he prepared to write a further book on another 'lost' composer – Domenico Scarlatti – to coincide with the 250th anniversary of his birth in 1935.

His reason for this apparent abdication as a poet was that he felt that the times were against him, 'and there seemed simply no point in going on getting kicked in the pants by the critics'.

Just before Osbert left Peking he received one of the rare letters Edith wrote to him. It was from Levanto, a small holiday resort between La Spezia and Rapallo on the Ligurian Coast, and scrawled across the top of the letter was the dire admonition: 'Do not let the Gingers know I am here.'

My Darling Old Boy,

... I came here a few days ago because poor Helen had to have some sun. The sea is lovely but the town is a triumph of ugliness, rather like the outskirts of Nottingham. However the hotel is nice, the English people haven't found out who I am – or if they have they don't care – and I can work quietly. I am at work at (besides that bloody Victoria) a small book on modern poetry in which I am simply going to 'let the devils have it!'

The other night I went to Cocteau's play on the subject of Oedipus Rex. I don't want to be unkind, but I should have thought *that* play *had* been written. But no, Cocteau comes along with his little bourgeois mind, and dwarfs it from a grand tragedy into nothing at all. The Queen isn't like a Queen, she is like 'la Patronne'. There is a ridiculous phosphorescent ghost that moans and groans, the sphinx is a dear little chorus girl with curly hair. And in the end, when the afflicted pair finally amble off the stage, I was more pleased to get rid of them than I was to see the last of any pair barring the Gingers. The play is what one might describe as 'mignonne'.

It is an interesting letter for what it omits as well as for what it says. It shows how little Edith actually confided in Osbert on the subject of her private life. There was no mention of her fears and bitterness for Tchelitchew, or the latest events in her miserable private life in Paris. For these affairs one turns to the letters she still wrote to Allen Tanner, all of which make it clear that one of her reasons for leaving Paris for Levanto was to cut her emotional losses with Pavlik, and forget him in her work:

Perhaps Pavlik did not realise, when he said goodbye to me, that he will probably not see me again for a year. For when I return to Paris he will be in England, and when I go to England, he will be in Italy, and I suppose he will be going to America in September before I return from England. But I hope that you will be there all the time.

In the same letter from Levanto she speaks of it being 'hot and drowsy' and says that she is so busy with the new book she is writing about poetry that she can think of nothing else. It is turning out to be a series of short essays on her favourite poems – and favourite enemies; 'and when it comes out', she told Tanner gleefully, 'I shall probably be lynched'.

Edith received one unexpected bonus early in 1934. Thanks to the advocacy – and, indeed, insistence – of her friend and admirer of her poetry, W. B. Yeats, the Royal Society of Literature awarded her its medal for poetry. At a time when it seemed as if her work was already out of fashion, and it appeared unlikely that she would ever find the time or peace of mind for poetry again, it was a reassuring act which meant a lot to Edith.

As for Osbert, he returned from China to a multitude of minor – and some quite major – irritations. William Walton had finally left Carlyle Square. By now his reputation was secure. His need for the Sitwell patronage was almost over, and it was clearly best for everyone that he should take his leave. But Osbert was incensed when he discovered later that he had started a romance with Lady Wimborne and gone to live with her. Osbert was jealous of his rich society ladies. To lose one to a protégé like Walton was upsetting, and Walton would never really be forgiven.

More damaging to Osbert's reputation – if not to his *amour propre* – was yet another literary torpedo that came homing in on him that summer.

Like most such missiles it was launched by an affronted former friend – this time the Keatsian war poet, the man whom Edith once compared with Dante, Robert Nichols. Since the great days of his *réclame* as the fortunes of the impassioned poet–hero, the author of *Ardours and Endurances* had been falling steadily. He was ill, and Osbert had been briefly rude about him when he joined the Squirearchy and contributed to the *London Mercury*. This affront – and the sight of Osbert's continuing career as a bounding social figure and successful man of letters – seems to have festered in poor Nichols's disappointed brain. Osbert clearly stood for everything that a failure most resented in the London literary scene – conceit, publicity-seeking, snobbery, the ruthless use of anyone to boost a reputation, and terrible ingratitude. Somehow this managed to sustain Nichols through an entire book of rhyming couplets, based, it would seem, on Pope's satirical 'Letter to Arbuthnot', denouncing and exposing Osbert Sitwell. He called him – and the poem – 'Fisbo'.

Like everything that Nichols wrote, there is too much of it, and he was emphatically no Alexander Pope. But as one wades on through this endless sludge of bitterness one gradually accumulates the debris of a case against Osbert. And some of it, as Walton said of Hernia Whittlebot, is not unfunny. Where Wyndham Lewis was profoundly critical of Osbert – and most of what he stood for – Nichols is merely mocking, but a good deal of the mockery strikes home. There is, for instance, the description of plump Fisbo's home:

> A Regency bed, wax fruit à la Victoria,
> Three chairs constructed from the bones of sauria,
> A poor Picabia, a worse Kandinsky,
> A caricature in waxwork of Nijinsky,

There is the hypocrisy of Fisbo,

> Who, vowed to register beyond dubiety
> Your attitude to an effete Society,
> Have, in the fury of your will to grieve it,
> Performed all possible outrage save to leave it,

And there is the status of Fisbo as a poet:

> To-day a poet is not read – he's seen
> And when unseen (too sick, say, to go out),
> He is not read – Lord, no – he's read about.

There is a great deal more. By the end of this exhausting poem it seems that there is not one of Osbert's weaknesses or sillinesses or past failures that Nichols has not managed to drag in. Even five years earlier Osbert would have exploded in a shower of law-suits, frantic letters and denunciations in the press – as Nichols evidently longed for him to do again.

But Osbert had learned a lot in the last five years. Nichols was met with total silence from Carlyle Square.

In one respect, at least, age was making Osbert and Edith increasingly like their father – in their obsession with their health. Edith had her abscesses, her back, her headaches and her occasional bouts of suspected appendicitis; and during August, after David had left for a holiday in Corsica, Osbert was soon complaining of insomnia, bad headaches, fears for his heart and shooting pains in his legs. He was also, as he still realised, considerably overweight.

To almost everyone's relief, Sir George decided not to come to Renishaw for August. Sacheverell and Georgia were off again in the Westminsters' ever-ready yacht, and so, to pass the time until reunion with David, Osbert decided to treat Edith to a brief therapeutic holiday – taking a cure at Brides-les-Bains in the Savoie. Fond though they were of one another, Edith was still no substitute for David. 'I began my cure on Tuesday,' he wrote in purple ink to the beloved on 16 August, 'getting here on Monday night, and am nervous and dispirited, and ready to cry at sight of a falling leaf.' His doctor, he went on, while wanting him to lose *some* weight, counselled against over-violent dieting which might easily bring on a nervous breakdown. Nothing was right – not even the hotel which was crammed with 'odious children and great swollen French bags'.

As for Edith, although they went one evening to the cinema 'to see Mr Paul Muni', she was not exactly lively company. As he explained to David: 'I hardly ever see her 'cept at meals, which are short: for she can't walk much, so I go for long alpine trudges by myself.' Edith, in point of fact, was desperately engaged in finishing a book Faber had persuaded her to write on 'that old bore', 'that bloody Victoria', and there was just one occasion when there was a flicker of the old rumbustious double act of years gone by. Osbert described it in a letter to David on 18 August:

Dearest Sambo,

Thank you for your letter from darling Toulouse. I don't know how you have the heart to go there without me.

An enormous Swiss woman has cottoned on to Edith and me. She is fifty, translates books, and is engaged to a young Englishman of twenty-six. She asked Edith if she wrote under her name. (An odd question, for she knows all about various of our contemporaries such as Robert Graves.) Edith said, no, we were both very shy: so she wrote under the name of T. S. Eliot and I under the name of Clemence Dane. And that when she, the Swiss, settles in England (which she is going to do shortly) she was to be sure to ring us up under those numbers.

There is an echo here of the occasion, several years before, when Osbert, asked to autograph a book for a villege fête, took great delight in signing the name 'John Galsworthy' in several copies of *The Forsyte Saga*. It clearly brightened up a dreary holiday, and he related the sequel two days later in a further letter to David.

> Terrific excitement here, now that it has come out that I am really Clemence Dane. The whole hotel is in an uproar about it.
>
> Edith told the large Swiss lady . . . that I did not like it to be known, as it was a different personality, practically automatic writing. I am sure I shall be arrested for false pretences.

Luckily there were only a few more days of spurious fame to live through. On 7 September Edith would catch the train to Paris, Osbert to Nice and David. 'I shall be glad to get out of this bloody place with my life,' [he told him]. 'I feel so ill always . . . but seeing you will no doubt put me right.' And the following day: 'Gosh what a place! But I've got back to my 1922 figure: but alas, not face . . . my clothes hang on me like a bear's.'

The 'lynching' Edith had expected on publication of her *Aspects of Modern Poetry* in November never quite occurred, although the book stirred up a lot of enemies – as she apparently intended. 'It is evoking an appalling storm,' she reported gleefully to Allen Tanner. 'All the whipper-snappers are rushing to each other's rescue, and Lewis is absolutely howling with rage.' This suggests something of the psychology behind the book – the love of a good literary brawl, the need she evidently felt to pillory her 'enemies', the way she had of pouring scorn upon them just as in old age she sometimes told her friends that, if anyone offended them, she would come and instantly 'pour molten lead on their heads as my ancestors did in Norman times'.

In fact, the book has two quite separate parts. The first consists of Edith blasting off at almost everybody who offends her, while the second is an attempt to praise the modern poets she approves of and adumbrate her theory of what poetry is all about.

Inevitably the blasting is more memorable than the praising, and it is important to recall the conditions under which she wrote. She had been miserably unhappy all that year and, to make matters worse, the two people most responsible for her troubles were two people that she nominally loved – Tchelitchew and Helen Rootham. They could not be attacked, so Edith seems to have diverted all her rage and her frustration on to those who could be. This had become her standard practice. 'When I feel cross, which is often, I tease Wyndham Lewis' – or Geoffrey Grigson, Gertrude Stein or anybody on her lists of 'habitual offenders'. It was her method of alleviating pain. In *Aspects of Modern Poetry* one can see the therapy she used during her wretched springtime at Levanto.

There is her inevitable party-piece on Wyndham Lewis in which she mocks his age, his poetry and what she still calls his 'sentimentality'. Then she officially declares war on another old enemy, Geoffrey Grigson, whose review of *Alexander Pope* in the *Yorkshire Post* was neither forgotten nor forgiven. By now he was editing *New Verse* and admired Wyndham Lewis. This was enough for Edith, and she scored several easy victories by selectively quoting from some of the sillier poems his magazine had published. (A good example of Edith's method is the way she quotes the line from a poem by Gavin Ewart – 'Deliver me from fornication and hockey.' and gaily adds, 'We will, Mr Ewart, we will!')

But the most spectacular of all the 'slaps' in *Aspects of Modern Poetry* is the one she delivered to a new *bête noire*, the sage of Downing College, F. R. Leavis. Leavis was rapidly emerging as the high priest of a literary tradition which stood for almost everything the Sitwells had rejected – 'maturity' against frivolity, reality against fantasy, the puritan tradition against the cult of the 'children of the sun'. Trouble was bound to come between them in the end, but it erupted with Dr Leavis's bland observation in his book *New Bearings in English Poetry* that 'the Sitwells belong to the history of publicity, rather than of poetry'.

This was not just upsetting: there was sufficient truth in it to make it damaging, and it began a feud that lasted until Edith's death. In the letter that she wrote to Osbert from Levanto while he was still in China, she was already working up her rage. A friend of hers had dined with the Leavises at Cambridge, and Edith reported: 'Mr L. is small and harassed-looking and does coaching. Mrs L. wears an emerald-green jumper and has a dyspeptic-looking nose and eyeglasses. They gave their guest Cydrax.'

Edith was not averse to dwelling on the supposed social inferiority of her enemies, and in *Aspects of Modern Poetry* she does her best, and in her most superior manner, to put the doctor 'firmly in his place'. His chief offence, one feels, has been 'impertinence'. She is sarcastic rather than witty, and the one brief flash of humour comes when she dwells on his criticism of Milton's *Paradise Lost*: 'As for the interpretation of the stressing, it is sad to see Milton's great lines bobbing up and down in the sandy desert of Dr Leavis' mind with the grace of a fleet of weary camels.'

Having put her prime offending trio in their places, Edith goes on to praise the poets that she does admire – Hopkins, Yeats, Davies, Eliot, Pound, and, of course, Sacheverell. This is done warmly and enthusiastically. One of the characteristics of Edith, as of Osbert, had always been a simplifying tendency to see every artist as a hero or a villain, and she writes well of poets she admires. But she is not a good creative critic. Her most interesting remarks are those of a working poet on the technique of other poets. She makes her preferences plain – for Pope and Dryden, for what she sees as the great tradition of English poetry which was betrayed by the late Victorians, for the 'lovely texture' of Sacheverell's poems, and the 'vitality' and acute visual sense of Hopkins. Where she fails is in establishing any clear criteria beyond her instinctive preferences, although she talks a lot about 'texture' and 'energy and speed': at one point she even borrows Berenson's celebrated hobby-horse of 'tactile values', as she laments the loss of 'tactile sense' in late Victorian verse.

But what was most important for her own status in the next few years was the way in which she seemed to turn her back decisively upon the younger poets of the thirties.

She makes it clear that she has no sympathy with any of them, and then proceeds, in her most superior manner, to mark down the grubby poets of the class of 1934: Cecil Day Lewis – 'a young poet with a certain spiritual vision', but 'the bodily visual sense and the tactual [*sic*] sense are lacking'; Stephen Spender – 'I hope to write of Mr Spender in a later volume' (but she never did); W. H. Auden – 'an able mind, but, unhappily, he writes uninteresting poetry, or, at least, his poetry nearly always lacks interest'.

A sorry bunch. They may have 'certain mental qualities, but they have not one touch of genius'.

CHAPTER EIGHTEEN

Rat Years

1935–1937

May every wish you have come true
In 1935,
May Pussy's eyes become more blue,
His tail still more alive.

May Pussy find a golden nugget,
With which to build a house,
And may he, crouching on the drugget
Procure his daily mouse.

THUS OSBERT started the new year – with a poem to the still adored but still determinedly independent David. During this year David would still be travelling on his own and insisting on his separate holidays abroad. The result for Osbert was an increase in the social round, as left to himself, he went to more parties and stayed longer in the stately homes he loved. April saw David in Geneva; and Osbert doing the rounds of Badminton, of Nether Lypiatt and Polesden Lacey. His friends the Duke and Duchess of York were among Mrs Ronnie's guests that year, and Osbert described the scene to David. 'I came back from Polesden yesterday night. It was like jazz night at the Palladium. All the butlers were drunk – since Maggie was ill – bobbing up every minute during dinner to offer the Duchess of York whisky.'

There were, however, none of these high jinks with the Beauforts. 'It is delicious here,' he wrote from Badminton, 'but I am suffering from guest constipation as you can imagine, and did one of my green-faced faints after dinner . . . perhaps from missing you.'

It might also have been caused by suppressed rage at reading Harold Nicolson's review of his book on Brighton – 'a very impertinent one, full of jealousy', as he described it to David.

In fact, Osbert's health was giving him much cause for genuine concern that spring. After the stately homes he travelled out to Paris, partly to visit Edith, but more importantly to consult the doctor who had treated him at Brides-les-Bains the year before. On 30 April he was writing to tell David his depressing diagnosis. (David was still in the Swiss mountains – '*so* middle-class' wrote Osbert disapprovingly.)

Alas! It was gout. A *cloche d'alarme* as he says. And nevermore is your Bertie to look a glass of cognac in the face. Above all, he says, avoid red wine, chocolate, and he elegantly puts it, 'insides' (ris de veau, etc.) He says it will recur whenever I do anything enjoyable. (But champagne is allowed.)

He adds that he has just received his invitation to the next Court Ball, 'and know I shall lose it. When Edith's *Victoria* comes out, I shall *not* be asked to any more I fancy'.

Come the beginning of July, and David was off again, this time to stay with friends at Quiberon in France. To while away the time Osbert decamped to Renishaw where he was soon in trouble once again.

You very nearly never saw your Bertie again as he fell right through the iron bridge at the bottom of his grandmother's garden at Renishaw yesterday evening at 7 o'clock. He thought he was done for. However, apart from a cut knee, a bruised thigh, a wrenched arm and jangled nerves, no great harm seems to have been done.

It wasn't my weight: please don't think it was: it was simply the rotten iron from acids in the coal seam.

But David was remiss that year, neglectful of his Bertie in the waves of Quiberon, and Osbert was soon despatching worried letters asking him for news, for postcards, for almost any indication of affection. Osbert was the most concerned and tolerant of lovers, although, as he confessed sadly, 'I am moping terribly without you, as usual.' Not even a grand dinner later in the month at Mrs Ronnie's house in Charles Street could raise his spirits. 'Londonderrys, Emerald, Philip etc.: v. boring'. The only bright spot on the whole horizon was that David would soon be back, and that he had just heard that the peaches at Renishaw were going to be 'very good that year'.

In fact, there was another great improvement in his life which should have cheered him up. He was financially better off than he had been for years. Indeed, by now he was managing to support quite an establishment at Carlyle Square. Apart from Miss Noble, who had replaced the lamented Mrs Powell as housekeeper, there was Herbert, who acted as his valet and his chauffeur, and the year before he had the assistance of A. E. W. Mason's secretary. This was Lorna Andrade, a diminutive and well-read young lady, who was a niece of the scientist, E. N. da C. Andrade, and who would play an increasingly important part in managing his life in the years ahead.

He was even buying pictures. While in Paris he bought two small canvases by the seventeenth-century Bolognese painter, Bibiena, 'expensive, but finest quality', he told David with apparent satisfaction. This was a strange choice for someone who once saw himself as one of the leaders of the Modern Movement in the arts, particularly when one thinks of all the modern paintings he could have bought in Paris for his money. But by now whatever sympathy he ever had with modern art was waning and his rising affluence reflected two important factors in his life. The first was that he had won his battle with Sir George, and his allowance was assured. More important, he had discovered a new source of income – popular Sunday journalism.

It was a shade ironic after the battles he had waged with journalists and press lords in the past, but the fact was that all the Sitwells had the makings of accomplished journalists themselves – and none more so than Osbert.

By 1934 the *Sunday Referee*, under the new and lively editorship of R. J. Minney, had begun a policy of publishing weekly essays by literary celebrities like H. G. Wells, George Bernard Shaw and Bertrand Russell. Osbert was invited to join them and almost instantly established himself as a regular contributor. From now on, until the beginning of the war, Osbert would be a well-paid weekly journalist, and these regular articles of his not only paid for the bills, the holidays, the chauffeurs and the occasional champagne, but also gave him a chance to do what he was particularly good at – airing his prejudices, talking about himself in print, and laying down the law with a mixture of outrageousness, originality and wit that guaranteed his popularity.

In 1935 he published the first collection of these journalistic essays under the title *Penny Foolish* (a later batch was called *Pound Wise*). Taken together, they provide a wonderfully readable self-portrait of the author. There are descriptions of his travels in America and the Orient, his views on privacy and snobbery, on friends and enemies and ghosts and gardening, on public schools and prigs and progress. Altogether the impression that he gives is of a garrulous, immensely civilised original, a high-brow egocentric who insists upon the luxury of saying and doing exactly what he wants. On 'Friends and Enemies' – 'Never forgive your enemies – it only makes them more bitter!'; on 'Games' – 'My fear and detestation of games is founded in the first place, I think, in love of my country'; and 'On English Food' – which 'inclines us to wish that Napoleon *had* in the end conquered England, so that French chefs could have reformed our food'.

But in these weekly articles for the *Sunday Referee* there were two topics Osbert wrote about which were of more than passing personal importance. The first was on the arms' race and the threat of war. Here Osbert makes it clear that he is still as unrepentantly a pacifist as ever:

It seems that the European nations *want* another dog-fight; if they do, it is our duty to watch it and not to join in. Every single principle for which we fought, or were said to be fighting, the last war, we have seen betrayed during the last fifteen years. . . . We must not be taken in again. No war is ever a 'righteous war'. A righteous war is only one degree less wicked than a wicked war.

This paragraph explains almost his entire attitude to foreign politics during the run-up to the war of 1939. It also partially explains his continued sympathy for Fascism – as he made clear in a *Sunday Referee* article of 4 March 1934, which he republished a year later with the title, 'Is Fascism British?'

By then the violence of Mosley's meetings had brought the period of 'respectable Fascism' to a close – but Osbert did not go back on his earlier support for it. His article takes the form of a review of a book published originally in 1917 by the eccentric and outspoken right-wing peer, Lord Harberton. Osbert is, more than ever now, the declared opponent of democracy, and once again harks back to the supreme disillusionment of the Great War when statesmen and journalists 'indulged in all the rest of that treacherous democratic claptrap which was consummated by a million British deaths'.

There is a clear alternative, and Osbert applauds Lord Harberton for being

the first to pronounce democracy a failure, and to argue that a great modern people may demand to be governed firmly and not to have ogres thrust on it; that it may not want millions of officials, nor the perilous oratory of democratic statesmen, but may prefer instead to be allowed to get on with its business, to be given a good bus service and punctual trains.

It is important to remember that Osbert was profoundly anti-Nazi from the start. For him the Nazis were unspeakable, the ultimate example of the Germanic barbarism he fought against in France. But Mussolini was, for him, a different sort of animal. As late as October 1935 he was writing in *Nash's Pall Mall Magazine* on recent rumours that the Italian leader had been ill. 'If so, I cannot help occasionally wishing that someone would inoculate our Mr Baldwin, much as I respect him, with the same germ. For the Italian leader, despite his present policy [against the League of Nations] has been a great benefactor to his country.'

With hindsight it must seem depressing that a man of Osbert's culture and humanity can have believed in someone who had more in common with Horatio Bottomley than with Stanley Baldwin. That he could do so points up an interesting fact about him – that for all his cherished love of Italy, and talk of it being his second country, he never really knew it. He knew its art and architecture and its food; he knew about Magnasco and rococo; he loved Venice, Naples and Amalfi: but he could not speak Italian, had no genuine Italian friends, never once attended an Italian political meeting, and dismissed Italian politics as something for the natives. He could thus ignore Fascism's debt to the theorists of violence like Pareto and Sorel, and the repression and the bullying it practised in his own beloved Tuscany. Such things had no part in the Italy he was concerned with.

For Edith the troubles that followed publication of her *Aspects of Modern Poetry* rolled on into the early months of 1935. The most damning of all the reviews was one in the *New Statesman and Nation* by G. W. Stonier which pointed out an embarrassing series of coincidences. Dr Leavis, in his earlier book, *New Bearings in English Poetry*, had written at some length on four of the major poets Edith dealt with – Hopkins, Yeats, Eliot and Pound – and, as Stonier made quite clear, Edith had plainly borrowed from his book. The parallels he cited were so apposite that there could really be no question but that Edith had quite cheerfully used the hated doctor's book as one of the major sources for her own. On top of Stonier's quotations, a later correspondent listed nineteen parallel quotations from the two books, and when one correspondent wrote defending her, the literary editor of the *New Statesman*, David Garnett, commented drily that, 'Disliking Dr. Leavis does not seem to us a good reason for borrowing extensively from his work without acknowledgement.'

Meanwhile, Geoffrey Grigson had accused her of plagiarising also Herbert Read's *Form in Modern Poetry*. Wyndham Lewis joined in. Osbert responded with alacrity, but it was Lewis who undoubtedly got the better of the correspondence that ensued.

People who live in glass-houses invariably throw stones, that is the law of nature; all the same, Miss Sitwell has built such a really enormous glass-house for herself in 'Aspects of Modern Poetry' that when that big stiff of a brother of hers butted in and discharged a brick at me . . . I was, for a moment, dumbfounded by his peculiar effrontery.

Fortunately for her, Edith had now discovered a less contentious champion than Osbert. At the beginning of January 1935 she wrote to Allen Tanner complaining bitterly of the way she had been treated in the press.

As Tennyson told a friend, the fate of a poet is, –
'When living the owls,
When dead, the ghouls'.

And she added:

> If it had not been for my new friend, John Sparrow, whom I met for the first time in London this summer, heaven knows where I should have been. . . . He is a barrister, and the way he 'takes on' my enemies in the papers is most amusing.

Sparrow was already considerably more than a mere barrister. In his late twenties, he was a Fellow of All Souls, a poet and critic of commendable erudition. He was a formidable young man, a close friend of Harold Nicolson, a successful lawyer with a flat in the Middle Temple; and in years to come his wry wit, learning and great charm would make him an important figure in post-war Oxford, and Warden of All Souls. Early in 1934 he published a short critical book on modern poetry, *Sense and Poetry*. In it he remarked of Edith that,

> Her happiest effects have been attained when she wishes merely to present a pattern; her verse is most pleasing when it is purely decorative; but like Mr Pound, she seems never to have thought out her aims clearly, and intellect is always trespassing in her poetry with the most unfortunate results.

In earlier years such outspokenness would have been classed as an 'impertinence', and might even have involved its author in a resounding 'slap'. But Edith had grown shrewd enough to spot a potential convert to her work. The letter that she sent young Mr Sparrow was as restrained as it was courteous. She said that he could not have read her *Gold Coast Customs* when he maintained that her verse was merely decorative. She would send him a copy, and went on, 'I am invariably extremely careful about the truth of my similes and there is nothing in any of my poems that I cannot explain absolutely rationally.'

In his book Sparrow had criticised her for saying that pigeons smelt of gingerbread. 'As for the trouble about pigeons smelling of gingerbread, I can assure you pigeons' feathers in hot weather *do* smell like gingerbread. I don't know if all birds do, but I can vouch for the fact that pigeons do.' During the row that followed publication of *Aspects of Modern Poetry*, Sparrow was really the only one of Edith's allies to put up a coherent case for her. While Osbert and Wyndham Lewis were busily insulting one another, and Edith was muttering darkly that she supposed that the literary editor of the *New Statesman* was 'going to allow every lower grade mental defective and protozoan in England' to give his opinion on her work, Sparrow deftly pointed out that 'her book is largely devoted to contrasting [Dr Leavis's] view of poetry with her own, and no two views could be more radically opposed; the Doctor thinks of poetry as a sort of moral hygiene, Miss Sitwell sees it rather as a means of creating beauty in language'.

A slight over-simplification, perhaps, but at least it enabled Edith to retire with dignity. What he could not do was repair the damage Edith had really done herself with her summary dismissal of all the young poets of the thirties. Edith, who had always passionately believed in the modern movement in the arts, Edith, who longed to lead and influence the young, had publicly proclaimed herself one of the old guard from the poetic twenties. If she found young 'Master Auden' a 'bore', young 'Master Spender . . . just a heightened and refined edition of Mr W. J. Turner' and that young Master MacNeice 'belongs to the bungalow school of poetry over which Mr Grigson presides', that was her loss, not theirs. The first poem Spender had ever published was a schoolboy 'Ode to Edith Sitwell' in his school magazine. But now, as he says, 'none of us read her poetry or ever really thought about her. As a poet she simply didn't seem to matter any more.'

To make Edith's sense of isolation worse, Tchelitchew was still in America – and still with Charles Henri Ford. His New York exhibition failed to be the great success he hoped, but even this had not convinced him of his overriding need for those who really loved him and who would further and protect his genius. 'Poor Pavlik,' she wrote to Tanner (who was still in Paris). 'I feel so dreadfully sorry for him. Oh when *will* he realise the absolute emptiness of these showy, second-rate people. He is always being deceived and it makes him so unhappy.' But to remind the erring Boyar of his real friends – and the invaluable influence they wielded – she persuaded Cecil Beaton, who had a great success now in New York, to send him letters of introduction to a number of potential wealthy clients. 'He knows crowds of influential people,' Edith wrote a little sourly, 'he must, considering the amount of money he makes.'

With Tchelitchew so firmly in America, it was high time for Edith to discover a new genius she could adopt, but they were few and far between. That June there seemed a chance that she had found one, and she wrote excitedly to T. S. Eliot about him. 'I have just read the unpublished poems of a young man called Thomas Driberg. They seem to me to show really remarkable promise, and, at moments, achievement.' So much for Edith's intuition if it led her to prefer the juvenilia of William Hickey to the work of Auden, Spender and MacNeice. (To be fair to her, she soon revised her opinion of his poetry, and through Beverley Baxter suggested Driberg as gossip columnist for her old enemy, Lord Beaverbrook. Here she was on firmer ground.) Then finally, towards the end of 1935, she found what she was looking for.

In May 1934, while she was still staying at Levanto, Edith had ended a letter to John Sparrow:

I have just got a book of eighteen poems by a youth called Dylan Thomas. Have you seen them? He ought to be dashed off to a psycho-analyst immediately before worse befalls; then if he would afterwards spend ten years in thought I am not sure that he would not produce some fairly good phrases.

Hardly a favourable reception for the first book by the twenty-year-old Welsh poet, and almost everything she can have heard about him must have encouraged her suspicions. The personal habits and appearance of this young man, whom his friend, the poet Norman Cameron, described as 'That insolent little ruffian, that crapulous lout', were hardly to Edith's taste and, as she must have known, his poetry had been praised by 'Master Spender' and her arch-enemy, Geoffrey Grigson. Indeed, even as Edith was writing these words to Sparrow, Thomas and Grigson were on holiday together, staying at a cottage on the west coast of Ireland. Grigson has described how they used to go down to the shore together.

There was no sand, no gravel, below these cliffs, only white pebbles shaped like eggs or heads by Brancusi. We drew faces on them with black crayon, we named them, set them against rock, and cracked them, with fling after fling of other huge white pebbles, into literary nothingness – since the faces were of authors – and literary oblivion.

Hardly surprisingly with Grigson there, the faces Dylan Thomas helped to smash included Edith's, and it is more than possible that word of this got back to her. Certainly her hostility to Thomas seems to have increased during the summer and, according to Constantine FitzGibbon, Thomas's biographer, she even went as far as 'mocking him in a parody that aroused his fury'.

Then came a dramatic change. In December 1935 Dylan Thomas had published his poem 'A Grief Ago'. Edith was impressed – and said so in an enthusiastic review in the February number of the *London Mercury*. She also said so in a glowing letter to the author:

It is no exaggeration to say that I do not remember when I have been so moved, profoundly so excited, by the work of any poet of the younger generation, or when I have felt such a deep certainty that here is a poet with all the capabilities and potentialities of greatness. I am completely overcome with this certainty and this admiration.

Edith was plainly thrilled with her 'discovery' of Thomas. She wrote to Lady Aberconway about him, saying that 'he stands a chance of becoming a great poet, if only he gets rid of his complexes'. She promised to review his forthcoming collection of poems – and duly did so in the *Sunday Times*, with even more enthusiasm. And finally, of course, she had to meet him.

This, given Thomas's past behaviour when faced with alcohol and polite society, was hazardous – but Edith was determined. Over Christmas she had been staying in a small village in Catalonia, to finish off *Victoria* in peace, and it was from there that she wrote to her friend Robert Herring, editor of *Life and Letters*,

I am coming to London (unless I die or something happens to prevent me of that sort) on February 15th, and shall be at my club, 49 Grosvenor Street, (the Sesame, (for about ten days. I do hope that you will come and dine with me there at 7.30 on the 18th. Mr Dylan Thomas is coming. . . . The more I read young Thomas's poetry, the more certain I am that he is the coming man. Is not the progress he has made during the last eighteen months extraordinary . . . ? I want to ask him some questions and give him some advice.

When he received the invitation, Thomas was understandably alarmed, and wrote instantly to Herring, whom he also knew, for reassurance. 'Can you tell me all about Miss S.? She isn't very frightening, is she? I saw a photograph of her once, in medieval costume.'

Edith's club, the Sesame (or, rather, the Sesame, Lyceum and Imperial Pioneer Club, to give it its full title), was an unusual and forbidding institution in those days. (It is now transfigured and transformed from the pre-war days when Edith held court there in the dark-green dining-room among the other members of this London ladies' club.) Surprisingly, the dinner went off perfectly, and according to FitzGibbon, Thomas even managed to wait until *afterwards* before getting drunk. Some years later, when Edith was asked in a broadcast interview how Thomas had behaved, she replied, 'Beautifully. I've never seen him behave anything but beautifully with me. He always behaved with me like a son with his mother.'

That is as may be – but as one knows too well, mother-and-son relationships can be extremely fraught. Edith's with Dylan Thomas most certainly were. In later years her early patronage of Dylan Thomas was to become part of the myth she spun around her life, and she would always claim, as one of her great achievements, the way she had discovered, helped and championed his lonely genius. She was extremely proud – as she had some right to be – of her review of Thomas's *Twenty-Five Poems* in the *Sunday Times* when it appeared that autumn, for it undoubtedly helped spread word of his achievement to a wider public than he could have hoped to have had before. But she did not 'discover' him – Spender and Grigson share that somewhat meaningless distinction – and whether the publicity and fuss that followed Edith's *Sunday Times* review was good for either Thomas or his work is open to debate.

Edith instantly turned him and his poetry into a fresh campaign in the battle against the philistine. For years she had been leaping to 'defend' her favourite poets – Wilfred Owen, Robert Nichols, T. S. Eliot – as if they were themselves beleagured and besieged in some doomed Rorke's Drift of literature.

An angry correspondence soon developed in the *Sunday Times*. Edith was in her element. As she wrote later, 'It was my privilege and pride to give the attackers, during two months, more than as good as they gave. The air still seems to reverberate with the wooden sound of numskulls soundly hit.' Edith loved cracking numskulls in the Sunday press, but all this publicity that she was the first to heap upon him was probably the last thing Thomas needed at the time. Indeed, for the rest of his life he suffered far more from his role as a celebrity than from anything the philistine could do. It was excessive adulation – not opposition or neglect – that helped to kill him in the end.

Perhaps he realised the danger of being taken up too eagerly by Edith even in those early days of 1936. He was certainly the most exasperating protégé. From the first meeting at the Sesame he seems to have made it quite clear that he was not accepting the 'advice' that Edith had told Robert Herring she longed to give him, on his poetry or on anything else. Edith never sent him the sort of magisterial letters on 'technique' which she despatched to earlier disciples such as Brian Howard or Harold Acton. On the other hand, he needed money – rather urgently.

One inevitable result of his being taken up by Edith was that he broke with Geoffrey Grigson. Grigson, who was now literary editor of the *Morning Post*, had been sending him regular batches of novels to review, and this had been a vital source of income. There was an argument, and the reviewing ceased. This caused a financial crisis, and early that spring Edith was writing to her old friend and Fleet Street supporter, Richard Jennings, with 'a most earnest plea to make. I've simply *got* (if it is humanly possible) to find a job for 22-year-old Dylan Thomas. That wretched boy is living – if one can call it living – on what he can earn from his poems and his stories – and he realises that he can't go on like it any longer'. He had been 'disgracefully rude' to Augustus John on the one occasion when they met. 'I have given him an awful ticking off.' Clearly Edith can have known little of the dramas that were beginning between the wild old painter and the wild young poet over Caitlin Macnamara, the turbulent young Irish girl who had been living for some years in John's household at Fordingbridge, and who would eventually become Mrs Dylan Thomas. Nor did her 'awful ticking off' prevent another row a few months later which ended with John knocking Thomas out in a car-park in Carmarthen.

Jennings, however, seems to have helped find Thomas work – as he invariably did when Edith asked a favour – and before long Edith was writing to him again. The mother-and-son relationship with Dylan Thomas was already under strain.

The last time we corresponded it was on the subject of Master Thomas, who is rapidly heading for having his ears boxed. I can feel the tips of my fingers tingling to come in contact with the lobes of his ears. And it would do him a lot of good, for he was evidently insufficiently corrected as a child. *What* a tiresome boy that is, although a very gifted one. Having given us all that trouble, caused me to pester you and to write dozens of letters to busy people who must now curse the name of Sitwell – he has now disappeared again – disappeared without leaving a trace. I have had reproachful letters from this person and from that, reproving *me* because *he*, after getting *me* to ask *them* to give *him* an appointment, hasn't kept it. But not a word can we get from him. This disappearance trick seems to be a habit of the younger generation. They ought to join Maskelyne and Devants. . . . I'm practically Master Thomas's secretary now, as everyone who can't find him – (and nobody can) – addresses his letters care of me and I have to readdress them.

The mystery of Dylan's disappearing act has been solved by Constantine FitzGibbon. Just for once, the poet was not being feckless, drunken or intentionally rude to Edith, but his situation was not exactly one that he could easily explain to her. He had caught gonorrhoea from a lady he had met at that year's international surrealist exhibition, and had retreated, chastened, chaste and sober, to his parents' house in Swansea. In those pre-penicillin days it took several months of gloomy living to recover from this unfortunate disease, and one can only wonder what explanation his fertile mind finally produced for Edith. At all events, it failed to placate her, for they were not to be on speaking terms for the following eight years.

1936 was the year of the Spanish Civil War. 'It had been rightly said', writes A. J. P. Taylor, 'that no foreign question since the French Revolution has so divided British intelligent opinion or, one may add, so excited it.' Most writers of the day inevitably became involved on one side or the other, but all three Sitwells stayed scrupulously aloof.

At that same surrealist exhibition that had such painful consequences for Dylan Thomas, Stephen Spender happened to notice Osbert. 'I was very young,' he says, 'and was trying at the time to organise writers to go to Spain. I recognised Osbert and asked him if he would care to go out to Republican Madrid. I have never forgotten the look of absolute horror on his face at the suggestion.'

Osbert explained his views on the war in an article in the *Sunday Referee*. It represented, he maintained,

a horrible and incredible gladiatorial combat, before our very eyes, and in it we can plainly discern the possibilities of similar conflicts elsewhere. . . . There seems, indeed, to be no end to the zeal, ferocity, energy and perverted heroism of each side.

And the chief lesson for the British race as our eyes are attracted, with a cheerful fascination, to the terrible tragedy enacted before us, is that we . . . refuse to take sides, or allow ourselves to be divided into two hostile factions, pro-government or pro-rebel.

Osbert, as usual, was advocating pacifism at all costs.

For anyone of Osbert's temperament it was difficult to steer clear of any fight – particularly one so fraught with obvious importance as the civil war in Spain. To follow his own advice meant standing quietly aside as Europe gathered speed towards inevitable destruction. This hardly suited Osbert, and throughout that summer he showed signs of restlessness and irritation. Even the consolations which his world could offer merely underlined the precariousness of the society he loved. That June he and David spent a weekend at Faringdon where the guests included Salvador Dali and his wife. A few weeks later, after David left for France, there was a special luncheon in Dali's honour (which he shared with André Simon) given by Curtis Moffat and his son. It was the sort of menu Osbert could appreciate:

Turbot en papillotes
Pâtes au gratin
Choux au lard
Jambon
Salade, fromage, fruits.

With which they drank:

Hugell's Riesling Réserve 1923
Château de Cheval Blanc 1921
Grand fine champagne Jubloteau 1906

While David was away Osbert did his best to keep himself amused – but everybody seems to have annoyed him. 'Saw Willie and Alice [Wimborne] walking down King's Road yesterday,' he reported bitterly to David. 'She looked old and footsore and slummy.'

He sent a bad-tempered notice to the Agony Column of *The Times*: 'Mr Osbert Sitwell regrets he is unable to read any more manuscripts gratis. None are to be sent to him, and none, if sent, will be returned. Otherwise his charge for reading is £1 a page.' Chips Channon noted in his diary what looks like the beginning of an old-style Osbertian stunt. He had, he said, 'been agitating for some time in a facetious spirit to get the Crystal Palace moved back to Hyde Park where it once stood'. It came to nothing: nor did a proposition about which Osbert wrote to his new publisher and former comrade from the Grenadiers, Harold Macmillan, in October 1936. It was a subject dear to him – a book of hoaxers, ranging from Lambert Simnel and the Holy Maid of Kent to Titus Oates and Anastasia – but it was also just the sort of book authors put forward to their publishers when they are drained of inspiration.

On 8 November he took a casual swipe at Noël Coward in the *Sunday Referee* – always a disturbing sign. 'Mr Coward,' he wrote tetchily, 'with his frisky tea-shop dialogues, has gained among nitwits a certain reputation for wit – but who in England is really witty, whether rude or civil?'

In contrast with Osbert, Edith for once had something really good to celebrate. In August Sacheverell mentioned, in a rare note to David, 'Edith is in transports today, having just received a cheque from her publishers for £356!! Isn't it lovely for her?' And all the lovelier for being unexpected. Edith's 'bloody book' on Queen Victoria had appeared in February that year. Far from the boring book she had predicted – or the embarrassment for his royal friends Osbert had half feared – *Victoria of England* was having, as she wrote, 'A violent success'. No book of hers had ever sold so well – or had such instant popular acclaim.

'A book of great beauty,' intoned the *Spectator*. 'A sincere and moving portrait of a great sovereign', hymned *The Times*. Most satisfying of all was the verdict of a critic in the *Observer* that 'her canvas invites immediate comparison with Mr Strachey's. But it is at once broader and warmer than his.' Edith was beating Bloomsbury's chosen son at his own game.

Paradoxically, the key to her success seems to have lain precisely in those moans of hers about her boredom with the book while she was writing it. She had not been in a rage as she composed it. It had no polemical intent like her books on Pope or modern poetry, nor did she feel the urge to identify herself with Queen Victoria – as she did in all her later work on Queen Elizabeth. Instead she read the published sources on the period and then produced a book of quite extraordinary charm and wit, with much of the colour and the freshness of her earlier poetry. And just for once it was exactly what the public wanted. The first printing of 4,500 sold out instantly, the book clubs bought it, the Americans were after it. No wonder that her publisher at Faber, Richard de la Mare, was, as she wrote, 'half off his head with excitement'.

1936 was even more of a year of achievement for Sacheverell. He had no less than three books published, books that in their sheer variety attest to his energy and range of interests. There was his sumptuous *Conversation Pieces*, 'A Survey of English Domestic Portraits and their Painters'. It was a subject dear to his heart, a study of the interaction of English social life and the greater and lesser artists who between them formed what Sacheverell claimed to be a clearcut 'school' of painting. It further established its author as an art historian, and showed his growing interest in the English scene. His *Collected Poems*, too, appeared that autumn. In a melancholy way they were to complete his career as a published poet – although fortunately this did not mean that he would stop writing poetry. Nothing could do that. But his most intriguing book of all that year was his extraordinary, *Dance of the Quick and the Dead*, 'An Entertainment of the Imagination' in which he brought together his talents as a poet, traveller and lover of the arts, to illustrate and comment on the oddity and underlying tragedy of life. It is a book of considerable originality, the first of several in which he expounded his own private, multi-faceted, dream-like vision of his private world.

If nothing else, these three books disproved the fears of those who had gloomily predicted that domesticity would be the end of Sacheverell as a writer, but it is hardly necessary to add that none of these books achieved a fraction of the sales of *Victoria of England*. They barely repaid the effort that went into them, and Sacheverell continued to be dogged by chronic money worries, particularly since the birth of his second son Francis the year before. At times it seemed impossible to maintain their life at Weston, let alone to entertain their friends and travel as they had in years gone by.

Throughout the autumn of 1936 there was one topic which increasingly intrigued and exercised Osbert – Edward VIII's affair with Mrs Wallis Simpson. For anybody 'in the know' it was the great unspoken subject of the year, but it was something more than this for Osbert. He had imperceptibly become a close friend of the Duke and Duchess of York, and when in England often lunched with them informally at their house, 145 Piccadilly. This meant that through all the dramatic and outlandish happenings that led to the abdication, Osbert knew rather more than most of what was going on. He was something of an insider to the whole affair, and very much the loyal partisan of the Yorks.

He also patently disliked the King. There was something almost personal in this. He had first seen him as a young fellow-officer with the Guards in France and, like so many others, had been impressed by all that boyish charm and royal glamour. This had made Osbert's disillusionment the more severe as he discovered that this superficially attractive, stylish Prince of Wales was, in fact, the philistine incarnate.

Worse still for Osbert was the company the young Prince kept, the racy, empty-headed gang who made up the so-called Fort Belvedere set, that glittering epitome of Osbert's original Golden Horde.

The Yorks were a happy contrast to all this. Not merely was the Duchess an intelligent and cultivated woman, who liked his wit, shared his taste in pictures and even read his books, but he had managed to amuse the Duke, who particularly enjoyed Osbert's stories of Sir George.

Inevitably, the arrival on the scene of Wallis Simpson confirmed Osbert's gloomiest prognostications for the Prince of Wales. When the Prince succeeded to the throne in January Osbert could sense the danger to the monarchy. Hardly surprisingly, he loathed the tone of the new King's court and the Simpsons seemed to sum up what was most offensive in this seedy bunch. The much put-upon Mr Simpson appeared to Osbert like an outrageous bounder, his wife was quite unspeakable. Everything that Osbert most disliked about them seemed confirmed by events that autumn. From the moment that the Simpsons were divorced on 27 October, leaving Mrs Simpson free to marry the King, Osbert had had a ring-side view of what was going on. The prospect of his feeble sovereign so gaily at the mercy of this cool American adventuress shocked and angered him. When the crisis broke on 1 December, and the question of Mrs Simpson hit the papers, Osbert saw it as the outcome of everything he had foreseen and feared for months. As a friend of the royal family he was all but personally involved. As a devoted monarchist he was appalled by what was happening to the monarchy. And as a long-standing enemy of Beaverbrook and Churchill he was enraged when they, of all people, threatened to support the King and Mrs Simpson with a new 'King's Party'.

Osbert was a man of action *manqué*. His thwarted love of power was always roused when he became involved in politics. If there was something he could do – even something as ephemeral as acting as a go-between in the settlement of the General Strike – he was happy. If he was powerless his impotence enraged him.

His anger had been brewing all that week, and on Sunday, 6 December he stayed in bed all day (as Lady Ida often did), overcome with rage and tiredness. As something of a release from anger, he began a poem. His old capacity for venom had not deserted him; and thinking over the events of the previous few days he hit on a splendid title. He called it simply 'National Rat Week'.

Where are the friends of yesterday
That fawned on Him
That flattered Her;
Where are the friends of yesterday,
Submitting to His every whim
Offering praise of Her as myrrh
To Him?

And it concluded:

> What do they say, that jolly crew?
> Oh . . . her they hardly knew,
> They never found her really *nice*
> (And here the sickened cock crew thrice):
> Him they had never thought quite sane,
> But weak and obstinate and vain;
> Think of the pipes; that yachting trip
> They'd said so then ('say when – say when') –
> The rats sneak from the sinking ship.
>
> What do they say, that jolly crew,
> So new and brave, and free and easy,
> What do they say, that jolly crew,
> Who must make even Judas queasy?

It also included scurrilous references to some of the better-known friends and hangers-on of the former King, such as Osbert's old Aunt Sally, Lady Colefax.

Having composed it in that fit of bed-bound anger, Osbert realised that it was too intemperate – and libellous – to publish. But he was proud of it, and it was just not in his nature to suppress it. Before long copies of the poem were all round London. It was theoretically anonymous – but everybody seems to have known who wrote it. Ten days after its composition, that determined gossip, Chips Channon, was writing in his diary; 'Much talk of rats. Osbert Sitwell has written a poem, not a very good one, called "Rat Week" in which he lampoons many deserving people. It is cruel, funny and apposite in spirit if not in the letter.' Osbert by now had left for France with David – anything to avoid an English Christmas. From Paris they went on to Vevey, on the shores of Lake Leman, and it was there that he received a telegram from a newly founded London magazine called *Cavalcade*, asking if he had written 'National Rat Week'. It was David who devised the splendidly succinct reply on the pre-paid telegram: 'SEZ YOU SITWELL'.

The editor rashly interpreted this to mean that Osbert would make no objections, and a few days later *Cavalcade* published a letter from a reader quoting a large part of the poem, and crediting it to Osbert. Osbert was in a tricky situation. If he sued for outright breach of copyright, he would be publicly admitting authorship of a poem, the unpublished parts of which he knew were libellous. So, acting for Osbert, Philip Frere applied for an interim injunction against *Cavalcade*. *Cavalcade* responded, as Frere thought they would, by trying to make Osbert admit authorship and disclose the remainder of the poem; but on appeal, Frere won his case that it was unreasonable to compel an author to disclose the contents of a poem which could make him liable to a further charge of criminal libel in the courts. This set an important precedent in law – and incidentally cost *Cavalcade* five hundred pounds. It also served to make it clear to editors in general that it was inadvisable to take on Osbert Sitwell in a court of law. Publication of the poem had another strange effect. The version in *Cavalcade* inevitably cut out Osbert's more scathing references to the Windsors' friends, making the poem appear simply as loyal denunciation of the unfortunate King's ungrateful friends, instead of an attack upon the whole circle including the King himself.

But it was Edith who gave the neatest verdict on the abdication when she wrote to David Horner saying that she could not help being relieved 'that Mrs Simpson has been given a raspberry by the whole nation'. 'I think she has been divorced twice too often for a Queen,' she added, 'and I don't think "Queen Wally" would sound well.'

Now that Edith had her royalties from *Victoria of England* she had what she had dreamed about for years – a chance of real independence. It was no fortune, but it was certainly enough to have allowed her to find herself a flat in London, and live in comfort and write poetry again. But the fates decided otherwise. At the end of September 1936 she was staying at the Sesame when she got news from Paris making it quite clear that she had no alternative but to return to the Rue Saint-Dominique at once. As she wrote to her friend Margery Beevers, 'On Saturday by the last post I got a letter saying Helen has *inoperable* cancer of the bone. I can't speak about it.' Helen had to be supported, and the household in the Rue Saint-Dominique maintained. Evelyn Wiel was there, of course, to do the nursing, but only Edith could provide the money that was needed now for Helen's treatment, and there could be no question of deserting her. Edith returned, and a few days afterwards was writing back to David: 'It is most dreadfully sad here. . . . The poor woman suffers a good deal of pain, and *never mentions it*. One just sees her going green, and a terrified look appearing in her eyes.' Although it was impossible to operate, radium treatment at the Curie Institute offered a certain hope. It would take at least a month, but when it was finished Edith decided she would take Helen to Levanto where she could get the sun and sea-bathing she needed.

She told John Sparrow that Helen seemed a little better, 'though terribly weak and she can't go out much'. 'She is', she added, 'very good and wonderfully uncomplaining.' Edith's biggest cross was now her sister, Evelyn:

That great – or perhaps it would be right to say big, spreading woman – is becoming more and more both my Mr and Mrs Watts Dunton. There is no smallest detail in my life now which she does not try to boss. But of course there are differences in the situation. The original Mr and Mrs W. D.'s victim was seventy years old and stone deaf when Mrs W. D. started blighting what part of his life was left after her husband's efforts. Therefore the effect cannot have been quite the same, or at least he couldn't hear the nonsense.

She might have added too that Swinburne did not have a Tchelitchew to worry him during those last gloomy years at Putney. For her prodigal painter had returned to Paris. He was still with Ford, and there could obviously be no chance of a revival of the romantic, sentimental days with Choura and with Allen Tanner. Pavlik no longer chased her round the room, nor did he tell her that she was the only person who could understand him and console him. But relations had been re-established – just. For anybody else there would have been brimstone and molten lead, but Edith was grateful to accept almost any terms he offered. She still refused to have anything to do with Ford, but at the end of October she was telling David,

Pavlik is painting a new portrait of me. This time it is a straightforward portrait, and I am frightfully pleased with it, and think you all will be. He is mournful but good. When I sat to him last, he uttered a long plaint, and for some time I was unable to discover the gist of it. But when boiled down it transpired to be the fact that he *cannot* work properly in Paris because of his brother-in-law's pince-nez. These flash, when he comes in to meals, tired from working; and they are, in addition, very large and round as to the lens, very thick, too, which gives him an owl-like appearance disturbing to a painter.

(The picture, the 'Sibyl' portrait, was bought by Edward James, and is now in the Tate.)

By mid-December Edith was writing, once again to David, with good news of Helen. She seemed 'really better, and the thing is disappearing at last. But of course, she is fearfully exhausted by the treatment which has had to be very severe.' By now, inevitably, Edith's own health was suffering as well. During a brief trip to London just after Christmas she had consulted Dr Dawson – now Lord Dawson and the King's physician – who said she was over-tired and that Levanto and some sun would indeed be very good for her.

And so, early in the new year, Edith and her one-time governess left Paris for the shores of Italy, leaving behind them Pavlik, Evelyn Wiel and all the other maddening distractions of the Rue Saint-Dominique. Helen was bravely hoping she was cured, while Edith had a brand new book to write, a book unlike anything that she had done before. It was to be a novel, a romantic novel. It would be experimental, with few concessions to the gentle reader, but it would help to exorcise the passion that still painfully obsessed her – her love for Tchelitchew. If Richard de la Mare and Faber did not like it, that would be just too bad.

With David's company, and the calm waters of Lake Léman at Vevey, Osbert was soon restored to equanimity after 'Rat Week.' He too began another novel, and with him as well it was rather different from anything that he had tried before. It was something of a memoir, a nostalgic evocation of his vanished life in London and in pre-war Florence. He already had a title for it – *Those Were the Days*.

Spring passed agreeably for Osbert. The house where they were staying, the Villa Beauregard, was quite luxurious, with a small terrace and fine views across the lake. David had started writing too – the story of *his* early life – and every day the two of them kept to an orderly routine, writing till lunch-time, walking each afternoon, and dining out and visiting each evening.

Edith's life, three hundred miles away on the Riviera di Levante, was less agreeable, and the letters she was writing now contain interminable laments. Levanto, 'which used to be heavenly', as she told Richard Jennings, was now swamped with 'the most awful people with legs like flies who come in to lunch in bathing costume – flies, centipedes'. She was being plagued with constant noise – 'One really does feel as if one's head had been shattered, one's spine abolished, and all one's nerves rent,' she told David.

The worst offender was a man she called the 'banging doctor' (apparently because he knocked all night on the wall of the room next door), and a man whose passport described him as 'Gentleman, Bank of England'; his wife was 'the sort of woman who would blow a toy trumpet in the Savoy Grill a few years ago'.

Helen was in constant pain with what Edith described as 'neuralgia of the spine': instead of sea bathing she spent almost all the time in bed. Edith complained that what with the noise, the other guests and Helen's sufferings, 'there were times when I felt my brain rocking in its orbit'. But she was writing – with immense determination – and by early May 1937 she was telling Geoffrey Gorer's mother that, apart from her criticism, the novel was her only book of prose that had ever really satisfied her.

To get the peace and quiet that she needed, she was getting up at six each morning, 'and when evenings come', she told Margery Beevers, 'after a day's furious work on my novel, I am honestly so tired that all I can do is to lie on my bed with my mouth open, – neither reading, nor thinking – just lying there'. By June the book was finished. As she wrote later to Raymond Marriott, she 'felt as if I had been through an earthquake, after I had written it'.

CHAPTER NINETEEN

War Cometh

1937–1939

In the spring of 1937, Osbert returned to London and his grandest social period of all. This was partly due to 'Rat Week' and the successful law suit, both of which enhanced his *salon* reputation, as an insider and a man of influence and rather dangerous wit. The lampoon had, of course, done nothing to upset his closeness with the new inhabitants of Buckingham Palace – rather the reverse – and more than ever now he looked like the uncle of a king. If not an uncle, he was a personal friend of the new King – and of his Queen – and during these last two years of peace Osbert the writer all but disappeared beneath Osbert the courtier, the wit, the polished social figure who went everywhere that mattered – and knew everyone who counted.

He made one attempt to link his writing and his socialising – as he explained that May.

My agent, in pursuit of her ten percent, has suggested that I should try to keep a diary. This is my first trial at it. My resolve is to keep it as far as possible on matters of no importance. I shall, however, when I have time, revert to conversations with Eden which I had at the weekend.

The Edens had to wait, but in the meantime Osbert made his début as a diarist.

Yesterday was rather uneventful. In the evening I went to the opera with Mrs Wynne, the other guests being Oliver and Lady Moira Lyttelton. The opera was *Don Pasquale*, which is being performed for the second time this year and has not been performed before for sixty years. It is a charming work, but my chief diversion occurred during the second entr'acte. The passages were crowded with people from the various boxes and just underneath our box was Lady Cunard. . . . I went to speak to her and to my brother and sister-in-law, who were sitting with her, and she rushed up to one in a coronation contingent of Nepalese.

'The Prince is wonderful,' she said. 'He knows everything. He's sure to have heard of you.'

The Prince, who was not more than three feet high, and who was, on account of his supposed knowledge, continually being asked questions, at once assumed the air of a Chinese conjurer about to produce a white rabbit.

'You know the Sitwells, don't you?' Lady Cunard said.

'Yes,' he replied. 'Poets. I know their work well.' Then he reeled off two of Shakespeare's sonnets and said to me, 'Them by you?'

Feeling it would be rude to contradict him, I agreed.

'Isn't he wonderful,' Lady Cunard exclaimed. 'I call it really touching of him to know that.'

This diary of Osbert's is one of the more tantalising might-have-beens of literature – or, at any rate, of very high-grade social gossip. He knew so many people, had such a natural ear for anecdotes and dialogue, and could be so amusing when he wanted to, that he could easily have been an important, and possibly a great, diarist. But he never wrote another entry.

There were two reasons for this failure. One was strictly practical: he wanted fame and money from his writing, and what literary energy he had was needed for his novel. The second reason is bound up with Osbert's character. As he must have known, the successful diarist has to reveal himself unguardedly – and this was something he was not prepared to do. He was still something of a dandy, and he arranged his life as carefully and stylishly as he arranged his hair, or his splendid suits, or the dining-room at Carlyle Square. Osbert was no believer in life's casual revelations.

Only in the letters that he wrote to David was he entirely at ease, and since David was away again for much of that early summer there is a record of at least a few of the events in his extremely social life. On 17 May, for instance, there is a paragraph about the visit of the Edens while he was staying with Sassoon at Trent.

The Edens as usual are here – and seem cheerful, I can't think why. I rather dread that fruity voice at Badminton on Friday. Lord Duveen motored down for tea. He is very infirm and gaga, but has modelled his social manner on Emerald's – and as they aren't very alike, it isn't a success.

P.S. Duveen says that he was offered all the pictures at the Prado about a month ago, but it fell through.

This contact Osbert had with Eden – and with the other politicians that he met at Trent or Polesden Lacey – confirmed his pessimism on the inevitability of war. It was becoming something of an obsession with him now, making him even more determined to enjoy life while he could: there was a certain poignancy in watching the approach of Armageddon from the houses of the very rich. 'I myself heard Eden say last night that he didn't expect a war at once, but did not see in the end how one could be avoided!! Rather alarming.' But there were other threats to life than war in Europe. Just three days later Harold Macmillan received a telephone message cancelling an appointment to discuss the jacket for an author's book. 'Mr Sitwell has met with a slight motoring accident. He will ring up about the drawings as soon as he is well enough to do so.' In a letter to the absent David, Osbert described what happened.

I am quite all right but had very nearly a nasty accident last night on leaving the Sutherland's Ball at about 1.30 a.m. We were going slowly away from the dance when a car driven by a drunken woman at about 80 m.p.h. crashed into the side where I was sitting. Unfortunately we were going too slowly to be able to get out of the way. I was knocked silly for the time and was much shaken, but otherwise apparently uninjured. . . . I am to stay in bed and see no one for a day or two.

P.S. The drunken woman was awfully troublesome when I came to and kept saying, 'Uncle's much worse. His nose is bleeding.' I had to get the police to remove her from my car.

Osbert was shaken and had mild concussion, 'like being knocked out in the ring'. He was particularly relieved that this was all it was, for had his injuries been worse, they could have seriously upset the most impressive social season of his life. Only a few days earlier, at Mrs Ronnie's, Osbert had met 'the Queen of Egypt and her clumsy fat-boy son [Farouk], Prince and Princess Chiclitz, and the Grand Duke and Duchess of Hesse'; and on the very evening of the crash Osbert had made a social conquest which was to lead to an important friendship. As he explained to David, 'I didn't see the King and Queen to speak to at the dance, but had a short "chat" with Queen Mary. Timmy Chichester, her secretary, told me I had made "a great hit" with her, I'm glad to say.' By early June Osbert, restored and de-concussed, was able to dine at Londonderry House. Queen Mary was there as well, and 'was too charming for words and sent for me afterwards – there was a dance – and had a long talk. I've never known anyone with more charm. I saw one or two people I *don't* like looking very cross.' And, to complete his happiness, there was a small news item which he had noticed in the papers. As he reported, dead-pan, to David, his old enemy Sir John Squire had been arrested 'in Regent's Street for being drunk and disorderly. But failed to turn up in court.'

In July there was fresh matter for concern which none of the family had bargained for – Lady Ida was unwell. Sir George had been playing the 'dear invalid' so long that his hypochondria had tended to obscure her failing health. Only the previous October he had been admitted to a Scarborough nursing home for what he insisted was a serious nervous breakdown. But now Lady Ida was rushed back to England. By the time

Osbert saw her it was already clear that she was seriously ill, and soon he was telling David: 'I am afraid that Mother is sinking. I saw Mother last night and thought I saw a great change for the worse. She recognised me but no one else. And she had now developed a horrible cough.' Pneumonia set in, and on 16 July the *Scarborough Mercury* printed the following front-page announcement:

We regret to report the death of Lady Ida Sitwell, which took place in a London nursing home early on Monday morning.

The news reached us from Mr Osbert Sitwell, who stated that his mother passed away peacefully after a short illness. We understand that Lady Ida, who had only just returned with Sir George Sitwell from Italy, was shortly coming to Scarborough for the summer.

The funeral was held at Lois Weedon church, near Weston. Sir George was apparently too unwell to be able to attend.

It was considerate of Lady Ida to take her departure with, for once, so little fuss. The family might be in mourning, but Edith could write that autumn: 'Renishaw was more perfect than ever this year. Not even the Baronet's brief descent could do a thing to spoil it. It was absolutely perfect.' Sir George was, of course, the problem now, for, however curious the marriage, Lady Ida had always acted as a sort of ballast to his eccentricity. In his strange way he had obviously grown fonder of her in old age and come to depend on her. What could be done now to look after him – and make sure that he did not plague his children?

Osbert and Sacheverell discussed it, and had a brainwave. One of Sacheverell's friends was Francis Bamford, an archivist, a keen amateur genealogist, and until recently librarian to the Duke of Wellington at Stratfield Saye. He had just edited a collection of eighteenth-century letters found at Weston, where they had been written by Sacheverell's great-great-great-aunt, Miss Heber. (Georgia contributed a preface, Sacheverell an introduction, and the collection was entitled *Dear Miss Heber*.) Still only in his middle twenties, Bamford was unmarried and unattached, a civilised, penurious, learned young man who with any luck would find enough in common with Sir George to get on well with him.

As he explains, 'the night after Lady Ida died, Osbert and Sachie rang me up to ask me to take on the job of Sir George's secretary. When I asked what the duties were, Osbert replied airily, "Well, you must just take Mother's place."'

Bamford's first meeting with Sir George might well have daunted any lesser man.

I turned up at the Washington Hotel in Curzon Street where Sir George had an apartment. The curtains in the room were pulled, and in came this extraordinary man who looked just like Don Quixote. He bowed. His manners were quite perfect – but of course didn't mean anything at all. We talked briefly, then he suddenly took off his jacket and climbed into the bed, pulling the cyclamen-coloured bed-cover up to his chin so that he looked exactly like one of the embalmed kings at Cintra in the royal mausoleum.

'You're a Derbyshire Bamford,' he said. 'Good yeoman stock.'

This rather irritated me, but he was perfectly right. He then remarked that a fourteenth-century Sitwell once got a John of Bamford to sign some title-deed. This seemed to convince him that I must be all right, and it was arranged that I would travel back to Italy with him a few days later.

Bamford was to stay as secretary to Sir George until war broke out. He came to like him and enjoyed his company, although, as he says, 'he was a cold fish and extremely strange and somewhat formidable and eccentric and all the other things people have said about him. But he could be wonderful company. He was amusing, with a puckish sense of humour, and had a fund of recondite information. I always found myself wanting to know what he would be up to next.'

Over money Bamford found him both mean and lavish. On the journey out to Italy they went through Switzerland and took their time, and Bamford had to pay all the small expenses of the journey. When it came to settling up, he found that Sir George had taken the precaution of asking the current rate of exchange at each hotel, quite forgetting that hotels invariably gave a slightly worse rate than elsewhere. He nevertheless insisted on paying only this precise amount and criticised the scale of Bamford's tips. 'You'll be bankrupt if you go on like this,' he used to say. Then at the end of the summer he suddenly gave him a hundred pounds for nothing.

Bamford once said to him, 'Sir George, you'd have made a good lawyer, you know.'

'I quite agree,' Sir George replied. 'I would have been an ornament to any Woolsack.'

After her 'absolutely perfect' time at Renishaw Edith returned to Paris. Even on the boat her troubles started, in the way they always seemed to with Edith.

I had an awful crossing [she told David], and an old girl, who had been carried on to Paris against her will, when she ought to have got out at the station for Paris-Plage, pestered me all the way to Paris, because her daughter, who had been left gibbering on the platform, had her ticket and her money.

Much worse was awaiting her in the Rue Saint-Dominique. Helen was more ill than she had ever been; neither the radium treatment nor the 'cure' at Levanto seemed to have done her any good, and she was obviously dying and in great pain. For Edith the galumphing Evelyn Wiel seems to have made matters worse – by her general bossiness and lack of tact. Meanwhile, on 27 September Gollancz had published Edith's novel. It was entitled *I Live Under a Black Sun* – a title Edith chose from a line by Nerval, referring to *le soleil noir de la mélancolie*. The first reviews did little to dispel it. 'Frankly,' she wrote to David on 3 October, 'I was horribly upset, both by the *T.L.S.* and the *New Statesman*. It seems to me that all the Pipsqueakery are after me in full squeak, and Raymond Mortimer's pet dormouse shouldn't have been allowed to bare its teeth at me.' After her success with *Victoria of England* – a book which had meant so little to her – the cool reception of this novel which she had produced with such intensity and suffering naturally depressed her. But this reception by the critics was not entirely surprising – nor can it be dismissed, as Edith tried to, by ascribing it to animosity. For *I Live Under a Black Sun* is a most unsatisfactory novel. It has pages of poetic prose of some beauty, but the plot, what there is of it, seems unnecessarily baffling and obscure. She would appear to have been influenced by Virginia Woolf's *Orlando*, but where that novelist achieved extraordinary switches of time and character, Edith's decision to tell the story of the love-life of Jonathan Swift against the background of the First World War makes the whole book extremely odd. At the time, though, none of the critics can have realised what made the book of such extreme importance to its author. Tucked away within its maze-like plot Edith had hidden the whole story of her love for Tchelitchew.

As if to demonstrate her limitations as a fiction writer, she used several real-life characters as figures in the book. There is a marvellous portrait of Sir George as the retired politician, Sir Henry Rotherham, pacing his enormous house 'like a procession of one person', worrying continually about his health, and being so obsessed with dodging all unnecessary risks, that although agnostic 'he said his prayers every night, on the chance of this proving to be a good investment'.

Wyndham Lewis is the villain, Henry Debingham, looking like 'a rather sinister, piratic, formidable Dago'. Swift – or, as she calls him, Jonathan Hare – is clearly Pavlik, while Edith casts herself in the dual role of the two women who are in love with him. The young and headstrong Anna is the Edith who went to live abroad to be near him; and the older, disillusioned but still faithful Lucy is the Edith at Levanto as she wrote the book.

There is a clear account of the romantic reasons that had first attracted her to France as the exiled genius begs the first heroine to join him.

For if you decide not to come [he tells her], I doubt – I think it very improbable – if you will ever see me again. But if you are brave, if you break away from this useless frittering life, we shall see each other every day, talk and work together. . . . And you will help me. For I need your help more than you know. I *must* have someone who believes in me, who has the faith in me that I have never found in anyone in the whole world excepting you.

But if she decides to join him in his exile, he tells her plainly that it will require 'a terrible and grave sacrifice . . . to give up your life to me, your thoughts, your heart, [and] make *me* your life itself'.

Anna says she is more than willing, but before she finally agrees the man she loves insists on spelling out the grim conditions of their love. 'For if you come, there are certain things for which most women hope – things which mean everything in life to nearly every woman, that you must give up for ever, for they will never be yours – hopes, wishes, which belong to one side of your nature.' He warns her that this sacrifice will mean that her romantic dreams will ultimately wither, but she will also have a clear reward. She will become 'the only woman in all my life, whom I will call my friend'.

It could hardly be much clearer. 'Sex is not love, no it is contrary of love – it is a sort of madness like gloutonnerie, like cleptomania', Tchelitchew once wrote, and Edith had gone to France plainly knowing that sexual love could never be for her. Instead she had been promised close and everlasting friendship with a man of genius.

And there are even closer parallels when Edith describes how Pavlik/Hare started to ignore the older woman, Lucy, and one hears continual echoes of the miserable letters that she wrote to Allen Tanner when Pavlik left them both for Charles Henri Ford. A despondent Lucy hears of Hare's success.

And though he made no reference to sharing it with her, it raised in her a strange feeling, one that she could scarcely analyse.

What did she feel? Her life was so uncertain. She said to herself: 'I must have *something* in life, some hope, something in which I can bury my thoughts, something to build.' And she thought: 'Soon, if this life goes on, I shall no longer even look in the glass.'

She had persuaded herself, by now, that his letters were all she had to lift her out of this dismal life of make-believe, of barely-hidden poverty masquerading as an easy comfort under prying eyes. And pondering over the ever-increasing rarity of those letters, she began to spur herself into jealousy.

Then all the bitterness comes welling up as the ageing heroine opens her heart to Hare/Pavlik's 'uncle', a character clearly based on Tanner.

'You know, you must realise, that I have devoted my whole life, the whole of my existence, *all* my thoughts to Jonathan, for years now, oh, yes, for long years.' And as she spoke, the realisation of the nobility of that selfless sacrifice, grew upon her. What might not her life have been if she had never known Jonathan? Remembering a phrase he had used when they were walking in the forest, she added: 'My life has been given up to helping him make *his* life.'

Reading this highly operatic rendering of Lucy's plight, with its nobility and sacrifice and saintly resignation, one can appreciate why the novel was of such deep importance to its author. For this was her attempt to make her love for Tchelitchew into a great romantic theme; this was her most ambitious effort yet to build a legend from the sadness of her life. No wonder she was so distressed by the 'Pipsqueaks' and 'pet dormice' who refused to see the point.

If the book failed, she was like her ageing heroine, with nothing left to show for her great love now it was ended. Like Lucy, her one reward was 'the little lines around the corners of her mouth', and as she gazed at her image in the looking-glass, 'Her body, her face, seemed blackened, charred, and twisted, as if some enormous fire had devoured her life and had then died in the daylight, leaving only this ruin, this empty hulk slumbering behind lidless and wide-open eyes.'

How enviable her brothers' lives must have appeared to her by contrast! Sacheverell should have been the happiest with Georgia and the two small boys and all the security of Weston. He has his roses and his pictures and the calm routine in which to finish off his book on that spring's travels in Romania. Osbert's life had little of this sort of domesticity, but it could boast excitements that Sacheverell's lacked. On 19 September 1937 he was scribbling a reassuring note to David from Balmoral, where he had been attending the ball given by the Royal Family for their employees on the estate. 'The ghillies' ball was such fun; a very pretty sight, and every kind of Primrose League dance; Veleta, Great Rushing Sergeant [*sic*] etc. Party included the Gloucesters, Princess Helena Victoria, the Carnegies, etc.' Despite his weight and gout and unconcern for women, Osbert remained a practised and enthusiastic dancer who enjoyed these jollities with this elevated clientele. Besides, that autumn he had finished off his novel, and felt justified in indulging in his social life. During October it became so all-absorbing that the idea of that wretched diary must have nagged at him again. Unfortunately, he never quite continued it – but he did leave a few unpublished notes which make one realise how good it could have been. It was now, for instance, that he met a statesman he always thought that he despised.

On Saturday I went to Cumberland Lodge to stay with Lord and Lady Fitz Alan and met there, for the first time, Lord Baldwin. Lord Baldwin I found to be a man of extraordinary charm. He told me two or three things which amused me. 'It was nearly a year ago that the Mrs Simpson business began, and the actual place where it began was Cumberland Lodge.' Lord Baldwin once or twice touched on this to me, and I asked him, 'Had he met Mr Simpson? What was he like?' He answered, 'I do not know if your knowledge of political life is sufficient, but he is the sort of man who would be chosen by a London borough as a Conservative candidate, because they could feel quite sure that he was a gentleman.'

Later on I asked him if 'he knew they had started a Warfield Museum (for Mrs Simpson) in the States in her old home.' He said, 'Yes, I had thought several times of sending them a signed photograph of myself.'

This was the sort of company and gossip Osbert loved, and it was maddening to be forced to leave the London scene now at a moment's notice. But there was news from Italy. Sir George again seemed bent on interfering to prevent his eldest son enjoying life. He had retired to his favourite sixteenth-century bed and was cherishing

a certain sense of grievance . . . at his children not being around him . . . telegrams were suddenly shot at me from all sides . . . to tell me that my father was ill and that it was now my duty, however disagreeable, to be at his side. I at once took the train to Florence.

Was Ginger dying? Osbert would have been a good deal less than human had the hope not struck him of arriving at Florence to discover that the old nuisance was already gone, and that *he* had inherited the title and the castle and the Sitwell fortune.

But there was no such luck. As Osbert reported back to David, 'I've seldom seen Ginger looking so well: and in high spirits. He's good for several years yet, though he looks old and frail.' Bamford attempted to explain that not even the local doctor had been certain whether Sir George was suffering from serious internal haemorrhage or from haemorrhoids. Probably the latter. Osbert failed to see the joke, and frustrated anger brought on an attack of gout. The only bright spot of the dismal trip was that he had lunch with Reggie Turner. The ugly little man was very old, but still amusing and still a source of all the latest Florence gossip, particularly about Norman Douglas. 'Norman's trouble is getting worse: he can't ever return to England,' Osbert told David.

Edith was ill again – 'I have collapsed completely, and have been in bed since Monday night with agonising pains in my head,' she wrote to Tanner on 3 November. As for Osbert, all the aches and pains from Florence seemed to pursue him through the winter. To escape them he and David took a villa at Hyères, between Toulon and Cannes. They were together there for Christmas, then had seven long weeks of splendid weather, but for the first time in his life neither David's presence, nor the sun, nor the South of France seemed able to restore him. Even a day or two at Monte Carlo brought complaints of rheumatics and lumbago in a note he sent to Harold Macmillan in London.

He suddenly appeared to have acquired another enemy – his body. He was forty-five and still considerably overwight. He ate and drank too much and cures and dieting scarcely seemed to help. Heredity was catching up with him and, like Sir George, his health would henceforth be one of the major interests in his life. He was rapidly becoming the 'dear invalid' himself.

One thing he could look forward to was publication of his novel. It was due out in March, and Osbert took unusual care with the peripheral activities of authorship, fussing over dedications, presentation copies, lists of possible reviewers. 'Will you please *not* send a copy to the Nation and Statesman [*sic*];' he wrote to his publishers: 'if they telephone and ask why, the answer is because they are impertinent to artists, if they can be. An awful paper I think.' In the same letter Osbert included a long list of friends who *were* to receive copies of the book. It included Somerset Maugham, Michael Arlen, Maurice Baring, Raymond Mortimer, John Sparrow and Her Majesty the Queen.

When *Those Were the Days* came out in April 1938 most of the reviews were quite polite – except for one by James Agate, who panned it in the *Daily Express* – but the triumph Osbert had hoped for never came. Sales were disappointing, so was his health, and it was a thoroughly depressed author who wrote to his publisher on 5 May.

My Dear Harold,

... I am only sorry that my first novel with Macmillans should not have had a whopping success. I am so tired of 'moral success'. I think the reviews have been excellent, and I believe it to be my best book: (author's vanity, I suppose).

The doctors now tell me that I am much *worse* after their dieting, so I revert to living as I like and hoping for the best. They say it's overwork and worry. And that Vichy (where they intend to exhaust me by their treatment into a vegetable state) will put me right!

Those Were the Days, which was to mark the end of Osbert's career as a novelist, is an interesting pointer to the future, demonstrating, more than anything else he wrote his strengths and his deficiencies as a writer. This was his great attempt to write an important novel, but he is simply not a major novelist. The lack of ideas, the failure of the characters to come alive, the weakness of the dialogue, make this the sort of book that with the best will in the world one never finishes. But buried within its pages there is something other than an ordinary novel. From time to time one seems to catch the rich, unmistakable tone of Osbert's voice reminiscing about the people and the places he has known. Standing behind the novelist is the embryonic presence of the memoirist to come.

The failure of *I Live Under a Black Sun* left Edith dispirited and at her lowest ebb. She spent the winter in Paris, but seemed to have lacked the will for fresh creative work. What time she had she spent on the new verse anthology she was preparing now for Victor Gollancz. Apart from this the dying Helen Rootham seems to have occupied her days – and nights. As 1938 began, she wrote to Sacheverell:

I am very despondent. I *had* thought that poor Helen was a little better, because there have been none of those dreadful shrieks for nearly a fortnight, and she has stopped moaning in her sleep. Let us hope to heaven that it will go on like that. But the nurse says it is only like that because she is under the influence of morphia the whole time; and new symptoms have turned up, and the nurse says it has not attacked the liver.

There could be no more trips for respite to Levanto. Helen could not be moved. Only her death could free Edith from the Rue Saint-Dominique.

In Europe it was the summer of appeasement. In February 1938 Eden had resigned, and two months later Hitler had marched unhindered into Austria: Czechoslovakia was threatened. For Osbert it was a time of gout and limbo as he waited for the European war which he had long predicted. Sacheverell and Georgia left for the Tunisian coast where they were sharing a villa with the wife of Edward VIII's former equerry, Major Edward 'Fruity' Metcalfe. According to Sacheverell, there was 'No place along the coast of Northern Africa, with such benevolence of climate and situation. The winter days are bright and clear; the summer . . . a paradise of peace and enervation.' Despite the enervation, he was able to continue with his long-projected book on Mauretania.

With Sir George still worrying about his health in Italy, Sacheverell in Africa, and Edith incarcerated in the Rue Saint-Dominique, there was no point in having the usual August party at Renishaw. David left for Switzerland, thinking it would probably be his last chance to see those middle-class mountains that he loved before war descended. Osbert stayed on at Carlyle Square, re-working his articles on Angkor and China into the full-scale travel book, *Escape With Me!* – and irritably correcting proofs of *Trio*. This was a joint book of lectures which he gave with Edith and Sacheverell the year before at London University. Edith had spoken on 'Three Eras of Modern Poetry', Osbert on 'Dickens and the Modern Novel', and Sacheverell on 'Palladian England' and 'George Cruikshank'. There was a slight crisis with Macmillan over Sacheverell's proofs, and Osbert reacted like an impatient elder brother: 'Really, Sachie is too difficult for words – he has now gone gallivanting off to the North of Africa, and goodness knows what he can have done with his proofs.' He suggested sending a fresh copy out to Tunisia, 'Or perhaps,' he scribbled as a postscript to the letter, 'I can correct his bl. . dy proofs for him.'

When war seemed imminent over the Czechs that autumn, Osbert was at Badminton staying with his Beaufort cousins. Queen Mary was there too. Osbert was flattered by the interest she took in him, but the life of a courtier – particularly with old Queen Mary – seems to have been demanding. There was barely time even to write to David in Geneva:

I haven't written before, as I've been kept on my gouty feet, day and night. At the moment am in bed resting with a bad attack of motor nausea after Louise's old Rolls.

I am afraid the position grows more desperate hour by hour. (It is now said that the Queen returns to London on Monday, instead of Thursday: which may mean she has information of developments.) I am worrying about you and how you'll get back, though I'm sure as long as there's a railway you'll be all right. Miss you frightfully; but I am glad to be here, where everything is peaceful, and yet we get all the news. Anyhow, you *won't*, will you, return through Germany.

Yesterday we went over to Savernake for tea. Lady Ailesbury hit me with a walking stick. She weighs twenty stone and says to her husband, 'Chandos! Don't contradict me please!'

Osbert was far too realistic – and too pessimistic – to share the euphoria which greeted Chamberlain's return from Munich. He had no faith at all in declarations about 'peace in our time'. It was too late for that. But nor had he any faith in Churchill – and still less in Eden. He was certain there would be a war within a year and, as he explained later, 'After the Munich crisis I found the mounting war tension in Europe suffocating, and I felt that I must go far away and obtain a draught of fresh air before war, so claustrophobic in its character, should break out again in the autumn.' Neither he nor David had ever seen South America, and this might be the last chance they would ever have. There was a luxury German cruise ship, the *Cordillera*, sailing from Hamburg for Guatemala just before Christmas.

It was to be a haunted voyage, for, apart from a few rich tourists like Osbert and David, the ship was crammed with German Jews fleeing from the Fatherland to South America. *Escape With Me!* was taking on a grimmer tone than Osbert had envisaged, and Jewish refugees were hardly the fellow-passengers he would have chosen.

But there was more amusing company aboard – Beverley Nichols and his friend Cyril Butcher, who, like Osbert and David, were intent on one last fling (in their case on some of Nichols's blocked German royalties) before the war began. They knew each other reasonably well, and soon Osbert and Nichols were discussing their work with the frankness that can sometimes come on long sea voyages.

He was [says Nichols] terribly depressed about his future, and finally confessed that he knew quite well now that he was not a creative, imaginative writer. He even said how much he envied what ability I had to string a story together out of nothing. I remember that we spent some time together discussing subjects that might suit him. I suggested a biography of Edgar Allan Poe. I told him he could come back through the United States and see Boston, and that Poe's life had all the ingredients that should appeal to him – the inspired writer against a philistine background, a passion for the supernatural, and that extraordinary gothic imagination.

The idea came to nothing, but the voyage brought further revelations about Osbert's character. According to Nichols, Osbert was most put out by the presence of the Jewish refugees, and made no effort to disguise the fact.

There was a swimming-pool and the swimming was divided into separate hours for the first- and second-class passengers. Osbert and I turned up for the first-class period, only to find the pool already occupied by several Jewish first-class passengers. He didn't care for that at all, and turned to me and in a loud voice bellowed out, 'Oh, Beverley, and when d'you think it's Aryan hour?'

The story has a sequel. The *Cordillera* arrived at Guatemala – but the refugees discovered to their horror that they were refused permission to land. There was nothing for them but to return to the persecution of Nazi Germany.

And suddenly, no one was more upset and hurt for them than Osbert. He did everything he could. He laid on parties for them – and even for their children, although normally he couldn't bear children – and I can still see him pouring out champagne for the parents, and playing party games with the younger ones. Some of the parents must have been the very people he had been so rude to in the swimming-pool a few days earlier.

Osbert and David spent three months in Guatemala, almost certain now that this would be the last of their winters of content. The gardens of the West were closing. The future seemed to hold no place for ageing children of the sun. In Antigua Osbert finished his book on China, while David began a novel. They were back in London for the nervous spring of 1939.

Edith was still in Paris. Helen had gratefully expired the previous October; and her death had inevitably produced a flood of contradictory emotions. From various letters and references that Edith left, it is quite clear that one side of her had grown to dislike her former governess intensely; though it is hard to tell how much of this was due to Helen's natural overbearingness, and how much to the ghastly situation in the Rue Saint-Dominique. Edith went in for ambivalent relationships, and none more so than with the unfortunate Miss Rootham.

Edith certainly admired her courage in the face of all her suffering. She also felt enormous gratitude and obligation to her for originally 'rescuing' her from Renishaw and making it possible for her to live in Moscow Road. She even appears to have depended on her almost to the end. But Helen *had* become a burden and a frightful tie. The household in the Rue Saint-Dominique had been a place of misery – for Edith almost as much as for Helen. Once Helen died, a new life could begin for Edith. Or could it?

Later she referred to the days following Helen's death as a time when she was 'unthinkably lonely and unhappy'. She wrote pathetically to David Horner that she had caught a chill 'visiting poor Helen's grave and have been rather seedy'. But the worst feature of this time of mourning was that she felt deserted by the one human being who could really help her: Tchelitchew. Later she wrote bitterly to Allen Tanner of how she had not been able to see 'how I was going on with my life, *and P. left me to it*'.

But Pavlik had made it fairly plain that he was completely past caring about Edith any more. Not long after Helen died he was in England for the triumphant first night of Massine's ballet on St Francis for which he had painted the designs. Earlier that June he had scored a great success in London with his latest exhibition at Tooth's Gallery. Now he was acclaimed by Edward James, Kenneth Clark, Lord Berners and Lady Cunard. Edith had come back briefly from Paris, but he had no time for her. Charles Henri Ford was with him, and on the one occasion when the excited painter met Edith he gazed disapprovingly at her mourning, then told her with more truthfulness than tact: 'It makes you look like a giant orphan.' Back in Paris, Edith exploded to the ever-faithful Tanner in a rush of stately outrage over Pavlik's continued friendship with Charles Henri Ford:

I could not believe *anybody* could be guilty of such absolute disregard of both your and my feelings as to take that person there. . . . I am so angry at the thought of this perfectly worthless parasite, with no gifts, and a thorough slum type, throwing his weight about in England, and I am powerless to do anything, because if I got him chucked out of every house because of that book it would ruin all prospects in England. It is nothing short of an outrage that he has been taken there. P. does not mind what pain he inflicts on one.

She was, she added somewhat unnecessarily, 'very angry and embittered'. But anger and bitterness were not much help in solving the real problem now with Helen gone. How, as she put it, was she going on with her life? She was very much alone, and free of obligations. Should she start writing poetry, or another novel, or fresh criticism? Early that spring it sounded like the latter. David Horner wrote to her saying that his friend Miss G. B. Stern was taking on one of her lesser enemies. Edith responded eagerly at once.

I am delighted to hear that G. B. Stern is going to launch an attack on Master Isherwood. But *where*? Do you mean in a paper, or just straight-from-the-shoulder. . . . Really I am so sick of the Auden Spender Isherwood clique. It is just a ramp. They all have a *small* talent, but how insignificant. If they hadn't all been at Oxford together, they would never have been heard of – I've got an attack waiting for them!

But she never launched it. Instead, she had suddenly become excited by the idea of America. Her agent David Higham had cleverly suggested that the time was ripe for her to hit the lecture circuit, but the idea ultimately foundered because of Edith's insistence that she had to *read* her lectures.

I have no memory at all, and my lectures are highly complicated, and, in the case of those on poetry, very highly technical. One slip would be fatal. . . .

They must think I'm not an artist but a trick cyclist. Next they'll want me to deliver my lecture cycling round and round the platform balanced on my nose with my feet in the air, and declaiming at the same time on the effect of texture on the caesura! So if they won't let me lecture from manuscript, then I cannot go to America.

Since America was hopeless, and she felt she needed to escape from Paris come the autumn, where could she go? She chose the last place in the world one would have expected.

At the end of May 1939 Sir George wrote to her from Montegufoni. He planned to visit Paris on 3 June and, since he was suffering his usual trouble, needed Edith's help. 'It is possible that the fatigue of the journey may bring on the haemorrhage again, and if so I would need a nurse to come in to give me twice a day injections of a certain drug I carry with me.' After a page of medical logistics, he ended up his note as follows:

It is delightful to think I may see you in Paris, and if all goes well that you may come for a long stay to Montegufoni in the autumn.

Best love, darling,
Your loving father,
GEORGE R. SITWELL.

The buffeting of life, the faithlessness of Pavlik and the death of Helen Rootham had made Edith turn for solace and support to the one retreat still left to her – the family. Before leaving for Italy that autumn she would see her brothers. As she wrote to Osbert on 16 August, 'I am *longing* more than I can say for Renishaw', but the summer schedule was a little late that year. Osbert and David had decided they would make the most of Europe in that last doomed summer, before the spas shut up for good, the roulette wheels ceased to turn, and the expensive restaurants closed their doors. August found them still at Vichy. 'The cure is so exhausting that I spend most of my time sleeping,' Osbert wrote to Harold Macmillan. Then as the German troops massed on the Polish frontier, Osbert and David went to Monte Carlo. Osbert was a great believer in fate's little repetitions. This was where he had celebrated the ending of the last great war. It was a good place to prepare oneself to meet the next. He gambled and, as usual, lost – but not a lot – and on 1 September was back in Renishaw. His friend Queen Mary was expected on the twelfth, but history intervened.

CHAPTER TWENTY

The Great Interruption

1939–1942

For Osbert war was simply 'the great interruption', both tragic blunder and appalling nuisance which put at risk, or put an end to, almost everything that made his life worth living. It was also, as he saw it, the quite senseless culmination of all he had campaigned against and warned against for years; democracy and militarism, cheap rhetoric, political incompetence. For him this present war was no more than a recrudescence of the same great war that he had fought in Flanders twenty-two years earlier. He had forsworn war then, and saw no reason to espouse it now.

He had no patriotic zeal and no enthusiasm for the great events his country was engaged in. He felt himself at liberty to contract out, and stoically and more or less bad-temperedly composed himself to wait until the storm blew over. In the meantime he had other things to think about.

Queen Mary's visit was inevitably cancelled. This was a pity, as it would have been the first royal visit since the Prince Regent came to Renishaw in 1808 and honoured Sitwell Sitwell with the baronetcy. More important now was what was to be done with Renishaw itself. It needed to be occupied: otherwise who knew what horrors would be perpetrated on that great, defenceless, empty, country house, with all its treasures? Houses were being commandeered for troops, for civil servants, even – perish the thought – for working-class evacuees from London. Much as he hated the idea, there was nothing for it but to go and live there permanently. Luckily the house was too uncomfortable for all but the most desperate Londoners. The haunted wing was derelict, and throughout the house there was scant heating and no electric light. Osbert had never stayed there for more than a few months at a time, and then always in the summer.

But Sir George had grown his potatoes there during the First World War. Osbert could do likewise in the Second.

Living there would solve another problem – Edith. She had been planning to return to Paris at the end of that September, but there was now no question of her going back. All her possessions – drawings by Tchelitchew, manuscript notes of poems, furniture and clothes – were in the flat, but Evelyn Wiel was there to look after them. Until they knew for certain what was happening, Edith and Osbert would stay on at Renishaw.

Osbert was forty-six. With his uncertain health there seemed no danger of his being needed for the war. David, on the other hand, was still in his late thirties, and had already volunteered for the R.A.F. He returned to London and continued living at Carlyle Square – looked after by Miss Noble – until he was finally called up.

Sacheverell naturally stayed at Weston, but what about Sir George? He was nearly eighty and, since Francis Bamford had returned to join the army, was now living on his own in the echoing great castle with the gardens and his *contadino*, Luigi, as his only company. Luckily he showed no sign of wishing to return and, not surprisingly, Osbert felt no great need to persuade him to.

It is strange how many problems war was solving – even the problem of the tortured and unhappy love Edith still nourished for her Boyar. When the war started he had been staying with Charles Henri Ford and Ford's mother at a villa on Lake Annecy. He sensibly decided that he owed it to his art to leave as soon as possible for America. Giraudoux helped them find a passage, and on 8 September Pavlik was writing:

> My Dear Edith,
>
> Here I am in New York, a terrible shock to find again such a sea of insensitivity and mediocrity – the women above all – who are dreadful – and the men no better.

It was the beginning of what may well turn out to be a historic correspondence. (As all the wartime letters between Edith and Tchelitchew are under lock and key at Yale until the year 2000, this will be posterity's privilege to decide.) At all events, the Boyar was transformed from a faithless presence into a faithful correspondent. Distance had made him the one lover she could cope with – a literary one. For the remainder of the war he would write every week: she likewise. As Camus once wrote of two of his characters, 'Separated, they write to each other, and he finds the right tone and keeps her love. A triumph for language and for good writing.'

After the shock of war came the anticlimax of the 'phoney war'. Osbert was still snatching at the last remnants of the peacetime luxury he loved before the war began in earnest. But David by now was already working at the Air Ministry as Pilot Officer Horner, and they could only correspond. In mid-October Osbert was writing miserably to Harold Macmillan in London: 'I long for a little news: and remain here, layered in white mist from morn to night, knowing *nothing*. . . . I do hope to hear that you will come over soon from Hardwick.' His politician publisher had more important things to do in London, and as the Derbyshire autumn turned to winter Osbert began to face the uncomfortable realities of life. He and Edith were together for the first time since childhood, and David had departed. He missed him terribly and wrote continually, although even that was difficult as their letters could have no endearments now – nothing that could possibly give a censor or a blackmailer grounds for making trouble. The only indication of his feelings that survives is a short poem that he wrote but never published:

> For me to say, 'I love you!' means 'I live!'
> To live without you is a waste of breath,
> Our love makes every day more fugitive
> And yet, without you, even life is death.

In such a situation it is surprising that two such bad-tempered characters as Osbert and Edith could have survived together. It says much for their underlying fondness for each other that they did. Also, in her way, Edith was a great accepter; and after the nightmare of her last few years, Renishaw, for all its cold and damp and lack of company, must have appeared a haven and a home. Part of the secret of the success of her relationship with Osbert lay too in a certain lack of closeness. She adored him but, as we have seen, she never confided deeply in him – nor he in her. He was her 'dear old boy', but there were curious taboos between them. A great deal must have stayed unspoken – particularly over David. Those perfect Sitwell manners had their uses, and in fact Edith would gradually rely on David as a confidant and intermediary with Osbert, rather as she had used Allen Tanner to get through to Tchelitchew.

Another big advantage they possessed was Renishaw itself. True, it was somewhat primitive – though not by any means as rough as the Rue Saint-Dominique – but it had one great asset – space. For much of the day they could live virtually separate lives. Edith had her bedroom at one end of the house on the first floor. Here she would stay in bed all morning (like her mother), reading and working. Osbert's bedroom was fifty yards away, and he had his separate work-room now where nobody dared enter. It had his favourite work-table (painted at Roger Fry's Omega Workshop before the First World War), his eighteenth-century gout stool to support his leg, and a view across the gardens to Sir George's ornamental lake. Soon he would have some favourite possessions sent up from Carlyle

Square – a shell sofa, a big Venetian mirror and a rare set of Chinese birds in wax. And gradually he would impose his own routine upon the place. Mornings were for work. Lunch was simple but extremely good, particularly the vegetables which came from the estate. (It was to be the boast of the Renishaw head gardener that he could produce fresh peas for Osbert from the garden for two hundred days a year.) The afternoon was devoted to Osbert's customary walk through Eckington Woods, and to the business of the estate which was carried on with the agent, Maynard Hollingworth, in the kitchen wing. (For the first time in his life Osbert, through force of circumstance, was now to become the Derbyshire squire his ancestors had been.) Not until evening were Osbert and Edith closeted together. They read the papers in the library, listened to the news at nine o'clock, and Edith knitted.

For two writers it was a life that had one great advantage – there were no distractions; no dying Helen Rootham, no Sir George, no Tchelitchew or David, and no dowagers. There was not even any real contact between Renishaw and Weston, and Sacheverell had grown increasingly reluctant to visit the house where he would never live. Gradually he and Osbert had pursued their separate lives and now had less and less in common. There had been no break, no open disagreement, but they met once or twice a year for lunch – and that was that.

At Renishaw the chief enemy, of course, was boredom, particularly during the early months of 'phoney war'. There was some talk of Osbert collaborating with Lord David Cecil and Evelyn Waugh to launch a new literary magazine which they would call *Duration*, but the idea was soon pre-empted by the start of Cyril Connolly's *Horizon*.

Osbert bore his burdens stoically – but not without complaint, particularly to his secretary, Miss Andrade. She remained in London, where she acted as factotum and sympathetic listener to his various laments. David apparently got leave just after Christmas, for on 7 February 1940 Osbert was writing to her:

> It's dreadful here. Rain dripping through old roofs. David in bed with influenza, and a temperature of 103. Three feet of slush outside in the park, and Lord Haw-Haw making propaganda out of a photograph of my brother Sacheverell in the Sketch. . . . Oh dear!

The reference to Sacheverell concerns a painfully unfair but most adroit piece of propaganda by the one-time admirer and henchman of Sir Oswald Mosley, William Joyce. On Christmas Eve the Duchess of Westminster had arrived at Weston to spend Christmas with Sacheverell and Georgia. Fuel was scarce, but to celebrate Christmas a log-fire was blazing in the drawing-room. The Sitwells and their friends all dressed for dinner, and one of the guests, the Greek photographer Costa, photographed them around the fire. A few days later the photograph was published in the society magazine *Sketch*. A copy must have reached Berlin, and Joyce made a point of referring to it in his next 'Germany Calling' broadcast to the British. This, sneered Lord Haw-Haw, was the way the English upper classes bore the sufferings of war, while the lower classes had to do the fighting.

Just a week after Osbert wrote his letter to Miss Andrade, there was more serious trouble nearer home. Gollancz had finally published Edith's big anthology to more or less favourable reviews, except for one by Hamilton Fyfe in the left-wing Sunday paper, *Reynolds News*. Fyfe quite liked the anthology – 'it would be a delightful bed book if it were not so heavy to hold' – but he used most of his review to elaborate on the subject of the Sitwells, with the patronising assurance of the young journalist he was.

> Among the literary curiosities of the nineteen-twenties will be the vogue of the Sitwells . . . whose energy and self-assurance pushed them into a position which their merits could not have won. . . . Now oblivion has claimed them, and they are remembered with a kindly, if slightly cynical, smile.

This talk of oblivion was too near the bone for comfort; Osbert was most put out, and sent the review to Philip Frere to ask if it was libellous. Cautious as ever, Frere replied that he would need to consult learned counsel before saying.

Gout, Renishaw and *Reynolds News* made Osbert long to be back again in Europe, far away from all these irritations. He was more or less resigned to his exile's life in Derbyshire when suddenly an opportunity cropped up to go to Italy. It might have seemed impossible, with Italy on the edge of a heroic war with Britain, but the British Council was still gallantly engaged in bringing British culture to the Italians. With France about to be invaded, and Mussolini in daily contact with the Germans, it was clearly in the best traditions of the British Council to invite Osbert Sitwell out to Florence to lecture on modern English literature.

Osbert was taken with the idea. Apart from seeing Italy again, it would enable him to see Sir George. The family was becoming understandably worried about the old man now. He was eighty. His memory was fading. Few of his letters made much sense, and with the likelihood of war something decisive must be done about him and his property – particularly his property.

The trip was fixed for early April, but Osbert was now delayed by that plebean malady – chickenpox. Edith had it, and he was forced to wait until the end of April to see if he had caught it too. In the meantime, with the alarming news from France and Italy, he thought it best to ask his influential publisher for reliable advice. Macmillan advised caution, and on 20 April Osbert replied:

I agree with you, that it is most doubtful if I get there at all in the end: and I will be most grateful if you would be so kind as to let me know anything you hear about the Italian attitude. I am most anxious not to get to Italy and then find myself interned. So for heaven's sake warn me if you think it too dangerous.

There was apparently no further warning – nor, for that matter, any sign of chickenpox – for in the second week of May Osbert crossed the Channel, stopped the night in Paris, then made his way by train to Milan. He described what happened a few days later from the Hôtel de Paris in Monte Carlo in a curiously stilted letter to David:

Mon Vieux,

The British Council wired to me in Paris, 'In spite of deterioration of situation, advise you to go, but it must be your own responsibility.' The chief military attaché said it would be madness to go, but I thought I had better. When I got to Milan, the chief of the British Institute advised me to deliver my lecture and return at once. The Simplon was out of bounds, Modane choked with returning Britons, and so, after advice from the British consul, who had received a message from the British Embassy in Rome, I got a sleeper here (it took twelve hours from Milan) the last one.

All the English were leaving Milan. The atmosphere was *indescribable*. People singing outside my hotel anti-British songs until 2 a.m. and constant demonstrations. . . . I am waiting here for a sleeper to Paris (II class) and hope to arrive by afternoon on the 23rd, this letter being taken home by Jim Barber, a friend of Jack Lysaght's. . . .

I only fear developments may make me set out with my grip before Tuesday. The Wops are certainly coming in against us. I miss you more than I can say, and constantly long to talk to you. I saw Boyer [his old doctor]. Am having blood test tomorrow. No money for gambling. Very bored. Am lunching w. Willie [Maugham] tomorrow.

This whole rash expedition seems untypical of Osbert's careful nature – particularly in the face of such extremely daunting warnings. Perhaps he could simply not resist the lure of an adventure. Perhaps he thought it worth the risk for one last lover's glimpse of Europe before the iron gates of war clanged fast. It was unfortunate he never saw Sir George. He and the family might have been saved a lot of anger and exasperation if he had.

During the first year of war Osbert and Edith both kept their links with London. David – and other friends – could visit them quite regularly, and Osbert stayed as regularly at Carlyle Square. Edith, for her part, used the Sesame Club as the centre of her London life. It was to be a place of great importance to her – base, refuge and campaign headquarters.

She treated both the club and the other members with off-hand disdain:

Last night [she wrote to David on 1 April 1940] an old girl died in the smoking-room. At first 'they' thought she was merely having a fit, so took no steps about it, and just left her to herself, but after a bit – half an hour or so – it struck them she was curiously quiet, and then all was discovered.

But the Sesame suited her and would finally become as much the natural background to her London life as Moscow Road had been. In the first place, it was an exceptional bargain, as she realised – a dining-room and bar and bedroom in the heart of Mayfair for less than a pound a night. The somewhat frowsty décor suited her as Moscow Road had done, enabling her to shine against the drabness of the background.

The summer of 1940, with the loss of France and the beginning of the Blitz on London, inevitably increased the sense of isolation now for all three Sitwells.

Sacheverell at Weston was occupied with Home Guard duties – and simultaneously enjoying one of his years of superheated literary fertility, with no less than three books published. The first was his *Mauretania*, the second a small book on a subject dear to Sir George, as well as to all his children, *Poltergeists*, and the third and most original of all, *Sacred and Profane Love*, the first of his 'travel books of the spirit, explorations in the main of art and artists and reflections that arise from these aesthetic journeys'.

Towards the end of that July Osbert made one of his few trips to London. He met David briefly to see Shaw's *The Devil's Disciple* at the Piccadilly Theatre – then it was time for Renishaw again. Even their letters now become attenuated. David became 'My Dear Old Boy', Osbert, less heartily, 'My Dear Osbert'.

David apart, one of the very few friends who found the time and energy to keep in touch with Osbert that summer was Rex Whistler, recently commissioned into the Welsh Guards and stationed at their camp at Sandown Park. He wrote a touching letter to 'Darling Osbert' asking him for help in selling a painting or two among his richer friends. There was, for instance, a small portrait he had done of Lady Aberconway's daughter, Ann.

If you ever see Sam [Courtauld] could you trick him into buying it? (Or is Sam still angry with you?) – You know so many people who *do* occasionally buy paintings. I have a little coming in, but am frightfully poor at the moment, which just means that I can't *drink* anything in the mess. You can imagine what *that's* like, when everyone is swilling all round and you've got the habit anyway!

Osbert could not do much to help: that summer the rich had other things upon their minds than buying pictures.

The fall of France was a particular blow, of course, to all three Sitwells, the amputation of a great part of their lives. Montegufoni was in enemy territory; there had been no further news of Sir George, and since the fall of France there had been no news either from Evelyn Wiel in the Rue Saint-Dominique. (Edith now began to worry about 'old Moby Dick', as she had nicknamed her.) 'My morale is *very* low,' wrote Osbert to Miss Andrade, who was staying imperturbably in London, 'I am quite miserable and helpless. . . . P.S. Would you be a super-angel, and order a hundred original Sub-Rosa cigarettes for me from Sullivan and Powell, to be sent here.' Later he was telling her: 'We have just received some wretched refugees; poor things, so lost and unhappy – but not more than I am.' He was not the person for the Dunkirk Spirit. Instead he wrote gloomily of 'the general *malaise* by which everyone sensitive and intelligent has been caught'. The loss of France was one more proof of what he had always known – that war was a disaster, and that politicians were incompetent to wage it anyway.

It was almost a relief to have some German air-raids in the district. They relieved the boredom, and were the equivalent, perhaps, of those acts of violence he had indulged in once when troubled and distressed. 'I'm glad that a good many bombs have fallen round here,' he wrote to Miss Andrade on 19 August, and from then on his letters give a running commentary on the Luftwaffe's efforts to attack the Sheffield factories, all related with a sort of quiet satisfaction:

Your prayers about bombs keeping their distance from Renishaw and myself have been answered in the usual way. The night before last there was a tremendous raid, though only by one airplane, at 11.15 and at midnight! I thought every window in the house would be blown in. And the rooms shook as though in an earthquake. The bombs fell about 1½ miles away.

P.S. Could you please ring up the herbalists, (Culpepper House) in Bruton Street, and ask them to send me three tins of Paris Gold for the bath, and also one box of tubes of it. I have an account there.

Then a few days later he reported: 'The bombs are raining down nightly. Several fell last night on our land. Sheffield has been badly hit.'

But despite the bombs and the difficulty getting his 'Sub-Rosa' cigarettes, there was one splendid happening that made all the disasters round him seem more bearable. On 21 August he wrote to the absent David:

Dearest Pilot,

I still can't get over your news. I woke up crying this morning, which I suppose shows what we have all gone through. . . . Edith is delighted at your being not so far away.

The 'news' was that David had been promoted Flight Lieutenant and posted as intelligence officer to the air force base at Watnall Chaworth, less than twenty miles from Renishaw.

In better spirits now, Osbert made a brief trip down to London at the beginning of September, staying in his empty house in Carlyle Square with only Miss Noble for company.

It was a relief to get away from Renishaw even for a few short days. His soldier servant, Robins, now the butler, seems to have been grating on his nerves. It was one thing to tolerate the servants' free-and-easy manners (which had once so intrigued young Evelyn Waugh) when you were with them merely once or twice a year. It was quite another to be forced to endure them all the time. 'The servants' hall', he told David, 'is just like a night club. I call it "The Cave of Harmony", but as a joke it has not gone down too well.'

In London he had business to attend to – what he called 'the gay round of Coutts, Macmillan, Pearn [his literary agent], Philip Frere'. He lunched with Alan Pryce-Jones and dined with Maugham. But the air-raid sirens started as he left the Dorchester, and Maugham was away to America in two days' time. '"Oh dear and oh dear!", as Xtabel says,' he wrote to David. 'And I believe the Huns are starting on us any day now.' This was the beginning of the invasion scare.

There seemed no point in staying on in London for more than a day or two; not because of danger, but because his friends were leaving or were already in the forces, and the city that he loved was closing down for the duration. Besides, David was no longer there. 'Your bedroom [in Carlyle Square] has a mournful, vacant look.' So after arranging with Miss Andrade to have more furniture and his favourite big Venetian books sent up to Renishaw, he came back to Derbyshire himself. The war would obviously last longer than anybody thought. Wars generally did. Renishaw appeared as good a place as any now to see it out, and David's presence twenty miles away did much to reconcile him to it.

With David at Watnall, Osbert's private life improved, but the outside world that autumn maddened him. Predictably, the rise to power of his ancient *bête noire* Winston Churchill did little to improve his temper; nor did that great man's handling of the French fleet at Dakar towards the end of that September reconcile him to his heroic leadership. 'Dakar made me ill with temper,' he told Miss Andrade. 'Blast Churchill! . . . What will that day of national prayer in the States do for us?' The popular enthusiasm Churchill roused throughout the nation depressed him further. Churchill had not changed. He was still the unforgivable Winston Churchill of *The Winstonburg Line*, the bloodstained demagogue whose natural habitat was war. If people rallied to him, that merely confirmed Osbert's profound contempt for the public judgement of 'the little man'.

Anti-Churchill, anti-democrat and anti-war, Osbert was withdrawing more and more into his mandarin seclusion in the winter mists of Renishaw. And that autumn, in the darkest days of war, he risked voicing his unpopular and very private credo in a short poem in his old magazine, *Life and Letters*, now edited by Robert Herring.

I hate the clamorous voices of the crowd,
Its call to sacrifice forever,
Abhor the dronings of its limpet leaders.
I love the quiet talk of those endowed
With reason – call it treason – their endeavour
To live and love. I hate the million readers.
(I love their money but shall see it never.)

As if the family had not enough to worry them already, Sir George was now plaguing them from Italy. There was not much that anyone could do – nor did he get much sympathy from Renishaw.

Early that October there was a brief message – the first since Italy had joined the war – simply announcing: 'Sir George is completely undisturbed.'

'That is so typical,' exclaimed an irritated Edith in a letter to Ree Gorer.

I suppose he thinks that this is just a slight international misunderstanding, which will soon blow over. If he had been in the Mont Pelée earthquake, he would simply have taken his pulse, and put himself to bed on a milk diet 'to rest the heart'. He is always 'resting the heart', though he has never given that organ any work to do whatsoever, from any point of view. . . .

Not even a desperate message from Sir George a few weeks later did anything to make her change her mind. 'My maddening old father is in Italy,' she told Geoffrey Gorer, 'and has sent through a message to say that if we cannot send money through to him by December, he will be starving. What an end and climax to a life in which one has been a constant nuisance to one's offspring!' Edith was more concerned about the fate of Evelyn Wiel – there was still no news of her from Paris – and it was left to Osbert to see what could be done to help their father. There evidently was not much, for just before Christmas further news came through that he was seriously ill. 'I never hit it off with him,' wrote Osbert in a note to his new editor at Macmillan, Rache Lovat Dickson, 'but am very sorry for him in such a country and at such a time, and over 80. . . . One can't help at all, and telegrams take so long.'

In February 1941 Edith, Sacheverell and Osbert were briefly reunited – in a court of law – and suddenly 'The Sitwells' hit the headlines once again.

According to Philip Frere, this was slightly unintentional. What had happened was that when he consulted the lawyers over the *Reynolds News* review of Edith's book, they had replied that there was a *prima facie* case for libel; and Osbert, who was feeling bored in that period of 'phoney war', decided he would take his chance and sue, refusing a settlement and apology from *Reynolds News*. The case had been delayed, but suddenly it came up again. The defendants now refused to settle. Osbert decided he would go ahead.

Osbert often acted as if the standard answer to life's aggravations were a lawyer's letter – even to his father; but it was all but unprecedented for three authors to sue in a court of law for libel over an unfavourable review. Superficially, at least, it looked as if the Sitwells had a feeble case. There was also a strong hint – which the defence was sure to mention – that this was one more Sitwell stunt to gain publicity.

Their strength lay in the fact that the offending words of the review were not about the book (which would have been covered by the legally accepted practice of 'fair comment' on books submitted for review) but about all three of them and their reputation as a whole. If they could prove to a judge's satisfaction that, far from being claimed by oblivion, they were still as important and significant as ever, they had a case.

They also, as it turned out, had a lot of luck. The judge, Mr Justice Cassels, was a man of culture who was not impatient, as some judges might have been, at the idea of the Sitwells bringing such an action in the middle of a desperate war. More important still, *Reynolds News* were ill-advised enough to use a famous criminal King's Counsel for their defence – G. D. 'Khaki' Roberts. Edith christened him 'The Buffalo' and called him – but not, of course, in court – 'a great blundering buffalo of a man'.

He should have known better than take on three such experienced old hands as the Sitwells in cross-examination and try to make them look absurd. They had no difficulty proving just how far they all were from oblivion, and Edith even managed to make legal mincemeat of the hapless Roberts. When he accused her of always having 'courted publicity', she slapped him down with the brisk reply: 'I have advertised my books in the way that all tradesmen advertise their wares.'

'Do you not consider that this action is rather a waste of time in war-time?' he asked Osbert weakly.

'It is a very serious thing to damage an author in this way,' Osbert replied with dignity.

And the judge agreed, awarding each of the Sitwells £350 damages and costs.

Sacheverell did not particularly enjoy this sort of knockabout notoriety, but it was just the triumph Osbert and Edith loved. The philistine press had been discomfited, a critic had been taught to mind his manners, and they had publicly proclaimed in court that, far from being in oblivion, they were all three on the threshold of the most important part of their career.

Certainly all three of them were working harder now than ever. Sacheverell, as Edith reported after a stay at Weston early in the summer of 1941, was 'writing the greatest book he has written yet, but is the prey of every kind of chattering magpie and busybody'. (This was *Splendours and Miseries* which came out in 1943.) Edith was 'doing what I was born for' and writing poetry again, with undiminished vigour; while Osbert was somewhat grimly finishing a new book of short stories. They were published later that same year under the title *Open the Door!*. In the introduction he described how he regarded his writer's role during the great disaster of the war. 'The art of writing (if you think about it – which fortunately the born writer seldom does) is comparable to the action of the band which played a hymn as the great ship the *Titanic* was sinking.' It was a simile that placed him firmly back among the Edwardians he once despised, just as the band on the *Titanic* symbolised the qualities he valued – stoicism, consistency, and a stylish affirmation of a forgotten civilised existence against the shipwreck of a foundering world. Osbert described his regular routine, and his attitude to life, in a letter that he wrote to David later that year.

I spend my days writing and walking. . . . I feel more and more sympathy with poor Nero who 'fiddled while Rome burned'. What else was there to do? He was, no doubt, good at the violin and due to give a concert in a few days' time, and didn't want to be bothered and badgered about a fire he couldn't put out. Of course, if everyone had done the same, what? The fire, I suppose, would have been over quicker.

In fact, of course, Osbert was exaggerating as he always did when he was feeling sorry for himself. Life at Renishaw was not that bad. Spring brought the first delicious young broad beans to the dining-table, the bluebells to the woods, and the clear northern mornings, in which Derbyshire became transformed into a sudden country paradise. It also sometimes brought him visitors he actually enjoyed.

Old-fashioned Tom Driberg turned up here for a night (last night) at an hour's notice [he wrote to David on 9 March 1941]: very louche and furious, quite seriously, when I told him I was an *anarchist*. He then saw he'd given himself away, and laughed. He has a servant of Italian descent, the illegitimate son of a conjuror.

Osbert and David Horner had known Driberg since he was at Oxford, and both were convinced – quite rightly, as it proved – that he was a member of the Communist Party. This did not affect their friendship for him, any more than did his work as William Hickey on the *Daily Express*.

Another visitor he liked was Somerset Maugham's secretary, Alan Searle. He arrived that autumn with news of the Master, now in California. 'Alan's here,' wrote Osbert once again to David '(complete with bleeding piles) and he tells me Willie M. is *not* returning till after the war. Really, these patriots!' He also managed an occasional long weekend at one of the big country houses that he loved. It was during one of these weekends at Badminton that he came face-to-face with another villain-figure from his past, Lloyd George. Just as with Stanley Baldwin, Osbert was evidently charmed by the old Welsh wizard who, like Osbert, was now profoundly pessimistic, defeatist and anti-Churchill. He reported back to David, somewhat alarmingly, 'D. Lloyd George is precisely of my opinion about the war.'

During that second year of war Osbert's favourite distraction was one of which he was extremely proud – his friendship with Queen Mary.

As he has explained in his posthumously published essay in *Queen Mary and Others*, that indomitable old lady had been packed off to stay with her niece in the relative obscurity of Badminton for the duration of the war. For Osbert's Beaufort cousins it was, of course, a signal honour to have the former Queen of England as their permanent house guest. It was also something of a trial. For the Queen and her private household took their right to be at Badminton very much for granted – while the long-suffering Beauforts tactfully withdrew to a cottage in the grounds. More trying still was the constant strain of keeping Her Majesty amused.

Much of the time she evidently spent keeping the nearby woodlands free of ivy – a plant she hated – and chopping down dead trees. But there was still the problem of providing her with suitably amusing company, and the Beauforts now relied on cousin Osbert for this elevated chore.

Osbert's manner, and his long apprenticeship as a 'tame cat' among the great ladies of Edwardian England and the twenties, naturally equipped him for his role as courtier to the Queen. They also had a lot in common – a passion for the European aristocracy, a fund of stories of the great names of the past, and a collector's mania for *objets d'art*. Osbert was apparently one of the few house-guests who could be relied on to keep Queen Mary amused at breakfast-time.

During 1941 he was twice summoned for long stays at Badminton. 'It is becoming a habit', Edith used to say – and in his accounts of what occurred, particularly in the letters that he sent to David, one sees his sense of reverence fighting a losing battle with his sense of humour.

My first day at Badminton was dull [he wrote to David on 10 May], (because Queen Mary was ill), my second unforgettably hysterical. In the morning we all set out to cut down trees. Already two chauffeurs and a woodman were 'out', two with concussion and several others with slings and black patches over their eyes. The Queen herself was attended by an unmistakeable pansy, lately conscripted and stationed at Badminton to whom she has taken a fancy. 'A nice boy, he always helps me.' A tree top has fallen on the Princess Royal (whom I very much like now.)

Having survived the royal woodmanship, Osbert was then roped in for a trip to Bath on one of Queen Mary's collecting forays which made her the terror of the antique trade throughout the land.

On the way the Queen saw an airman waiting for a lift and sent her detective to kidnap him. He was brought into the motor and suddenly realised his fate. Only about 19, he was gibbering with fright. The Queen was splendid with him, talking to him and making him happy. . . . A wonderful moment came when she asked him what he did before the war, and he replied, 'I worked in a hospital, Ma'am, in the *Maternity* Ward.'

('But what *could* he have been doing; I can't make out.' The Queen and Princess Mary fretted away at the question later.)

Osbert was a great admirer of the Queen. She was the sort of very grand Victorian lady that he loved, and he liked her regal imperturbability and royal aura. 'I arrived two hours ago. Queen Mary looked really wonderful in white,' he reported on his second visit in August. But Badminton was really not the place for him, and he had more important things to do than play the flunkey. However great the honour, it began to pall.

It's raining cats and dogs (and horses) here [he wrote on 13 August], and everyone is black with rage. I ate kippers and mushrooms and bacon for b'fast. Q. Mary is being very cagey about how long I am to stay, and very stingy about petrol. (Incidentally, they've got *no* petrol in Eckington.)

By the end of August he was grateful to be back at Renishaw, for now that he had finished his short stories he had work to do.

One of the best accounts of Renishaw about this time is given by Sir Alec Guinness, who in 1941 had just become Lieutenant Guinness, R.N.V.R. He had a slight acquaintance with Osbert from before the war, thanks to a performance in a stage adaptation of *Great Expectations* which Osbert had admired. Also his wife, Merula, was a distant cousin (which always helped on a friendship where the Sitwells were concerned). Soon after he joined the navy he met Edith, who added him to her long list of correspondents and insisted on knitting him a pair of seaboot stockings – 'very long and curiously shaped with two left feet'.

Then one night [Sir Alec says] I found myself on leave and stuck in Sheffield with my wife and three-month-old baby son. I rang Edith, and she instantly invited us all to Renishaw for the weekend. She made a great thing about Osbert having to be kept from knowing there was a baby in the house. Babies were supposed to make him ill, so we had to go through the motions of smuggling the infant in. In fact, of course, Osbert knew perfectly well what was happening, but it was the sort of elaborate fiction the Sitwells loved.

Then, inevitably, during dinner, Robins the butler came in and solemnly announced: 'Young Master Guinness is crying upstairs, sir!'

'Oh!' said Edith brightly, 'I expect the baboon has just looked in and frightened him.' And suddenly, in that extraordinary house with its oil lamps and creaking stairs and miles of corridors and haunted rooms, one really *could* imagine that there well might be a baboon or two roaming around.

From what he saw of them at Renishaw on this and several later visits, Guinness came to think of Osbert and Edith as rather like a comfortable, long-married couple when they were together. Increasingly Edith signed her letters 'with love from Edith and Osbert', and she described how 'Osbert and I are living a sort of Robinsoe Crusoe existence, with one very aged, fearfully bad-tempered good Man Friday.' (This was the inimitable Robins, who was tending more and more to order the daily running of their lives.)

Almost the only evidence of any disagreement at this time is a short note to David Horner – one of the very few apologies which Edith ever brought herself to pen to anyone – which must have been composed for Osbert's sake.

Dearest David,

I am horribly ashamed at having made that maddening scene when you had come over here to have a rest, dead tired as you were. . . . Also, poor Osbert gets all Sachie's worries and it was too bad of me to make a scene. . . . It is to be ascribed partly to the fact that I get terrified, as I cannot understand business. I won't do it again. You shall have a tearless Thursday and Friday I promise you.

Money – or the lack of it – was once again a source of misery to Edith, and in the autumn of 1941 she was writing to her agent:

My financial position is about as bad as it could be, – indeed, most urgent, as I have lost all my possessions and most of my income. I *cannot* go on being supported by Osbert; it simply is not fair on him, and it is most painful to me.

That was one thing that no longer worried Osbert, and by the second summer of the war he gave a firm impression that he was coming to accept his full-time life at Renishaw, and had begun to build an organised routine that suited him completely.

Of *course* you can come to Renishaw [he wrote to Lovat Dickson in June 1941]. Merely name the day. And I shall treat you as one of the family; which means that I shall leave you to bathe or walk and work myself as though you were not there. . . . We'll simply meet at meals and by the bathing pool if it's warm enough.

Osbert was making Renishaw unique, with everything subordinated now to work and his writerly routine. As he put it in a lecture on George Meredith he gave later:

I have always bitterly hated the existence led in any country house but my own – and that, in its life, resembles no other, being, in so far as I can make it, a place of seclusion in which to work. The real country house was a place where bored but very healthy men could just get through life with the aid of estate work, port wine and rural pursuits.

Writing about 'the artist' (by which he invariably meant himself), Osbert insisted that 'to be able to work at his best, it is necessary for him to have an endless vista of hours and days, within the space of which he can write or paint without any interruption, except those which are carnal or which he makes himself'. This was a boon which Renishaw and war between them had conferred upon him.

In the summer of 1941 he had made fresh contact with an important figure from the past – Adrian Stokes. He was transformed from the troubled golden youth Osbert had loved at Rapallo and Amalfi during the vanished playtime of the twenties. He was now living at St Ives in Cornwall, writing, painting and fathering an endless progeny, but his letters were sufficient to revive Osbert's memories of an inaccessible lost Europe they had known together. During that summer of 1941 an escapist and nostalgic Osbert wrote to him:

Yes, I've been working hard, but I don't much care for the world I find myself in at the moment. I daren't let my mind dwell at all on *places* I like; except to think that, after all, tomatoes will still be growing on the outside walls of flat-roofed houses in September whatever happens.

I've enjoyed working here because for the first time I can surround myself with books of reference. It's too difficult when travelling. Also I read a great deal, except I can never get what I want out of the London Library.

I've just finished writing a book of short stories, and am engaged on an endless autobiography, but it has interested me tremendously, and I'm trying to do something new in it. . . .

I don't personally believe the war will ever end, and I know so well the familiar stages. It's high hysteria just now – of course to a new tune. And that awful old heroic oaf, and the brand of high-faluting crooks around him and the maniacal crooks opposite him, make sad thoughts.

This was the first news that he gave of his latest project, which was destined to become the most important book he ever wrote – the story of his life and times as told in the five volumes of his autobiography. It was a massive undertaking from the start, and one can see how it grew directly out of the life he led at Renishaw. He was approaching fifty, and at the height of his powers as a writer. At this crucial moment of life, the war had cut him off from almost everything he loved and valued – from society, from Italy, from art and travel, food and friendship. It had also left him, for the first time, with no distractions and a regular routine; so he began to recreate the rich, nostalgic dream-world of his past to set against the drab reality of Churchill's war.

He had an untapped capital to draw on – the range and richness of his early life, the number of his friendships and acquaintances, the battles he and Edith and Sacheverell had fought, the story of their rise as writers in the twenties. But he could also see now that there was a pattern to it all. Immured as he was in Renishaw with Edith, he was particularly aware of their shared ancestral past, that familiar, genetic world which had helped make the trio what it was.

At Renishaw this aspect of the past was an inescapable reality, and gradually he saw how the rise and the disaster of the family mirrored the rise and fall of the society he knew. He formed a great symbolic theme, matching in its way the theme which Proust had re-created in his attempt to rediscover the 'lost' time of his own youth. Osbert was very much aware of Proust's example, and those books which the London Library never seemed to have when he wanted them were the twelve volumes of *À la Recherche du temps perdu* – in the translation of the long-dead but still unforgiven Scott Moncrieff.

Sacheverell too was building his literary escape-routes from the war. That same summer of 1941 saw publication of his *Valse des Fleurs* – a marvellously vivid and poetic re-creation of a day in St Petersburg in 1868 and a ball in the Winter Palace of the Tsars. For where Osbert was delving in his memory, Sacheverell was more intent than ever now in using his poet's visionary imagination to construct a world of more bearable fantasy and happiness to set against his anguish at the destruction of the Europe that he loved.

Edith too was on the edge of fresh achievement but, unlike her brothers, she had no intention of escaping or forgetting the full horror of the war. In January 1942 she was writing once again to her friend Ree Gorer about her life at Renishaw:

> Here I have stayed, for a whole year, with the exception of one week spent at Sachie's – without moving. For what is there to move for? . . . One tries to go on with one's daily life, and regard the whole time – excepting for the unspeakable disaster and horror of it – as a railway journey which has to be undertaken. But when one thinks of the thousands of millions of wretched lives, it is a little difficult to think of any life as continuing at all.
>
> I have sent, or rather Macmillans have sent, my new book of poems to you, and to Geoffrey. I hope they all arrive safely.

Thus she announced the proudest moment of her life for many years, the re-emergence of her full poetic talent in her first important book of poetry since her *Gold Coast Customs* thirteen years earlier. As she wrote grandly, 'My time of experiments was done.'

These poems of her full maturity show an extraordinary leap from her earlier work. In these 'prodigious hymns', as Stephen Spender called them – poems like 'Still Falls the Rain' or 'Any Man to Any Woman' – we have Edith the prophetess of doom, using her full poetic power and Christian imagery to express the horror of the war. The fantasy and prettified conceits of the earlier poetry had gone. The rain of the bombs is compared with the dripping of the blood from the crucified Christ's side, for as she proclaims in 'Poor Young Simpleton', 'the season of red pyromaniacs, the dog-days are here'.

She had suddenly established her claim as the first of the major poets of the Second World War and, as John Russell put it in the *Sunday Times*, she 'now belongs to the greatest tradition of English religious poetry' as well.

CHAPTER TWENTY-ONE

Victory

1942–1945

'I COULDN'T have believed it possible to be so cold,' wrote Osbert miserably to Miss Andrade early in February 1942. Despite the free coal and the enormous fires, Renishaw was particularly exposed to the icy north-east winds, but he had managed to complete the first volume of the autobiography, and after he had sent her off the green-inked manuscript to type he was relieved by her reaction.

I am really so very enchanted you like the book. Any criticisms to offer? Do you think the first ancestral part *sounds* snobbish? It is not: but it might *seem* so. In any case I think I shall leave it as it is, though I think in a month or two to have even a father may be a capital offence.

He was still concerned about his own, and just as the whole autobiography is overshadowed by the constant presence of Sir George, so the distant baronet remained a source of worry even as Osbert wrote it.

Sir George was eighty-one and genuinely ailing. Despite his desperate earlier letters asking for money, he seems to have now arranged some sort of income for himself. But letters out to Italy had to pass through neutral Portugal. The post was at best erratic, and at the end of 1941 rumours trickled back from the Red Cross via a certain Lady Crump that Sir George was beginning to feel lonely and aggrieved at the lack of letters from his children.

Edith replied for all of them.

My Darling Father,

Osbert, Sachie and I are terribly distressed because we gather from a letter recently received by us from a Lady Crump, that none of our letters have reached you for some time. How this can have happened, we do not know. We can only hope that this letter will reach you.

Darling father, you are in all our thoughts, always. We are dreadfully unhappy that we are separated from you, and most unhappy knowing how ill you are. We long for the time when we shall all be together again. . . . Osbert sends you his best love. He asks me to tell you that there are difficulties now about his writing to you. And we all beg you to believe that a lack of letters from us means no lack of affection or thought for you.

Sir George, in fact, was by no means as lonely and neglected as he seemed. The Italians barely troubled him. For a while he had been nursed by the Blue Nuns of Fiesole, and he had even found an English-speaking couple to look after him – the Woogs. Herr Bernard Woog was Swiss, a former manager of the very Zürich bank where Sir George had kept his money; his wife Olga was English, a large, blonde, pallid lady who before her marriage was a Chandos-Pole.

This was a coincidence that set Sir George's mind at rest, for the Chandos-Poles of Radburne Hall, near Derby, were neighbours of the Sitwells, and distantly related. The Woogs, who normally lived in Switzerland, had been stranded in Florence by the war; they took pity on Sir George, and he was grateful to allow them to adopt him. Osbert was unquestionably grateful too when he first got news that there was somebody in Florence who could keep an eye on his father. He seems to have heard from Woog direct – to judge by the reference to him in a letter now to David. 'Woog's last letter about Ginger is really enjoyable. One recognises the touch in the distance. And I think he (Woog) is as bored with his foster-father as I should be!' Sir George's health continued to decline. An operation that he had that autumn had revealed an internal tumour, but his health precluded further surgery. On 6 January 1942 Osbert sent further news to David.

Today I've had a letter from Woog. Very nice, saying that Ginger is growing worse, and is not likely to recover from the next bronchial attack. His mind is odd (because of the increasing poison, I suppose), and the tumour, poor old boy, is worse. He thinks the chef is putting ground glass in his spaghetti and is giving away all his money as fast as he can get it. Woog says he thinks Otello is dishonest, as well as the nurses, but that if they were to go there would be no one to look after him. Father is still at Fiesole, and won't go to Switzerland, having been persuaded by the nurses to stay on, because they want his money.

For the Woogs had thoughtfully suggested that it might be best for all concerned if they took Sir George to neutral Switzerland when they returned themselves. The medical treatment would be better there. Osbert appears to have agreed, and finally Sir George did too.

That claim that the Sitwells were on the threshold of the most important part of their career had seemed a shade far-fetched when it was made at the

Reynolds News trial the year before. Now it was coming true, and suddenly they were attracting notice once again. The clearest sign of this was the reaction now to Edith's latest poetry in *Street Songs*. As John Lehmann says,

These poems were published before Dylan Thomas's *A Refusal to Mourn*, Louis MacNeice's *Brother Fire* or Cecil Day Lewis's *Word Over All* had appeared, and their impact was inevitably profound at a time when complaints were being voiced (not without encouragement from the Philistines) that no war poetry worthy of the world-engulfing catastrophe was being written.

The times were on the Sitwells' side again. The left-wing poetry of the thirties had in its turn been superseded by the rhetoric of war. And Edith's poetry was receiving the sort of critical acclaim that she had never known even in the twenties. The sweetest praise of all must certainly have been that which came from those who had previously hitched their wagon to 'Master Auden'. Apart from Stephen Spender, whose praise naturally delighted her, there was Cyril Connolly, who had abruptly switched from his dismissal of the Sitwells and their works in 1938, and was now comparing Edith's latest poetry with Eliot's *Four Quartets* and Yeats's post-Byzantine manner.

Edith still regarded Auden as her prime poetic enemy, the leader of those left-wing boy-scouts who had stolen all her thunder in the thirties, but he had lost out now by his defection to America, and Edith was immensely tickled by the reports about him that she got from Tchelitchew, who had met him in New York. As she told Robert Herring,

I've just had a letter from Pavlik, who seems to have taken a dislike to Auden. To read his poetry, he says, is 'Choos (just) counting mackintoshes in large warehouse.' Oh dear, it *is* a good description, isn't it? Simply brilliant. Nor does Mr A's face escape censure: 'Choos large disaster, badly carved Roquefort cheese.'

A further indication of the reviving interest in the Sitwells and their works was a new production of *Façade* at the Aeolian Hall. This time it was rapturously received by an audience most of whom were too young to have been aware of the earlier storms and tribulations it had caused in 1923. Edith was naturally thrilled to have this new mass audience acclaiming her, but Osbert was already just a little wary of what was happening, and stayed at Renishaw. The autobiography was taking all his energy and, as he told Lovat Dickson, he had his doubts about these sudden swings of popular enthusiasm.

As you know, *Façade* was the scandal of the century when we produced it twenty years ago, and now that the audience are enthusiastic about it, and it has to be repeated twice, I believe it is only in order to prevent our doing new work by fixing our attention on the old. It is a sort of booby-trap really, but I don't intend to be taken in by it.

How typical of Osbert to reject those 'clamorous voices of the crowd' now that they wanted to acclaim him. But he was suffering from gout again, 'brought on by these infernal worries one has the whole time', and seems to have spent a crotchety early summer, with part of July actually in bed. 'I enjoyed reading Hitler's speech very much,' he told David. 'His descriptions of our statesmen, however exaggerated, are always enjoyable reading. What, I wonder, is the German for nincompoop – his description of Eden?'

Apart from the seeming endlessness of the war, which he regarded as a personal affront, Osbert's 'infernal worries' had the same infuriating source as ever – Sir George. That bronchial attack which should have carried the old gentleman peacefully away had failed to do its work. As Swiss citizens, the Woogs had managed to persuade the Italians to allow him to join them in Locarno, and at eighty-two Sir George was as much the master of the unexpected as ever. According to a letter Edith wrote to David in June 1942, Osbert had just received a telegram from him. 'He wants Haig's book of food rules to be sent out to him. He is apparently wildly and ferociously well.' So well, indeed, that at the beginning of September there was further news for David, this time in a note from Osbert: 'Ginger wants to marry his German nurse (born 1896) and give her a permanent £500 a year. (All three children to keep it up after his death.) Edith is furious, Sachie beside himself. Personally I am amused.' Osbert's amusement was predictably short-lived, and on 8 September he was telling David, 'I am getting cross at the prospect of a German step-mother and having to pay her 500 quid a year.'

But Osbert's rage was temporarily diverted by an alarming prospect nearer home. At fifty-one he was suddenly informed that he was liable for work of national importance. The horror of it was, as he explained to Lovat Dickson, that 'on industrial call-ups, as opposed to Military, the interviews take place at once and you are posted, often, within a few days. In all the ones that I have observed here, the posting to various factories has taken place within a week of registration.' If this happened he might even have to cancel the trip he was making to Balmoral at the end of September as guest of the royal family. But there was more than this that worried him. As he told Adrian Stokes, 'I am being driven nearly mad at the idea, or even the possibility, of being stopped writing, for Moloch has had his seven years of my life before, and now I am at the height of my powers.'

While Osbert was increasingly oppressed by Moloch and his father's marriage plans, Edith's own private star was rising. That summer of 1942 she had had an unexpected windfall, as Osbert announced in a letter to Miss Andrade. 'Edith is just buying a lovely small old house in Bath. The old lady of 96 who is living there (it is the house of Wm. Piozzi in Gay Street) is lending us all the furniture for the period of the war.' As he went on to explain, Edith had no immediate intention of moving in herself, and wanted to let the house. But it was a good investment for the future and the money for the house had come from an unlikely fairy-godmother – the mysterious, immensely rich sister of the shipping magnate Sir John Ellerman, who wrote historical novels under the nom de plume of Bryher. Osbert had met her years before when she was married to the poet Robert McAlmon; she had become the patroness of *Life and Letters* and, now that the war had exiled her from her home on Lake Geneva, she was a frequent visitor to Renishaw.

This tiny millionairess in her old blue beret seems to have been entranced by Renishaw and its inhabitants, and the account she gives in her wartime memoir, *The Days of Mars*, gives one a clear idea of the romantic picture Osbert and Edith could create when on their better behaviour.

> Whenever I think of Edith it is summer at Renishaw. We seldom obeyed Osbert's instructions to sit out of doors but stayed instead, just inside the doorway that led to the terrace above the formal beds that I had once seen bright with flowers but now were full of vegetables. There Edith used to read to me Shakespeare, the old ballads, and if I pleaded very hard, her own poems. Her voice, the air, an ornament on her dress, all seemed golden, while the landscape beyond us hovered in the cool Derbyshire sunshine as if a giant butterfly were protecting the grass with its coral and amber wings.

There seems a certain justice in Bryher's gift to Edith, for, despite her own lack of money, Edith undoubtedly possessed a generosity and a concern for others which Osbert lacked. There was her patriotic knitting ('will you bring your ration card, and I'll get wool to make you another pull-over *without coupons*', she told David). The shortage of wool was a matter of some concern in wartime Renishaw. In *The Days of Mars* Bryher tells of how she bought a sackful of camel-combings from a keeper at the London Zoo and had it spun and made into cloth. 'I shared it eventually with Osbert. He had a coat made which he gave later to a gamekeeper.' And Edith took endless trouble in trying to get money through to Evelyn Wiel in Paris. She had finally heard that Evelyn was now marooned and all but penniless in the Rue Saint-Dominique, and started sending her £10 a month. At the end of that September, when Edith heard that the money had not been getting through, she wrote complaining of the Foreign Office for all the world as if Wyndham Lewis were in charge of it.

> The person responsible for this wicked, heartless muddle has done his best to commit a slow murder by starvation and want. That woman is utterly destitute.

She has not a farthing. She is 62, has had two major operations of the worst description, she is lame with open wounds on the feet and legs. . . . Ought I to come to London to raise Cain at the Foreign Office? I do *not* want to if it can be avoided. But certainly there must be a frightful row. I will not let this matter rest. This horrible thing *must* be righted.

The threat of Edith's presence was enough, and Edith's small allowance did more than anything to enable Evelyn Wiel to survive the war. The strange thing is that Edith did not really like her.

The new year of 1943 found Osbert increasingly concerned about his health. In January he wrote to Thomas Mark at Macmillan, thanking him for his suggestions about the first volume of the autobiography, *Left Hand, Right Hand!* He used a postcard he had evidently saved from Italy – showing the papal palace at Viterbo – for, as he explained, 'I find myself forbidden for a while to write letters! I am not well; nerves and heart. What a bore! But I may perhaps have a rebound. . . . How are you? I thought you might like this view of Viterbo. Such a lovely place.' This lowered state of health had at least saved him from a possible fate in a Sheffield factory, but he was entering one of his periods of irrational depression. Even the supernatural was disturbing him and, as he wrote to David at the end of January,

The ghost has got into my room. Last night I woke up at three, looked at the time, fell asleep peacefully. Then was suddenly woken by a violent rap on the forehead, a real good slap. I lay quite still wondering what would happen next. Nothing. Then I thought perhaps a piece of plaster had fallen. Nothing.

David was back in London and inevitably his private life had drawn away from Osbert's, who was clearly worried, and it was now that his doctor diagnosed a murmur in his heart and ordered him to spend at least one day a week in bed. Despite this, he had managed to complete the introduction to his book of essays, *Sing High! Sing Low!*, and despatched them to Lovat Dickson, who in his reply inquired anxiously about his heart. Thanking him, Osbert wrote:

Yes, I fancy my heart would murmur with no uncertain voice, 'Peace! A warm climate! Pictures! Food! Roulette Tables!' And the rest: but it does behave in an odd way. And how singular the mechanism between heart and head. For example, several nights in recent years, if I have lain on my left side, I've experienced, just as I was getting to sleep, an almost psychic feeling of impending catastrophe and that unless I turned round I was 'in for it' in some way.

Then at the end of February the black mood seemed to lift, his health improved, David had reassured him that they were as close as ever, and Osbert and Edith were caught up in a sudden bout of energy like that with which they faced the world a quarter of a century before. As he told Miss Andrade, 'I am in a flutter of proofs today – for *Horizon*, for my book, for *Life and Letters*. Yesterday was one of the most beautiful February days I ever remember.' It was a freak early spring and for a few extraordinary days winter departed. A fortnight later he was noting that it was 'Hot and fresh and cool and sunny, with everything bursting into flower. Peonies in a Chinese abandon and perfection'.

Osbert revived as well. He had sat passive and alone in Renishaw too long. The time had come to act. A few days later he was in London. One of the people he met there was Bryher; she describes how they walked across the Park and Osbert suddenly announced:

'The time has come to do something to keep the arts alive. I have decided to organise a poets' reading. . . . The Queen has accepted to be Patroness. . . . Edith will read and at least a dozen others. . . . Eliot has accepted and I've asked Masefield. . . . Morale is low at the moment.' Osbert looked so regal when he made this pronouncement that I expected him to be in the gold lace and ample cloak of an 18th century general concerned about the winter rot that had set in among his officers. 'It is time for the arts to assert themselves.'

He might have added that *his* morale was low as well and that he felt the time had come for him and Edith to assert *themselves*.

Despite – or possibly because of – its royal patronage, Osbert's great Royal Poetry Reading turned out to be one of the more innocently hilarious artistic happenings of the war.

The idea was admirable, if just a little dated, echoing as it did those drawing-room affairs at Lady Colefax's or Mrs Arnold Bennett's at which the Sitwells nervously recited their earliest poetry – to the cynical delight of an Aldous Huxley or a Lytton Strachey. And, as one would imagine, there were some fraught preliminaries before it started. Osbert and Edith chose the poets and rehearsed them, an experience which brought forth a typical lament from Edith to her latest friend, John Lehmann, who was now editing the influential quarterly *New Writing and Daylight*.

The Reading! – the *Reading*!! The letters to poets about their collars, the threats to faint on the part of one female poet – the attempts to *A* read for 20 minutes instead of 6, *B* change the time of rehearsals, *C* be rehearsed by *me just* before reading, *D* read a funeral eulogy of Yeats – on the part of another female poet. And other troubles of this sort. Life is very difficult.

The threat to read the eulogy of Yeats was made by his old friend Lady Gerald (Dorothy) Wellesley.

Lady D.W. (Dotty) [she wrote to David] is being beyond any words tiresome. She truly pesters me beyond endurance. Practically every day I get letters worrying me about something. She sends me all the tripe she writes – (and she never stops) – she expects me to time her poems . . . and now she expects me to spend the afternoon of the Reading listening to her doing her stuff – because she says she can't come to the rehearsal. How *did* Y. put up with it? Honestly, the old man's mind must have been going for him to think her any good at all as a poet.

The reading finally took place on 8 April at that legendary Sitwellian battleground, the Aeolian Hall. It was a very grand occasion in its way, with Osbert sitting in the front row next to his friend the Queen, and the two Princesses in their seats beside them. When Denton Welch asked Edith later how they had behaved, she replied that 'they sat very still in the front row and stared at one'. For both it must have been, if nothing else, a useful training in royal deportment under awkward circumstances, as several of the poets were a great temptation to unseemly mirth.

Eliot was impressive as he read a portion of *The Waste Land* in that High Anglican asbestos voice of his. Osbert read several of his eclogues from *England Reclaimed*, and Edith was undoubtedly the star of the afternoon – as she intended from the start – with her dramatic declamation of her newly written 'Anne Boleyn's Song'. Unlike the other poets, she was a genuine performer, and her appearance and delivery had immeasurably improved since the time of those excruciating evenings at the Arnold Bennetts'.

Some of the other poets were less fortunate. Little Walter de la Mare was unable to reach the gigantic lectern Osbert had provided. W. J. Turner – once a bitter Sitwell enemy but now restored to favour – went on and on until the chairman silenced him. And, after the Queen had left, Dorothy Wellesley hit Harold Nicolson with her umbrella, and had to be restrained by Beatrice Lillie.

It was, in short, a good example of the hazards of attempting to project poets as performers on a grand occasion, although Osbert himself was gratified to get a letter afterwards from Lady Cunard saying that his 'distinction and dignity were outstanding in a sloppy age of mechanised disorder'. He was less pleased with a letter he received from Dorothy Wellesley inviting him and Edith to a reading she was organising for some sailors at Tunbridge Wells. 'I shall read "Yo-ho-ho and a bottle of *rum*",' he told David.

Although the reading was at best a partial success, it demonstrated Osbert's own importance in the arts. No one else in Britain could have mounted such a grand event for poetry. He was apparently resuming his original rôle of champion and impresario of the arts – but no longer as a rebel. He had imperceptibly become a distinguished figure in the establishment – almost Lord Badgery himself – and he was now intent on using this new power and influence to help poetry to 'reassert' itself. One can see the seriousness of his intentions in a memorandum he dictated shortly after the reading. It was headed, most importantly, 'Private and Confidential', and outlined the setting up of an impressive committee with royal backing – he suggested Masefield, de la Mare, Eliot, Edith, Blunden, Spender and Lord David Cecil – 'to encourage poetry, both the writing of it and the reception'.

For a start the committee would finance promising young poets, sponsor lectures on poetry throughout the country and hold further readings. It could be the start of that great revival of the arts against the blight of war that Osbert dreamed of. But nothing came of these plans – perhaps because the work was already being done by the Committee for the Encouragement of Music and the Arts, which later became the Arts Council.

There was certainly a great revival of the Sitwells now – and particularly of Edith, who was beginning to attain the role that she had tacitly assumed for years as the acknowledged *grande dame* of contemporary poetry. The potent combination of her poetry and personality caught the imagination of the young as they had never really done before. The authoritative role became her, and in the spring of 1943 her book *A Poet's Notebook* was received with reverent notices from the critics. Age was now on her side. With the success of *Street Songs* she was no longer Edith the rebel or the controversial 'modernist'. After her ten years' hibernation she could assume her place as one of the great survivors of the twenties, the friend and fellow-pioneer of Eliot and Pound, and a poetic icon for the young.

The most striking evidence of this new image she had gained is shown in the journals of the young novelist Denton Welch. In September 1942 *Horizon* published an article he wrote after meeting an important ancient ally of the Sitwells, Walter Sickert. It attracted much applause, and Welch received letters of congratulation from a medley of distinguished writers including Vita Sackville-West, James Agate and E. M. Forster, but

> the last final plum jewel, diadem and knock-out was a four-page letter from Edith Sitwell, telling me how much she and her brother Osbert enjoyed my Sickert article. How they had 'laughed till they cried'. How one thing was clear, and that was that I was *a born writer*! It was such a beautiful, generous, deliriously exciting letter. . . . It is so thrilling to have such warm-hearted praise from a great genius.

By the early summer of 1943 life, despite the war, was smiling on the Sitwells, and it seemed that their revival now had come to stay. Sacheverell had sent off the proofs of *Splendours and Miseries* and Osbert had completed *Left Hand, Right Hand!* Edith was still muttering about 'that bone-head, Connolly' for the way that his review of *A Poet's Notebook* had mentioned 'the passion of the great poets of the past for their art'. 'What', she asked majestically, 'does he suppose *I* feel?'

She could afford such carping now, for her own position was assured. Her ancient enemies seemed to have subsided with the war, her latest poems were being published as she wrote them in journals like *The Times Literary Supplement* and Lehmann's *New Writing*.

Osbert's gout had vanished with the summer weather, David's conversion to Roman Catholicism had made no difference to their love, and to crown his happiness he was now in funds. Mrs Ronnie Greville, who had died the year before, had stuck by her intention to leave her money to the deserving rich. Her jewels went to the Queen and £10,000 to Osbert Sitwell.

And yet, despite all this, something of that 'feeling of impending catastrophe' with which he had begun the year persisted. Reading his letters now one gets the feeling that he was waiting – as he had done at Renishaw so many times before – for some disaster to occur. And, sure enough, in mid-July it did.

Despite the family's relief that Sir George was now in neutral Switzerland and being cared for by the Woogs, they had begun to be a shade uneasy over what was happening. There was his mysterious telegram soon after he arrived at the villa at Locarno: 'Safe at last in this haven of good food and drink. Love Daddy.' As Sacheverell says, his father had never shown much interest in his food, still less in drink, and had never in his life signed himself 'Daddy'. Then came unlikely rumours that he had been drawing heavily on his bank account in Zürich. Finally they heard that he had given power of attorney to the mysterious Herr Woog.

But there was not a great deal anyone could do. Switzerland was inaccessible, and Sir George himself appeared to have lapsed into complete senility. His children could only trust the Woogs and hope for the best, taking their letters about 'the dear old man' at their face value. At least he no longer seemed to want to wed his nurse.

Then, suddenly, on 8 July, in the white-walled Villa Fontanelle, facing Lake Maggiore, Sir George Reresby Sitwell, fourth baronet, expired.

Sir George's death was, in its way, Sir George's masterpiece. So much of his lifetime had been spent plaguing his children, using his money – or the absence of it – to play those complicated power-games he loved, and now in death he played the subtlest game of all. As one reads his will and understands the consternation and suspicions that ensued, one almost hears the old man's thin, elusive voice as his grey shade insists: 'They may think I shall – but I shan't!'

Those celebrated words of his must have been in the minds of many of the congregation in the village church at Eckington the Sunday after as they assembled for a memorial service for the departed Lord of Renishaw. His ashes were in Switzerland (and would remain there for many more years until his grandson, Francis, dutifully collected them) but Sir George's presence must have been extremely real to his family.

Osbert, of course, was now fifth baronet. Not even Sir George was able to deprive his son of that, and he enjoyed the honour unashamedly. 'Being a baronet is a delicious experience, like a new toy,' he told Miss Andrade. But Sir George left little else to cheer his children. A recent codicil to their father's will revealed that an original bequest to Edith of £1,000 had been revoked in favour of an income of some £60 a year (which, as she bitterly but over-pessimistically lamented, would bring her in around £30 a year after tax). Sacheverell's bequest was cut as well, and there was no further money for Osbert. (Sir George had cut him out of the settlement on the grounds that Osbert had joined the Labour Party, and was spending all the money he had given him on subscriptions to the Party. 'A fearful lie', said Osbert who, as usual with Sir George, failed to appreciate his little joke.)

The will was, of course, unable to affect Osbert's original tenure of Renishaw, which was still governed by the terms of the existing entail, nor could it deprive him of Montegufoni (a somewhat hypothetical possession at the moment), as it was originally purchased in his name. Apart from this, Sir George's prime intention seems to have been to shift whatever wealth he had straight to his grandchildren, Reresby and Francis. For years now he had made no secret of the fact that he preferred them to his children, and the way that he insisted on regarding them as the true mainstay of the Sitwell line made Edith call them bitterly, 'The Hopes.'

But on the day of the memorial service there was not only bitterness about Sir George's will but also much confusion. Nobody knew for sure how much of his estate remained in Switzerland, and there was some suspicion that Sir George had somehow been able to disinherit Osbert of everything – Renishaw included – in favour of Reresby who was now the Lord of Renishaw. This was entirely untrue, but Osbert was certainly extremely angry, and Edith was already worrying that she would soon be leaving Renishaw for good.

She was beside herself, and next day wrote to David: 'Sir George – (I am going to call him that in future, *not* Father, as I don't see why I am to give him the credit for producing me!!!!). . . .' Sacheverell, in Edith's words, 'could not have been more sweet about Osbert, and realises absolutely how outrageously he has been treated. . . . He really showed at his *very* best.'

But family relationships had not been improved by what had happened, and one feels that for once Edith had understated what had happened when she wrote, 'The weekend was *most* trying.'

It took Edith several weeks to recover from her father's death – not out of sorrow, but from overwhelming anger. 'It hasn't been possible to work at all,' she told John Lehmann on 20 August, apologising for the non-arrival of an article that she had promised, 'owing to the really ghastly, *foudroyant* worries and disgust from which I have been suffering. – My father died a few weeks ago, and I dare not trust myself to speak of what he did before he died.' Her shock and fury made her so ill that for several weeks she stayed in bed; even now, a whole month later, she still had dizzy spells whenever she began to work. Osbert inevitably succumbed to gout again – a 'paroxysm for a week', as he told Mark at Macmillan. 'A foretaste of Hell, I suppose . . . devils with prongs.' Sacheverell remained distressed, and Georgia, so Edith said, was understandably put out that, 'as the mother of his only grandchildren', Sir George had left her nothing.

However, the first crisis fears were soon allayed. Osbert was certainly not dispossessed of Renishaw – nor Georgia and Sacheverell of Weston, for that matter. And the money was thought to be safe in Switzerland where Sir George had providently placed it twenty years before. As autumn arrived it appeared as if Sir George and all the troubles he had left his family could be quietly forgotten.

Osbert was getting an enthusiastic response from people who had seen the manuscript of *Left Hand, Right Hand!* 'I need praise as a salamander needs flame', he told one of them, and he was more convinced than ever now that this would be his masterpiece. This made him particularly demanding over the sale of the book's foreign rights. None of his books had previously earned more than a pittance in America: he was determined now that this one would. As he wrote to the foreign rights department at Macmillan, 'This book, I hope, will be a classic; it is the book of a lifetime, and the Americans, who appreciate success, will only despise me if I do not obtain the best possible offer.' He was also very much aware of the reviving literary fortunes of the Sitwells, and anxious to cash in on them, as he showed in an admirably realistic letter that summer which he wrote to Lovat Dickson at Macmillan.

I don't propose to publish for charity which is what American publishers like. . . . I think the American public do not realise the position that Edith, Sacheverell and myself occupy here in literature, or the difference the last ten years have made. They are very liable to take a valuation, if handed to them, and to be interested; but, chiefly owing to the influx of refugee English left-wingers who have lost touch with the world of today, no one has told them how we stand. . . . Could it not be done? It sounds an interesting subject for an essay in one of the papers if one could find a writer who would like the task. What do you think?

Edith too was obviously enjoying this new 'position' Osbert referred to. In the twenties, what following she had had been a narrow one, and it had not lasted. Now she was enjoying a regal role, a wider audience, and something she had always loved and longed for – power.

Cyril Connolly was proposing to devote an entire number of *Horizon* to the Sitwells, while John Lehmann at *New Writing* had by now become one of Edith's closest friends and allies, one of the favoured few she wrote to regularly, like some devoted, very anxious maiden aunt. 'And *do* put Bengue's Balsam up your nose. It is wonderful,' she told him when he had a cold, and even started knitting him one of those talismanic pullovers. But through all the letters that she wrote him one can also see the firm, assertive Edith laying down the law on poetry and poets as only she knew how.

On the 23 November, for instance, she wrote to Lehmann about a rival publication, Tambimuttu's celebrated wartime magazine, *Poetry*.

He needs a member of the sterner sex – a horrid white mem-sahib – (myself for instance) after him. There are, of course, one or two beautiful poems – Stephen's for instance. But, oh, ouch! Mrs Ridler! We read about people being allergic to flannel. That poor little woman has the same effect on me. And *need* she be *quite* so confidential about the more intimate aspects of her life?

More than ever, any living female poet was a personal affront to Edith, but her female bore-in-chief and permanent Aunt Sally was not a poetess but Arthur Waley's mistress, the eccentric translator and writer on exotic dance and drama, Beryl de Zoëte. For almost all the rest of her life Edith would use this tactless, gushing lady as a source of endless gossip in her letters, and build her into a running joke as she had done with 'Ginger'. When Arthur Waley brought her to Renishaw that autumn, Edith was rather stodgy about meeting her. 'If Arthur wants to inflict her on people,' she wrote to David, 'he really ought to marry her. That, any how, would explain her presence in a civilised house.' But she was already making fun of her as well: 'Edith says that Beryl's nose is like a tin-opener,' Osbert told David.

Renishaw was now becoming quite a haunt for 'lunatics and bores'. 'Dear Gerald [Berners] is here,' Osbert reported gloomily to David. 'God, he is a bore now, bless him, with all his little *diableries*. But I am very fond of the old boy, very sweet and pathetic, but those stories of his are like flies circling round one on a summer day.'

With Miss Andrade still in London, Osbert had been obliged to hire a 'new and expensive' part-time secretary. She annoyed him too. 'Her trouble is B.O.; a veritable *badger*. What is one to do? I have to burn incense and open all the windows.' But Osbert thought he had found the answer to the servant problem with another employee. 'We have a new, heavenly deaf woman to work in the house,' he told David in September. 'No gossip, and absolute peace. I always said deaf-mutes were the thing.'

Although he complained about the tedium of writing – 'I'm sick of it, and would like to enter a non-stop dance competition in Brazil instead, he told Miss Andrade – he allowed nothing to divert him from this 'book of a lifetime' he was now composing. Interestingly, he even turned down a chance to go to Italy again. The writer Lionel Fielden had just joined the Allied Control Commission which was administering the 'liberated' parts of Italy, and wrote to Osbert:

I am en route for Italy dressed up as a major!!! Are *you* coming? Because I strongly feel you ought to. It will be hell unless people like you have a hand in it. . . . Perhaps this is impertinent, but I do feel so very strongly that the time has come for all Italophiles to pack their bags – you perhaps most of all. Do come.

Osbert was one Italophile who had no intention of packing his bags until the mess in Italy was thoroughly cleared up. 'Lionel Fielden wants me to be a Wop gauleiter,' but 'I don't want to go,' he told David, adding in a heartfelt postscript to his letter, 'I'm gouty, oh dear! – I work all day and pine for my friend.'

After the bores and lunatics and gout, it was time for influenza, winter, streaming colds and days in bed. Renishaw was at its winter worst. 'No gas to cook with, no electric light, no door or window that fits, no coal, – only damp wood that hisses at you like a serpent when you put it on the fire!' moaned Edith to John Lehmann. Osbert had sinus trouble and depression on top of his more bearable afflictions. Dosed to the eyes with M. and B., he stayed in bed. Edith succumbed to influenza too. It was nearly Christmas before either Osbert or Edith was allowed up again. 'I must look rather extraordinary, too,' Osbert told David. 'One eye is bright red and almost invisible; the other *couleur de rose*; and my wrinkles are like your Aunt Frances's, and I have to go about in a cap! Do stop this war soon. It *has* been on a long time.'

This atmosphere of strain and sickness up at Renishaw helps explain a curious episode which indicates how little Edith had forgotten or forgiven all the dramas following her father's death. Just before Christmas David spent a weekend's leave at Renishaw. A few days later Edith sent him a letter explaining the atmosphere that he had evidently found within the house. 'You may, or may not, have noticed that I had a slightly strained look in the eye the last day you were here. This was due to the fact that I was in a passion and was trying to keep the fact dark.' The cause of this 'passion' would have been unthinkable a year before. It was the arrival of an advance copy of Sacheverell's latest book, *Splendours and Miseries*.

> Most of it is magnificent as usual, to a degree. But the last chapter, 'Songs My Mother Taught Me', is devoted largely to my mother's admirable qualities as a mother – her mother-love etc!!! She is represented as a gentle sweet woman, brimming over with mother-love. . . . I really haven't any words to say what he has done to *me* by that chapter. I do not need to tell you that by this, he has . . . succeeded in conveying it was all *my* fault. (I suppose *I* maltreated *her* when I was a child of 8 or 9, a child of 13!). . . .
>
> He has raised up the whole of my childhood and youth by that chapter, and has succeeded in infusing into my mind the whole of the way in which Father, as well, has behaved. . . . I had succeeded in forgiving her – which took some doing, and now he has brought back all my old feelings. I suppose I shall have to forgive *him* some time, remembering his misfortunes.

Reading the offending section of Sacheverell's book it seems inconceivable that anybody – even Edith – could take the mildest offence. Sacheverell had had a very different childhood from Edith's. He had adored his mother then, and touchingly recorded some of his childhood memories – the 'agonies of separation' from her when he went off to school, her beauty and 'her calm and dignity during, and after, the extraordinary and unlikely tragedy that befell her'. That was all – and there was no mention of Sir George.

But Edith was not rational on the subject of her parents – and the events that summer had revived such suffocating bitterness towards them that she went so far as to say that Sacheverell's words had 'succeeded in flaying me alive with the hell of my childhood'.

This was the language of hysteria. Sacheverell was one of the very few people she really loved, and she must have known that he had no intention of upsetting her – as David hastened to remind her. He calmed her down, and managed to prevent a row developing – a service which she later gratefully acknowledged when she dedicated her next poem, 'Green Song', to him. But her great bitterness towards her father remained unassuaged. It would soon discover a fresh outlet.

During the months following Sir George's death his children gradually discerned an additional catastrophe to blame him for. Early that autumn Osbert was already forming his suspicions. 'I believe', he told Miss Andrade, 'that my father has been cheated of three quarters of his fortune: the end to which suspicious people come. Please do *not* mention this.' Then in the months that followed the suspicion turned to certainty, as Philip Frere made contact with the British consul in Zürich to discover what had happened to Sir George's famous *Stiftung* in the Zürich bank. Those unlikely rumours Osbert heard about his father 'giving his money away' seemed to have been too true. Of the Sitwell money which should have been there for his dependents, less than a tenth remained.

It took, as such things do in wartime, many months and many letters to discover what had really happened. For some time it seemed as if some of the money was recoverable, but by the spring of 1944 the facts were undeniable. As Osbert told the chairman of Macmillan, Daniel Macmillan, 'My father seems to have spun through £75,000 in two years; after having spent the forty previous ones in "calling" (as the servants used to say in the days when there *were* servants) me, for extravagance! . . . And now we have to bring lawsuits in Switzerland. It is the most appalling bore.' This talk of lawsuits naturally involved the character who, until now, had remained firmly in the background, the mysterious Herr Woog, Sir George's ultimate protector and the man who held his power of attorney. According to Sir George's Swiss will, he had been left £10,000. This in itself was strange enough. In his right mind Sir George was simply not the man to waste the financial life-blood of the Sitwells in such a way. Alone, unwell, and with none of his family in reach, Sir George in his dotage would have been fair game for a determined fortune-hunter.

But how had it occurred? As Edith and Osbert asked themselves this question during their evening chats in draughty Renishaw, they gradually produced what must have seemed a satisfying theory for them both, a wish-fulfilling fantasy to crown all the conspiracies and plots they had been spinning round their father almost all their lives. Edith appears to have hit upon it first – and hastened to elaborate on it in a letter that she wrote to David on 27 February.

> Personally I think that the sulphoral played its part in gathering the old gentleman to his forefathers – not self-administered. That is my belief. . . . The heir is a charming man. Thick brown double-breasted suit (tobacco-brown and moustache to match, *pince-nez* on nose). And when I think what I endured from his mother-in-law!

This heavily sarcastic description of Herr Woog makes it clear that Edith had already met him, presumably at Renishaw before the war, and the reference to his mother-in-law concerns the fact that old Mrs Chandos-Pole had been one of the local ladies who had befriended Lady Ida and helped to plague Edith's girlhood. As for the sulphoral, this was the drug which Woog himself had mentioned in his earlier letters as prescribed for Sir George's breathing troubles and bronchitis. For anyone of Edith's powerful morbidity, it was clear that, if the sulphoral *could* have been used to kill Sir George, it obviously *had* been used.

Murder fascinated Edith. As an old lady one of her favourite London drives would be to Rillington Place where Christie the mass-murderer immured his victims, and patently Sir George's death had all the classic elements of murder too: motive, opportunity, and all the hallmarks of the perfect crime. By brooding on her father's death, she more and more convinced herself that he had found the fate that he deserved. She and Osbert spent much time now going over all the letters they received from Woog. Soon it was all quite horribly, excitingly clear.

There are [she wrote to David on 4 March] horrible passages in Woog's letters about George's 'Terrible agitations', and about how, on one occasion, he succeeded in getting into the garden. What was he trying to escape from? There then is a horrible thing about his false teeth, (a new set ordered) coming after the old man was cremated. 'Strange,' says Woog, 'to think of somebody's false teeth coming *after* they *were cremated*'. . . . Gloating? Hysteria? Horror at what he had done? The 'who could have thought the old man had so much blood in him?' touch. Or simply morbidity?

In any case I hope that we have the creature on toast. Osbert says that he wouldn't be in his shoes. . . . It is precisely like having a pair of relentless bloodhounds after one.

Having decided that Sir George was murdered, they could now have the satisfaction of bringing his murderer to justice. It would assuage whatever guilt they felt themselves about Sir George's death and possibly help recover the lost Sitwell fortune. But how could this pair of bloodhounds sniff out a murder in a far-off neutral country in a time of war? How, for that matter, could they ever know for certain what had really happened in that white-walled villa on the lake? A few days later it was Edith once again who hit on a curiously Sitwellian solution.

On 13 March Osbert wrote a letter, which he headed 'PRIVATE AND CONFIDENTIAL', to his secretary in London.

My Dear Miss Andrade,

Can you do a very strange job for me? I become more and more doubtful as to the circumstances of my father's semi-captivity in Switzerland, and the disposal of his fortune, and even to the nature of his death. Edith suggests that we should send a letter of his, and one of his host's, to Mrs Nell St John Montagu, the celebrated medium and psychometrist . . . just to see if anything should occur to her. . . . If we made an appointment with her, would you give her the letters and give us a shorthand report of what happens at the interview?

I don't [know] if you may have a prejudice? I have none, only, combined in equal proportions, superstition and disbelief. . . . And if you say you can and will do it, I wonder *when*, for I know your days are busy. And I don't want to harass you.

Miss Andrade reassured him that she had no 'prejudice' and was quite willing to do as asked; and five days later he sent her several letters from Sir George and Woog to show to the 'celebrated medium and psychometrist'. Armed with these letters and her shorthand pad, the indomitable Miss Andrade duly visited Mrs St John Montagu in her small house in Belgravia.

Miss Andrade is unable to remember much about the visit except that the medium – a blowsy lady in her sixties with dyed black hair – took a lot of trouble looking at the letters, questioned her about Sir George, and then, by re-examining the letters line by line, started to reach some startling conclusions. She seemed to think Sir George had certainly been in danger when he wrote the letters; also that Herr Woog had been intent on stealing his money, and that the nurse referred to in the letter had been hand-in-glove with him. Miss Andrade took all this down in shorthand, and sent her report to Osbert that same evening.

The report itself is not with Osbert's other papers, but Osbert's letter of acknowledgement exists. In it he makes a show of being open-minded and still unconvinced about the evidence.

> Thank you so much for your letter and enclosure. *What* do you make of it? . . . I'll write fully to you when I have thought over it: I confess I find various points she makes useful, possible and interesting [here he has added later 'and horrible'].
>
> A full report from the Consul General concerned is on its way to the Foreign Office: and Swiss lawyers and chartered accountants are already being consulted. . . . The psychic is only an extra, so to speak.
>
> I find it depressing that N. St. J. M. evidently thinks no money will be forthcoming.

But the 'horrible' evidence of the medium in fact made more of an impression at Renishaw than his letter might suggest. Edith was particularly excited and wrote to the medium herself, asking for further details. On 3 April she received them – in a florid hand in purple ink. The letter was not exactly calculated to calm anyone of Edith's temperament. After mentioning that Lady Ida had been to see her years before, Mrs Montagu went on: 'The whole tragedy of Sir George has been so gruesome and appalling – and honestly haunts me. One of your father's nurses was absolutely hand in glove with your father's horrible "host", and as I told Sir Osbert when he phoned, he helped him in every way.' Edith and the medium wrote to each other several times over the weeks that followed. In one of her letters Mrs Montagu expressed her fervent hope that 'that monster will be brought to justice'.

Edith was soon in a state of wildest agitation and near collapse. 'Forgive a tiny note,' she told John Lehmann, 'I'm now ill as a result of shock – It is something most dreadful about my father, who died last July. It is too bad and too serious to write about in a letter.'

And a week later she was still telling him that she had to 'write rather slowly, as I get very unpleasant fits of giddiness, due to this miserable affair'.

For by now Osbert had received the promised report from the British Consul-General in Zürich. This too is missing from his papers, but from what he wrote about it to Miss Andrade it would appear to have been an inconclusive document. This, however, did not stop Osbert using it in confirmation of his theory – which for him and Edith was now a certainty – that Woog had undoubtedly killed Sir George and absconded with the money.

The enclosed – which please return to me at your leisure – is the official and restrained account of the villain of the piece, from H.M.'s Consul General in Zürich: I fear that the last sentence leaves little hope that my father was kindly treated. The middle paragraph is a perfect *croquis* of a potential murderer. The information is all new to us. I never met the man – but was always given to understand that he was manager of his bank. I had no idea that he had been retired – or had lost his memory (which must have occurred about '28). I am horrified beyond measure. I would like to have your opinion. Don't hesitate to tell me if you disagree. I do hope I'm wrong.

Two days later, in a further letter to Miss Andrade, Osbert continued the strange story.

The visitor to my father was the local British Vice-Consul. My father sent for him, or rather, managed to get round his nurse to send for him one day when Woog was away. It was that, I imagine, which sealed his death warrant. They became frightened. . . . It is *so horrible*, that I feel as if I had shell shock. As you know, the old boy was the plague of my life; but he was a highly organised and nervous human being and I find it frightful beyond words. I still can hardly believe it, though a day or two after he died, I had a haunting feeling and spoke about it to my sister.

One of the oddest things is that Woog called himself Chandos-Pole. It was his wife's name: but her brother is alive and in the Grenadiers – and a most charming man. The Sitwells and the Chandos have been friend and neighbour and distant cousin for the last 200 years!

I never saw her – Olga [Chandos-Pole] – after her marriage: but I saw her constantly when she was a child. Her mother was a Prussian: and Olga was like a hollow being, but with a flighty manner. . . . They were related distantly to Cecil Pole: his family, like Woog, were foreigners, Van Nottam, and married a Pole and took the name.

I think I told you that Woog was always represented to me as the manager of the Bank, and a nephew of the Director. I have asked for inquiries to be made.

And made they were – although, in fact, there was not all that much for the inquirers to go on: Osbert's psychic 'feelings' two days after Sir George's death, Edith's determination to believe he had been murdered, Mrs St John Montagu's corroborative evidence from the great beyond – all would have fitted perfectly into one of Osbert's best short stories in the modern-gothic manner, but it was not the stuff on which the Swiss police could act.

The case went rumbling on all summer, with lawyers and private detectives 'buzzing away' in Zürich, the accountants trying to decide just where the money had all gone, and Philip Frere in London doing his best to sort out the final details of Sir George's will. From time to time the two old bloodhounds at Renishaw really believed that they had Woog 'on toast'. 'I have a feeling that the great Ginger problem will come to a head,' Edith told David in July. 'Really we have got into an odd circle. Murderers (probably) embezzlers (certainly) and blackmailers (possibly).' An exciting prospect, but it came to nothing. Despite a great deal of activity not the slightest real evidence was ever found against Herr Woog, with his pince-nez and tobacco-brown double-breasted suit, or the Wagnerian Olga with the flighty manner; and they retained the legacy Sir George had given them in gratitude for tending his last days and rescuing him from Italy in time of war.

As for the remainder of Sir George's fortune which had disappeared, much of this had vanished before he even met the Woogs. The Woogs could well have been given some of it during that period when Osbert had reported that his father was giving his money away, but the real responsibility for the losses lay with Sir George's bank in Switzerland – and when the war ended Philip Frere adroitly managed to recover a considerable sum in compensation.

No evidence was ever produced that the Woogs really *had* killed Sir George. Like Osbert and Edith, they may have had an eye on the wealthy old man's money, and Woog himself may well have looked to Osbert like a *croquis* of a potential murderer. But even the most uncharitable interpretation of their actions must make murder seem unlikely. Sir George was, after all, extremely old and ill and feeble, and the Woogs had helped to prolong his life for nearly two years after Osbert originally expected him to die.

The Woogs spent the remainder of the war in Switzerland, where Bernard Woog helped organise an escape route for Allied prisoners across the Italian border and down to Spain. He died in 1948, his wife in 1969. Her brother – the 'most charming man' Osbert referred to in his letter to Miss Andrade – Major J. W. Chandos-Pole, knew Osbert well, and when the war was over lunched with him several times. But neither of them spoke about Sir George's death. 'I knew quite well that Osbert thought Bernard had murdered him,' he says, 'but somehow I never liked to bring the subject up, and Osbert never mentioned it.' Major Chandos-Pole insists, however, that the idea of the murder was:

quite utterly absurd. There was never a scrap of evidence, and Bernard, whom Osbert never met, was simply not that sort of man. He was intelligent and cultivated, while my sister always said that she had really been extremely fond of poor old Sir George. His chief trouble was his constipation, which with the wartime shortage of castor oil in Switzerland had become quite a problem.

As for the 'evidence' of Lady Ida's medium, it would be hazardous to take it quite so seriously as Osbert and Edith evidently did. For while Mrs St John Montagu was so emphatic over all the details of Sir George's violent end which she extracted from the ether, she was, it seems, unable to discern her own. Early that autumn – not long after Osbert himself had called on her to hear if there was any further news – a German flying-bomb destroyed her house. She was killed instantly.

While Osbert and Edith were so busily constructing their strange murder plot around Sir George's death, they were both rapidly progressing to the heights of their success and fame, and were now in the middle of their most important creative period of all. It was a curious situation for this most unlikely couple, now in the fourth year of their exile from the world they longed for; she writing poetry in bed and knitting sea-boot stockings in the evenings, he plagued by gout and melancholy and a murmuring heart, and reliving, page by page, his life in a vanished society.

Left Hand, Right Hand! was serialised in the spring of 1944 in the *Atlantic Monthly* – and the first American public reaction served to encourage Osbert's own estimate of the book's classic status. The second volume, to be called *The Scarlet Tree*, was complete by April 1944, and he immediately embarked on volume three, *Great Morning*, with its nostalgic picture of his life as a young officer on the brink of the First World War, of Renishaw as he remembered it, and the golden, distant paradise of Italy. Most interesting of all now is his treatment of Sir George, who dominates this portion of the book just as he still dominated Osbert's life. For this is where he tries to make his peace with the old man's memory, by re-creating him as a comic character, an almost lovable eccentric. It is as if Osbert is himself attempting to forgive him, and reading these pages one must look hard for any hint of the hatred and the lifelong bitterness he felt towards him. There is one reference, however, which he could not resist, a scarcely noticed echo of those bitter poems he wrote at the end of the First World War in which Sir George had been symbolic of the hated father-figure who had, Judas-like, betrayed the younger generation in the war. At the end of the book Osbert described himself leaving Renishaw for the trenches: 'I said goodbye to my father – who offered to lend me thirty shillings in silver for the journey, as my allowance had ceased.'

'Did I tell you,' Osbert wrote to David early in July 1944, 'Edith and I are both being elected *fellows* of the Royal Literary Society.* I feel extra important.' It was a further indication of their rising status. Honours were just beginning to pursue them, and when Edith's latest poems, *Green Song*, were published in August, they were greeted as the work of an established major poet. Kenneth Clark would pronounce her later work unreservedly as the greatest poems of the war, a judgement that was more or less accepted by the press.

At the same time she was resolutely marshalling her allies and supporters just as she used to do in Moscow Road. Since then her techniques of public and of personal relations had immeasurably improved. Certainly no living poet could have hoped to muster anything like the personal support which Edith could command from influential members of the literary establishment. One can be very cynical about it all, seeing the way she operated as proof that Dr Leavis had been right. But to say this is to miss the essence of her particular achievement. For Edith was not concerned with merely writing poetry. She was born, as she reiterated, to be a *poet*, and over the years she had been learning to live out this dedicated role. Everything in her life – her childhood miseries, her unhappy loves, and her extraordinary appearance – had helped to build the unique personality she was. During the twenties she had been serving her apprenticeship – in those Thursday gatherings at Moscow Road, those battles in the press, those personal appearances and poetic friendships – and she had learned a lot from Osbert. During the thirties she had suffered obscurity and decline. But now in her own late fifties she had the confidence at last to be a public poet. Consummate actress that she was, she lived the part of the celebrity that she had made herself, using her charm, her wit, her friendships to further and promote her work.

She was quite shameless and quite single-minded in the way she did this. (One of her great complaints was that Sacheverell never bothered, and that his sales suffered as a result.) But she took endless trouble with the teas and dinner-parties she began holding at the Sesame to build a whole new network of allies and admirers she could rely on.

John Lehmann was more than ever at the centre of this private network. During this period he became one of her closest confidants and friends, and her immense correspondence with him shows her at her best – and worst – witty, funny, endlessly concerned, and also bitchy, snobby, indiscreet and sometimes verging on hysteria when she felt threatened or oppressed.

While many of her letters consisted of bulletins on her private life at

* The Royal Society of Literature.

Renishaw, underlying everything she wrote was one consistent purpose – to promote her poetry. And Lehmann, an influential editor who believed in Edith's work, was naturally delighted to oblige. In May that year, for instance, he arranged to publish an article on Edith's poetry by Henry Reed, and wrote to her about it. Edith instantly replied, making it quite clear *exactly* how she wished to be presented.

Oh, about Mr Reed's article on my poetry. What I understood from your letter disturbed me a little; but perhaps I misunderstood it. I am extremely anxious that the public, at this moment, should be led to contemplate my *latest* poetry (written since this war), until, in the course of time, and through getting to understand this later poetry, they are able to understand the earlier.

Heavens, what I have been through in the past! In the first place, neither the sadness nor the gaiety of those earlier poems has been understood, nor any underlying meaning. It is always so, I think, with any new poetry, unless there is a layer of meaning on top – which *can* happen with very good poetry; but it is not the only way of writing good poetry. – Then there are the technical problems. I practised for years, like a pianist, meticulously, with infinite patience. I am very anxious that people should get accustomed to my later poetry; and then, when they realise I am not in the least the person they supposed me to be, they will at last understand what I have done in my earlier poetry.

Lehmann was an invaluable friend to Edith now (as he had been earlier to Virginia Woolf),

For one thing, in order to be able to work [she told him], I do need sympathy and comprehension. And you are one of the very few people on whose help I rely in that way, as in others (technical problems, clarity, being among the other ways in which I would most deeply value your advice).

A kindly, tactful man, he not only gave Edith advice but also helped her with her friendships. This was a time of reconciliation. Stephen Spender, who had 'never, I think, cared much for my poetry before', had been much impressed by *Street Songs* two years earlier, and was by now a close friend and had fallen under Edith's personal spell. Another reconciliation was with John Hayward, the crippled but extremely influential critic and editor who shared a flat with T. S. Eliot. In 1936 he had written an unappreciative review of *Aspects of Modern Poetry* for the *New York Sun* – and had been consigned ever since to the frozen realms of Edith's enemies. Now, thanks to Lehmann's pacifying influence, Hayward made restitution, and Edith at her most Plantagenet gratefully accepted it.

I am writing to you for my brother Osbert and myself. I have long wanted to tell you that I feel towards you exactly as if there had never been any interruption of the friendship, and Osbert . . . joins with me in asking if you don't feel, with us, that the whole episode should be forgotten and treated as if it had never existed.

Henceforth, whenever Edith made her queenly progress down to London

– as she did now with increasing frequency – Hayward would be among the honoured few she summoned to her court for dinner at the Sesame.

Osbert tended to keep clear of London. The steady re-creation of the past appeared to satisfy him, and what friends he saw now came as guests to Renishaw. David, recently promoted Squadron Leader, was still in London, but generally managed to get to Renishaw at weekends.

An important new friend was the painter John Piper. Osbert had admired his work ever since he published a set of aquatints of Brighton, and he decided that he wanted him to illustrate his latest book with paintings of Renishaw. It was a fresh example of the old Sitwell flair for patronage. Piper would stay at Renishaw for short breaks from his work as official war artist, and, like Alec Guinness, he has glowing memories of the extraordinary house, and its increasingly reclusive owner.

It was quite wonderful to get there – a complete respite from the war, and the house appeared marvellously rich with its enormous fires and splendid food from the estate. Osbert was deep in the autobiography, and in the evenings he would simply talk. Later I realised that he was trying out his stories on me for the book, but he seemed infinitely knowledgeable and wise, with endless anecdotes about a world that I had never known – the Asquiths and the Keppels and the hostesses and artists of the time. He had a particular way of speaking and such a lapidary style that almost anything he said sounded like an epigram. He also had an extraordinary memory, and a fund of situations, so that something very simple – like the way the salt was passed at dinner – would remind him of another time with somebody important when the same thing happened.

As for Sacheverell, life at Weston was becoming arduous. Edith told John Lehmann: 'Poor Sachie has the same H[ome] G[uard] adventures. Every night of the week he is up till the middle of the night, lying in a bog or something of the sort; and he seems never to have a decent meal, all his work is interrupted, or he is too worn out to do it.'

Osbert's health was troubling him again. In the summer of 1944 he told Miss Andrade that his heart was 'valvular'. 'I really feel ill,' he added, 'especially from 4–8 a.m. and from 10 to luncheon-time, dithery.' And that same summer came a further blow – the death of his friend Rex Whistler, that perfect painter and most amiable of souls, run over by an army truck in Normandy. 'I heard last night, and tried to ring you up,' he wrote to David. 'It is a great grief to me.'

It was an appropriate moment for him to publish *A Letter to My Son* – a pamphlet (previously printed as an article in *Horizon*) which courageously enunciates his total opposition to the war. It was his personal testament, couched in the form of sage advice to an imaginary son at the beginning of a literary career – and it contains the essence of those mandarin ideals of the supreme importance of the artist and his art which were his religion and his private way of life. The war – like democracy, and poverty and the modern state– is something that the artist must use all his cunning to survive. 'Had Mozart been a modern Englishman – or for that matter a modern Austrian – he would have spent the last four years training to fight. . . . Conceive the loss to the world had conscription been in force!' Conceive the loss, he might have added, had Osbert Sitwell too been forced to donate those seven more years to Moloch.

But, apart from its element of special pleading *A Letter to My Son* is an important pamphlet, more relevant today than ever, with its reminder that the modern state, with its restrictions, mass taste and demands for the general good, will always be a threat to the heretical freedom of the artist. At the time he wrote it sentiments like these were hardly tactful, and Osbert soon discovered that he had summoned up a fresh antagonist, his former friend James Agate, who weighed in against him in the *Daily Express* as the defender of the common man. 'By what law of God or man', demanded Agate, 'must the common people do the dirty work of saving the world for the artists, while the artists look smilingly on?' He also challenged Osbert to 'come out into the open and tell us what he believes is reasonable compensation in terms of plays and books and paintings and symphonies for the duty to put an end, for example, to the murder of 600,000 common people' which the Nazis had just performed in Poland.

As Osbert knew quite well, it was impossible to answer, and for once he attempted no reply – even when Agate went on to compose a pamphlet of his own entitled *Noblesse Oblige; Another Letter to Another Son*. Agate was transferred to Osbert's private list of enemies, and it was left to Edith to explode about conspiracies and plots hatched by 'that hound' – that 'dreadful creature', Agate. 'I absolutely *hate* Agate,' she exclaimed to David, 'and would stick at nothing to do him in. Poor Osbert! What a mean, base persecution!' While to Lehmann she was more determined still:

> I am so furious with Agate that I shall certainly cut him dead under circumstances of the greatest ignominy, for him, if I ever see him again.
>
> It really is a filthy trick, made filthier by the fact that Osbert has a very large, warm-hearted, simple character that simply *cannot* understand dirty actions of this sort. He never suspects them.

A better reply to Agate was perhaps the famous reading which all three Sitwells gave that autumn at the Allied Forces' Churchill Club at the height of the V1 raids on London. A warning sounded just as Edith started reading her 'Still Falls the Rain'. She took no notice, but as she read on everybody heard the unmistakable roar of an approaching V1.

The rattle grew to ominous proportions, and it was impossible not to think that the monstrous engine of destruction was poised directly overhead. . . . Edith merely lifted her eyes to the ceiling for a moment, and, giving her voice a little more volume to counter the racket in the sky, read on. . . . Not a soul moved, and at the end, when the doodle-bug had exploded far away, the applause was deafening.

By a curious coincidence, in the very midst of the controversy with Agate over the war effort versus works of art, Osbert discovered he himself had been unknowingly involved in one of the most successful operations to protect the heritage of Western art from destruction in the war. During that summer he had been anxiously following the news of the advancing Allied armies blasting their way to Florence. In July he wrote to Adrian Stokes, 'I suppose Montegufoni is in the battle line this week. We shall probably see the usual photographs of "Montefugoni Liberated" and the usual heap of stones and Guido offering a lemon to the forces.' But for once his pessimism was misplaced; and soon he was writing proudly to Miss Andrade with his extraordinary news:

I heard yesterday from an old comrade-in-arms, General Alexander, that Montegufoni, my Tuscan castello is untouched. And it has been used to put all the great Italian pictures in by the Italian Government. So I have had, as distinguished guests, Venus rising from the sea, and Primavera and the Duke and Duchess of Urbino (by Piero della Francesca) and many saints and Popes. I am so happy about it. It is just the use I would like the house to have been put to.

Guido Masti, Sir George's old agent at the castle, far from presiding over the destruction of Montegufoni, had preserved the great house intact. The famous paintings from the Uffizi, secretly placed in the castle for safe keeping, had miraculously survived and, as Osbert soon heard from Eric Linklater, the novelist, who was with the liberating army, the place was in quite reasonable repair, and the furniture 'about as much damaged as it would be by very bad tenants'.

If Montegufoni could survive, perhaps the carefree pre-war world of Italy would soon return as well.

This was of no particular concern to Edith, and her reaction to the war by now had become quite different from Osbert's. He was attempting to avert his gaze from it and live in his re-creation of the past until peace and sanity returned to Europe. Edith had no such refuge and made no attempt to shut her mind to the suffering and death of war. She was very much aware of death in the summer of 1944.

Oh, the war and the misery and horror [she wrote to Lehmann in the middle of June]. All those young men; and the young women at home whose lives will be broken. My young maid (a village girl from here) who came when she was 16 and left at 21, to go into the W.R.N.S., got married here . . . to a boy of her own age in the Air Force. They had three weeks leave for their wedding honeymoon, and a week after they both went back to duty, he was posted missing – it is simply frightful to see that poor child, to whom I am devoted and who was devoted to me. She was given compassionate leave, most of which she spent in my room, not speaking except to say from time to time, 'Oh, Miss Edith. Miss Edith.' She is back at work now, which I think is the best thing for her. I think there is still *some* hope; but not much.

Not long afterwards news came of another death in action: this time of the twenty-two-year-old son, David, of her favourite cousin, Veronica Gilliat, and once again Edith witnessed the terrible effect of grief. As she told John Lehmann,

If I could hear she [Veronica] had died in the night, I could be almost happy again. But I think it is quite possible she will go mad.

That is why I have to wait to get on with the poem until the *first* shock of the horror has left my mind. It shrivels one's emotions, and, too, in a strange way, one's language.

It was no coincidence that the poem she was working on was her great testamentary poem of suffering and death, 'The Song of the Cold'; and it had an appropriately tempestuous gestation. The letters that she wrote in the autumn of 1944 show her more than ever in a state of turmoil. 'The wind has started one of its non-stop rages, howling like a universe of wild beasts,' she wrote to Lehmann that September. 'We are near the Wuthering Heights country, and that is the atmosphere at the moment.' Then finally, at the beginning of November, 'The Song of the Cold' was finished, and as she sent the poem off to Lehmann she accompanied it with a note explaining how she wrote it. '*Now* I am satisfied with the poem. I hurled myself on it like a maniac, biting, tearing, kneading and kicking it into shape, and doing what I can only describe as a blood transfusion act. I think it is all right how. I like it.' Reading the poem, one can understand this violence, just as one hears the raging of the elemental winds across the fells and shivers at the cold of the approaching Derbyshire winter. For there is a close connection now between the fury which disfigures so much of Edith's private life, and the extraordinary rage and passion which she can pour into her poetry.

Just as she could work up her hatred of her father into an imaginary symbolic murder, so she could use a similar anger to cry out against the suffering of man. This poem in particular reveals a powerful fusion of her feelings – her sense of suffering and grief, her bitterness at ageing, her horror at the slaughter of the war. It is her rage and passion that gives this elderly and sheltered noblewoman the unquestioned strength to speak for suffering humanity as her resounding rhetoric proclaims:

But the great sins and fires break out of me
Like the terrible leaves from the bough in the violent spring . . .
I am walking fire, I am all leaves –
I will cry to the Spring to give me the birds' and the serpents' speech
That I may weep for those who die of the cold –
The ultimate cold within the heart of Man.

The poem finished, Edith's rage went marching on. Her fury at the behaviour of 'the hound Agate' found another target – E. M. Forster's great friend, J. R. Ackerley, the editor of the *Listener*. In the spring of 1944 she had been deeply upset at the death of the young Greek poet Demetrios Capetanakis, who was also a close friend of Lehmann. She praised him generously, but at the beginning of December an article in the *Listener* had criticised his poetry. For Edith this was a personal affront, and she wrote instantly to Lehmann of her

day of really *passionate* rage. . . . *The Listener* had reached me in the morning and as I write of it my hand trembles with anger. What do these *mean* attacks portend? I am extremely displeased with Mr Ackerley and shall most certainly turn my back on him in future should I see him. He has, in his anxiety to be impertinent to me, allowed a dead boy of great gifts to be underrated – simply because *I* praised him.

A few days later, in a brief letter, she took Ackerley himself to task for 'impertinences which should not be offered to a poet of my standing'.

Worse followed that same week when another enemy, Julian Symons, criticised her poetry for being 'removed from life'. This seemed to be the start of something Edith always feared – a conspiracy behind her back; for, as she knew, Symons was an admirer of Wyndham Lewis. It was unbearable, uncalled-for and unfair. Just before Christmas she bewailed the fact to Lehmann in a phrase that only Edith could have used: 'The dregs of the literary population have risen as one worm to insult me.'

There was one source of solace by now. The war was clearly drawing to an end, and with it was ending Osbert and Edith's time of exile and confinement at Renishaw. For both of them the war had been a time when they had done their major work, a time when their reputations had recovered and advanced to reach extraordinary success. In different ways, the war had remade them both.

But the loneliness and boredom of those five long years at Renishaw had left a mark. Their youth had gone for good, and they seemed to realise that with it they had lost the leisured, easy Europe they had loved. The war had also broken up the trio.

You ask me about Sachie [Osbert wrote to Adrian Stokes early in 1945] Alas, I hardly know more than you do. I never see him and he never answers a letter.

I don't even know what he is working at. And he has invented a system by which all the faults are mine. And when I do see him he first of all lays down an intensive barrage of small grumbles, (for half an hour); then opens up an hour-long bombardment of questions without listening to my answers. This absolutely exhausts me and protects him from any real contact with me. It makes me horribly sad.

There had been too much bitterness about Sir George as well, and with the enemies they knew were lurking in the world outside.

When the war in Europe ended on 7 May 1945, they met the peace without excitement.

'I *cannot* yet, can you, realise that part of this war's over,' wrote Osbert to Miss Andrade. 'Will one ever?' And a few days later Osbert added, 'I'm getting ready for the next war – I suppose in August or September, with an election in between for a diversion.'

As for Edith, she wrote to Lehmann saying that she could hardly comprehend the peace, 'simply because I am *beyond* registering anything . . . if one once began to think of the young lives! Believe me, one would go *raving mad*! Yes, quite mad. I walk about with tears in my eyes.'

CHAPTER TWENTY-TWO

'*Dibs and Dignity*'

1945–1948

The more I think about the National Gallery, the more I LIKE IT, (though I'd never take it.) But it would mean a chauffeur, car, £1,800 a year: dibs and dignity. Travel abroad. Red carpets. It's them I feel the want of.

Thus Osbert in a letter that he wrote to David – now the resplendent Wing Commander Horner, R.A.F.V.R., but soon to be demobilised – in August 1945. In it one detects the longings of the solitary writer, the hankering for easy prominence, success and power – in short, for 'dibs and dignity' – which he knows quite well he will reject if they are ever seriously offered.

But this idea that Osbert evidently had that he could actually become director of the National Gallery is an intriguing daydream. Lord Clark – who as Sir Kenneth Clark was just retiring from the job – expresses himself quite 'flabbergasted' at the whole idea. He never heard Osbert's name mentioned as a candidate at the time, and says that 'Osbert never had the faintest chance of the directorship. Equally fantastic is the idea that the job would provide a chauffeur, and red carpets. He should have known better.'

Perhaps he should, and one can only speculate on the source of the original suggestion. But the red carpets and that chauffeur-driven car seem to have haunted Osbert during those grey, uncertain, post-war days. Even the publication of the first part of his autobiography, *Left Hand, Right Hand!*, that summer failed to satisfy him.

The book had more than justified his expectations. Its tone was right. A public jaded by the austerity of total war was ready for his endless sentences; and readers longing to return to the remembered happiness of peace responded to his own gargantuan nostalgia, the sheer excess and splendour of his sense of loss. For Osbert the real 'dibs and dignity' lay in the memory of his own ancestral past – the dynastic power of the Beauforts and the eighteenth-century Sitwells, the wealth of the Londesboroughs, the lost bucolic paradise of Renishaw, and a society of 'characters' unmarred as yet by democracy and war and the rise of the blighting middle classes. The very breadth and detail of the book were in its favour. The verbally deprived could gorge themselves upon this convoluted prose, and relish the extravagance with which he re-created a suddenly familiar world which everyone had lost. With this one book, Osbert had become a sort of national remembrancer.

But the profound distaste for almost everything about the present world, which still inspired him as he wrote, was slow to lift. He showed no eagerness to get to Italy, nor, for that matter, to return to Carlyle Square. Even the spring of 1946 failed to disperse his fears of almost instant Armageddon, although he felt that the defeat of Churchill in the general election had possibly postponed a war with Russia for a while. 'I think the Labour Government will delay it by about a year', he wrote to Thomas Mark at the beginning of March. 'If Winston were here, we should be in it by now. . . . Oh for the German army to be in existence to help us! A labour camp in Omsk will be cold, but not much colder than here – and there will be reindeer to look at and even eat.'

Edith's reaction to the threat of war, and the gloom around her, continued to be very different. Whereas it drove Osbert into an ever-deepening nostalgia, it inspired Edith to protest and project herself into the role of the delphic prophetess of doom. In the autumn of 1945 the time had come for her most sibylline utterance of all. As she wrote:

> On the 10th of September 1945, nearly five weeks after the fall of the first atom bomb, my brother Sir Osbert Sitwell and I were on the train going to Brighton, where we were to give a reading. He pointed out to me a paragraph in *The Times*, a description by an eye-witness of the immediate effect of the atomic bomb upon Hiroshima. That witness saw a totem pole of dust arise to the sun as a witness against the murder of mankind. . . . A totem pole, the symbol of creation, the symbol of generation.
>
> From that moment the poem began. . . . It passed through many stages.

Osbert gave his reaction to Hiroshima in a note to Miss Andrade. 'The atomic bomb, final present from Churchill, has just finished me!'

But for Edith it would ultimately result in her impassioned 'Three Poems of the Atomic Age' – her 'Dirge for the New Sunrise', 'The Shadow of Cain', and 'The Canticle of the Rose'. She laboured over them all winter, and in late April 1946 was writing to John Lehmann that she was 'nearly *dead* with fatigue': she used one of Sir George's favourite gardening expressions – used to describe the transplanting of a tree – to account for the heavy labour of poetic composition. 'The hauling and dragging operations on the poem, the muscling of it, were fearfully tiring.' But Edith was now developing another role than that of the prophetess of the atomic age. Early in 1945 she had been telling Lehmann of a new prose book she had been working on:

It has been held over my head for years, and now it is descending. I wrote it ages ago, because I *had* to make money. What makes it so embittering is that I now make quite as much money out of poetry as I did out of the hateful prose books I had to grind out, so needn't have done it. When I say it is a terrible book – well it is *all right* you know; but I am *not* a prose writer. I am a poet.

The book was her *Fanfare for Elizabeth*, which Edith described in the blurb as 'the story of the childhood – terrible, but romantic and dramatic – of one of the greatest women ever born in England, Elizabeth Tudor'. It was the sort of dramatised historical account she did so well in her earlier *Victoria of England*, but it was obviously written – the disclaimer to John Lehmann notwithstanding – with far more sympathy and feeling of identification with the young princess than she had managed with Victoria. Reading the book, one cannot miss the emphasis she puts upon the parallels between the childhood of Elizabeth Tudor and of Edith Sitwell – a powerful father who ignores her, a beautiful and flighty mother, unhappiness and disinheritance in magnificent surroundings, and an awareness of a great role still to come.

But there was more than mere coincidence to bind Edith to Elizabeth I. There was that Beaufort blood as well, and one can see how potent was its spell for Edith in a little incident she relates to Lehmann, soon after publication of her book. The comedian Gillie Potter had arrived at Renishaw to open the flower show, and Osbert did his duty and invited him to lunch. Potter was something of a national celebrity, thanks to his radio broadcasts, and also a very bumptious and infuriating man. From the moment he congratulated Edith on having just produced 'a nice little book of poems', storm signals were in evidence.

Osbert did his best to observe the gentlemanly rule against insulting any guest beneath his roof, whatever the temptation. But Edith was no gentleman. And Potter courted fresh disaster by saying she was very brave to write another book on Queen Elizabeth after so many others had done so.

I replied, quite politely, that I was *not* brave, considering that all the people who had written *personal* histories of Henry and Elizabeth were vulgarians in the worst sense, (I consider Strachey's book on Elizabeth vulgar and *odious*).

This was the beginning of Edith's grand Elizabethan role. From now on she would rely upon that dash of Beaufort blood within her veins to stress her affinity with Gloriana. Actress and myth-maker that she was, she gradually began to live the part. It became annexed to her public image of herself, to emphasise the grand poetic role she felt she had to play.

For more than ever now she had to have this living legend of herself. Without it, as she knew quite well, she was a fifty-eight-year-old spinster poetess living in her brother's house in Derbyshire, and her non-legendary self was poor, shy, often bored, and terrified of being treated with indignity. But as a legend she could be invulnerable, and the legendary Edith had no problem in asserting her inherited and justified superiority to lesser mortals. Now in the middle of composing her most ambitious poetry, she finally achieved her best-known public face of all. To outsiders she was the dangerous ageing lady with the gothic face and the Elizabethan clothes and an acute awareness of exactly how she wanted to appear before the world. Just how aware she was of this one sees from the desperate complaint she made to David Horner about the short – and entirely sympathetic – introduction Harold Acton wrote to the French edition of her poems. 'It is of an almost unbelievable vulgarity – (part of it) – very familiar and personal. I shall have to have all the personal part cut out. None of it *ought* to appear . . . no serious artist's reputation could survive the personal part of it.'

Her complaint was made too late. The book was published with its introduction still intact – and no harm was done. But at all costs now the public image had to be preserved, and one can see how she defended it in the early summer of 1946 when her friend Maurice Carpenter was ill-advised enough to invite her to discuss her poetry at the Unity Theatre with Roy Fuller and another poetess, Anne Ridler. Edith replied with majestic rudeness that it was quite impossible. Mr Fuller had apparently been 'impertinent', and Mrs Ridler was not qualified to discuss poetry in her presence. 'I added that to ask me to discuss poetry with them was like asking Sarah Bernhardt to discuss the art of acting with two nice – or nasty – young people who had appeared twice in some remote, sparsely peopled, and dusty provincial theatre.' Edith's technique when 'dealing' with people now was to combine the manner of the Edwardian *grande dame* who dresses down a servant for 'impertinence' with the even grander manner of the great offended artist – the Sarah Bernhardt touch. There was an element of over-kill in the way that she pursued her often inoffensive victims. In mitigation one can plead her chronic sense of insecurity, but there seems something more than this at stake, almost as if these outbursts were dictated by that same assertive, histrionic instinct which shows so clearly in her later poetry. For the essence of her grand poetic rhetoric lay in the assumption of a queenly right to speak for the whole of suffering humanity in her verse, and this in itself required a working up of all her imaginative and histrionic powers. She who had never known physical love would speak for lovers, she who had borne no children would declaim the power of mother-love, and she who had had seen nothing of the first-hand savagery of war would lament the slaughter of Hiroshima.

The first full year of peace shows Edith stepping up her entertaining at the Sesame. She would have hated the suggestion, but there seems something of her mother's reckless passion for inviting guests for dinner in the way that she insisted on her dinner-parties every time she came to London. She had the Sitwell love of quality in food and drink – smoked salmon, lobster Newburg, good white wine – and most of her income went in paying for it, for this was her principal treat in life. It accorded with her queenly functions to surround herself with clever and distinguished friends, and she was always ready to invite poets and artists – and even critics – on whom she felt she could depend for potential allies. Her main Sesame dinner in April 1946 included John Lehmann and the Henry Moores, the young poet George Barker, and that one-time member of the 'bungalow school of poetry', Louis MacNeice himself. The following month she held an even more ambitious dinner – which misfired.

It followed a verse-reading given by T. S. Eliot, and was intended as a celebration of a great event in Edith's life. After the eight-year break with Dylan Thomas – during which time his name had been anathema – Edith had exercised her sovereign's privilege and finally made peace with him again. He and his wife Caitlin had been at the Eliot reading, and Edith magnanimously invited them to a lavish dinner at the Sesame, together with John Hayward and John Lehmann. To start with all went well. Thomas was civil, Eliot rigidly polite, and Edith intent on playing the literary *grande dame* with her favourite poetic protégé. And then, as so often in the rumbustious social history of the Thomases, the alcohol began to do its work.

Dylan Thomas was good as gold to *me* [she told David Horner afterwards] but he suddenly felt an inescapable urge to fly at Tom Eliot on the subject of Milton. (This was not due to any outside or *inside* cause, but was the poet in him.) He muttered across me to Tom – 'I'm surprised we were allowed *Milton* this afternoon; I thought he was dead.' Then in a tone of scorn, '*Dislodging* Milton with very little fuss – Dislodging Milton with very little fuss.'

Tom said mildly: 'I can't be held responsible for what Leavis says.'

'Well, you ought to *stop* him,' said D. 'And look here. Why does a poet like you publish such *awful* poetry. *You know* it is bad.'

Thomas had recently 'discovered' Milton. He had been learning long passages of *Paradise Lost* by heart for a Third Programme reading in which he took the part of Satan, and seems to have resented what he felt was Eliot's condescension to his latest hero. Apparently Eliot received the young man's rudeness with his customary aplomb. Edith, by now convinced again of Thomas's poetic 'greatness', was obviously anxious to avoid another *contretemps* with him, and so blamed Caitlin rather than her husband for the subsequent collapse of her dinner-party.

It then transpired that Mrs Thomas had spilt some ice-cream on her bare arm and had ordered John Hayward to lick it off. John H. refused. She insisted. He said he would lick it off any other part of her body anywhere else but *not* in the dining room of the *Sesame*.

'Mother of God!' she replied. 'The insults of men! You great pansy. What for are you sitting in that throne [his wheel-chair] and twisting your arms like that!!!' She afterwards seemed to take a fancy to him, although she perpetually, throughout the evening, addressed him as 'Old Ugly'.

Edith was determined not to lose her young Welsh genius again, and although one feels that she was not as angry as she might have been at his boorishness to Eliot, her chief concern was that the disastrous dinner might have caused another break in relations with the Thomases.

Oh, wasn't that dinner party after the Poetry Reading *Hell*! [she wailed to John Lehmann]. I was never so despairing, so worried, or so unhappy for everybody. I cannot imagine *anything* more unfortunate. One never knew what was going to happen next. . . . Dylan has, as a result, disappeared off the face of the earth as far as I am concerned. I mind very much because I consider him a wonderful poet, and (although there was a misunderstanding for eight years) am fond of the boy. He has been good as gold with me – that was why I put him next me at dinner, in order to keep the goodness well in evidence – and it is very trying that he should go off like this because of his maddening wife. Either, *A* she has told him I am a demon in human form and insulted her (in this case she would have been backed up by Roy Campbell with whom they were staying) or, *B*, he thought he would have to apologise for her and is waiting until he thinks the storm has blown over.

Edith had too much capital invested now in Thomas to allow him to behave like this. She believed he was the greatest poet of his generation, and she had no intention of allowing anybody to forget that she had fought his battles for him in the *Sunday Times* so long ago: hence her extraordinary tolerance of his behaviour and desperate insistence that he was always 'good as gold' in her maternal presence. But this was not exactly true.

After the dinner at the Sesame a fresh break *did* occur, despite all Edith's efforts to the contrary, and on 26 July she was writing miserably to William Plomer asking if he had news of him: 'He has wandered away again, and renounced me and won't have anything to do with me.' But Thomas was too smart to waste a friend as powerful as Edith. He had gone off to Ireland still angry at the reports that he had heard of Edith's attitude to Caitlin. But on his return he managed to forget his pride, and tenuous relations were restored.

His generosity of soul was soon rewarded. Bryher had recently set up a travelling bursary for promising young writers, and Edith had no difficulty in persuading her devoted millionairess friend that there was nobody more promising than Dylan Thomas. So it was really thanks to Edith that the Thomases set out that autumn for a three-month trip abroad. They even followed her advice on where to go, and ended up near Florence at Scandicci – within walking distance of the Villa Mirenda, where twenty years before the Sitwells had visited the D. H. Lawrences.

By then Osbert had already made the first of his post-war visits to Montegufoni. This was in early May 1946. He had just finished his corrections to the proofs of the second volume of his autobiography. David was now demobilised and back full-time with him again, and Osbert had finally agreed – it seems with some reluctance – that the time had come to bestir himself to visit that appalling old white elephant of Montegufoni.

Edith was worried for him, and penned David a touching note to say, 'How much I wish you were not going to Italy. Also to beg you, *A*, to see all water is boiled *with your own eyes*. *B*, avoid bridges, and *C*, avoid crowds that look as if they were going to make a row.'

But not even Osbert's legacy of wartime gloom was proof against Tuscany in May. Within a few days of arriving he wrote to Adrian Stokes with all his news.

> Since I arrived here on Wednesday, I have done nothing except moon, and can't work or write a letter (I hadn't realised how much seven years in England had tired me): but I must at least write to you and tell you that I'm here and that I have loved every minute of being in Italy again. Florence is very broken and much proletarianised (if such a word exists). The damage is frightful. *But* everyone in Italy seems just the same.
>
> Here there is very little damage, but it is due to Guido Masti who guarded my interests with a fervour I had done nothing to deserve, and who incidentally really saved the Uffizi (Ufizzi?) pictures by haranguing the Germans (the pictures were in the house) and telling them when they wanted to destroy them, that they belonged, not to the Italians, but to the world.

Osbert was now 'Il Barone Sitwell', and for the first time in his life was able to enjoy the castle without the plaguing presence of his parents. At one time he had even thought of selling it, but he was delighted and surprised at how much he liked it now. There were no rows over money with Sir George, no endless luncheon parties with poor Lady Ida. Instead, he and David could enjoy their first holiday abroad for seven years. The gardens were in bloom, the valley gleamed, the house was his.

'Only the ghost of my father, walking about a little at night,' he wrote 'disturbs the tranquillity and makes the dogs howl.'

The return to Montegufoni meant the restoration of Osbert's old escape route to the sun. Edith had no particular desire to go to Italy even if Osbert had invited her – which he didn't. Nor, for that matter, did he invite Georgia and Sacheverell. Wartime confinement to the shores of Britain had turned Sacheverell's attention to the art and architecture around him: the result had been his monumental *British Architects and Craftsmen, A Survey of Taste, Design, and Style during Three Centuries 1600 to 1830*, with its timely exhortation to 'consider, rather, our own glorious past and draw profit from it for the future'. Once again, Sacheverell was acting as a 'trail-blazer' in the appreciation of our artistic heritage, and this book would make way for many subsequent more detailed studies of the art and culture of the period. It also came as something of a counterblast to the grimness and the horror of the times. 'It was' wrote Cyril Connolly, 'one of the books that preserved my sanity in the last war.'

Edith continued her strange hermit's life at Renishaw with Osbert and David for company when they were there. Despite her work, despite the weekend guests and growing fame, she clearly found the life extremely boring, but there was no alternative now that she had no Helen Rootham to take charge of the practical details of setting up her own establishment. She never visited her house in Bath.

Her main alleviation and escape was still her monthly trip to London. Then she would come alive, enjoy the gossip and the lionising and the attentions of her court around the dining-table of the Sesame. Then, when the few days of her London jaunt were over, she would return to Renishaw, mull over everything that she had heard, and carefully compose her letters to her chosen friends, most of them written late at night on Osbert's fine Renishaw paper with Whistler's vignette of the house, while the oil-lamp by her bed gave off as much light as a cigar.

Edith's letters contain much of her finest – and certainly her funniest – prose, and in them she did something that Osbert in his carefully constructed memoirs never quite achieved: she totally revealed herself. Reading these letters in that energetic, forward-sloping hand, one can almost hear her voice, by turns querulous, sarcastic or concerned, and start to understand this angry, very funny woman who always seems to teeter on the edge of some extraordinary excess: sometimes so common-sensical and humorous, sometimes so admirable and brave, and then so shrill and vitriolic and self-pitying that she becomes embarrassing.

But from her letters now one starts to see the curious pattern of her life and her imagination. It is a sort of *commedia dell' arte* world which she constructs – without particular concern for factual accuracy – from the characters around her as a back-drop for the great role which she set herself to play. First come the groundlings, her omnipresent regiment of lunatics and bores – old women at the Sesame, who breathe cold-germs over her, workmen who disturb her peace, fans who write in to ask her to correct their poetry, Robins's daughter endlessly practising the piano. Then come her perennial Aunt Sallies. Throughout this period Robins was firmly at their head, himself the incarnation of Osbert's hated 'little man', closely followed now by Miss de Zoëte, or 'Beryl the Peril' as Edith christened her. 'One of these days,' she wrote darkly to Lehmann, 'Beryl's nymphomania will really land us up a gum-tree!' In their wake follow numerous editors, working-class children, journalists and female poets. She had her heroes – headed by her brothers, and including 'geniuses' like Dylan Thomas who could do no wrong. Finally there are the main contenders in this running drama of her life – the adversary–victims she did endless battle with, dread successors of her parents who, like them, would mock her and insult her by belittling her work. Her great enemy, Wyndham Lewis, had become too blind and too infirm to be much use in such a role, while D. H. Lawrence was dead and buried now in Mexico. This, however, did not mean that Edith's fighting days were over. There was still Noël Coward, and the memory of his mockery of *Façade* a quarter of a century before could still set her quivering with rage. Julian Symons was an ever-present threat, and late in 1946 yet another potent villain crept once more upon the scene – the poet Geoffrey Grigson, once 'G.G.' of the *Yorkshire Post* and, like Symons, one more follower and friend of Wyndham Lewis.

Only that autumn she was certain he had been insulting her – not by name, it is true, but that made hardly any difference – in a talk broadcast on the B.B.C. She moved into action in swift retaliation. She had heard rumours that the Sheffield Literary Group had invited Grigson up to lecture to them on poetry and, as she told Lehmann, she had no difficulty persuading Osbert to send the secretary 'a really appalling rebuke'. When Grigson heard he had offended Edith, he wrote suggesting a gentlemanly truce. 'You will not expect me to recant in my criticism of the past: I cannot expect you to recant in yours. But if we can agree upon that expectation, we may perhaps agree on keeping, for the future, our judgements relating to each other's work to ourselves.' There was small hope of this, particularly as Edith would increasingly believe that Geoffrey Grigson was now behind a plot to undermine her reputation and her work. For one of the main themes of this constant drama was that her poetry was a 'cause' – just as the Modern Movement had once been *the* cause in the twenties. One was compelled to be for it – or against it. She would allow no neutral ground. And any critic of her work was automatically regarded as an enemy, just as the faintest criticism was seen as the vilest insult to her person.

This was the source of all the most dramatic action in the letters that she wrote, and the people who received them formed the essential audience to the living drama of her life. The composition of this audience changed from time to time as fresh appreciative members joined and old ones were discarded. Two important new members were the poet William Plomer, and the Master of Wadham College, Oxford, Maurice Bowra. But the people she relied on most were still John Lehmann, Kenneth Clark, John Hayward and Stephen Spender. She took enormous trouble with them all, dedicating individual poems to them, sending them copies of her poems which she wrote out for them by hand, and never failing to invite them to the Sesame. But, most importantly, she treated them as confidants and allies as each fresh act and drama burst upon the scene.

In May 1947 *The Shadow of Cain* was published to generally awed and enthusiastic reviews, but the tension before its publication seems to have upset Edith. 'Not very well, and get the most dreadful *angoisse*,' she told Lehmann, 'for which nothing can be done until it decides to depart. It is a horrible nervous thing which I inherited from my beloved Papa.'

As usual at such times, the lunatics and bores were soon acting up. Her long-suffering fellow-members of the Sesame were, she told Lehmann 'absolute *hell* . . . and getting like the Yellow Peril and threaten to overwhelm and swamp my civilisation completely'.

They were a sort of comic turn that she could joke about continually. 'You know,' she told John Lehmann, 'there are moments when I feel that in some past life I must have, *A* imprisoned and martyrised an old lady, *B* driven somebody to suicide through sheer boredom and, *C* committed the most horrible aesthetic crimes. No other theory will meet the case.' But besides the comedy there was time for melodrama, and in July 1947 Edith discovered a new minor villain for a short entr'acte of anger – Lawrence Durrell. She was incensed to hear that he had apparently suggested that, at any other time, the work of the Sitwells 'would have been only printed in private presses and country houses' – and as usual Edith took this criticism personally. Durrell had been, she wrote to Lehmann, 'most caddish as well as grossly impertinent'.

Shortly afterwards, Durrell was ill-advised enough to write to Osbert asking his help to publicise an exhibition of the paintings of his old friend Henry Miller, only to receive a firm Osbertian refusal and rebuke for the 'gross public insult' – as Edith called it – which he had given all three Sitwells by his earlier remarks. But this was not enough for Edith, and one can see her *angoisse* suddenly exploding in the letter that she wrote to Lehmann on the subject.

I for one am sick of being attacked by persons of no talent, of a small talent like Mr Durrell's. In 1923, Mr Coward began on me in a 'sketch' of the utmost indecency – *really filthy*. I couldn't have him up, as I had no money and didn't know how to set about it. Nobody helped me, and I had to put up with having *filthy* verses about vice imputed to me and recited every night and three afternoons weekly for *nine months*. They weren't just dirty, they were filthy. The woman who recited them went so far in her persecution of me as to have a photograph of me in her dressing room. These attacks, in one degree or other, have gone on since then, and I have always had to put up with them.

This event had occurred a quarter of century earlier, when the Lord Chamberlain would not have permitted the slightest impropriety on the stage. She had not witnessed it herself and she had accepted Coward's apology in 1926. This extraordinary letter, with its frantic underlinings and exaggeration, shows more than anything she wrote the cruel force of her persecution fantasies. Depending on her mood, a criticism like Durrell's could become a 'gross public insult', which in turn could set off this hysterical reaction from the past at her insecurity and fear of being mocked. Nothing could be forgotten or forgiven. The pain was still as great as when she was a girl, and it was the self-same cry of anguish that she put into her poetry.

The summer of 1947 was a time of strain for Osbert too. Gout and the effort of finishing the fourth volume of the autobiography – which he would finally entitle *Laughter in the Next Room* – had had their predictable effect, so much so that when John Lehmann saw him in London in July he wrote asking Edith what was wrong.

Don't give an instant's thought to Osbert's having appeared strange in London [she replied]. It was illness. He made exactly the same impression on me. (I mean I thought that he was angry for some reason with me), so that I thought, 'Oh, God, what *have* I done now?' And it worried me a good deal. But this demeanour was simply due to pain, being unable to sleep, and having taken a medicine against gout called Béjean, which makes him feel extremely ill.

But 'a month's gruelling treatment of electricity, etc.' from his Sheffield doctor seems to have eased his gouty foot – although the doctor kept on telling him that he must realise that he was now fifty-four. 'But just don't I!' he exclaimed to Miss Andrade. A blazing August helped on his recovery. Soon he was swimming twice a day and managing to walk with a stick. 'Only Robins' face and your absence prevents this week from being paradisiacal,' he wrote to David now. For David too had not been well. He had been struck with shingles and was in Vichy for a cure. Parted from him, Osbert was as lovelorn and concerned as ever. 'I do hope that you are feeling better, bless your heart, Your devoted O,' he wrote.

Then, by the middle of September, happiness returned. The manuscript of *Laughter in the Next Room* was safely with Miss Andrade for typing. Osbert spent a week alone at Carlyle Square, and then was off for Italy, and reunion with David at Montegufoni. And, as usual, Italy and David soon restored his spirits. When they arrived the castle was superb in the golden days of late September. The zinnias were 'tall as trees', the gardens a bower of plumbago. All day white bullocks plodded up the valley bringing the grape harvest to the high-arched *cantina* underneath the house, and the cool rooms held the acrid aroma of young wine.

This year Osbert had brought another visitor – John Piper. Osbert commissioned him to paint the house to illustrate his book, and Piper's memories of Osbert now are in pointed contrast with the angry, gouty figure Lehmann had seen in London just that spring. Away from England, Osbert became gentle and relaxed 'and quite the ideal patron in the way he would suggest a subject but never interfere'. Every morning Piper worked, but in the afternoons Osbert would often hire the local taxi and take him and David sightseeing to places like Volterra, Lucca and Siena. 'He was always putting you in the way of things he thought you would enjoy. He was extremely good like that, and when I left he sent me off with a great batch of old master prints from Alinari, and some Ferragamo shoes for my wife. He could be very generous when he liked you.'

Life must have seemed particularly benign for Osbert now, for while in Italy he heard that he had won an unexpected honour – the newly instituted *Sunday Times* book prize of £1,000 and a gold medal (since, alas, discontinued) for the first two volumes of his autobiography. He had always previously made fun of literary prizes, but now he was delighted. As he told his publisher, 'being vain and avaricious, both honours and money are dear to me. Very dear. . . . Now all I *need* is the garter.' Perhaps this mood of strange benevolence at least partially explains the historic act of reconciliation which now occurred. In the middle of December Miss Andrade, who was dealing with corrections to the typescript of *Laughter in the Next Room*, received an urgent note from Osbert. 'In the *Façade* chapter, I shall, owing to tactical considerations which I will reveal to you later, have to cut out much of my vituperation of Mr Noël Coward.' She wondered why – and on New Year's Eve, when Osbert returned the now pruned manuscript, she got her answer. Writing of Noël Coward, Osbert explained,

The brute apologised to me the other day, and as I was in a good mood, I forgot to kick him downstairs – it was on the staircase at Buckingham Palace – and as my sister magnanimously says she'll let bygones be bygones, I suppose I had better cut out the insults. . . . I really am loath – or is it loth? – to cut; but there it is!

The real author of this reconciliation seems to have been the Queen herself. Both men were personal friends of hers, and by inviting them together to a reception at the palace she tactfully gave Coward the chance to meet Osbert under conditions which it would be hard for even Osbert to refuse. But even so it was a bitter pill to swallow and, as Osbert evidently feared, it soon caused repercussions in the family.

For Osbert knew how passionately Edith hated Coward, and was so worried at the prospect of upsetting her that, like any frightened male, he shied off mentioning a word of what had happened. But news leaked out about it from the palace. Sacheverell and Georgia heard, and it was Georgia who inadvertently broke the news to Edith. The results can be imagined – and when the uproar had subsided, Edith described the conclusion of the whole affair in the inevitable letter to John Lehmann, complete with extensive underlinings.

I am terribly sorry for poor Osbert who is not in the least to blame, and who has been wretchedly unhappy because he thought I would mind. He sees very badly, it was in a dark passage, and at first he didn't see who it was speaking to him and asking to be on friendly terms. Osbert didn't sleep all night after it. But actually I think it's a good thing, *for my family*, that this has happened.

She went on to say that despite poor Georgia's 'blabbing' of the news, she refused to upset Osbert by indulging in a full-scale family row. She also made it plain that, while she had agreed to a reconciliation with Coward for the sake of the family, she never could forgive that brute, Coward, in her heart, and ended up by reiterating just how 'unspeakably obscene and filthy' those far-off verses had been – '*so* bad that it would have been hard to discuss their meanings in a court of law'.

As if to reward her for her act of magnanimity, fate now stepped in to offer Edith an unexpected prize which gave her great pleasure. The first hint of it is in a postscript to a letter that she wrote to Lehmann early in January 1948. She had explained how she had been happily immersed in Blake's prophetic books, then added, 'On May 7th I am going to be given the honorary degree of Litt.D. by Leeds – and from then onwards shall call myself Doctor Sitwell.'

Edith was already taking her doctorate as seriously as she could. And five months later, after she duly received it at Leeds, she wrote telling him exactly how she wished to be addressed.

Oh, will you please address my letters to Miss Edith Sitwell, D.Litt, because I am having to train Robins to call me Doctor Sitwell. It has been a fearful shock to everyone – because until now writing poetry was supposed in my case, to be on a level with *knitting*, and they can't *think* why I should need quiet . . . tempers here are past bearing. It has been a terrible shock to the shouters and disturbers that I have been made a doctor and shall soon be two, and they have been very disagreeable ever since.

The reference to her soon being two doctors concerned the fact that, not to be outdone by Leeds, Durham University had offered her a doctorate as well. Oxford would follow, giving her the right, which she made the most of, of signing herself D.Litt., D.Litt., D.Litt., to anyone she thoroughly disliked. William Plomer said he would register with Doctor Sitwell on the National Health Service.

Just before Christmas 1947 Osbert had received his medal and his cheque from the *Sunday Times*. Edith naturally attended, and at the luncheon party afterwards she was infuriated by the flippant remarks of an unknown guest. As soon as she was back at Renishaw she wrote to her friend Plomer complaining about what had happened.

There was a man there who began talking to me by saying he knew you. I think I have got his name wrong, but it was either Frazer or Fleming, I think the latter. I thought at first, from his manner with her, that he was Lady Cunard's social secretary, but realised afterwards that she would scarcely have employed him in that capacity. He told me he was 'very amused' to see that you had put my 'Shadow of Cain' first on your *Horizon* list. He ran no immediate risk in saying this, as we were both guests at a luncheon party given in honour of Osbert. He then, until I was at last released from him, expatiated on other 'amusing' and 'surprising' choices.

Ian Fleming was then thirty-nine and Foreign Manager of the *Sunday Times*. It would be some years before he married Lady Rothermere and began the first of the adventures of James Bond, but he was already liable to make this sort of breezy, Bond-like conversation in unappreciative surroundings. Plomer was an old friend from his time with Naval Intelligence in the war, and Plomer replied to Edith promising to convey the gist of her extreme displeasure to the insensitive young man – and tell him he should mind his manners. It is some measure of the extraordinary lengths to which her outraged dignity could carry her that more than a week later the remarks of this unknown and as yet unimportant philistine still managed to incense her; and she thanked Plomer 'for so kindly saying you will reprove Mr Fleming. As I say, he was lucky to have been rude to me under those particular circumstances.'

But this was not the end of the affair. Edith could be unnaturally swift to anger, but she was just as speedy with her queen-like powers of forgiveness when she felt inclined. (Indeed, these sudden swings of people in and out of Edith's favour were more dramatic and pronounced than ever now. They served to keep her friends and followers on their toes, flattered her sense of power and theatre, and stopped life getting boring.) Prompted by Plomer, Fleming wrote back a letter of appropriate apology and obeisance. Edith was delighted and, to show just how richly she could reward the truly contrite, invited him to something of a gala luncheon at the Sesame – where the guests included Maurice Bowra, John Lehmann, William Plomer and T. S. Eliot.

Quite how Fleming made out is not recorded, although he did tell Edith he was horrified by the array of talent she had asked. But the interesting thing is that the friendship prospered. For, apart from his good looks and charm, he was also something of a power on the *Sunday Times*. Like Edith, he knew all the tricks of self-promotion and publicity, and just a fortnight after the luncheon at the Sesame he made his usefulness quite clear to Edith. *The Shadow of Cain* was printed in its entirety in the *Sunday Times*. For a while there was even talk of Edith and Ian Fleming collaborating on a short book on the unlikely subject of the sixteenth-century alchemist and mystic, Paracelsus. (Edith had become interested in Paracelsus through Tchelitchew, who was obsessed by him; Fleming while a student had translated a short paper on him by C. G. Jung.) Although the book came to nothing and Fleming never made it to the inner circle of the Sesame, the two of them remained good friends. Edith always made the most of this persuasive ally on the *Sunday Times*, Fleming would make a point of seeing that her books were favourably reviewed by people she approved of, and in return the creator of James Bond was always gratified to count the *grande dame* of English poetry among his gallery of friends.

A more important – and even more unlikely – friendship started for Edith early in 1948, with the Australian Marxist poet and critic, Jack Lindsay. It started with a long review he wrote of *The Shadow of Cain* for a left-wing magazine, *Our Time*, analysing the poem in elaborate Freudian–Marxist terms, and then comparing Edith and her work with the development in France of such major left-wing poets as Tzara, Eluard and Aragon.

It must have been a little disconcerting – to say the least – for anyone with Edith's attitudes towards the church, the common man and the sanctity of birth and breeding to find herself acclaimed as a potential hero of the Left, indeed, as one of 'the poets who stand at the pivotal points of a vast and decisive spiritual change in man, bound up with social, political and economic revolutions', as Lindsay put it, but so it was. And indeed, Jack Lindsay, who is a subtle and extremely learned critic with interests far beyond the usual run of Marxist dialectic, made out a fascinating case, comparing her effect on English poetry with the development of Tzara through the Dadaist experiments of the twenties into a full-fledged poet of the Left.

Edith was instantly excited by the whole idea – a little like the character in Molière who was so thrilled to learn he wrote in prose – and wrote back enthusiastically, thanking Lindsay for his 'profound and wide understanding of all the implications of the poem'. 'You understand what was in my mind when I wrote "The Shadow of Cain" in a manner so extraordinary that you might have been the poem's author.'

Nevertheless, one may reasonably doubt whether she really had worked out the politically loaded implications of her poems that Lindsay was saddling her with. Certainly it is hard to credit her with that 'true dialectical insight' Lindsay so admired. But, on the other hand, one can understand why Edith was so inclined now to accept this elaborate interpretation of her work. In the first place, Lindsay did what few other critics could be bothered to do – he read her poems seriously and based his estimate of her importance on a profound analysis of her work. This was, in itself, a welcome change from the glib acclamations of her 'greatness' coming from reviewers wary of offending so powerful and termagant a figure as Edith was now. (Cyril Connolly, for instance, praised the 'pure honey' of her verse, yet privately asked Stephen Spender whether he really liked her poems. 'I wish to God I could,' he added.)

And in the second place, it was most exciting finally to have the chance to steal the clothes from her old enemies, the left-wing poets of the thirties, by having herself proclaimed the greatest left-wing poet of them all. For she was still exceedingly aware of them and their annoying reputation. Auden was due to return to England from America in the spring of 1948 and Edith made a point of asking the Spenders to be sure to introduce her to him when he came.

At all events, Jack Lindsay now became for several years her favourite critic, as well as a devoted friend, for, as he says, 'in her heart of hearts she always feared that, despite her fame, nobody else was really interested in her poetry – or took it seriously'.

For Osbert, 1948 brought more sciatica, a number of maddening reviews of *Great Morning* which appeared that spring, and a row with the local council over a dead cow in a nearby reservoir. He was, as he had told Miss Andrade, by now fifty-five, 'already somewhat lame, very forgetful, and living in a vile age'.

He was particularly incensed by a review in the *New Statesman*, so much so that he took the injudicious course of penning a heartfelt protest to the editor, Kingsley Martin. According to C. H. Rolph, Martin's biographer, Martin wrote back saying he was very willing to publish the letter since it was 'good copy' but that he thought that writers lost their dignity 'when they show that their feelings are hurt by adverse criticism.' Osbert Sitwell 'awkwardly accepted this' and the letter was not published.

Edith too was at her most 'crocodilous', as she called it. The nuisances were bad in 1948 – including an unknown woman who forced herself into Edith's presence and asked her if she realised how wonderful time was. 'It goes on and on, Miss Sitwell,' she remarked. 'Never again – *never*,' Edith exclaimed to John Hayward, 'will I be kind to anyone unless they can produce a certificate signed by a magistrate, clergyman and two doctors, to the effect that they are not a nuisance.'

She seems to have felt the lack of Noël Coward's presence as her major demon, and had to make do with Geoffrey Grigson now, who had produced a book of criticism which she was sure 'by implication' had insulted her. (There is, in fact, no mention of her in the book.)

I really do not know what to do [she mused to Lehmann]. Osbert keeps on telling me that these attacks made by these little people can do no harm to me professionally – but it is very injurious to my pride to find that – whilst many reviewers protect the other people attacked, – not *one word* has been said in *my* defence. Mr Grigson's gifts really do not entitle him to insult or instruct me. He is an egregious little poetaster, and his poems are debilitated pastiches of Hymns Ancient and Modern. Only he does not write of, or invoke God; he examines the nature of groundsel and the sex-life of the winkle.

This latest feud with Grigson was particularly complicated by the presence of John Piper in the house. For Piper was a friend of Grigson and she considered that in some obscure way Piper had 'betrayed' her by keeping up this friendship with somebody she disapproved of. Piper had 'tied her hands' and so had prevented her replying as she would have wanted. To show how annoyed she was with Piper, she stayed in bed throughout his stay. He wondered what was wrong with her.

Then, in the middle of this most bad-tempered summer, there was suddenly a chance of fresh excitement. Her correspondence with Tchelitchew had continued steadily after the war, and now, incredibly but sensibly, she had made peace with her once hated rival, Charles Henri Ford, who was more than ever Pavlik's inseparable companion in America. At the beginning of 1948 Ford had even written to her to say, 'How I wish, how we both wish, Pavlik and I, that you were going to be here for the exhibition of the "Paris Period" paintings done between 1925 and 1933 – paintings that you know so intimately, most of them.' There was no chance of that, but then in June Ford wrote to her that he and Pavlik were off to Sarasota, Florida, that coming winter, where the Ringling Museum was holding a series of celebrity lectures. Why didn't she and Osbert come too?

CHAPTER TWENTY-THREE

America

1948–1949

THEY MUST have made a most impressive trio that grey October as they sailed to America on the old *Queen Mary*, Osbert and Edith and David Horner, bound for the lecture circuits of the Middle-West at the invitation of Colston Leigh, the lecture king of the Americas, the man who, just before the war, had so infuriated Edith with his refusal to allow her to read her projected lectures on her poetry, and who had wanted her instead, she joked, to deliver them extempore when balanced on a bicycle.

Since then Edith had grown enormously in self-confidence and poise and she no longer had the faintest qualms about her public appearances. Leigh admits that at this stage he was not particularly enthusiastic over the idea of the Sitwells. That year his star lecturers were mainly American and included Mrs Eleanor Roosevelt. But Colston Leigh had built his business up by trying everything, and his advance publicity had promised that 'the famous brother and sister – two of England's most celebrated literary figures', would provide 'in their contrasting voices and methods of recital, a program that gives a splendid insight into their brilliant intellects and unusual personalities'. And he had already been agreeably surprised by the response. From Montreal to Buffalo, from Florida to Boston, assorted women's clubs and student bodies had been clamouring to hear them at a standard fee of 1,750 dollars for an evening. (Half the money would go to the Sitwells, and Leigh would pay all expenses and publicity.)

Osbert had not been particularly keen on going. The memory of his last unhappy visit to the States still disturbed him, and he did not really need the money any more. But Edith did, and she was also drawn by the same force that had brought her to Paris in the twenties – Tchelitchew. It was nine years since she had seen him, but thanks to those weekly letters they were still very much part of one another's lives. Pavlik had shared in the excitement of her gathering success, just as she had in his; for in New York Tchelitchew had finally achieved the prominence and the acclaim that were his due. His masterly technique had made him a fashionable painter of the rich and famous in New York, and at the same time his profounder work, while never getting the success which life had lavished on his old and hated rival, Picasso (or Pot-cassé as he called him), had a devoted public all its own. His masterpiece, entitled 'Hide and Seek', which he had worked on in the war, had a proud place now in the New York Museum of Modern Art.

Through Pavlik's letters Edith had all but collaborated in its creation – he wrote to her about it endlessly, and throughout this period of their enormous correspondence Edith appears to have grown in imaginative importance to the painter. 'I have no friends besides you,' he told her. 'I have no real friends.' As he became increasingly obsessed with mysticism he came to believe that their continuing rapport, via their letters, 'formed some "astral" physique important to life'. By 1948 Tchelitchew was undoubtedly as anxious to see Edith as she was to visit him. He had actively encouraged the reconciliation with Charles Henri Ford. (Allen Tanner, in the middle of a nervous breakdown, was undergoing electric-shock therapy in a New York hospital, and was quietly erased from the list of Edith's correspondents.) Pavlik encouraged the idea of Edith's visits among the New York socialites he knew, as well as with a group of artists and intellectuals broadly associated with the Museum of Modern Art. He even managed to persuade Mrs David Pleydell-Bouverie and Mrs Vincent Astor to provide tax deductible funds for a suite at the St Regis, and had put Edith and Osbert in touch with Colston Leigh. As the ultimate incentive, he was promising Edith that he would paint a great new portrait of her when she came, to go beside the 1937 painting of her as a sibyl, which had been bought by Edward James. 'When I paint your next portrait it will be a "Muse" not a "Sybill", an inspiration and not a destiny,' he wrote.

But once the long-awaited visit turned from a dream to inescapable reality, everything began to change. According to Tchelitchew's biographer, 'Edith Sitwell exists for him most truly as the great Sibyl he painted in 1937 and it has been to this image, a purely spiritual quantity, that he has addressed all his letters.' Certainly the arrival of the *Queen Mary* and the prospect of confronting Edith in the flesh seems to have panicked him, and, but for Ford's insistence, he would have funked turning up at Pier 19 to greet her as she disembarked.

As it happened, Osbert's sociable presence and the excitement of the grand reunion saved the day, and everything appeared set fair for the happiest of stays. True, Pavlik had aged badly in the ten years that had elapsed since his days in Paris, and the once handsome Boyar was going bald and had the features of a melancholy horse. But Edith had aged as well, and there was something touching in the almost girlish joy of this stately, sixty-one-year-old poetess in the presence of the exuberant Russian who had brought her more suffering and joy than anyone alive.

But that very night, at dinner at the St Regis, came the first hint of trouble. It was a distinguished little gathering of a number of the visitors' most eminent well-wishers in New York – among them Mrs Astor, Mrs Pleydell-Bouverie and her mother, Lady Ribblesdale, Monroe Wheeler of the Museum of Modern Art, Tchelitchew, Charles Henri Ford and Lincoln Kirstein and his wife, Fidelma.

Edith was tactfully wearing a spectacular dress made from the black and gold striped material Tchelitchew had sent her specially for the visit, but instead of being pleased by this Tchelitchew objected to the way the dress had been cut by her Sheffield dressmaker 'and neither his face nor conversation can hide his irritation'.

Indeed, irritation seemed to be his dominant emotion now with Edith. He was irritated by the way that she insisted on referring to herself as 'Doctor' Sitwell. He was irritated by the red turban that she wore. But the height of all his irritation came when he took her to the Museum of Modern Art to confront his painting, 'Hide and Seek'.

This was the pregnant moment Tchelitchew had awaited for years, for Edith was his muse, his confidant, his closest friend who had always championed his genius and passionately believed in all his art. He had written to her endlessly about 'Hide and Seek', but now, as she stood before this canvas of the young girl with the butterfly surrounded by so many foetal heads, something was seriously amiss. Edith, for once, was at a loss for words.

One can sympathise with her. It is always hazardous to give an off-the-cuff reaction to a painting in the presence of the artist, particularly one as touchy and extreme as Tchelitchew, and 'Hide and Seek' is certainly a most alarming work. But Tchelitchew was panting for reaction from his muse. Edith still stayed silent. Finally she told him that her emotions at his painting were so profound that to do them justice she would have to write them in a letter – which she did next morning. By then it was too late. For Pavlik, nothing could make amends for her behaviour. From that moment Edith lost her role as Vittoria Colonna; and Pavlik ceased abruptly to be her Michelangelo.

Tchelitchew's damaged dignity was not helped by the enormous interest Edith and Osbert obviously aroused from the moment that they landed in

New York. They were *his* friends, and he had arranged their visit, but now, within a few days of arrival, too many other people were becoming interested in them too; indeed, they were gathering a fame and notoriety which he in all his years in America had never managed to achieve.

It was a curious success, part of that same mysterious phenomenon of fashionable Britishness which had ensured a similar American triumph for Oscar Wilde, and which would do the same for the Beatles and Ian Fleming. Colston Leigh's publicity combined with the Sitwell legend and their public presence to make their lecture tour a sell-out from the start.

The most important moment in building this celebrity occurred when Leigh's publicity machine arranged for a photographic session with *Life* magazine in Frances Stellof's Gotham Book Mart on 47th Street, early in their trip. This was an extraordinary occasion, and something of a classic exercise in elegantly engineered publicity. For Miss Stellof's tiny bookstore had long been a famous meeting-place for New York's poetic underground. Edith had corresponded with her in the past, and on the afternoon that she and Osbert called for tea some careful staff-work by the *Life* photographer had assembled an impressive group to welcome them. Among the older generation of American poets there was Horace Gregory, William Rose Benet, Marianne Moore and Randall Jarrell. Auden and Spender had turned up, and so had the fresh-faced Gore Vidal and the youthful Tennessee Williams. Charles Henri Ford arrived. Pavlik didn't.

The resulting photograph has something of Aquinas's Vision of the Heavenly Host, with poetry's angels and archangels grouped around their virgin queen. One had no need to have read the Sitwells to appreciate that here in Edith Sitwell there was a major modern poet of awesome dignity and presence, respected and revered by famous fellow-poets. When reproduced in *Life* with a photograph of Edith reading and a reverential article – 'we are the only two people to whom the magazine *Life* has ever been polite' crowed Edith – the photograph proclaimed this instant message, and everyone knew who they were.

A celebrity has been defined by Daniel Boorstin, as a person who is known for his well-knownness, and by this truism of the higher ballyhoo the Sitwells had become celebrities, and naturally loved every minute of it, Edith in particular. For here, with one article in *Life*, they had achieved what thirty years of dedicated work and struggling in Britain had not given them – unquestioned literary status, and an enthusiastic welcome from all sides. There were no Geoffrey Grigsons in New York, and no *New Statesman* critics with their impertinent reviews. Edith had reached the paradise that she had always longed for. 'Dear me, how much I *do* like the Americans,' she told Lehmann. 'Anyone who doesn't must really be mad.'

That autumn, as they hit the lecture trail, Edith and Osbert found that they were rivalling Mrs Roosevelt herself. Osbert spoke on 'Personal Adventures' and 'The Modern Novel, its Cause and Cure', Edith on 'Modern English Poetry', but they were most successful when they read their poetry together, for it was the Sitwells *themselves* their audiences paid to see. There was a crash audience at Yale, a sell-out in Boston, and in New York ten thousand stormed the Town Hall to hear them.

But this public triumph was only part of their American success. Back in New York a more select and dignified acclaim awaited them, as high society took them to its brittle bosom. The welcome of the women's clubs and student gatherings was more decorously re-echoed in the great drawing-rooms off Central Park. Again they really owed a lot of this success to Tchelitchew, for many of their introductions came originally from him. It was thanks to him that they had the entrée to a number of his clients in the solid upper-class New York society and among the old guard intelligentsia round the Museum of Modern Art. As visiting celebrities they were custom-built for such a world – regal and spectacular, gracious when necessary and arrogant when not. They didn't miss a trick. Since the days of Whitman, poets had never made much impact on American polite society, but here in New York the Sitwells managed to combine what many English critics secretly resented in them both – the assurance of the aristocrat and the glamour of the dedicated artist.

It was Edith's triumph in particular. Osbert pursued his favourite public role of royal uncle, but Edith was royalty herself, and welcomed as a reigning celebrity. For here in America she could play Gloriana to the hilt. This was her peak period as a Plantagenet, and it gave her an invaluable role.

But Edith's grandest hour was yet to come. It was Monroe Wheeler, the ex-advertising man who ran the special exhibition side of the Museum of Modern Art, who first suggested a celebrity performance of *Façade* at the Museum. Edith was thrilled at the idea. Her great success as lecturer and reader of her poetry had boosted her self-confidence, and the more inclined that Wheeler was to turn this performance of *Façade* into a gala social evening, the more the prima donna in her rose to the occasion. It was superbly handled. With all the tickets sold out at thirty-five dollars a head, Wheeler could afford first-rate musicians, and all the rehearsals even Edith wanted. Mrs Astor arranged for Edith to be fitted by her own couturier – in a regally imposing gown of gold – and on the night of the performance the magic of her voice and presence brought her the sort of public triumph she had dreamed of all her life. The only real crisis came at rehearsal when she discovered she could no longer get her tongue around the staccato verses of the hornpipe. But even this was turned to her advantage. David Horner stepped in to recite this one poem in the programme with great *éclat*, and the change of voice provided Edith with a useful break in her delivery, as she gave the performance of her life and polite New York shared in the incantatory high mass of the Sitwell cult.

A few days later an exultant Edith wrote to John Lehmann in wintry, post-war London.

I must admit that the Americans seem to have taken to us almost with violence. *Façade* was a wild success, and at the dress rehearsal, to which all the poets, artists and pressmen were invited, the whole audience rose to their feet when I came to the hall after the performance.

In the same letter Edith enclosed Philip Hamburger's review from the *New Yorker* for 29 January 1949:

To me it was wise and wonderful art. . . . Dr Sitwell's voice is an orchestra in itself, and it has a haunting quality, a range and power that are quite staggering. Dr Sitwell made an appearance at the end of the performance, looking just about the most elegant woman of our time. She was wearing a great gold something over her shoulders – a perfect complement to that monumental face, which is certainly one of the marvels of the world.

At the bottom of the cutting Edith had scribbled: 'You must admit it's the stuff to give the troops.'

All of this adulation should have guaranteed Edith's happiness. Osbert was delighted, and quite captivated by New York. Despite a bout of influenza and trouble with his leg which left him 'in constant pain and rather dopey', he unashamedly enjoyed a little lionising, saw all the galleries, made several visits to his beloved Frick Collection, and cemented some important friendships, including one with James Fosburgh, and another with Lincoln Kirstein, then at the beginning of his great career as co-founder of the New York City Ballet.

But Edith was clearly not content. After the triumph of *Façade* she seemed increasingly irritable – even with close friends. She was drinking very large martinis, and the novelist Glenway Wescott remembers visiting her one morning at the St Regis, and finding her immersed in gloom.

'I can't stand myself,' she told him.

'Why not, Edith?' he replied in his slow, mid-Western drawl.

'Because I'm so ugly. Oh, how I just wish somebody would put his head in my mouth so I could bite it off!'

'Anyone in particular, Edith?'

'No, almost anyone would do.'

But, as Wescott knew, there was one person in New York responsible for

Edith's billowing ill humour – and, as so often in the past, it was Tchelitchew. After the disastrous episode in front of 'Hide and Seek' she had seen less and less of him. Theoretically he was at home, suffering acute colitis, but in fact he had been observing every move of Edith's social triumph – and getting increasingly put out by it. There was no longer any mention of the great new portrait he had planned of her. Instead he complained continually now of her pride and the unforgivable way that she was 'basking ... in the public eye'.

'There are two people in Edith which I can't put together,' he told Ford. The greatness of mentality in her work was not in her life, and he no longer could communicate with her. On past form some sort of outburst was inevitable, but Pavlik saved it for the very eve of Edith's departure from New York.

A farewell dinner for Osbert and Edith had been laid on at Voisin's. Kirstein was there, and Ford, Kirk Askew, David Horner, W. H. Auden and, of course, Pavlik, who had recovered now from his colitis. It started off as something of a celebration, for the Sitwell visit had succeeded beyond anybody's dreams. Tchelitchew alone stayed silent, then finally, towards the end of dinner, he threw the inevitable scene. White-faced with anger, he began shouting at her that she was self-obsessed, that she had let herself become corrupted by the vulgar social figures who surrounded her, and that she had finally betrayed the poet in her that he had always loved and cherished. Cruellest of all, he coldly told her that everything that there had ever been between them now was over.

It was particularly embarrassing in the smart, crowded restaurant, and Edith was completely at a loss for words. She looked stunned and vulnerable and infinitely old, while Tchelitchew was quivering with anger. Some of the guests around him tried to calm him down, and then pretended that nothing untoward had happened. Osbert looked down his nose, was very dignified and said nothing. Auden appeared profoundly bored. For the remainder of the disastrous dinner, Tchelitchew, his anger purged, stared icily ahead. When Edith rose to leave he could barely bring himself to say goodbye. Afterwards he told Lincoln Kirstein, with evident self-satisfaction, 'I was as hard as iron – as hard as *iron*,' and to Ford he still complained, 'She's been ungraceful – ungracious and selfish; she has neither intelligence nor heart. I'd like to slap her face and have her kneel at my feet and crawl like a worm.'

That was something Edith would never do for anyone – not even for her once beloved Boyar – and for the last hours of her stay she gave no sign of how she felt. Royalty controlled their feelings before lesser mortals – and so did Edith. Consummate actress that she was, she managed her farewells impeccably. But thanks to Tchelitchew, what should have been a triumphant return to Britain had become a thing of ashes and of total misery. Once aboard ship, she took to her bed, and stayed there for the rest of the voyage.

She must have hoped for a letter of apology, or at least of explanation, from Tchelitchew, but he was determined to keep up the role of man of iron. When a letter from him did arrive towards the end of March it was so bitter and accusing that she was torn between replying and ignoring it. Finally she wrote to him.

Dearest Pavlik,
Your heartless and callous letter reached me this afternoon. I should not reply. But your accusations against me are entirely untrue. And there are certain things you force me to say to you. I thought from the beginning that mischief – the result of spite and great envy – had been made by your 'artist friends', as you call them.

She went on to tell him that, on her return, she had been in hospital for an operation on her throat. There had been a growth. She had feared cancer. Luckily it was not malignant, but she was still in bed and in great pain.

I don't tell you this because I think it will be of the slightest interest to you. There is no reason for your treatment of me, excepting your wish to rid yourself of me. I was aware of that wish from the evening after my arrival in New York. You have indeed succeeded. . . .

I have never made any claim upon you, I have neither the right nor the wish to do so. But you should be ashamed for ever of the letter you have written me, and your behaviour towards me.

Having written the letter, Edith then decided not to post it.

CHAPTER TWENTY-FOUR

'Gid and Dizz'

1949–1952

AFTER THE adulation and excitement of New York, the return to Renishaw soon produced the inevitable sense of lassitude and boredom in them both. For Edith there was the worry and discomfort of the operation on her throat, and the misery she felt at losing Pavlik. Neither improved her temper: nor, for that matter, did the work she was engaged on now, selecting poems for the personal anthology of American poetry and prose which she would publish two years later under the title *The American Genius*. Her lunatics and bores received short shrift that spring. (A particularly insistent Indian Ph.D., attempting to interview her on the state of English poetry, was promptly packed off to Dr Leavis at Downing College, Cambridge, with her blessing.)

Osbert's health was more disturbing still, and he was mystifying all his doctors now with what exactly could be wrong with him. His damaged leg – he tore a cartilage when he slipped in snow at Boston – still gave him trouble and the nerves appeared to be inflamed. But this would not explain the sense of curious malaise that troubled him. Perhaps, as he half suspected, it was all in the mind, just as Sir George's illnesses had been. But he undoubtedly did feel very sorry for himself – so much so that, when Adrian Stokes visited Renishaw that year, Osbert could not face seeing him. In a letter of apology he explained that he had been feeling

> very middle-aged: fatigue, pains in the legs, a deep-seated cataract in both eyes, inoperable but not at present too bad, and so *ad infinitum*. I think the strain of composing my book told on me – but it is nothing like the strain of not writing. As for my leg, on a good day I can walk for miles, but at times am lame.
>
> I wish I had seen you when you were here, but I am so conscious of being under the weather that I took no steps. You know the dread – I hope you don't – of seeing someone you're fond of when you are not well, stupid, feeling a hundred and so on.

Summer weather brought no improvement – rather the reverse – and his puzzled doctor now forbade him to appear at any sort of public function. 'He finds me too *nervous* after the U.S.A.,' Osbert explained regretfully to Lovat Dickson. There was much talk of nervous fatigue and some sort of aggravation of his heart murmur. Osbert still blamed it all upon the strain of finishing the autobiography – the fifth and final volume, *Noble Essences*, was sent off to Macmillan now – and it was to his editor, Thomas Mark, that he attempted to explain his thoroughly disturbing symptoms.

This last week has been very strange [he wrote on 13 August]: I have been feeling very giddy, in the worst sense. (Do you remember Connie Ediss's pre-1914 song, 'I'm gid and I'm dizz'?) Really it's rather a delicious, not of this world sensation: and I feel anchored to this globe only by the noise the servants make.

Cyril Connolly defined the natural qualities of the homosexual writer as 'combativeness, curiosity, egotism, intuition and adaptability', and one can see how much Osbert had relied on all these literary advantages to complete his monumental work. One can also see that in his private life he had made the most of all the benefits of his condition. He had been free to travel as he willed, to act on whim and please himself entirely, and to pursue a rich and varied social life as only the most resolute and selfish heterosexual could have done. Marriage and children would have anchored him quite fatally to family bothers and affairs and loyalties. In the unlikely role of married man and father, he could never have managed to afford the travel and the pictures and the friendships that had made his life so rich and so rewarding. David had given him the love and the stability he needed while leaving him the twin essentials he required both as a writer and a man – artistic freedom and detachment.

All this helped to make the autobiography almost the 'classic' he intended it to be. It is a unique document of an extraordinary family, a quite unrivalled picture of a vanished era, and an exceptional account of the battles and successes of the trio in their outstanding career. It is also, as he intended, a polished work of art. But it is not quite the masterpiece his great abilities and varied life had fitted him to write.

Osbert once described how as a 'professional observer' of human beings he had learned to conceal himself behind 'the blank, and I hope bland, mask and massive frame that I have inherited'; and this is half the trouble. During a lifetime as a homosexual, he had learned to be too blank, too bland in his relations with outsiders. The mask goes on, the terrible discretion starts, the inner self is not revealed. Ironically, his sexual status, which conferred so many incidental benefits upon him as a writer, seems to have deprived him of his masterpiece.

It must have been extremely difficult for him of course. At the time he wrote it would have been impossible to have confessed to anything – or even to have hinted at the truth – and this in itself produced a mood of reticence and tact incompatible with any great work of autobiography. In Osbert's case the damage to his work does not stop there. The vulnerability he must have felt extends to almost everything he wrote. He can be witty, lyrical, nostalgic, but he cannot begin to tell the truth about so many of the friends he had: and this fatal reticence is not confined to other homosexuals, but to almost everybody he describes. All his life Osbert had been a man of passion, a great hater, fighter, and a devotee of social scandal, but not a hint of this is permitted to invade his pages. He has a few clear-cut villains – Field, Kitchener, Horatio Bottomley and J. C. Squire – but apart from them he writes with carefulness and tact, as if determined at all costs to avoid offence which might have brought retaliation.

This sense of reticence must account also for yet another loss – the way the autobiography peters out in 1930. David, as we have seen, is mentioned once, and then the 'blank, bland mask' descends, and nothing more is said about the most important person in his life and all those fascinating years he spent with him.

During the summer of 1949 Edith laboured on with her American anthology, and one can see her disapproving of the poetic world in general, as she wrote testily to Lehmann:

> Heavens! There is practically not a living poet, excepting Tom, Ezra, Marianne, E. E. Cummings, Wystan, and Jose Villa. . . . There are also very few dead poets either!! . . . I shall be in fearful trouble with all the Americans for not putting them all in. I shall therefore say, in my Preface that this is the First Volume only, and that they will all come in the Second. And then there will be *no* Second!

She went on to ask him if he by any chance possessed the poems of Robert Frost and could lend them to her. 'I think all Frost's poems I have seen really *stinking*!' And she ended simply, 'Must I have all the old bores? You know, they will spoil the book.'

But despite such rigorous criteria, there was one new American poet she was determined to include – Charles Henri Ford. Although she was not on speaking terms with Tchelitchew, Ford was now a good friend and a vital link with him. She included two of his poems on the grounds that they were 'curiously interesting'.

But the *contretemps* with Tchelitchew had left her very bitter, and all that summer one can see her boiling up to further irritated outbursts, so much so that Philip Frere was moved to write a trifle wearily,

My Dear Edith,
I seem to spend so much of my time telling you that statements which appear in the Press, and which cause very justifiable annoyance to Osbert or you, as the case may be, are not libellous, that I am beginning to doubt my own judgement.

Everybody seemed to be offending her. Early in September she invited Ian Fleming to the Sesame, and asked him to use his influence on the *Sunday Times* to retaliate on a certain poet who had annoyed her. Fleming, like Philip Frere, decided that discretion was the better part of literary valour where Edith was concerned. 'I will certainly do all I possibly can to get the poems panned,' he told her, 'but the one that you described really does not seem strong enough meat.' But if Philip Frere and Ian Fleming were too faint-hearted – or too prudent – to take on the role of Edith's champion, there was a most unlikely candidate who was not – that typhoon in a beer-bottle, as Edith had once christened him, the South African poet, Franco-ite and friend of Wyndham Lewis, Roy Campbell.

Always something of a brawler, Campbell had been drinking heavily that year, and had recently assaulted Stephen Spender during a drunken argument. Then Edith heard some news that thrilled her to the core. Campbell, she heard, had 'slapped' her arch-enemy, Geoffrey Grigson, in the B.B.C. canteen. She was revenged for all those years of insults and impertinences, and Campbell was promoted to be one of Edith's closest friends and heroes, a true knight of Our Lady, and, as she described him finally, 'one of the very few great poets of our time'.

In fact, the whole incident had been nothing like as chivalrous and stirring as Edith liked to think, and the row, such as it was, had not been over her at all. What actually happened was that Grigson and Campbell had both been members of the B.B.C.'s Literary Advisory Committee under the chairmanship of Desmond MacCarthy. Campbell was something of a protégé of MacCarthy's, and took offence at what he thought was a rude reference to the older writer in a broadcast talk by Grigson. A few days later Grigson was walking along Upper Regent Street with Anthony West when, as he says, he met Campbell 'prancing unsteadily towards the B.B.C., in bronco-buster's hat, and carrying his knobbed stick. He raised his stick and said, "Why were you rude to my daddy?" – Daddy because MacCarthy had befriended him, and persuaded the B.B.C. to employ him. I said, "Don't be a fool, Roy," and after a moment or two of nothing that was that.'

But for Edith that emphatically was *not* that. This was the stuff to give the troops – and before the week was out this small affair had been expanded and had become a heroic part of Edith's personal mythology. She wrote excitedly to Lehmann and to Osbert – who was now at Carlyle Square with David – giving them the news. She seems to have taken it for granted that the whole row had been on her behalf, and in her note to Osbert added a further piece of make-believe. The B.B.C., she said, had 'asked Roy to promise he wouldn't hit him again. Roy replied that he would rather throw the whole staff out of the window than give such a promise'. Osbert's advice was to forget the whole affair as soon as possible, but there was no chance of that. Campbell was instantly invited to the Sesame, Edith could talk of nothing else, and the rumour soon got round that she had actually incited Campbell to attack her enemy. This caused trouble – even among her most devoted allies. As she explained to Osbert,

William Plomer came to lunch with me. I can only imagine in order to insult me. He was quite *odious*. He said – I can't remember his words – that it was nothing for me to have had Grigson slapped as Roy had slapped Stephen. . . .

I said, 'I have never asked anyone to defend my poetry.' He replied, 'Well, you *couldn't very well, could you!*' . . .

I said, 'I am sorry that you are not pleased that this man was slapped for insulting me.' He replied, '*One cannot take both sides*'!!!

William Plomer was the gentlest and courtliest of men, and it is some indication of the dangerous way the row was escalating that he of all people should have felt obliged to challenge Edith in her wrath at her own table. For more than Geoffrey Grigson was involved. There was Stephen Spender, who had had the earlier row with Campbell, and there were also several former left-wing poets, including Cecil Day Lewis, who had been very harshly treated in a recent article by Campbell. Plomer mentioned this, and Edith instantly defended Campbell's article. It was, as she told Osbert,

the only harsh criticism any of these little pipsqueak poets have ever had – I mean softy Cecil and so forth. Well! I think Plomer is very right to take the side of a great poet like Day Lewis instead of a *woman* who is the *poet I* am! And yet, I think I know whose poetry will be remembered.

Essentially the row was over her status as a poet. As insecure as ever, Edith had inflated the non-event of Campbell's tiny threat to Grigson into a *cause célèbre* in which she envisaged once again her friends and allies rallying around her to defend the great cause of her poetry. One sees this in the excited way she reported the latest war-news in a letter to David at the beginning of October.

The Campbell–Grigson affair has caused a sensation in wide circles. Fr. D'Arcy's eyes (I met him at the Campbells') gleam with an unchurchmanlike interest – (the Campbells are Catholics) – the Spanish ambassador is enthralled. Mr Daniel George, who hates G. is spreading the news that he has told Lord Beaverbrook.

And so on. It is difficult to credit the idea that any of these gentlemen was quite as thrilled as Edith liked to think. Certainly Father D'Arcy, that worldliest of Jesuits, was far too shrewd a man not to have recognised a few of Edith's symptoms, and it is interesting that he was present when this new friendship with the Campbells was already bringing Edith closer to the Catholic Church. But at the time Edith seems to have believed that London was applauding Campbell's gesture for a sadly used great poetess.

She was tormented by the suspicion that her enemies were intent upon consigning her and all her poetry to some carefully planned oblivion, and the conspiracy was now so wide that all her energy was needed to combat it. Her so-called friends were not above suspicion. Plomer had refused to take her side. Spender was suspect too. Even that old and trusted Sitwell ally, Alan Pryce-Jones, the influential editor of *The Times Literary Supplement*, seemed to be infected by 'disloyalty'.

For that September she had published a new book of poetry, *The Canticle of the Rose*, and he had promised, some time earlier, that he would, as Edith put it, 'hit the town' with a full-scale review of it in the paper. In fact, he had been under something of a misapprehension when he promised this, believing, as he says, that the book would contain a body of new poetry. It did not. As so often with Edith's books of poetry, it was made up of a selection of her verse, some of it revised, but most of it published and reviewed before. There could be no question of a long review – but Edith was very angry. 'Alan Pryce-Jones . . .' she wrote to David, 'has given it a review which might have been – and probably was – written by a dear old moss-grown, low-church clergyman, with beard dripping with rain, living in a Manse outside Aberdeen.' Her wrath was soon diverted by a graver insult still, when to her horror she perceived that *The Canticle of the Rose* had been reviewed in *Time and Tide* along with the latest verses of another poetess, Kathleen Raine. The review was one that Edith might have comfortably ignored, considering her fame and her assured position as a poet. Instead she took it as the final insult. For the review, by Norman Nicholson, extolled Miss Raine at her expense. It failed to call her Doctor Sitwell. Most unforgivable of all, it criticised 'Still Falls the Rain' on the grounds that Edith had contemplated the atomic bomb but only to make a myth about it.

The great siren call of Edith's wrath went wailing out from Derbyshire to Kensington in an indignant letter to John Lehmann. With the *Time and Tide* review an ungrateful nation had proved itself unworthy of her poetry.

> I am so indignant at the way my poems (my work of nearly thirty years) have been treated, that I will never, *never* allow the British public to see one of my poems again. Never. They shall all be published in book form in America, and I will give them to my friends here. But they shall *never* be published here.
>
> I ask you to look at the enclosed [a copy of Nicholson's review]. Is it not disgraceful? I resent nobody's success. But it is a little much that I should be insulted to bolster up this lady's-maid poetry – underbred, undervitalized, undertechniqued, messy and *déplaisant*.

I am so angry at the whole way I have been treated that I am going to shut myself up for ever more and only see a very few close friends: you, naturally, Rosamond [Lehmann], the Clarks, Maurice [Bowra] two or three more, and the John Russells.

The *T.L.S.*, whilst giving me that short damp review, has just given its middle page to the important subject of fireworks.

Fireworks! They don't know them when they see them.

Edith's own firework display went sputtering on thoughout that autumn and into 1950, for she had reached the point where rows and arguments seemed to have become her most intense – and certainly her most dramatic – form of contact with the outside world. They were addictive, and she could not resist the lure of further battle.

But while her battles staved off boredom, they did little to improve the winter atmosphere at Renishaw. Miss Andrade was invited up just after Christmas 1949 to help Osbert with corrections to the proofs of *Noble Essences*, and found the whole *ménage* unlike anything that she had ever seen before – the flickering oil-lamps lighting up the great house in the evening, the miles of draughty, ill-lit corridors, the statues in the gardens draped against the frost, and dinner every night with David, Edith and Osbert served in the shadowy dining-room. One night she listened as the three of them discussed the after-life: David and Edith believed in it, Osbert emphatically did not.

She found the way of life there 'beautiful and strange and most depressing'. Osbert had little of that easy, bland assurance of the past. He was tense and obviously ill with whatever strange malady it was that now afflicted him. (He had developed a continual tremor in his arm, and wrote to Lincoln Kirstein saying he had been 'doped all day long for months'. 'I enjoy the dope,' he wrote, 'it seems to go well with alcohol, only it slows one up.')

Edith she found 'very prickly', and when she took offence at something she would stay in bed for days on end. When she did this, Osbert made no reference to her absence. Miss Andrade found her 'very strange at times', and finally she put this strangeness down to her desperate fear and ignorance over money. For one night at Renishaw Edith stopped her on her way to bed and asked, 'Miss Andrade, can one still go to prison in this country for writing cheques the bank won't meet?'

There was one event that should have eased her anguish now – she had made peace, of a sort, once more with Tchelitchew. This had been thanks entirely to Charles Henri Ford. Through him she had resolutely kept in touch with Pavlik, and had heard that he was planning to come to London late in 1949 for an exhibition at the Hanover Gallery. In fact, he never came. He was unwell and his nerve failed him, at the final moment.

American homage to the Sitwells
Left seated, William Rose Benet; *behind him,* Stephen Spender; *behind him,* Horace Gregory and his wife; *behind the seated Sitwells,* Tennessee Williams, Richard Eberhart, Gore Vidal, Jose Garcia Villa; *on steps,* W. H. Auden; *standing,* Elizabeth Bishop; *seated,* Marianne Moore; *seated at right,* Randall Jarrell; *in front of him,* Delmore Schwartz; *squatting in centre,* Charles Henri Ford

Osbert and Edith in front of the Sargent portrait at Renishaw Hall

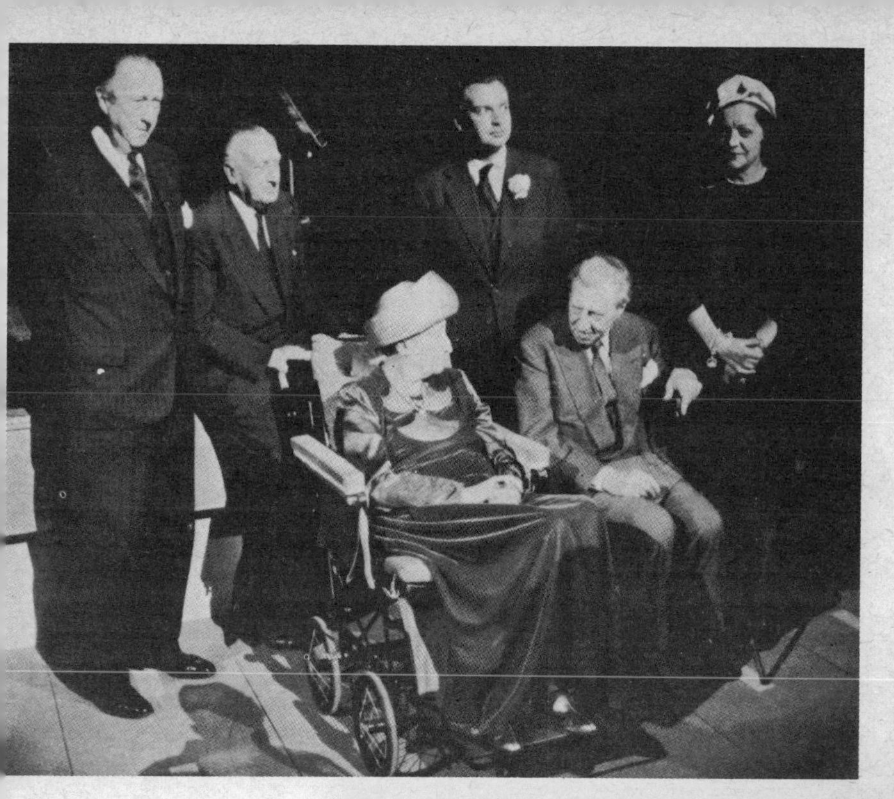

Above: 'Something macabre out of Proust'. Edith's seventy-fifth birthday celebrations at the Royal Festival Hall; Sacheverell, Sir William Walton, Edith, Francis Sitwell, Osbert, Georgia

Edith reading at the Royal Festival Hall

The Castello di Montegufoni: Lorna Andrade and Cyril Conn

Below: Frank Magro at Montegufoni w the Dobson bus

'Frankly,' he wrote to Osbert, 'I feel that I am almost completely forgotten in England.'

This exhibition, though much praised among the critics, was a commercial failure, and this was serious for the painter. For, like Edith, he had his money troubles. The colitis, which had begun with Edith's visit, still assailed him, and to escape the various pressures of New York he fled to Paris. From the letter that Edith wrote about him to an anxious Lincoln Kirstein in December 1949, it is quite clear that she had once again forgiven him – and was still as worried for his material well-being as she had ever been in Paris, twenty years before. Her concern for the few she really cared for was as instant as her anger for those she did not.

I can only tell you what I have heard from Charlie. . . . He has been taken very ill in Paris, with, apparently, some kind of tropical germ. He had violent giddiness, etc. Then he went to the Pasteur Institute where they gave him very drastic treatments, which seemed to make him, for the time, more ill. Poor boy, I think a large part of the illness derives from the ghastly non-stop worries about money and the future. I really think it is *shameless* the way he is battened on.

Then in the spring of 1950 Ford and Tchelitchew moved on to Rome – and it was from there that she finally heard from Tchelitchew in person. He was still sick, and there was now a sort of battered reconciliation. All Edith's pity and protectiveness were roused on his behalf, and she wrote desperately to Kirstein of how Pavlik was 'living *in terror* financially', frightened he would lose his New York studio, and incapable of work. She even quoted from the letter he had written her: 'If only some of the rich people in New York could get an idea to give me a yearly sum on which I can rely and live in peace a few years and get some work done in peace and not in fear and misery.' It was a cry she could have made herself, for her debts – and her extravagance – still plagued her. But she was touched that Pavlik in his extremity should turn to her again for help. He wanted her to go to Rome that Easter to be near him. Sensibly, for once, she refused, but the note that she sent Kirstein explaining her refusal shows how her bitterness had left her.

I feel I should not have spoken harshly about Pavlik. One's hurt feelings get, sometimes, the better of one. Not, *really*, that I *am* hurt any longer. But one remembers, sometimes, that one *has* been. I shall always think of him as one of the greatest living artists, and shall always do everything I can to help him. But I shan't go to Rome, because the really appalling second-rate, Comus-rout he has been mixed up with for all this time would turn him against me.

But, in fact, she was to see him later in that spring of 1950, not in Rome with his hangers-on around him, but on her own home ground of Montegufoni, with Ford and Osbert in attendance.

It was the hope of such a meeting that had prompted her to join David and Osbert there at the end of April. She did not enjoy the journey, and 'left a trail of nervous breakdown and desolation from Victoria to Pisa', she reported to John Lehmann with grim satisfaction, and Osbert viewed the whole idea of meeting Tchelitchew again with understandable alarm. 'Tomorrow, TOMORROW,' he wrote to Kirstein at the end of May, 'I regard with some apprehension, for Pavlik arrives with Charles Ford for the weekend. I do hope, oh I hope, that he will be in a good mood.' But the reunion proved to be a gloomy contrast with those stormy, unpredictable meetings which had starred Edith's love for Tchelitchew for over thirty years. The long horse face was pale and thin, the bear-like vitality all but quenched. Nor was there any of the rancour which he had shown to Edith that disastrous night at Voisin's.

Merlin [as Osbert called him in his letter reporting the weekend to Kirstein] was fairly good here: Charlie came too. He [Pavlik] was always very remote, consciously, from life, only a moment later to show himself immensely mixed up in it, knowing all the tricks and one extra. He has enough personality for 2000 in his work: but what would happen if he put the rest, the remaining 2,000,000 into it, I can't think. What explosions and eruptions on canvas there would be!

Edith reported back to Kirstein too – but, unlike Osbert, she had no doubts at all about her hero, and was as starry-eyed about him as she had ever been.

Pavlik and Charlie spent the weekend here. Pavlik seemed *like his old self*. But he was so sad – his face was so terribly sad, his speech was so terribly sad that it was really heart-breaking. I could hardly bear to see it. He was very good and sweet. He spoke with the greatest affection of you and said that you are one of the only people in the world on whom he can rely.

Beyond this there was little to be said, although she ended up her letter with a fond reaffirmation of her faith in him. 'He is a *very* great genius.'

Tchelitchew was not the only invalid at Montegufoni during May. Osbert had been hoping that the warmth and rest would do their work and cure him of his dizzy spells, but if anything they were worse. The tremor in his arm remained, while the drugs continued to depress him and to slow him down. When he returned to England, early in July, he finally arranged to see one of the top physicians in the country, Sir Henry Cohen.

Lincoln Kirstein happened to go with him. He had been in London with his fledgling New York City Ballet, which was having its first London season – to dubious reactions from the press – at Covent Garden. He says that to this day he is grateful for the way that Osbert seemed to understand his problems, and for the support and sympathy he gave the ballet in almost every way he could. But it was all too clear that Osbert was not well, and he recalls how casually Osbert broke the news to him when he came out of the consulting rooms. The mystery was over. He had Parkinson's Disease.

Both of them knew quite well what this implied. Parkinson's is an incurable nervous disease – it was originally called 'Shaking Palsy' – which produces shaking and rigidity in all the muscles of the body. It can be stabilised, but in most cases it becomes progressively worse with age, until the face becomes affected, speech inaudible, and movement uncontrolled. It would be hard to think of any crueller long-term illness to torment a writer and a social figure such as Osbert.

But Kirstein says the way that he received the news was very much in character. Instead of feeling sorry for himself, he acted just as if the whole idea of Parkinson's were a sort of social affront. and that the doctor had been guilty of bad manners merely by diagnosing it. 'Then as we walked down Harley Street we found ourselves agreeing that the whole idea of the illness was simply a sort of insult.'

Osbert knew how to deal with insults, and for some time this would be the way he would regard his malady.

Sir Osbert is very highly strung and nervous – like most authors – and fusses a great deal about matters both important and unimportant. He has suffered from occasional attacks of gout which started about 1937, but to my knowledge he has never had more than two fairly severe attacks: but he is obviously gouty.

In the past he has eaten and drunk a good deal, but for the last few years he drinks practically nothing until he has two rather strong cocktails before dinner, with half a bottle of wine during the meal and nothing much afterwards. He usually smokes two cigars a day and very few cigarettes.

Considering his weight and age he walks very rapidly and for quite long distances, as he likes walking very much. I have noticed that recently he is inclined to be slower in his movements, and stoops a little, rather as his father did as he grew old.

Regarding the shake in his left arm, I have observed that this varies very greatly. When he is bothered or nervous it increases. When he travels by train or by motor it decreases, and very often after two cocktails, it appears to be entirely calm.

I have heard from his G.P. that his heart is not too good.

He is excessively nervous about his health and especially fears catching colds from other people.

Sir Osbert will be 58 in December.

This was the report that David wrote for Sir Henry Cohen, and it provides some insight into Osbert's habits and his state of health during these early days of illness. But the nervous shaking must have been distinctly more pronounced in public than this suggests, for Aldous Huxley, writing to Christopher Isherwood in July 1950, mentioned that he had seen 'Poor

Osbert', and that he had 'started to tremble. It is a very depressing look-out.'

This was something Osbert himself was stoically refusing to admit. He had an inborn sense of how a gentleman behaved and saw no reason now to change his habits or his way of life. Indeed, he still insisted on regarding the disease as something of a solecism and best ignored, and David tried to do the same. Edith was more positive in her approach, appearing to attach great weight to the fact that Sir Henry had diagnosed what he called 'A Parkinsonian Condition', rather than Parkinson's Disease itself. In Edith's eyes this meant that Osbert should be able to be cured by rest and drugs and expensive doctors. There was no call for untoward alarm.

But Huxley, who saw them both again later that summer, paints a gloomier picture in another letter to Isherwood. 'Edith a monument and Osbert suddenly rather old and tired.'

But, tired or not, Osbert still had a role to play – that of the distinguished man of letters he had finally become. *Noble Essences*, the fifth and concluding volume of his autobiography, was published that September to almost universally benign reviews. His greatest work was finished and, by the sort of irony that he appreciated more than most, life suddenly appeared to offer everything he had longed for just as his body had begun to let him down. Through all those years of self-denying work at Renishaw he had created this success. Now was the time when he should by rights have been able to re-enter the society he loved as a distinguished man of letters, carefully taking on the role that he had seen Gosse or Arnold Bennett adopt in their prime. It was a role that would have suited him. Instead, the Parkinson's Disease got worse.

For Osbert and for Edith there was one bright ray of hope for them that dreadful summer – America. Edith had been quite serious when she had told John Lehmann that she was thinking of cutting herself off from her native land and rigidly adopting New York as her spiritual and literary home. For in America it seemed she could live out the exalted literary role she longed for. She would be universally fêted and admired, and free from the 'pipsqueaks' and the demons who oppressed her in England. Osbert was less ambitious – but just as keen to go. What he wanted most of all now was a rest from writing, and to see as much as possible of life while there was still the chance. He was grateful for the kindness of the Americans, and his last trip had taught him to enjoy the travelling, the comfort, and the sights. Despite his tremor and his bouts of lethargy, his mind was sharp as ever; so were his curiosity and appetite for people. He was still treating his illness as an 'impertinence' of life – and in New York there was the chance of seeking further specialists and perhaps even getting cured.

New York always seemed to bring out the Plantagenet in Edith. Here, more than anywhere, she could act her royal role, and Lincoln Kirstein cleverly appreciated this when he suggested what appeared to be the starring part of Edith's lifetime. On 14 November 1950, at the City Center of Music and Drama, the New York City Ballet had been asked to stage a gala night in the presence of a bevy of assorted diplomats and the Secretary-General of the United Nations. Ashton had agreed to create a ballet based upon Britten's setting of *Les Illuminations* (translated originally by Helen Rootham), but the great role of the evening would be Edith's.

When the curtain goes up, the stage is huge and quite dark: a light faintly illuminates what seems to be a sort of idol-thing; lights glint on the edges of a gold stiff ruff, on jewels, on a sort of throne, on you. You are an Elizabethan Queen figure, partly conceived as a chess piece, partly out of Zuccaro; you are seated on a rather high chair with a canopy over it supported by standards and plumes; it is a very royal setting.

And in this 'very royal setting' complete with jewels and ruff – and obviously a crown – Edith would solemnly intone the poem Kirstein was all set to commission. It was to be entitled 'The Great Seal', and, in Kirstein's words, it would describe 'the present position of England, in poetry which only you can write'.

Edith responded with predictable delight.

I was so excited to get your letter, with its truly wonderful suggestion which has filled me with enthusiasm, as you may well imagine. It could be *terrific*. I feel that very greatly. The whole idea is on a very great scale. . . . I have a passion for reciting the right things.

Kirstein had Walton in mind to compose the music, with the hope of re-creating the successes of *Façade* some thirty years before. Edith had somewhat grander plans. 'If it is done, I do not think Willie Walton would be the right person to ask, because I know he is frightfully busy at this moment with an opera. . . . Would it be impossible for Stravinsky?' Edith as Elizabeth I declaiming her own poetry to music by Stravinsky in the presence of *le tout New York* – that *would* be something to give the troops.

Alas, it would remain a daydream, nothing more. The cost of staging 'The Great Seal' proved prohibitive, and the idea failed to fire Stravinsky. Edith would have to play the part of Gloriana now in simpler surroundings. But this did not affect her visit to the States. Colston Leigh had booked her for an even more ambitious round of lecturing that year. Monroe Wheeler wanted her for a repeat performance at the Museum of

Modern Art, and Hollywood was sniffing at the Sitwell magic. George Cukor had read *Fanfare for Elizabeth* and wanted to discuss its adaptation for the screen. With his direction, Edith's vision of Elizabeth I could even take its place beside that earlier triumph Cukor had worked on – *Gone With the Wind*.

They sailed in the *Queen Elizabeth*. Osbert and David both disliked the ship, but Edith felt its name was something of a tribute to her ancestry. All three of them were grateful to be away from Renishaw with winter coming, and Osbert and David had no sooner landed than they were both delighted with New York. 'The city', Osbert wrote to Lovat Dickson, 'teems with English authors,' several of them personal friends. He was amused by Evelyn Waugh, who was busy acting out *his* private fantasy, complete with bowler hat and bright check suit, of some crusty Edwardian literary gent. Osbert had written him an enthusiastic letter of congratulations over *Brideshead Revisited*, but told David that he had found the book unspeakably vulgar. In New York 'He has even taken to leaving cards instead of telephoning', Osbert reported. And there was Aldous Huxley too, an older, gentler Huxley than that young predator, 'William Erasmus', who had borrowed Osbert's anecdotes so long ago, and given such offence with his portraits of Sir George and Lady Ida. Osbert was willing to forget all that – or so, at any rate, it seemed – although he became impatient when Huxley kept exhorting him and Edith to 'keep well and don't exhaust yourselves'. He wanted to forget his tremor, not be constantly reminded of it now.

And in New York it seemed as if he could really forget it for a while. The trio had their favourite suite again at the St Regis, and Osbert was being gently lionised once more by all his New York friends. He gave some public readings of his poetry, recorded several favourite passages from the autobiography, and kept his eye on Edith.

But for the first time now he also had a chance to see a fresh side of New York, thanks to a new and most unlikely friend. From the moment he arrived it was obvious he needed looking after. His movements were becoming clumsy, and his walk was dangerous. David was sympathetic, but he was plainly not the one to cope with Osbert's illness, and it was Lincoln Kirstein who suggested somebody who could. This was Daniel Maloney, a young Irish painter of formidable charm and great good looks, who had been a medical corps man with the U.S. Navy. Unlike the usual run of Osbert's friends, he was completely unconventional and classless, with contacts ranging from the higher reaches of society to the outlandish outposts of the New York underworld. During this period of Osbert's stay he acted as a discreet companion, and Osbert was very taken with him, both for his stories – he is a virtuoso raconteur – and for the people that he introduced him to. For Dan Maloney's world was very different from the world of the Astors and the Fosburghs and the Askews. For the first time in his life, Osbert was meeting burglars, pimps and all-in wrestlers, the whole forgotten other-world of downtown New York. He was fascinated.

Edith, meanwhile, was busily preparing for her great performance to repeat the triumph of *Façade*, but since the collapse of Kirstein's project there was considerable uncertainty over what exactly she would read. Monroe Wheeler asked her finally what she had chosen. She replied 'Lady Macbeth'.

'But why her, Edith?'

'Because she amuses me. Because the part suits my voice. And because she was one of my ancestresses.'

Wheeler was not convinced that this was such a good idea, but Edith was adamant. It was Lady Macbeth or nothing, and the performance was irrevocably fixed for 16 November at the Museum of Modern Art. But once this was settled there was the question of who was worthy – or would actually dare – to play opposite Edith as Macbeth. Several names were mentioned. Basil Rathbone? Edith shook her head. Jose Ferrer? 'But surely he's a film-star?' she replied, opening wide those pale-grey eyes in mock-innocent surprise.

Monroe Wheeler is convinced that Edith now had no intention of being upstaged by any famous Broadway actor, especially as she must have known quite well that Macbeth, not his wife, had really the most impressive lines. It was finally agreed that, instead of an actor, Glenway Wescott would read Macbeth. This seemed to reassure her. Wescott was, after all, a writer and a friend.

But once started on this extraordinary project, Edith was most professional – and most demanding.

Wescott recalls rehearsing the performance several times, 'just as if it were a full-scale Broadway musical' – lights, make-up, costumes, everything – and for the dress rehearsal Edith once again invited all the writers and the artists she could think of, including Tennessee Williams, Carson McCullers and Marianne Moore.

By then it was clear that she was not going to repeat her triumph with *Façade*; indeed, for Edith to have even risked so difficult a part for her full-scale acting début in New York shows something of the confidence she must have felt in her American admirers. Perhaps she was over-confident. Some of those who saw her remember the performance as one of the most bizarre Shakespearean evenings of their lives. Edith in spectacles at sixty-three did not resemble any ordinary vision of her blood-stained kinswoman and, as Westcott says, 'with her voice like a viola, she might have done better as Cleopatra'. It was certainly quite unforgettable, and it was saved from bathos by her overwhelming dignity and presence. But this time there *was* criticism from the press. John Mason Brown, in that week's *Saturday Review of Literature*, panned it very hard, describing the performance as an example of the more outrageous sort of English amateurism put over on the Americans in the name of art.

Luckily for Edith, Hollywood was calling. Edith had suffered more than usual from the cool reception of *Macbeth*. She told Jack Lindsay that she had influenza now with 'every kind of complication', including bronchitis and a poisoned gland which had to be removed. Her eyes were troubling her, and she had sprained her ankle, hurt her back, and thought that she had pleurisy. To get some sun, she went to Mexico shortly after Christmas, but got amoebic dysentery instead. 'I . . . really thought that I was dying,' she told John Lehmann.

But Edith, like Sir George, was most indomitable when most afflicted, and nothing on earth was stopping her from reaching Hollywood. Osbert was exhausted. The drugs that he was taking slowed him down, and he told Lincoln Kirstein that he was 'in a trance after Austin, Houston, Mexico and now Los Angeles. I have forgotten almost how to write, how to shape letters.' By mid-January 1951 he had relapsed into a state of grateful idleness in the comfort of the Fairmont Hotel on Nob Hill, San Francisco. But Edith dragged him on to Hollywood for a scheduled reading of their poems. Here once again their legend had preceded them, as they began their now familiar poetic double act. 'We had a tremendous reading of our poems,' Osbert reported to Lovat Dickson afterwards. 'Very enthusiastic audience of about 2,000 people including among them Harpo Marx, Aldous and Dorothy Parker!' Then there was a party at which Edith got into 'a Laocoon entanglement with Miss Mary Pickford, that lasted for ¾ of an hour. She discoursed to me on her role as Little Lord Fauntleroy, and said she always regarded herself as a Spiritual Beacon.'

Osbert, less fortunate, was face to face with Ethel Barrymore, who 'was breathing heavily', and picked up laryngitis. But Edith seems to have felt instantly at home in Hollywood. Something of a star herself, she found this city of the stars a natural habitat. She also formed an instant liking for George Cukor, who gave her the sort of deferential treatment few Hollywood directors naturally accord their writers. They talked of *Fanfare for Elizabeth*. He was immensely optimistic over the chances of the film, and was already talking of signing Olivier for Henry VIII, and Vivien Leigh for Anne Boleyn. In the meantime, Edith would return to Renishaw and write an outline treatment for which she would be paid a cool ten thousand dollars.

The only person who was out of all this lionising now was David, who, not unnaturally perhaps, was hankering for a little life himself. Osbert was a demanding man to live with, and throughout their time together David had always felt the need to claim a certain freedom. Time had been kind to him as it was cruel to Osbert. At fifty he was tall, elegant and slim, with all his hair and teeth and youthful looks, and all his considerable appetite for life. While Osbert and Edith moved on from Hollywood to Sarasota, Florida, to lecture at the Ringling Museum, David was staying with friends in Kansas City. Osbert had Dan Maloney to look after him but seems to have felt dispirited, and worried both by David and by the Korean War. 'David pursues his career,' he wrote to Kirstein, 'but that is more for talk than for letters. . . . Time seems very foreshortened now . . . a few weeks, then London, and I suppose WAR. I don't feel so keen on the next as the two last. And I didn't care for them.'

'When I arrived [at Carlyle Square] in the morning,' wrote Miss Andrade in her diary for 21 March 1951, 'OS was there, just as if he had never been away. He is looking fairly well and his tremor seems less or about the same.' He may have seemed unchanged, but this was because he was coming to terms now with his illness. The New York specialists could offer little hope of cure, and confirmed Sir Henry Cohen's opinion that the condition was almost certainly the aftermath of those bad attacks of Spanish influenza Osbert had suffered in the spring of 1919. There was the possibility of new discoveries that might ultimately cure him. But in the meantime all they could suggest was that he learned to live with it. And this, with courage, he began to do.

He knew his weaknesses. The tremble in the left arm was the worst, his walk was increasingly affected. But his speech was clear, his brain was unimpaired, and with people that he knew and who were used to him he could forget his illness. One of the worst things was the way it made him so conspicuous in public. He who had always been meticulously clean and something of a dandy was starting to be messy as he ate, and this made meals an ordeal, particularly when his uncontrollable left hand began drumming with his fork against his plate. It drew attention to him, which he hated, but there was nothing he could do to stop it once it started. His drugs were a confounded nuisance too, worse, he often felt, than the disease itself. They made him feel tired and old, and although he had suffered from insomnia since childhood he suddenly slept ten and twelve hours a night.

His secret dread, of course, was a deterioration in his condition, but his doctors had assured him that if this came it would be very gradual indeed. This did not stop him watching himself anxiously as each day passed. All the strength and energy with which he had once attacked the generals and the fathers and the philistines was needed now to fight his final battle against this insidious disease. It called for cunning and determination. The greatest danger was despair; the prime objective was to keep the brain alive. His surest way of doing this was work; so that writing, which throughout his adult life had been his one vocation, now became something more – his best hope of survival.

Even in New York he had been writing to Harold Macmillan about a 'wonderful new idea' he had for another book. He would call it *The Four Continents*, and it would take the form of reminiscences of his travels throughout Europe, Asia, North and South America. Dan Maloney had already promised he would illustrate it for him.

Once he began, he found an unsuspected problem. Although it did not show yet, his right hand was beginning to be affected, and his writing, never particularly legible, was now impossible. For several weeks this brought him near to despair, but by early April he had made a start, dictating the book to Miss Andrade, and soon found it easier than he had imagined; once at Renishaw he picked up his old routine of working every morning in his study with its view across the gardens. By mid-May enough was done to justify a letter to Harold Macmillan saying that he would be in London for serious discussions some time around the middle of June, 'but hardly before, as I am having new treatment for my tremble. ... About health, I am "better in myself", but still very shaky and there is little prospect of getting better. My friends will have to get used to seeing me tremble like a leaf. It is a nuisance.'

Edith had mercifully calmed down after her rages of the year before. Ten thousand very welcome dollars from Columbia Pictures, plus all she earned through Colston Leigh, had eased her nightmares over bouncing cheques. She had the outline of her film to write, and Lehmann pressed her for another book, long toyed with but never seriously begun – her autobiography. In her reply she made it plain that she could never do it now. It 'might upset Osbert. His autobiography is in one way his greatest life work. And I couldn't bear for him to feel I was trying in any way to rival him over that.'

There was another reason too. Her ghosts were still pursuing her.

> I *can't* tell the truth about my sainted mother. If I had been a slum child, I should have been taken away from her. But I wasn't a slum child, and motherhood is a *very* beautiful thing! I often wonder what my poetry would have been like if I had had a normal childhood.

This was a pity. One would give a lot for a full-scale account of Edith's life in place of the attenuated, bitter old-age memoirs that she finally composed.

'Roses most perfect,' wrote Miss Andrade on 30 June, 'and big blowing peonies, irises of every colour, flowering bushes and trees of great beauty. Walked round lakes in afternoon. It seemed very hot and I was tired. In the evening the three of them tore everyone to pieces as usual.' It was like old times, and just a few weeks later Osbert was inspired to make one final public blast against his ancient enemies, the critics. The occasion was the sudden death of Constant Lambert, shortly after his last ballet, *Tiresias*, was savaged in the press. He had always looked on Lambert as something of a protégé, as well as a survivor from the heroic brotherhood around *Façade*. His death was both a personal shock and a call to arms. The obituary he wrote for Lambert in the *New Statesman* ended with some of the bitterest words ever directed by a writer against that group of men whom Osbert always felt to be the artists' natural enemies.

I must record my opinion that he would be alive today, had it not been for the savage onslaughts of the critics on 'Tiresias'. Such is the rage of the uncreative against the creative that nearly the whole body of critics, ignoring all Lambert had done, both for music and for British ballet, jumped at him as if he were a criminal. And they felled him just at the point where he was unwell and seriously overworked. By this sole act will the critics responsible be remembered in future.

This was unfair. Lambert died from drink and diabetes, not from unfavourable reviews. But Osbert was shaken by the tragedy. Death was suddenly too close for comfort, and he could understand too well the irony of 'poor Constant's' death.

It was a relief to get away to Italy, but even there it seemed to Osbert as if everything was coming to an end – Constant dying, his own tremor getting worse, the war scare looming.

I have a feeling of Armageddon being near again [he wrote to Kirstein early that October from the autumnal fastness of the castle]. The vintage is in and the tractors rage over the countryside. My hand is bad today, as I've temporarily abandoned palliatives to see what happens and how bad I am.

In keeping with his mood of imminent mortality, Osbert had begun a brand-new project which appealed to him. 'I've been working for a change on the chapter of my posthumous memoirs. It is odd to address people you can't see – like being blind.'

But Osbert was not to be defeated by despair. He was miserable and ill and Armageddon was at hand, but he was also wealthy. Thanks to judicious purchases of farming land on Maynard Hollingworth's advice (he used the compensation money from the nationalisation of the coal industry), and a subsequent rise in land values under the Labour government, he had more money than he needed, or than he could ever hope to spend. Rather than feel sorry for himself, he would behave like the essential hedonist he was. He and David would indulge themselves; and he would travel comfortably, drink well, eat richly, and stay in the best hotels for as long as possible.

And so, with winter coming to the valley of the Pesa, and Edith ready to depart for London, Osbert and David hired a car and once again began the great escapist journey to the south. They stayed a night in Rome – at David's favourite Roman hotel, the old Minerva – and by a strange coincidence, as they strolled through the old part of the city, they saw Stravinsky 'sitting on the steps of Sant' Ignazio, facing the baroque stage set [of the houses opposite] waiting to be photographed. He looked like a cross between a beautiful cicada and a punchinello.'

It was a strange glimpse of a man who was still a living link with part of the artistic history of the Sitwells – *The Firebird* at Covent Garden before the First World War, the Ballets Russes, Diaghilev and *The Rite of Spring*. He too, Osbert was reliably informed, was ill.

Next night they reached Amalfi. The bay was wonderful as ever, but now possessed for Osbert 'a tremendous tragic quality in its beauty; something one can never quite seize'. David seemed little changed from the first winter he had spent with him in 1929, but *he* felt infinitely old. 'Osbert wobbles a bit, but is no worse,' wrote David in a note to Lincoln Kirstein.

They stayed several weeks and, when the weather broke, moved south again, this time to Taormina, where the flowers were still in bloom and the hotels emptier than they had ever seen them. Each morning they would read, each afternoon they walked beside the sea. They must have made a curious couple now, the big old man shaking as he walked, the younger one beside him with the unlined face and gold hair still untouched with grey.

In December they were joined by Edith, who had lost her luggage in Montreux. Her arrival seemed to be a signal for the hotel San Domenico to be suddenly inhabited with lunatics and bores. The hotel lounge, she wrote, 'was like a den of sea-lions with the coughs and sneezes, and at last they got me down, and I had to take to my bed'.

Then it was Amalfi once again for February 1952. The Waltons called to see them. Osbert was still wary of Walton and had never quite forgiven him for going off with Lady Wimborne; but he liked Lady Walton, and was grateful for their visit.

He felt that he had benefited from his winter holiday. As he told Kirstein now, 'I am better, but not my shake: but I grow more used to it (will others?) and feel for the first time the return of some vitality. I put that down to sleep. I sleep all the time.' In April, when he and David finally returned to Carlyle Square, Miss Andrade recorded that Osbert was 'looking very well and brown'. 'Life', she added, 'has usual horrors of a return of OS.'

The chief horror was that it was now quite clear that his illness was advancing, and there was really nothing he could do, apart from accept it and make still more adjustments to his life. He thought of death a lot now and offered his views upon that sombre subject in a brief note he wrote to Lincoln Kirstein, offering his condolences when Kirstein's mother died.

> The end is always awful, awe-ful in the original meaning of the word, and ties are rent that one had never examined but taken for granted. It takes time to grow fully isolated again. I am so sorry to think of you undergoing this remorseless business – an inevitable process in one's own way to the same place.

It was a stoical view of life and its extinction, together with the view that one must bear one's isolation here as best one can. This was the aim that he had set himself, and at Renishaw that summer he did his best to follow it.

His only task now was to select the best of his short stories for the collected edition which his publisher was anxious to bring out. This pleased him, and he enjoyed re-reading them, for they were like old friends who would neither die nor change with age; and, as he wrote in the preface that he now composed, 'my own journey can be traced through' them. The selection made, his own artistic record and account was in a sense complete.

But one thing troubled him – the feeling that these stories did him less than justice as a human being, and made him seem much colder than he really was. To correct this he finally sent Lovat Dickson a brief note asking if an extra paragraph could still be slipped into the preface.

> It has sometimes been objected to one or two of my short stories, such as *Low Tide* and *Staggered Holiday* that I have shown no sense of pity. This emotion is for the reader to feel and for the writer to make him feel; and the very complaint shows that the reader has felt it. As to the writer, it is difficult to write between sobs.

Once he had got that straight, it was time to pay attention to the ever-pressing problem of his life and how to live what still remained of it. Perhaps the solution was quite simple, and in the month that his *Collected Stories* was published he wrote another note to Lovat Dickson: 'I think that now I am obliged to do less work, I have more time to fuss in, and look forward to making everyone's life real hell before I've finished.'

CHAPTER TWENTY-FIVE

Hollywood

1952–1955

IT WAS Ernest Hemingway who, from grim experience, formulated the ideal relationship between Hollywood and the writer. The writer and the producer, he said, should both stand on opposite sides of the California State line; at a given signal from their lawyers the writer should throw the producer his book, and the producer in return a cheque. Then they should drive off in opposite directions, never to meet or think of the transaction ever again. It was a procedure Edith might have followed once she had finished the first treatment of *Fanfare for Elizabeth*; for Cukor was delighted with it, and was all set to go ahead and use a tried and trusted professional script-writer, Walter Reich, to finish the film's final script.

But Edith had no intention of abandoning the scene – and Cukor, a most civilised and literate Central European, seems to have been flattered at the chance of working with a great and famous European writer. It was a risky undertaking for them both, but they had set their hearts on making the film a worthy masterpiece and Edith was in love with Hollywood. Why should she stay in dreary Derbyshire when she could be fêted and admired as a visiting celebrity in California at Columbia's expense? And so it was that, during the early 1950s, *Fanfare for Elizabeth* made Edith, Osbert and David regular winter visitors to the States, with far-reaching results for all of them.

The first of these visits – late in 1952 – followed the pattern of their earlier trips, except that there were no more invitations from the Museum of Modern Art. There were a few weeks in New York, where they stayed at the St Regis, more readings, and then Hollywood, where Edith and Cukor got on famously, and Walter Reich was introduced. Osbert met Aldous Huxley once again and found him 'much kinder and more benevolent than ever before, without being less entertaining and interesting.

He is now a grandfather, which makes me feel older than ever, for I remember so well his rushing in to see me to tell me he was engaged.' And then, while Edith worked with Reich and Cukor on their script, Dan Maloney took Osbert off for another holiday in Florida and David returned to New York where he had a new friend now, the painter Brian Connelly. It all seemed most relaxed and civilised; the only outward sign of umbrage Osbert gave came when he visited the Huntington Hartford pictures in San Francisco which, as he wrote to Lincoln Kirstein, 'made me ashamed of my own country. What fearful pictorial tosh! And how odd to find such Gainsboroughs, for surely he could be a good painter. At any rate, I'm grateful to Duveen for getting them out of my country.'

Then on 9 April 1953 Miss Andrade was writing in her diary, 'Well, they have come back. OS seemed very much as usual, neither better nor worse, just moderately shaky in the morning and hardly at all in the evening.' On 1 May she was recording, 'Renishaw with OS. Everything as usual and grim. ES was there and she was as usual amiable and pale.' Six days later David arrived to spend the summer with them, and life appeared unchanged. That night at dinner the talk turned to one of Edith's favourite subjects, John Christie, the mass-murderer of Notting Hill, and she asked how so many bodies could have stayed undetected for so long behind the wall of the tiny kitchen. 'Well, of course,' David replied, 'the lower classes have no sense of smell.'

But Christie's crimes were insignificant beside the enormities Edith was soon complaining of from Hollywood. Early in May Walter Reich's script arrived. Edith was horrified. Her book was desecrated, the script so brimful of the most outrageous vulgarisms that she could never possibly associate herself with it. The film's suggested atmosphere was 'like a weekend in Surbiton', and as Edith struggled for the right word to describe the mutilation that had been inflicted on her book she hit on a phrase that Nancy Mitford would have relished. It had been, she spluttered, 'serviettted'. Just what she meant by this she explained in a letter she wrote a few weeks later to Michael Stapleton, when she had had a chance to rewrite some of the offending portions of poor Walter Reich's script. By then she was even admitting that he

> did get a certain *movement* into things. So what I have done is to keep the movement, whilst throwing out the 'serviette rings', 'cookies' drunken rolling on the floor, and general Laughtonisms. . . . I have told George [Cukor] that Charles Laughton (who is entirely responsible for my woes) will undoubtedly go to Hell and stay there.

But it was hard to exorcise the ghost of Laughton's rumbustious performance as the King in *The Private Life of Henry VIII*. She did her best, and by the end of summer had finished her revisions to the script and was all set for the next round with Hollywood, George Cukor and his vulgarising minions.

While Edith was working on her *Fanfare for Elizabeth* the First, Osbert had a chance to hear a real-life fanfare for Elizabeth the Second, with the new Queen's coronation at Westminster Abbey. As a friend of both her mother and her grandmother, he was naturally invited – and thoughtfully provided with a seat in the Queen Mother's box where he could watch in comfort and in blessed isolation from the mass of the congregation. It was his last appearance as a courtier, and nothing – neither gout, nor lameness, nor 'the shakes' – would stop him attending.

But writing was such an effort now that it was several weeks before he could summon up the energy to describe the scene in a letter that he penned, in quivering hand, to Lincoln Kirstein.

I have wanted to write to you every day; but my writing is now so difficult and execrable, that I have waited till the Coronation was over, in order to take an afternoon off and write to you about it. But now you have probably seen it as well as I did; that is to say, perfectly, for I had a wonderful seat, poised in the air, in a box built over the three thrones. . . . I started dressing at 5.30 in the morning as I had to put on knee-breeches and coat and jabot of lace and silk stockings, all unfamiliar to me for 15 years and with my tremble far from easy to put on: I had to be in the Abbey by 7.30. However a motor whizzed me along the Embankment, and I made it in record time. In Hyde Park, mackintoshed crowds had camped for 48 hours, in great good humour. The waiting for the processions was rather long – 7.30 to 10.30 – but even that was rather fun, watching the peers and the peeresses arrive in their extraordinary robes. The Queen was marvellous with her slow hieratic walk, and the ceremony seemed exactly to fit her as its centre. Even the Bishops were good. I was horribly tired after the day, and have now got *gout*!

There was another less spectacular ceremony that summer. Sacheverell's eldest son, Reresby, married the beautiful and lively Penelope Forbes quietly at a civil ceremony in Paris, while his parents were in America. Since Sacheverell, driven by his chronic lack of capital, had sold the reversions of Renishaw to the Sitwell trustees, Reresby was heir to the property, but Osbert remained determined to do nothing positive about the rest of the Sitwell fortune, or the succession to the ownership of Montegufoni, which he could leave to whom he liked. He never had been particularly close to Reresby, and while he rarely saw Sacheverell now his inclination was to keep clear of further family entanglements and please himself entirely over the whole question of his now considerable property. However, he lunched amiably with Reresby and his bride at the Sesame shortly after the wedding, when Edith tactfully invited several other guests including Stephen Spender and his wife, the Portuguese poet, Alberto de Lacerda, and one of Tchelitchew's old lovers, Peter Watson.

Osbert was charming and polite. A few days later he and Edith and David sailed for New York.

They arrived in time for Edith to become involved in yet another uproar over Dylan Thomas – the final one. The eighteen whiskies that he claimed to have consumed after a reading of *Under Milk Wood* at the New York Poetry Center had done their work, and he was lying dead from that final 'insult to the brain' in St Vincent's Hospital, surrounded by an all too familiar furore of recrimination, accusation, and high melodrama from those who loved him, loathed him and had helped to kill him. The shock to Edith was very real. As she lamented to John Lehmann, 'I *cannot* believe my dear Dylan, whom I loved as well as knowing him to be a really great poet, is dead. I just feel a dreadful numbness.' Her numbness and her grief apart, there was a very real danger that she would now be dragged into the dog-fight that had started round the corpse, and some of the letters that she wrote show just how bitterly she felt. Osbert urged caution and for once she managed to control her feelings. Even so, she told Lincoln Kirstein that she was having 'absolute *hell* about Dylan's death. Everyone is throwing accusations at everyone else and I am really seriously alarmed. They are all trying to involve me, and I can see the thing may lead simply anywhere.' There was another danger too. If things went on like this, she might be forced to change that mythical, endearing picture that she had always managed to maintain of Thomas as her loving poet-son. But luckily this was not necessary, and she could write to Maurice Bowra with the calm sadness of a grieving mother. 'And he was a most endearing creature, like a sweet and affectionate child. I can't bear to think how we shall miss him.'

Thomas's death was bad enough, but as soon as she reached Hollywood there were fresh worries to contend with. The enemies at home, emboldened by Edith's absence, were becoming active. That old enemy of all the Sitwells, the *New Statesman* (*and Stagnation* as Osbert liked to call it), had been impertinent again – more than impertinent this time, positively outrageous in their lack of reverence for Edith's royal presence, with an unflattering profile which they had actually entitled 'Queen Edith'. The anonymous article itself was bad enough; the cartoon by 'Vicky' was a positive affront.

And that same week Peter Ustinov had produced his new play at the Lyric, Hammersmith, *No Sign of the Dove*. Not for the first time in their lives, the Sitwells found themselves the subject of a satire on the stage. Edith was soon predictably enraged by the reports that filtered through to Hollywood. As she wrote irately to her biographer, Max Wykes-Joyce,

Osbert and I have been most gravely libelled by a creature called Peter Ustinov in a play called 'No sign of the Dove'. This is about a famous writer who is also a baronet, his sister who is 'a famous poetess' and wears a turban (I always did, before anyone else) and their eccentric old father. Three papers identified us *by name*. The witty Mr Ustinov gave the family the name of *D'Urt*!!! Both Osbert and I are represented as sex maniacs, and one theatrical paper that identified us by name said I was shown hunting an unwilling gentleman through bedrooms, 'looking for a bed'!!!!!!! What makes it especially disgraceful is that the part of me was taken by Beatrix Lehmann, whose brother is a great friend of mine and to whose sister Rosamond I have been most kind. . . . What *filthy* people to do that to a poor crippled man who has never harmed them.

Had Edith seen the play she might have realised that it was little more than a fairly harmless romp, which soon found the extinction it deserved. But Edith never changed, and this was the anguish she had felt at *London Calling!* all over again. The result was the inevitable letter of complaint to Philip Frere – which in turn produced the one reply which he had learned to write from a lifetime as the Sitwells' lawyer.

Libel actions, as no one knows better than yourself and Osbert, are wearing things to stage and bring to fruition; wear and tear on nerves, anxieties, time occupied, and a whole host of other irritations are their inescapable accompaniment. Are you prepared to face these in order to show up Ustinov more than he has already been shown up by the recoil of his own misdeeds?

Edith reluctantly agreed. And by now she had recovered her resilience; and when Anthony Hartley criticised her poetry in *The Spectator*, she cheerfully cabled to the editor, 'Please have Anthony Hartley stuffed and put in glass case with moth balls at my expense. Finest specimen in your whole collection.'

The editorial staff at *The Spectator* were not amused, according to Edith. They were, as she told David Horner,

furious at my cable, having never, I presume, heard of the phenomenon known as a joke. No doubt they have all been too busy playing yo-yo with their adam's-apples in belated Browning societies.

But for Edith, far from home and isolated in the Sunset Tower, the whole phenomenon was not a joke at all. Many less paranoid than she might easily have felt this sudden rudery and ridicule as more than mere coincidence; for Edith it was, of course,

quite obvious that the Ustinov attack and the *New Statesman* attack and the *Spectator* attack are all part of a most carefully thought out plan, and spring from the same source. . . . I think – and it is only my instinct – that the whole thing was cooked up at the Savage Club. Can you tell me [she inquired at the end of her letter to Wykes-Joyce], if B. Ifor Evans, Grigson and Ustinov are members. *If* they are, I think we know everything.

Edith reacted as she always had – by making a bold front of her public presence and, as one columnist wrote of an earlier visit, 'Hollywood, which thinks it knows all there is to know about publicity, is being taught a lesson by a little old lady in an ankle-length fur coat and black sandals.' Six-foot Edith enjoyed the phrase about the 'little old lady', and in fact her whole performance in Hollywood was a virtuoso exercise in totally effective self-projection. As she knew quite well, her fame and her extraordinary appearance both guaranteed that anything she did was news, and almost her first move had been to inform the head of Columbia's publicity that she intended spending most of her time in Hollywood in bed, working there all morning, taking a short walk after lunch, then going back to bed each afternoon. There would be no interviews with journalists, but in the evenings she might deign to venture out to parties if invited. 'I have been told that there are some fabulous parties over here,' she regally informed him. 'I expect I shall attend them.'

'Hollywood, which has thought up most stunts in its time, just hadn't thought of that one,' wrote Michael Ruddy, the veteran Hollywood reporter, admiringly, and the invitations were soon showering in on her. Bangles clanking, aquamarine ring flashing from her gothic fingers, she queened it quite unmercifully, playing the part she loved of the great theatrical celebrity, and moving with ease among the stars and outsize personalities of Hollywood.

She was delighted to meet Rubirosa, and was girlishly amused by the great super-stud's reputation as the supreme seducer. She called him 'Porfirio the Persecuted' and compared him to Lord Byron in the way that he was always suffering from lovesick women and threats from jealous husbands. She was also fascinated to see Zsa-Zsa Gabor in action from a nearby table as she dined at Romanoff's. She had, she said, never seen any woman so successful, and described La Gabor's mesmerising glance as 'a sort of slow-motion, deep-sea diving into the eyes of her companion – a considering look as if the passing moment and the companion were of supreme importance'.

But the one star she was determined she would meet was the reigning queen of Hollywood herself – Marilyn Monroe. And as a sort of tribute now to Edith's status, Miss Monroe called on her at her apartment. Edith found her, she said later, 'extremely serious-minded', 'pleasantly shy' – which was not surprising in the circumstances – and they apparently talked mainly about Rudolf Steiner.

But this was by no means Edith's most spectacular encounter during these weeks in Hollywood. This honour must unquestionably go to that redoubtable old journalistic battleship, Miss Hedda Hopper. She seems to have declared war on Edith before she had even seen her – possibly detecting here an undefeated rival in the gentle art of self-promotion – and in her syndicated show-business column early that January she related an unflattering story about Edith and the abrasive Harry Cohn, the head of Columbia Studios.

Edith retaliated, as only Edith could, by describing her one and only encounter with the lady.

I was at a party, when something or other propelled itself towards me. The man talking to me declared that it must be a badly done-up parcel that had been discarded after a Girl Scouts' encampment – half shielding from the gaze (though insufficiently) some outworn ironmongery and the dear energetic girls' outgoing laundry. But I am a trained observer. I was not, for one moment, deceived. I knew *at once* that it must be Miss Hedda Hopper. I said so. And I was right.

She seemed to have an extraordinary amount of clothes in some places and – it being the evening – none whatsoever in others. The same system – if the word may be used in this connection – applied to her hair, which seemed liable to sporadic outbreaks in unlikely places.

Edith demanded a retraction and apology for the story about Harry Cohn. When one was not forthcoming, she calmly issued a statement saying that, as she would hate to see Miss Hopper disgraced by publishing untruths, 'I here and now withdraw her statements on her behalf.' At the same time, less charitably, she issued her own private warning – not just to Hedda Hopper but to the world in general.

It must be remembered that it is far less dangerous to stir up a black mamba in the mating season, than it is to irritate me. It is advisable to remember also, that there is a great deal more where this comes from, and that I have never found the slightest difficulty in getting the Press to allow me space for anything I have to say.

This was fair warning, for by now Edith's long love-affair with Hollywood was drawing to an end. Much as she loved the lionising, the script sessions she was having to endure with George Cukor and the excitable Walter Reich were clearly showing that no compromise was possible between Edith's vision of the Tudor England of her ancestors and the demands of the Hollywood dream machine. She did her best to make a joke of all her troubles, especially in letters to John Lehmann.

Walter (my collaborator) was determined from the beginning that the atmosphere of Henry's Court should be a cross between that of Le Nid and Mon Repos in Surbiton, and that of the sixth form dormitory at St Winifred's. Anne Boleyn eats chocs behind a pillar, and pinches Jane Seymour's bottom behind Cardinal Wolsey's back. She is frequently addressed by her brother as either, 'Sister mine' or 'Little sister'. . . . When contradicted [Walter] . . . shrieks so piercingly that, although I am on the 10th floor, he can be heard on the 4th. It is exactly like a battle picture from Gericault transformed into sound. . . .

'Dr Sidwell,' said Mrs Pastor, the sweet elderly Hungarian who cooks for me, 'dose shrieks dey heard down the lift shaft on de fourth floor, and doze people dey crowd round de lift wid their eyes raised to heaven.'

But after a couple of weeks, the time for jokes was over. Edith was angry and exhausted, the film obviously doomed. She wearily confessed to Lehmann now,

I am beside myself; the whole damned film has got to be done again – because of dear Walter's unconquerable obstinacy.

You see, he has a mania for that regal personage Charles Laughton, so dwarfs everything down to his image.

I have told George that I *have* to return to England with Osbert. My life here is really too miserable for words. I am shut up here alone all day, with these attacks on me going on in England, and I cannot sleep at night. When Walter comes for a conference (which lasts for three hours) he shrieks so deafeningly that it splits one's head.

Since David was away with Connelly, there had been no point in keeping Osbert hanging on in Hollywood while Edith worked and mingled with the mighty. So for the early days of January Dan Maloney took him once again to Florida, this time to the small resort of Hobe Sound where they had what Osbert called 'a lovely lazy life'. Then he met David, and they went off for still more rest and sun in Honolulu. One gets some idea of how impaired he had now become from the letter from there to Kirstein on 11 February, apologising for the 'pathological laziness and dullness' which had prevented him from writing earlier.

This place has been lovely and I am better except for my gait, which is frightful. I do no work – though I keep thinking that peace may lead to some – and behave rather like a retired business man. . . . I heard two old ladies talking in the loggia here. They had been talking for some time, and I paid no attention until one remarked to the other, 'My husband is the same. He has a stroke regularly every fifteen months.'

This seems to have amused Osbert, who was miraculously managing to enjoy himself in spite of everything; and Lincoln Kirstein, who saw a lot of him in New York now, firmly believes that Osbert had begun to love America, 'because nobody bothered him, and he could quietly be himself and not be subject to the conventions and restraints he needed to observe at home'.

Back in New York again in March 1954, Osbert, with Dan Maloney as his friend and guide, could happily continue with his explorations of the city's counter-culture. It seems to have delighted him, and once again excited all his curiosity. While David was away and Edith was recovering from Hollywood in the suite at the St Regis, Osbert and Maloney were continuing their expeditions into the outlandish quarters of New York. One man who fascinated him was a body-building midget who, according to Maloney, 'was married to a model and lived with a burglar'. Most of his friends were criminals, and Osbert went on asking him about his life until the tiny man replied with dignity: 'I would have thought that you, as a titled Englishman, would know the answer to all *that*,' a reply that Osbert repeated with delight to Kirstein when they met.

Osbert's capacity for enjoying the bizarre had clearly not deserted him. One afternoon Maloney and Ted Weeks of the *Atlantic Monthly* took him on a tour of Coney Island in a wheelchair. Osbert loved it, and as they were passing one of the fish wharves they saw an expiring giant sting-ray on the sidewalk. Maloney was smoking at the time. 'Give him your cigarette, Dan,' said Osbert, and Maloney placed the end of the cigarette in one of the fish's gills. As the great fish puffed away, Osbert remarked that he had never seen a smoking fish before.

But certainly the most impressive of the unlikely friends whom Osbert met through Dan Maloney was a black male nurse called de Witt Harrison. Maloney had found Osbert too demanding, and too heavy, to manage on his own, and since Harrison was immensely tall, and a very competent and gentle soul, he felt he was the answer to their problems. He was quite right, and Osbert soon enjoyed being pushed around New York by this enormous man.

Osbert liked him and Harrison soon became devoted, always addressing Osbert deferentially as 'Sir Osbug'. Shortly before Osbert left, Harrison said to him, 'Sir Osbug, as a man of culture who is interested in the arts, perhaps you would care to see my dancing?' Osbert replied that he would be most honoured, and a date was fixed at Harrison's West Side flat. David and Dan Maloney went with him, and they sat together as Harrison's record player boomed out Stravinsky's *Firebird*. A moment later the door from the bedroom opened, and out came Harrison, stark naked and with feathers in his hair. For all his great height he was a ballet dancer *manqué*, and for the next half-hour the enormous man leapt, danced and pirouetted round the room.

Osbert used to say that this was the finest ballet that he ever witnessed in New York – and whenever Harrison appeared to flag Osbert would clap his hands and in that quiet, commanding voice of his exclaim, 'I think a little more. It's quite, quite wonderful.'

After her gruelling five months in America Edith was exhausted, and all she wanted was to leave for Italy with Osbert and David and quietly forget her public role and Anne Boleyn and everything associated with 'that damned film'. And this she did. For her and Osbert spring and early summer were a time of calm recuperation in the solitude and peace of Montegufoni. They rarely ventured into Florence now. It was too far, and there was little they desired to see there any more. Besides, as Osbert wrote testily – but all too accurately – to Thomas Mark, 'Florence is full of Germans, come to see what they destroyed and what is left for next time.'

Then on 8 June Osbert returned to Renishaw, and two days later there was exciting news. The Queen had finally confirmed what nature had long made evident. Edith was proclaimed a Dame. 'In the evening', wrote Miss Andrade in her diary on the twenty-first, 'ES arrived full of beans about her damehood etc. We had champagne for dinner to celebrate it.'

It was a signal honour for a poetess, and for Edith, given her obsession with titles, it was to be a constant source of pleasure and of satisfaction for the rest of her life. But the most benevolent effect it might have had did not occur. It failed to calm her chronic insecurity and anger with the world. It should have settled all those fears she had about the machinations of her 'enemies' and the constant threat of rival poetesses, and placed her high above the battlefield. Instead, her title was added to those three resounding doctorates as one more defensive weapon she could wield against the impertinences of the philistine.

'Why do you call yourself "Dame" Edith?' an uninstructed newsman was to ask her during her next trip to America.

'*I* don't call myself that,' she replied with full Elizabethan dignity. 'The Queen does.'

And if anybody knew the way to use a title to put the unwary firmly in their place, it was Edith.

'Actually, I'm Daming this year,' she replied quite genially to a fellow-member of the Sesame who had called her 'Doctor Sitwell', while to one wretched member of the public who had committed the enormity of sending her a manuscript to read, and the still greater one of addressing her as 'Dr Sitwell', she replied as follows!

Dear Sir,

Dame Edith Sitwell (that is how she should be addressed, not as Dr Sitwell) . . . has made it abundantly clear, over and over again in broadcast after broadcast, book after book, article after article, that any unsolicited manuscript sent to her is an intrusion into her working time, and that under no circumstances whatever will she read it.

It seems clear now that Parkinson's disease slowly changed Osbert's character. It was a strange affliction, for beneath the tremor and rigidity which gradually advanced upon his face and limbs his actual health was relatively unaffected. His mind was as sharp as ever, the condition of his heart no worse, and he could even get around, although his walk was now reduced to the so-called *marche-à-petit-pas*, which is one of the most distressing symptoms of the disease. This consisted of a headlong rush of small steps forward, and once started it was difficult to stop. (Dan Maloney still has nightmare memories of trying to hold Osbert back as he went staggering uncontrollably across the rush-hour traffic of Fifth Avenue. 'That was a damned close shave,' Osbert said afterwards.)

But inevitably the illness made him more of a recluse, and more and more dependent on the few close friends he had. He became gentler and more accepting of his lot, and very tolerant and thankful for the affection that he could command – particularly from David. He seems to have been very wise about him now. He knew that it was only natural for David to have other friends and something of a life away from him. This always had been a condition of their life together and it still applied. But David was still fond of him, still spent the greater part of every year with him, and he was grateful.

That autumn he was once again in Italy, while David took a two-week cure at Vichy and then went on to Switzerland before rejoining him and Edith at Montegufoni. The few notes Osbert wrote him, laboriously scribbled in an all but indecipherable hand, show how much he still depended on him.

On 2 September 1954 he wrote,

> This is just to thank you for your letter which made a great difference to me and put me in a contented state of mind. You say so much in a few words, and, as you know, I have always loved you. Can't write more at the moment. Longing to see you.
>
> Yours devotedly,
> OSBERT

Ten days later he was writing once again from Montegufoni.

> It is lovely weather, but I am very lame: can just potter in the garden. E. arrived safely. Readers' Union have taken *Four Continents*. Minimum circulation, 30,000.

And on 20 September there was a final letter to explain arrangements for David's eagerly awaited arrival from Geneva.

> Just a line . . . I've ordered Bonini's car to meet you at Florence at 5.40 on Saturday; so you'll find him outside the station.
>
> The weather has been very hot: only hope it will continue.
>
> Edith sits out in a blanket. Add a few feathers and she'd be a red Indian.

The calm and the good weather were exactly what she needed, and early in October she was writing excitedly to Lehmann, 'I am just on the verge of writing poetry again, after ages of nothing happening. (Indeed, – but I scarcely dare say this because it is still in germ – I think I am going to get a new atmosphere.)' She also mentioned that she was now completely reconciled to Tchelitchew (who was still in Rome) after what she described as their 'estrangement' over the last four years; 'But like the row between

Yeats and Arthur [Waley] (Whom Yeats said had behaved disgracefully, but he couldn't remember why) that, now, seems to be over.'

This happy interlude in Italy could not last for ever. Edith was obliged to go to Hollywood to make one final attempt to save the film, while Osbert had high hopes of treatment from the New York specialists. Through Anita (*Gentlemen Prefer Blondes*) Loos, a Californian friend of Aldous Huxley, he had an introduction to the acknowledged world expert on Parkinson's Disease, Dr Irving Cooper, and there were exciting rumours now of some new miracle-working drug for the disease. Failing this, there were developments in a new form of operation for the brain. There seemed a chance, at least, that he could still be cured.

Osbert, David and Edith sailed for New York, this time aboard the French ship *Liberté*, in mid-November, and soon after they arrived David went off to Wilton, Connecticut, to stay with Connelly, who was now painting his portrait. Edith was lecturing – one of her most successful offerings was an evening of her reminiscences of Hollywood and its personalities, including Hedda Hopper – and Osbert began to make the melancholy rounds of all the different doctors he could find. A few weeks later he was telling Miss Andrade, 'I'm having the various treatments here – vibration, massage, and heat which seems at last to be doing some good. I still tremble dreadfully.'

David was back with him for Christmas, and the note he wrote Miss Andrade makes it clear that Osbert was benefiting now to some extent. 'Sir Osbert has had success with his new treatment, which consists of vibration of the spine and massage of the neck. The treatment has completely cured his ankle and he walks well apart from the actual symptoms of his unfortunate malady!' As for the 'unfortunate malady' itself, there was still no sign of any cure, and with the new year of 1955, Osbert departed once again for Florida with Dan Maloney, while David disappeared to Santa Barbara. Edith went off to battle on her own in Hollywood.

There was little she could do, apart from belatedly discover that Columbia had now decided they would shelve the film indefinitely. It was a decision she accepted with surprising equanimity. Perhaps she was secretly relieved to know her battles with excitable Walter Reich were over, and her chief cause of irritation was not the film but persistent rumours which would plague her in the months ahead.

I have not the faintest idea what you and your editor are talking about [she wrote imperiously to Ruth Braithwaite of the *Sunday Chronicle* early in February 1955]. I am *not* bringing Miss Marilyn Monroe to England. Is it supposed that I am a publicity agent or a film agent or a press agent? I am an extremely busy person, and have been caused grave annoyance by members of the press thinking they are at liberty to disturb my work because of this quarter-witted story which *has no foundation whatsoever*.

Miss Monroe, like a good many other people, was brought to see me while I was in Hollywood. I thought her a very nice girl, and said to her as I said to others, that if she came to London she should let me know and should come to a luncheon party. There the matter began and ended. I shall certainly hope to see her at luncheon if she comes to London. . . .

I can now only suggest that the people who are interested in this world-shaking subject, should dig up William Blake or Percy Bysshe Shelley and enquire whether *they* are 'bringing Miss Monroe to England'.

But Osbert had more serious troubles to contend with now. Gossip had started about him and David, and while he was still in Florida he received a disturbing letter from David saying that there was talk that they were breaking up. This brought a painfully written but resigned note in reply.

Dearest David,

Thank you for your letter. . . . I return to New York on Sunday the 13th.

About what you tell me, what can I say, what can I do. . . . There was bound to be talk.

The weather here has been volatile; one fine day, then storm.

I wonder if the *United States* has cabins for us? Do send Mr Kinch [the New York travel agent] a line. . . . Sorry to be so illegible. We went to see a wonderful orchid farmer.

Best love,
OSBERT

For most of March Osbert and Edith were together in New York, and it was not until the second week of April that David joined them. Osbert was glad to see him back, and, impassive as he always could be where his emotions were at stake, seems to have given not the slightest sign of jealousy or resentment. David was in the very best of spirits.

They got the first-class cabins that they wanted on the *United States*, and it turned out to be a perfect voyage for April, 'like a jaunt on the Mediterranean in July', as Osbert described it in a note to Lincoln Kirstein. The Windsors were aboard, and Osbert was amused and rather gratified to receive an invitation to drink cocktails with his former monarch. The Duke was evidently oblivious of the extent of Osbert's disapproval of him during the Abdication. The whole incident passed off quite amicably. But in a letter to Kirstein Osbert showed he had not really changed his mind about them, and that he still possessed a sharp eye for the grotesque, for he 'found that they had exchanged voices – he now talks with a strong American accent, and she with the sublimated cockney which was the legacy to him of his nurse'. Edith was not included in the invitation. A trifle ominously, she spent the voyage in her cabin. David, on the other hand, was full of talk and bonhomie for everyone.

CHAPTER TWENTY-SIX

The Convert

1955–1956

FATHER MARTIN D'Arcy was one of the legendary figures of the modern Roman Catholic Church in England, one of the select few who kept alive the old tradition of the Jesuits as men of preternatural subtlety and social influence. He was an ascetic, bright-eyed little man. In his youth he was one of the very first to hail the poems of his fellow-Jesuit, Gerard Manley Hopkins; at Oxford in the thirties he was the Master of the Roman Catholic Campion Hall and lectured on philosophy; later, in *Vile Bodies* his friend Evelyn Waugh depicted him as Father Rothschild, the priest who remembered everything that could possibly be learned about everyone who could possibly be of any importance'.

During the five years after the war he was appointed Provincial Superior of the English Province of the Jesuits, and he had helped to bring a stream of most distinguished converts to the Jesuit church in Farm Street (just opposite the Connaught and around the corner from Claridges). Edith had met him, as we have seen, with Roy and Mary Campbell, and she encountered him again in the winter of 1954 in America when he was lecturing at the Catholic University of Notre Dame.

They corresponded briefly – and in April 1955 he wrote her a short and, in its way, historic note, which ended: 'I cannot say how happy it has made me.' This happiness was caused by what Edith called 'the most momentous decision of my life'. She had made up her mind to join the Church of Rome.

On the surface it could hardly have been an easier conversion, and certainly there have been few speedier acceptances into the arms of Mother Church. Since Father D'Arcy was unable to instruct her from America, he immediately put her in touch with another London Jesuit, Father Philip Caraman. It was a good choice, for Father Caraman is a man of letters on his own account. His excellent translation of the autobiography of the seventeenth-century English Jesuit and martyr, John Gerard, was published with a preface by his friend and co-religionist, Graham Greene, and for many years he was a valued friend and spiritual confessor to England's other major Catholic novelist, Evelyn Waugh. In Father Caraman Edith was getting an instructor who was every bit her intellectual equal, and thoroughly at home among the souls of literary lions.

Edith was plainly an important convert, and Father Caraman gave her all the zeal and dedication for which his Order has been justly famous. That April, on her return from America, she was staying at the Sesame, and he had lunch and several sessions with her there. When she retired to Renishaw he came and stayed for several weeks at the nearby Jesuit college at Spinkhill so as to carry on her spiritual instruction every day. There was no doubting the seriousness of her intentions. As she wrote to him at the beginning of May, 'I believe and trust with all my heart, that I am on the threshold of a new life. But I shall have to be born again. And I have a whole world to see, as it were for the first time, and to understand as far as my capacities will let me.' Indeed, she seemed a model convert. Most intellectual converts naturally have doubts and misapprehensions which need coping with, foremost among them is the question of authority. Not so with Edith. For so powerful a personality she showed a quite unusual eagerness to subordinate herself to the absolute authority of the Church, so much so that Father Caraman soon formed the opinion that she was already a Catholic in all but name.

> She was well-read in the mystic poets, she was eager to enlarge her knowledge of the theological basis of the Faith, and she possessed a natural understanding of the need for the authority of the Church which is often such a stumbling block to intellectuals. Indeed, she often told me that she could not conceive of Christianity as other than an authoritarian religion.

Because of this, her period of instruction went off smoothly, and Father Caraman soon met Osbert. The priest was impressed at the courage with which he was fighting his disease, and remembers how he would insist on struggling to help him with his coat each time he left. Agnostic though he was, he gave no sign of opposition to his sister's new Catholicism – rather the reverse – and, as an expert in such matters, Father Caraman wondered if he too might soon follow Edith.

As for Edith, she was rapidly developing a deep affection for her priest.

'When I baptise you, I shall have to call you Edith,' he informed her, early in their meetings.

'I hope you will always call me Edith,' she replied, and soon her letters to him changed from 'Dear Father Caraman', to 'Dear Philip'.

Her reception into the Church was fixed for 4 August at Farm Street.

This was to be an interesting ceremony in several ways. As she was being baptised she required two Catholic godfathers: she chose Evelyn Waugh and her old enemy and new-found champion, Roy Campbell. Campbell's wife Mary was to be her godmother.

Waugh took his unlikely duties as Edith's father in God with the extreme seriousness with which he approached everything to do with his religion, and this led him to make a concerned suggestion to Father Caraman which shows how well he knew his future god-daughter.

I am an old friend of Edith's and love her. She is liable to make herself a little conspicuous at times. She says she will be received in London. Am I being over-fastidious in thinking Mont St. Mary's much more suitable. What I fear is that the popular papers may take her up as a kind of Garbo Queen Christina. I was incomparably less notorious when I was received and I know that I suffered from the publicity which I foolishly allowed then. There are so many malicious people about to make a booby of a Sitwell. It would be tragic if this greatest occasion in her life were in any way sullied. Can you not persuade her to emulate St Helena in this matter?

Waugh's suggestion that the service be held outside London was turned down on the grounds that it would be unreasonable to expect the rest of the invited congregation to make the journey up to Spinkhill for the ceremony, but his advice was heeded over relations with the press – who were carefully excluded, so that much the best account of what occurred is found in Waugh's own diary. He had been working at Folkestone, and caught an early train to London on the fourth.

From Charing Cross I walked to White's buying a carnation on the way and drank a mug of stout and gin and ginger beer. Then at 11.45 to Farm Street where I met Father D'Arcy and went with him to the church to the Ignatius Chapel to await Edith and Father Caraman. A bald shy man introduced himself as the actor Alec Guinness. Presently Edith appeared swathed in black like a sixteenth-century infanta. I was aware of other people kneeling behind but there were no newspaper men or photographers as I had half feared to find. Edith recanted her errors in fine ringing tones and received conditional baptism, then was led in to the confessional while six of us collected in the sacristy – Guinness and I and Father D'Arcy, an old lame deaf woman with dyed red hair whose name I never learned, a little swarthy man who looked like a Jew but claimed to be Portuguese, and a blond youth who looked American but claimed to be English. We drove two streets in a large hired limousine to Edith's club, the Sesame. I had heard gruesome stories of this place but Edith had ordered a banquet – cold consommé, lobster Newburg, steak, strawberry flan and great quantities of wine. The old woman suddenly said: 'Did I hear the word "whisky"?' I said: 'Do you want some?' 'More than anything in the world.' 'I'll get you some.' But the Portuguese nudged me and said: 'It would be disastrous.'

This luncheon at the Sesame was very much in character for Edith, and also an expression of the absolute importance she attached to the occasion. Indeed, the letters that she wrote to Father Caraman throughout this period reveal a desperate longing to believe, as if she was willing herself towards it.

> The wonderful writings of St Thomas Aquinas, and Father D'Arcy's *The Nature of Belief*, make one see doubt – perhaps I am not expressing this properly – as a complete failure of intellect. Then again, I see that purely intellectual belief is not enough: one must not only *think* one is believing, but *know* one is believing. There has to be a sixth sense in faith.

Her chief problem, it appears, was over prayer. 'Prayer has always been a difficulty for me. By which I mean only that I feel very far away, as if I were speaking into the darkness. But I hope this will be cured. When I *think* of God, I do not feel far away.'

But the greatest blessing of her new religion was, as she thankfully told Father Caraman, the sense of peace it gave her. 'What a fool I was,' she told him, 'not to have taken this step years ago!'

So why hadn't she, and what had finally impelled her towards this momentous step?

One gets something of an answer from the turmoil in her life during the spring of 1955, for it was then that she had found herself in a sudden situation she could not control. The cause of this was Osbert – 'that poor, crippled man', as she had called him – and the rage she felt for the way that she believed David was treating him. Remembering the violence of her emotions at the time when Tchelitchew left Allen Tanner, one can understand how passionately she must have felt throughout that previous winter in New York when David was away with Connelly. And then in the following March – just before she wrote to Father D'Arcy – she was alone with Osbert for a month at the St Regis, waiting for David to return from Santa Barbara. During this time she must have seen how much her brother missed him and how pathetically he welcomed his return. One will never know how much Osbert told her, but she was no fool in such affairs and knew exactly what was going on. It was now that the friendship she had previously evinced for David turned to hatred. It was now too that her real troubles started.

For, as she knew quite well, there was absolutely nothing she could do about the situation. David would not change, and Osbert still adored him and still depended on him. If she said what she really felt, if she gave the faintest hint of the anger in her heart, total disaster would ensue. Not only would her beloved Osbert suffer in the inevitable uproar, but it would end with his being forced to choose between her and David. She can have had few illusions over how the choice would go, and she would lose not only

Osbert but her home at Renishaw and her holidays in Italy. She could not possibly afford the risk.

For almost anyone it would have been a tricky situation. For Edith, who could be raised to frenzy by an impertinent remark or an unfavourable comparison with Kathleen Raine, it must have been continual mental torture, particularly when she witnessed David, sun-tanned and confident as, ever blithely return to stricken Osbert for the summer. Small wonder that she spent the voyage back to England in her cabin and avoided seeing him. Small wonder too that she felt an overwhelming need for some all-powerful authority that would assist her to control her feelings, and teach her to turn her anger to acceptance. Without such help she would either be destroyed or give in to a storm of rage which would inevitably sweep away her world and the person that she loved.

Naturally it took a little while for all this to emerge in her dealings with Father Caraman, although it does explain that eagerness she showed him from the start to submit to an authoritative church. Otherwise she was concerned to show her true credentials as a Catholic convert. But during most of May she was alone with Osbert and David at Montegufoni, and in the almost daily letters that she wrote to Father Caraman the burden of her anger soon appears. On 7 May, a bare four days after their arrival, she was writing:

I am under daily temptation to great anger because of something terribly cruel that has been done to my dear Osbert, and by his greatest friend, and has made Osbert who has now even to have his food cut up for him see himself as a hopeless cripple dependent entirely on the ordinary kindness that has been denied him. His friend deserted him after telling him that he did not care if he lived or died. But now he who owes everything in the world to Osbert is back in this house for his own convenience and intends to live with us again. It is a great difficulty to me. I have so far shown no anger but I do feel it. I must, of course remember my own grave faults.

Edith was appealing for the Church's answer, and it was this that Father Caraman offered in reply – that it was wrong of her to question Osbert's illness, that it might well be God's will to lead him to salvation through his suffering, and that it was not for her to judge anyone in such a situation. Judgement was God's prerogative alone.

It was advice that evidently worked, for Edith replied with one of the few genuinely humble letters she ever wrote.

You are very kind to me. I am very grateful. What you say about Osbert is, I feel, deeply true. He trusted this poor man but twice in my life I have seen a saintly goodness grow from an unspeakable calamity. One is Osbert. All the human faults he seemed to have before his illness seem to have been burned away. He behaves with the greatest gentleness under this frightful burden. (He fell down again two days ago: he is always doing that now.) Yesterday I almost pitied the man of whom I spoke. He is a catholic convert and had been to church. He said he had prayed for a friend of his. I thought that so sad. He hasn't, I think, any idea of the terrible thing that he has done. Osbert always tries to find excuses for him.

But Edith had a long way still to go before she could patiently accept the situation. The ever-present provocation to enormous anger must have been very real to her, especially when all three of them returned to Renishaw. Thanks to her new-found belief and Father Caraman's support, she managed to control it – but evidently only just. It was soon after this that she told Stephen Spender that without her Catholicism she would have murdered David, and Miss Andrade, who was there that summer, wrote in her diary for 14 July: 'How ES hates DH, and yet I believe her to be so wrong in much of what she thinks.' Which, in a sense, she was. For Edith saw the situation solely from her own passionately loyal and pro-Osbertian point of view. What she could not understand was that, during the twenty-five years that David and Osbert had been together, David had given Osbert years of happiness and peace of mind. He and he alone had rescued Osbert from the loneliness that followed Sacheverell's marriage, but it had always been quite firmly understood between them that David should retain his freedom. Also it was clear that Osbert's illness had put a strain on their relationship. They may have had their bitter arguments, but in the last resort it was for them alone to decide what happened.

For Edith the important thing was that the Church and Father Caraman between them carried her through this time of crisis, and she undoubtedly made sincere and heartfelt efforts to believe. The crucial question now was whether her beliefs had given her a lasting antidote to that extraordinary 'temptation to anger' which was the greatest hazard in her life. It seemed as though they might have done. After her reception she appears to have been more devout than ever, and in a mood of almost mystical acceptance of the Faith was duly confirmed at Farm Street by Archbishop Roberts on 2 October. This time there was no need for discretion. What Waugh called 'the cream of Catholic London' made up the congregation. Three days later she and Osbert left for Italy.

David did not go with them. Instead he was off, as usual, for America. There was no formal break between him and Osbert, but as is quite clear from a note to Lincoln Kirstein, Osbert had decided it was best for him not to go to America but to spend the winter peacefully in Montegufoni, then see what happened in the spring.

Alas I shall not see you until March. I've had to abandon the idea of coming before, as I must try to work out in quiet how much I can work – and here I have a genius called Luigi who can help me. . . . I see doctor after doctor – witch doctors, real doctors, sham doctors etc.; but my walk is terrible, blast it.

The 'genius' who could help him now was the devoted Luigi Pestelli, the *fattore* at the castle, who had increasingly taken over the running of the place in the absence of *Il Barone*, and who had also taken on the chores of ministering to Osbert when he was there. For he had reached the stage where he required help for almost every movement in his life – to eat, to bath, to shave himself, to dress.

But at the same time he had also set himself to work. Whatever anxiety he felt for David he kept strictly to himself, and with extraordinary self-control applied himself to an unfinished task. In 1927 he had begun his series of 'eclogues', portraits in verse of characters, most of them remembered from his childhood at Renishaw. In 1954 he followed this with a second volume in the series, *Wrack at Tidesend*, which was devoted to the far-off world of Scarborough, with such nostalgic characters as Sir Fiddler and Lady Sparrowebank, the Misses Eurydice and Alberta Colbert, and what he called 'A Frieze of Doctors' Wives'. Now he was determined to complete the trilogy with a poetic portrait gallery of the Florence he remembered from the early twenties and before the First World War. It particularly pleased him to think he would be finishing a work which he had first planned thirty years before.

This was to be the first full winter Osbert and Edith ever passed together in the castle. Apart from her 'Elegy for Dylan Thomas', Edith did little serious writing. Most of her creative energies were taken up with the enormous task of compiling the most onerous of her anthologies, *The Atlantic Book of British and American Poetry*, which appeared in 1958 as something of a personal counterblast to Auden's even larger *Poets of the English Language*. Her incidental writing now reveals a new benevolence in line with the demands of Christian charity. She wrote a glowing acclamation of the poetic works of her godfather, Roy Campbell. She also gracefully conferred her accolade upon another writer who had once annoyed her. On 20 October 1955 Cocteau was elected to the French Academy, and in the piece that Edith wrote for the *London Magazine* to mark this singular event there was no hint that she could ever have despised him for that 'little bourgeois mind' of his. 'The Académie Française', she wrote, had 'honoured itself and France by bestowing its membership upon one of the only living writers of genius, Monsieur Jean Cocteau'.

With reconciliation in the air, she even showed the Christian spirit in her dealings with her enemies. William Plomer was forgiven, Grigson and Ustinov were quietly forgotten, and there was even an armistice with other female poets. When a carpet dealer from New York wrote reproaching her for joining the Church which still bore the bloodstains of the St

Bartholomew's Day Massacre, she simply sent him a postcard in reply with 'Don't be silly, Edith Sitwell.'

Altogether the impression that she gives is of a chastened spirit, doing her best to tend her brother and face the world with fortitude and faith. She and Osbert seem to have been closer than they ever were, and Edith was courageously attempting to create the Christian life for both of them as winter started in that draughty castle.

Inevitably she hoped for Osbert's conversion too. Father Caraman arrived for a brief stay, *en route* to Rome, where he was visiting his old friends the FitzHerberts (Mrs FitzHerbert was Evelyn Waugh's daughter); and he was so concerned for Osbert that he sent him back a rosary blessed by the Pope himself. Edith was duly grateful, and wrote devoutly of 'the greatest joy' which it would give her brother. Whether he really did appreciate the gift, except as an act of kindness, one must take leave to doubt. Certainly he showed no sign of any inclination to abandon that stoical agnosticism which he had lived by all his adult life. But as Edith was attempting to accept her life as preparation for a nobler one to come, she was patently attempting to persuade Osbert to do the same. Indeed, the picture that she gives in her Christmas letter to her confessor is of a touchingly united Christian couple.

Osbert and I do wish you with all our hearts a very happy Christmas. We shall think of you so much. I am sure the service here on Christmas Day will be very lovely and touching. The Feast of the Immaculate Conception was deeply so. It was held in the small white chapel leading out of the main church and was decorated with crowds of yellow roses the colour of candle flame. The young priest who celebrated the mass speaks English so that next day, with great kindness, he came to visit me so that I could confess to him. He is going to be a missionary in Africa and before that he must spend six montns in England so that I hope to see him there. I was greatly moved.

But what of Osbert? Was he really as resigned as he appeared to be behind that show of self-control? Had he given up all hope of being cured even now of Parkinson's Disease?

He described himself that winter in a note to Lincoln Kirstein as 'infernally wobbly . . . and a great bore to myself and everybody else'. But there was still a chance of some alleviation of the symptoms. As Edith, wrote to her friend, Lady Lovat, 'There is now a faint hope for the sufferers from this illness. There is a young surgeon in New York who *may* be perfecting a *safe* operation. (The old former operation was very dangerous.)' This was Dr Irving Cooper, the doctor Osbert had already seen through Anita Loos's introduction, and he was determined to consult him once again. If there was any chance he would take it. He was also quite determined to see David and so was sticking to his plan of visiting New

York in March, despite the fact that Edith had firmly told him that she would not go.

The New Year of 1956 brought at least one unexpected bonus. Osbert received a C.B.E. It was, as Sacheverell quite rightly said, 'rather a measly' decoration for a major English writer, but in the eyes of British governments living writers rank far below those generals and sportsmen and fifth-rate politicians Osbert had always so despised; yet he had also, as he readily admitted, loved honours, praise and flattery. A C.B.E. was proof, if nothing else, that he was not entirely forgotten.

Just before he and Edith left for England something occurred that might have been more serious than it was. Even so, it had an ominous ring to it. As Edith explained in a note to Father Caraman,

> I had a wretched accident. I fell the whole way down a flight of stairs. There are large holes in some of the stairs, and I caught my heel in one of them. Why I didn't kill myself I simply can't think. I picked myself up somehow and crawled upstairs. I didn't dare tell Osbert what actually *had* happened because I knew that it would have terrified him, so I lied and told him I had slipped up in a passage. But I was in bed for ten days and now can only walk if Luigi supports me. However I haven't broken anything.

She blamed the fall on overwork. She had been 'working so appallingly hard that by the time the day's work is over I was hardly making any sense'.

Osbert went on to America in March 1956. Meanwhile, Edith's unwonted mood of Christian resignation was already wearing a little thin, and she was on the warpath once again for upstart poetesses – always a disturbing sign.

> Oh the Lord have mercy on us! [she lamented to John Lehmann] *Who* is Gillian Freeman? Don't tell me we have another poetess looming? The whole lot of them ought to be told to shut up. They are such a bore. No passion, no fire, no interest, no beauty, no technique. . . . Miss Jennings ought to undergo treatment for this flow of verse.

Her letters to Father Caraman too were showing signs of trailing off. In one of them she actually apologised for the delay in writing, and excused herself because her American publisher, Ted Weeks, had recently been plaguing her, insisting on interminable discussions about her anthology. Because of him her days, she wrote, had been 'swallowed up by despondency, my life having been a cross between the ancient routine of the treadmill, and that led by a poor, worn-out electric hare pursued by particularly ferocious greyhounds'. But the reason for her heaviness of heart was not really her anthology at all but the news she had been hearing from New York.

In the first place, none of the American specialists seemed to be able to offer Osbert any immediate alleviation of his illness. Even that miraculous operation on which they both had built such hopes was not considered right for him at present. Then came an even greater shock. David and Osbert seemed to be reunited. The friendship with Brian Connelly was over, Osbert was overjoyed, and he and David were both coming back to Renishaw at the beginning of May. Edith was prostrated by the prospect, and seems to have resorted to her customary place of refuge. As she wrote gloomily to Lehmann now from Renishaw,

> I have been so got down by the return of that creature for good and all, (which is having a dreadful effect on Osbert) and by general fatigue and despondency, that I thought the best thing I could do was to come back here and go to bed for a few days. That man comes here on Thursday – I suppose for the summer and indeed for ever – and I really hate him terribly.

Osbert's investiture that May as a Commander of the British Empire was one of the now rare Sitwell family occasions. Edith was very proud of him, but Sacheverell, who had seen little of his brother for many months, was shocked at his physical deterioration. He wrote afterwards to Nancy Cunard – now very sick herself and rather mad, but anxious to hear about this old friend from her youth – describing how sad it had been to see him hobbling along the palace corridors, led by a royal footman.

And back at Renishaw Osbert soon had to bear the full spate of Edith's passionate resentment at the prodigal's return. She managed to contain herself in David's presence, but when she was alone with Osbert there was little sign of the Christian tolerance she had been trying to adopt the year before. Miss Andrade, who was there again, recorded simply in her diary, 'ES had an outbreak of vituperation against DH. I believe OS dislikes him. No, it can't be possible.' For the remainder of the summer David was at Renishaw, and Edith spent an awful lot of time in bed. Her ankles, always weak, were causing her considerable pain, and once in bed she was safely out of range of David whose offending presence was so patently delighting her beloved Osbert. After her valiant efforts to save her brother from despair when 'that creature' was away, by winning him to the Faith, it must have been particularly frustrating; and she still had that ever-present problem – how to control her anger at the presence in her home of this man she hated. True, she had her religion, but religion on its own was no longer quite enough.

Luckily, there were, as usual, various distractions at Renishaw in the summer of 1956. At the end of June Osbert invited Reresby and Penelope to stay. Relations were improving now. Reresby's charm and Penelope's vivacity are hard to resist, but despite the parlous state of Osbert's health he still made no reference to the obvious need to take appropriate long-term action against death duties by making over some of his possessions and his fortune to his heir. The subject was not mentioned.

Osbert could still swim, as he did each day throughout the summer in the Renishaw swimming-pool. He had his favourite cronies he would see, particularly his old friend Malcolm Bullock, the novelist L. P. Hartley, and his extraordinary near-neighbour, the penurious, bowler-hatted, spaniel-loving Catholic bachelor, Lord Grey de Ruthyn. Moreover, the world was far from consigning all the Sitwells to oblivion. In September, shortly after Edith entered her seventieth year, there was a highly successful one-night gala performance of *Façade* in which Edith and the singer Peter Pears took part.

'All the Sitwell world there,' Miss Andrade wrote with customary understatement.

After *Façade* it was time for Italy again. Osbert felt he could no longer stand the bleak cold of the Derbyshire winter, and Edith could hardly stay at Renishaw alone, even though David, much to Osbert's satisfaction, was going too. Osbert was now determined to begin another book and asked Miss Andrade out to join them for the winter.

And so this strange quartet set out for Florence. It was the first time that Miss Andrade had seen the castle. At first sight it delighted her, but she soon found it curiously like Renishaw in its combination of great beauty and a sort of ingrained gloom. Soon the inevitable signs of strain and tension had begun to show. The grapes were harvested, the short Italian evenings started to draw in. During the mornings she would work with Osbert who was starting to dictate his final reminiscences about Sir George, which would make up the last book he would write, *Tales My Father Taught Me*. Osbert could not escape him even now. And in the evenings after dinner all four of them – David, Osbert, Edith and Miss Andrade – would sit in the immense *salone* with the painted ceiling. Most of the conversation was between David and Osbert. Edith invariably looked grim and paid attention to her knitting. Around 9.30 Luigi would help Osbert off to bed. Edith would usually go next, followed by Miss Andrade. David would stay on alone, reading and drinking the Italian brandy which he liked.

This pattern of their days continued until nearly Christmas, when it was interrupted by a fresh event in the enveloping catastrophe. Edith had what she herself described as 'a really terrifying accident'. She was apparently walking from her room into the corridor outside when, as she wrote to Father Caraman:

I slipped and crashed onto my face onto the stone, coming to to find myself lying in a pool of blood, my dress absolutely soaked in blood, Miss Andrade, Osbert's secretary holding my head on her knee and towels soaked in blood to my face, and Luigi fanning me and holding ice to my forehead. Why I did not break my nose and both my cheekbones we shall never know. As it was I had two black eyes, a completely blackened face, and my nose swollen right across my face. . . . I am just up again.

This was the second time that she had had a bad fall in the castle. She blamed the servants. 'The maids', she told Father Caraman, '*will* wax the stone floors'.

CHAPTER TWENTY-SEVEN

Edith Agonistes

1957–1960

EDITH WAS still confined to bed when she received another shock – this time to her spirit rather than her battered body. The Italian newspapers for 12 January carried the news that her oldest friend in the pantheon of living poets, T. S. Eliot, had suddenly remarried. She, more than almost anybody, knew the truth about his wretched life with Vivienne, and might have been delighted that the great man had discovered a devoted bride to cosset and protect his final years. Instead his action left her speechless, and nearly a week elapsed before she vented just a fraction of her wrath in an outburst to John Lehmann.

> Oh, what a *beast* Tom is!!! . . . You wait! I'll take it out of that young woman! I'll frighten her out of her wits before I've done. As for Tom – he will, of course, be punished. He will *never* write anything worthwhile again. And indeed hasn't for a very long time now. The *Four Quartets* are, to my mind, infinitely inferior to his earlier work, completely bloodless and spiritless. It makes me quite sick to think of the pain John [Hayward] has endured – that waking up at 5.30 in the morning . . . to be told that his greatest friend, on whom he depended in his unspeakable physical helplessness and humiliation, had done him this sly, crawling, lethal cruelty. I feel I never want to see Tom again!

One's first reaction to this outrageous letter is that Edith must have been overcome with jealous rage. She often said that Eliot had been in love with her (she even told one friend that she had been the unlikely original of the Hyacinth girl in *The Waste Land*) and certainly appeared to feel that her long and slightly jealous friendship with the poet gave her a certain status in his affections. But a second reading makes it evident that this was not the cause at all. Her rage appears to centre not on Eliot himself but on the plight of her crippled friend John Hayward and the way he had been left by the man who had looked after him. And even this is not the whole truth of the letter, for Edith's certainty that Eliot had 'deserted' Hayward is in fact ridiculous and quite beside the point. Hayward, who, like Eliot himself, was strictly heterosexual, was in no way dependent on Eliot and had no reason to feel in any way betrayed by his marriage. (He may have felt envious, but that was quite another matter.) But Edith had a curious habit of transferring her hatred from its real object to a safer target on which she could then expend her wrath. She did this countless times. When she was writing *Aspects of Modern Poetry* Leavis and Wyndham Lewis both became convenient whipping-boys for the anger that she really felt for Tchelitchew, while Grigson and Noël Coward performed similar functions for her in her later rages at Sir George's will. Now it was Eliot's turn. As she had been in Italy she had had no chance of knowing the true facts surrounding Eliot's remarriage. She was still obsessed by David's treatment of her brother, and patently transposed her version of the David–Osbert situation on to what she imagined the 'abandoned' Hayward must be feeling. It was no coincidence that the words she used to describe John Hayward's wretched plight repeat almost word for word the harrowing description that she gave to Father Caraman of Osbert's misery when he was left by David. They were both crippled and dependent men of letters, and both, in Edith's eyes at least, had been deserted for another by their 'greatest friend'.

The real significance of her explosion over T. S. Eliot was that it allowed her to express her untrammelled anger to everyone on a subject that was otherwise forbidden – David's behaviour to Osbert – and by castigating Eliot she could say everything she felt about the pity that one owed a crippled friend, the weaknesses of older men for younger lovers, and the unforgivable unfaithfulness of lifelong friends.

Seen in this way, her outburst against Eliot boded ill for Edith's peace of mind and the continuing tranquillity of the trio. Evidently the peace the Church had given her no longer functioned. That brief forgiveness she had managed to accord to David thanks to Father Caraman was no longer possible, and she was locked inside this triangle, hating as much as ever. When David returned, far from forgiving him, she watched with her anger unappeased as his influence increased over her 'dear old boy'.

Hardly surprisingly, her health began to show signs of the continual strain that dogged her. The previous November Stephen Spender had urgently invited her to join a group of Western writers on a hurried trip to Hungary to show solidarity with the Hungarians during their rising. She refused. 'I shall be seventy on my next birthday, and am now extremely lame. I have arthritis in both knees, acute rheumatism in both feet, and often am only able to walk – and slowly at that – with the aid of a stick.'

Then, to complicate things further, it appeared that David's friendship with Brian Connelly was not completely over. Early in the spring of 1957 he was in London and David left Montegufoni earlier than planned especially to meet him. Osbert apparently acquiesced quite cheerfully. David had reassured him that he would never leave him now, and it was arranged that early in March they would *all* depart together for America – Osbert and Edith, Connelly and David. Osbert had his friends and specialists to visit, Edith had arranged to read her poetry and lecture, while David and his friend were planning to enjoy a holiday.

Osbert was clearly very reasonable and realistic now about the situation. He had his work, he had kept David in his life, and on those terms he was prepared to tolerate whatever else was necessary. Certainly the brief note that he scrawled to him in February from Montegufoni shows not the faintest sign of bitterness or fear of losing his beloved.

Doodlums,

I haven't written as intended but have been working.

Don't forget to let me know when you are coming here. Longing to see you.

Osbert

P.S. The terracotta pots look lovely.*

But Edith was the one who really felt the strain of what was happening. Osbert was sufficiently in love with David to be wise and understanding, and even to accept some sort of provisional *ménage à trois*. Edith was not – although she really had no possible alternative. By now she had made Father Caraman her confidant, and that spring Sacheverell was drawn into the secret too. By an uncomfortable coincidence, he and Georgia had arranged to travel to New York on the same boat at the beginning of March, and Edith naturally explained her plight to him. But for Sacheverell, just as for Edith, there was little to be done except suppress his feelings and avoid an outright row.

For Osbert begged him to be polite to Connelly and act quite normally to David. Otherwise, he told him, there would be atrocious scenes and everyone would suffer, he, Osbert, most of all. Sacheverell was every bit as angry now as Edith and felt that his brother was being dominated and exploited by David. (The fact that Osbert seemed quite cheerful about the situation was neither here nor there.) It was, he felt, a hideous predicament, but all they could do was to accept the situation as painlessly as possible.

Thanks to those courtly manners of the Sitwells the fraught voyage to America passed off without a hitch. The stay was in some ways a success as well. Edith enjoyed fresh acclamations at her readings, for her public

* These were big decorated flower-pots which David had had specially made by a local potter as a gift for Osbert.

presence now was at its most impressive. 'The recitals', as she reported proudly afterwards to the poet Charles Causley, 'were a huge success, and I had "standing audiences" everywhere. (The Americans don't cheer: they get up and stand, to honour one.)'

Osbert seems to have borne David's temporary absence now with equanimity. He made the rounds of all his New York friends. There were fresh expeditions with Maloney and de Witt Harrison, and, as an additional excitement, his friend James Fosburgh had begun to paint his portrait. (The original, depicting Osbert as a grey-haired, rather ghostly figure with a walking-stick, hangs today at Renishaw, but one gets a picture of a more insouciant Osbert in the sketch that Dan Maloney did of him at one of Edith's readings.) The doctors were encouraging as well. They still advised against an operation, but were ready to prescribe a new drug which might well reduce his tremor quite significantly.

Osbert was optimistic, eating and drinking well and miraculously enjoying life. The one who seemed to suffer most, despite her continuing success in public, was Edith. Much as she adored the acclamation, the readings were an inevitable strain upon a highly strung and highly temperamental poetess of nearly seventy in doubtful health. And as so often in the past, the role of the assured celebrity was masking an agonising medley of self-doubt and private agony. Apart from all her troubles over Osbert, she was feeling oppressed about her work. It was, as she was painfully aware, a full three years since she had written any poetry, apart from her short elegy for Dylan Thomas. The anthology continued as an endless burden, and she was also overdue with the latest prose book she was writing – her interminable study of the lives and times of Queen Elizabeth I and Mary Stuart which she entitled *The Queens and the Hive*. It seemed impossible to finish – 'the more I write, the less complete it seems', she lamented.

These problems with her work would not have mattered quite so much had she been solvent, but by the spring of 1957 her overdraft at Coutts Bank had reached five thousand pounds. Nor was this all. During the previous years of her success and heavy earnings in America, she had consistently ignored the one grim follower no author can evade – the tax man. He was now clamouring – not violently yet, but already with enough insistence to arouse those nightmare fears which still lay dormant from the days of Lady Ida's bankruptcy. She was already in a state of mortal terror over money.

The more she worried, the less she ate and, all too predictably, the more she drank. The falls at Montegufoni might have warned her of the danger to a person in her state of health, and it did not take too much to break that constant self-control she needed in her relations with Osbert over David. Those 'outbursts' on that forbidden subject, which Miss Andrade had seen at Renishaw, seem to have increased, and there was certainly one major scene with Osbert now when Osbert said things which, as she admitted later, had hurt her 'most terribly'. Whatever Osbert actually said, it made her realise that if she went on like this he would not tolerate her and she would certainly be left abandoned and alone, with no one but herself to blame.

This was an important lesson which explains how the potentially explosive situation managed to continue much as usual throughout 1957. David returned to Osbert, and the trio, superficially devoted and united, sailed back to Britain aboard David's favourite ship, the luxury S.S. *United States*, at the end of April. On their arrival at the house at Carlyle Square Miss Andrade recorded in her diary, 'OS looks, as he usually does, remarkably well, but the tremor has now reached his mouth.' The facial twitch was partly the effect of new drugs he was taking to improve his speech, but his sense of balance was affected and his walk grew still more dangerous. At Renishaw that summer he fell several times, and tiny Miss Andrade has frightful memories of running after Osbert and trying to halt him in the middle of a headlong Parkinsonian run.

As for Edith, she was plainly doing everything she could to be discreet and keep her feelings and her troubles to herself. She went to Weston for a few days at the end of May, but even then she seems to have been terrified that Osbert might begin to be suspicious. Somewhat desperately she wrote to Father Caraman on 4 June, 'Osbert must on *no* account know that anything is being discussed anywhere. It might easily lead to my losing him altogether and I simply could not bear it. He has been a wonderful brother to me.' She even added, towards the end of this letter, that she was beginning to 'understand' the situation as she had not previously. 'It's more than platitudinous to say that there are two sides to every question but there *is* something to be said on both sides.'

This new attempt at tolerant acceptance coincided with fresh bereavements which must have left her lonelier than ever. She had been in New York holding a press conference after one of her readings when 'an oaf of a reporter' blurted out the news that Roy Campbell had died. She managed to control her feelings, but she was terribly upset.

'He was one of my dearest friends and was, as you know, one of my godfathers,' she told Father Caraman, and added sadly, 'When I think of that noble creature and great poet being dead, it is too much.'

She had seen him – however optimistically – as her own faithful champion, a strong right arm forever ready to chastise the impertinent and smite her enemies. No one could replace him.

Then on 1 August she suffered a still more grievous loss. A telegram arrived from Rome. 'Dearest Edith, This message brings tragic news but also our deepest love. Last evening at eight Pavlik's great heart beat no more love again dear friend Choura Charlie.' Worn out and ill and full of rage, Tchelitchew had died at fifty-eight. All she could do was make a legend of his memory – and this she duly did, beginning with his *Times* obituary in which she recorded her devotion to 'that tragic, haunted, and noble artist – one of the most generous human beings I have ever known', as she described him later. But nothing she could write could truthfully describe the agonies and disappointments he had caused her, nor could it lay the ghost of the man whose 'power of living' was, as she wrote, 'so great that I cannot believe I shall never see him again'.

Later it transpired that he had left her a painting in his will – the 'Rite of Spring', one of his most sombre, most foetal pictures, much in the style of that other painting, 'Hide and Seek', which she had so disastrously failed to appreciate at the Museum of Modern Art in 1948. When it arrived at Renishaw she asked Maynard Hollingworth what he thought of it.

'Well, Dame Edith,' he replied, 'I think if I had it I would use it to cover up my bee-hives in the winter.'

'Very well, then,' she replied. 'I shall leave it to you in *my* will, so that you can.' Fortunately for the bees, but unfortunately for the Hollingworths, she never did. It went to Reresby at her death and hangs today at Renishaw.

Now that Osbert's face and diction were beginning to be affected it was all too evident that he was entering the secondary stage of Parkinson's Disease. This is the stage at which the majority of sufferers begin the miserable retreat into themselves, becoming more and more dependent and retiring and losing hope in everything. This Osbert resolutely declined to do. Against considerable odds he had hung on to David. He had compelled the family to continue to accept the situation, and he had kept his independence. Now, as full summer came to Renishaw, he made it plain that he was making no concessions to his illness. He had fallen badly in the bathroom and bruised his face, but he refused to let this worry him, and in the second week of August, with the beginning of the agricultural show, Osbert insisted on entertaining the customary crop of house guests for the occasion. There were L. P. Hartley, Malcolm Bullock, Lucien Freud, the painter, and the Duchess of Devonshire. But for Osbert the star guest that summer was undoubtedly Evelyn Waugh. Cyril Connolly had somewhat spitefully maintained that Waugh had a snobbish admiration for the Sitwells, and Waugh had certainly picked up from Osbert something of the gentlemanly art of suing enemies – and making on the deal. That summer it had brought him considerable success, and Osbert wrote delightedly to Lincoln Kirstein,

This year he has won £5,000 in two libel actions and is stated to have said about them that the results were the direct answer to prayer; that he had been to a convent and asked the abbess and her nuns to pray for the success of the enterprise and that he would give them 10% of any money he might win. He not only won £5,000, but in addition at the outset a very great enemy of his was killed in a motor accident.

Another source of regular amusement to Osbert now was a wealthy widow called Mrs Alice Hunt, who was in love with him and had set her heart on marrying him. Throughout that summer she was still in passionate pursuit, bombarding him with gifts. He described the unlikely climax to the courtship in another note to Lincoln Kirstein.

I received a letter from her saying, 'I have left you my head in my will,' and I was on the point of writing to her in reply, asking for the name of the best Mexican head-shrinker, when I suddenly realised that she was referring to the head of Christ painted by Graham Sutherland and not her own. But the idea of it still haunts me – that fine old Gainsborough mug unpacked from a parcel labelled THIS SIDE UP WITH CARE would be a shock to anyone – and then there are the cats to think of.

Autumn came, and it was time for Italy and work again as it had been for nearly forty years. Nothing could be allowed to change the accepted pattern of Osbert's life. This year, in addition to Miss Andrade, he decided he would take Miss Noble, the housekeeper from Carlyle Square, and the entry in Miss Andrade's diary for the day that she arrived shows Osbert still enjoying life. 'DH met me at Florence and we went to the Uffizi to pick up Osbert and Miss Noble. Lunched at restaurant high over Florence, then OS took us all to see San Miniato, the Cascine, and the old Medici Farmacia.'

It was a cloudless autumn, and throughout November they were still dining out each evening on the terrace. Now more than ever, Miss Andrade was entranced with the uncanny beauty of the place – the gentle valley with the shimmering olives, the scent of the plumbago in the air, and then the way the stars and fireflies came out together.

According to what Edith wrote now in a note to Father Caraman, the weather and surroundings suited Osbert. He was, 'if it's possible to say so, a *tiny* bit better. He walks better but varies from day to day. It's very terrible. He bears it with the most wonderful patience and goodness.' Sacheverell, who had seen him just before he left England, gave a similar picture of his brother in a further bulletin to Nancy Cunard. The latest drugs were having their effect. He was feeling better and walking more steadily now, but Sacheverell wondered pessimistically just how long the improvement would last. This was, of course, the dreadful question. By changing treatment there was always the possibility of these brief improvements, but then the disease began to creep insidiously back and it was time to find some further tactic to defeat it.

But none of this appeared to worry Osbert overmuch. He was still working on his final reminiscences about Sir George, and correcting proofs of his poems for *On the Continent*, with their wonderfully urbane and loving portraits of characters from that long-vanished Anglo-Florentine society, most of whom existed in his memory alone. Umberto, the gardener, *Il Capitano*, the bewhiskered fencing master, and Archdeacon Sawnygrass and his 'flannelled, Prussian' spouse, who played the harmonium and hated Latin. There were also portraits of his more famous friends – Orioli the bookseller, Reggie Turner (he called him Algernon Braithwaite in the poem) with his 'ugly sallow face' collapsing into 'a thousand wrinkles' when he laughed, and, most memorable of all, Norman Douglas (given the alias of Donald McDougall) the 'complete realisation, spiritually, Of that ideal of ancient China, The Old Scamp'. There were also several very short poems which he wrote describing his impressions of the castle – the tower at night against the stars, the sunlight quivering inside the grotto, the wooden sound of shutters being opened in the morning. They are not great poetry, but when one remembers how they were composed they become, like the whole book, strangely moving.

Osbert's real achievement is that he resolutely stops one feeling sorry for him. Edith, on the other hand, for all her fame and passionately pursued success, had reached a stage of life which all but her direst enemies would have pitied had they really known her situation. It was very odd. Outwardly it seemed that she had gained the very summit of success. She was a great celebrity. For her seventieth birthday in September 1957 the *Sunday Times* arranged a special luncheon in her honour, and published a full-page symposium in which a group of her distinguished friends – including Lehmann, Plomer, Frederick Ashton, Raymond Mortimer – hymned her praises and descanted on her fame. Early in 1958 a greater honour was accorded her, when she was elected one of three new vice-presidents of the Royal Society of Literature, along with Maugham and Winston Churchill. The B.B.C. was avidly pursuing her for readings of her poems.

Ted Weeks [she wrote to Father Caraman] is trying to drive me into the lunatic asylum and my poor secretary is hovering on the verge. I have asked him if he can to confine himself to writing me *one* letter plaguing me per day.

I am now not going to America at all in March, but shall probably go in October. If I saw Ted I think my mind might collapse completely. Amongst other things he says that I mustn't 'put in too many personal friends'. What he doesn't seem to realise is that as I am a poet, naturally all the top English poets *are* friends of mine. It stands to reason.

Elizabeth and Mary Stuart were demanding ladies, and the readings that she gave at the Royal Festival Hall and then at Oxford were crammed to overflowing.

But the gap was widening between this unique public image and the private self that painfully inhabited it. 'I was born extremely arrogant and vain,' she confessed to Father Caraman, 'but now, although I have not conquered my arrogance, I have I think entirely conquered my vanity, and do not care *what* lower grade mental defectives think about my poetry.' The very way she wrote this showed how untrue it was. She still cared passionately. Almost from childhood she had been building and maintaining her façade out of her poetry and her public presence as a poet. She had refined and stylised her own appearance to accord with it. She had subordinated and immersed her private self within it. She had publicised it, fought for it, sought allies for it, and bitterly defended it. And now in her seventies this public vision of herself was almost all she had. With growing bitterness she found that it was not enough.

Perhaps the greatest contrast now was with Sacheverell. Edith had always criticised his shyness, and the way that he refused to push and publicise his work. To some extent she had been right. 'Shyness', as he says, 'has always been the absolute bane of my life'; and, partly because of this, the extraordinary range and originality of his literary achievement had undoubtedly been undervalued by the critics and the reading public.

But with his temperament he had also been extremely wise to keep clear of the battles and the personal publicity for which he had so little taste. The private life that he had led was rich and varied and his marriage was particularly happy. His books had never made a fortune: indeed, he admits to having been pursued by 'quite appalling financial worries' for most of his working life. But these had not prevented him and Georgia from building up their home and garden and a circle of devoted friends.

Nor had these worries ever stopped them travelling, and his energies and curiosity had grown with the years. 'Surely no one else alive', writes Kenneth Clark, 'has seen and heard so many beautiful things in so many places, from Paris to Lagos,' and in 1959 Sacheverell published a travel book based on his longest journey. Thanks to financial help from Bryher, he and Georgia had visited Japan, and the experience had been second only to his discovery of Italy in his youth. Its gardens and theatres, art and aesthetes appealed to him immensely, making his reminiscences of the country in *The Bridge of the Brocade Sash* a true enthusiast's account of the art and artists of Japan.

But in the same year he had also published the first volume of his most ambitious prose work yet, which he entitled, echoing Céline, *Journey to the Ends of Time*; and here, as nowhere else, Sacheverell reveals that sense of private anguish which he had always shared with Edith, and which lay beneath the polished surface of his life. 'The Sitwells', as Cyril Connolly wrote, 'were not really happy people', and this book on which its author laboured for seven years shows that those early mental wounds left by his mother's trial and the First World War had never really healed.

He based his book on an unfulfilled idea of Berlioz's in which a mock Judgement Day staged by Antichrist was interrupted by the real thing, and there is nothing else in modern literature quite like this non-believer's meditation on the loneliness and horror of the human condition. C. P. Snow has written of Sacheverell that 'Under the surface of connoisseurship, he is wildly romantic in the most desperate fashion', and nothing could be more desperate than this literary nightmare so lovingly constructed from the macabre images of a lifetime's reading and acquaintance with art. 'In the middle decade of the twentieth century,' he wrote, 'the news is always bad and more often than not, terrifying and alarming.' But whereas he could temper his alarm by the many consolations of the life he led, Edith had few such palliatives. She had no real home of her own, no children, few genuine close friends – only a restricted private life which she had to share with Osbert and a man she hated.

At the same time, the public Edith was increasingly demanding of that 'worn-out electric hare' which she had now become. And, like an old electric hare, the private Edith wearily obeyed.

The public Edith also remorselessly insisted on her duty to appear in public, however silly the occasion and disruptive the experience might be. In July 1958, on top of all her interviews, appearances and public readings, she let herself be

> dragged off to Manchester to be televised, being asked questions by schoolboys and schoolgirls aged sixteen and seventeen. They were positively terrifying in their imbecility, poor children, and two boys would have liked to have been extremely impertinent. One *was*. He asked, 'Do you think your poems are going to last?' I replied, If they were not going to last, I shouldn't have written them, but I am sure you are not meaning to be impertinent. This threw the radio reporter of the *Daily Herald*, who is obviously suffering from acute hysteria and an attack of class hatred into an absolute frenzy. He produced a long diatribe against me under the heading, in large letters, 'You must apologise, Dame Edith!' Really.

But, as always, the most gruelling demand the public Edith made upon herself was to defend her reputation as a poet, for she emphatically *did* care what that whole gang of 'lower grade mental defectives', the hostile critics, dared to impute about her work.

Some of her friends realised the danger of her appalling sensitivity. 'You *musn't*, MUSTN'T let the silly little pipsqueaks who review poetry in the papers make the slightest difference to your writing poetry,' wrote Geoffrey Gorer. But part of the trouble of her outsize public ego was its sheer vulnerability. Her poetry and fame were all she had to take the place of the lovers, children and home life she had never known, and since the beginning of the war her reputation and the fear that she undoubtedly inspired among the middle-aged critics of the time had guaranteed her almost absolute immunity from criticism. For all her *Angst* about them, the older generation of her enemies like Grigson, Leavis and sick old Wyndham Lewis had been remarkably restrained throughout this period, while almost all the thirties poets had become her allies or refused to get embroiled with her. ('Edith was really such a dreadful bully, and so *very* powerful', Connolly sighed wearily.)

It was impossible for such immunity to last for ever. Poets pass out of fashion, and a new generation of younger poets and critics was emerging for whom Edith Sitwell was not the untouchable *monstre sacré* she had been for so long. The publication of her complete *Collected Poems* at the end of 1957 had been the inevitable signal for a new and irreverent evaluation of her work.

Cyril Connolly's review in the *Sunday Times* compared her favourably and at length with Eliot and Yeats, but in the *Observer* there was a dissenting voice. While at Oxford in the early fifties, the young critic Al Alvarez had become increasingly annoyed by what he felt to be the Edith Sitwell cult there led by Maurice Bowra. As for her poetry, he thought 'her talent had been evident in *Façade* but that since then her pretensions as a poet had become so overblown as to be faintly ridiculous.' And in his *Observer* piece he said so. Several other younger poets followed suit.

Edith was congenitally incapable of seeing this as part of the natural process of evaluation and rejection which she herself had boldly practised in her time, and which is essential if fresh poetry is ever to emerge. She had been battling all her life to get her poetry accepted, and now that she had won her honours she could not let the battle lapse. Any criticism of her work was unacceptable and was seen as she had always seen it – as some sort of lower-class impertinence against her regal presence. Indeed, she identified so absolutely with her poetry that she still could not see any difference between criticism of her work and insolence towards her person. And insolence required retaliation.

'John Wain', she told John Lehmann, 'ought to be skinned alive. It would be a nasty messy job, but I am not sure I shan't do it.' Then it was Thom Gunn's turn for treatment.

Mr Thom Gunn was most impertinent to me the other day in a review of somebody or other. But I 'prefer', like Dr Johnson, 'to avert my attention and contemplate Tom Thumb', (a giant in comparison). Mr Thumb or Gunn seems to think great poetry consists of understatement. It doesn't. It 'speaks always somewhat above a mortal mouth' as Ben Jonson said. And if Mr Gunn is ever pert to me again, I am going to make of him the mincemeat (or thick soup) that I made of Mr Alvarez.

The mincemeat-making of Alvarez was performed by Edith in a retaliatory review of his first book, *The Shaping Spirit*, in the *Sunday Times*. As she had boasted once to Hedda Hopper, Edith had 'never found the slightest difficulty in getting the Press to allow me space for anything I have to say'. But her hostility no longer had the same effect as in her prime. Alvarez shrugged his shoulders philosophically. 'With enemies like Edith Sitwell, who needs friends?' he told himself.

Betrayal and treachery were in the air. Even old friends were not immune. Her friendship with Stephen Spender ended now because in his magazine *Encounter* he had allowed

a young gentleman named Levine, of whom I had never heard previously, to say about the later poems of my 'Collected Poems' that 'my language lacks any sense of immediacy' and that I have to rely on my reading rather than on the small acts of experience that gave the immediacy to my earlier poems.

She ended this letter to John Lehmann by saying that Spender had been 'very treacherous' and that she would never see him again, despite the fact that she was devoted to his wife, Natasha. What was so nightmarish for Edith was that, however hard she tried, there was no possible escape from her predicament. Even in Italy that autumn there was no respite from the constant call to arms.

Her great anthology finally appeared, and harvested its automatic crop of criticism. One critic in a Sunday newspaper complained that, whereas she had allocated Robert Burns three pages and Browning only four, Osbert and Sacheverell had fourteen between them. 'Really,' she exploded to Father Caraman, 'what odious malice! Not a mention of all the wonders! I have sent the review to the solicitor of the Society of Authors, but do not expect he can do anything.' The American reviewers were even more impertinent. Early in January 1959 Edith was writing once again to Father Caraman of how she had just given two of them 'a good slap'.

I have just finished writing about the second to the American Review of Literature. I was reviewed by a gentleman called Winfield Townley Scott. I said that as a humanitarian, I am glad that Mr Scott is not dead, I mean more dead than usual. I had thought he was. But I am also an eccentric. I prefer to be taught my job by someone who is in a position to do so. I end my diatribe thus. 'My anthology stands somewhere near, but not after, all the best in the field.

The Auden–Pearson anthology (I said) is a great one, admired by all poets, but good heavens, why has Mr Scott omitted to mention Little Annie's School Treat Reciter?'

By the way, who *is* Winfield Townley Scott?

On this level one can perhaps admire the gameness with which she was managing to head off her 'enemies'. But while she could still achieve some flashes of her ancient wit, her real situation was far bleaker than a letter such as this suggests. This was no easy badinage, no light, amusing riposte to a stuffy critic, but part of her earnest struggle for survival as a poet. And simultaneously there was an even grimmer struggle which assailed her while she was still in Italy. At the end of January 1959 she was confiding in Father Caraman that she had just had

a complete collapse due to nothing but nervous exhaustion, work and the ceaseless and merciless demands of one of my hangers-on. For over a week I could eat nothing but mashed potatoes and couldn't sleep. However, I am now recovering, and am reminding myself that the old lady in question is in many ways a very heroic person.

This 'old lady' was the dependent Evelyn Wiel, who was still in Paris, still hard up, and still badgering Edith for money. In payment of the obligation which she felt she owed the memory of her sister, Helen Rootham, Edith had continued all these years to help support her, however tiresome she was, and had always faithfully entertained her at the Sesame during her occasional trips to London. The payments in themselves were not excessive, but they had served to aggravate the desperate state of Edith's own finances, and the true cause of Edith's panic was not old Evelyn Wiel at all but a far more potent and disturbing figure who was now entering her life in earnest.

Two things in life are certain, death and the tax man, and of the two the tax man suddenly appeared by far the more unpleasant. A few weeks after complaining of her exhaustion to Father Caraman, Edith was writing of more upsetting troubles, this time to her new friend, the novelist Elizabeth Salter, who was acting as her part-time assistant and secretary back in London.

I have been very ill indeed, though better today. For three days I couldn't keep any food down and got no sleep at all. The nights were an absolute nightmare of retching and a kind of mental horror. This was simply brought on by the Income Tax badgering me to send them money I have not got.

Like Lady Ida, Edith had no money sense at all, and her tax liability was one more burden which the public Edith had incurred. For almost all the money she had earned had gone on what she had thought appropriate expenditure for a major literary figure – those lavish luncheons and enormous dinners at the Sesame, the travelling in style, and above all the entertaining of her followers and friends and docile critics which was essential for the promotion of her art.

Much of this could then have been tax-deductible, but she had kept no accounts, and she dreaded making herself more of a burden to Osbert by telling him her troubles. One of the few she did confide in was Elizabeth Salter, and in a memorable passage in her book on Edith she describes how she gradually discovered that she was really living on two levels – one the extravagant public life, and behind it her panic-stricken private world ruled by the horrifying fear of debt.

As always in this private world of hers, there was no defence, no one to turn to, no one who could help. Behind that powerful facade of what Wyndham Lewis himself once called 'Edith, brave as a lion', there lurked the private, very simple woman tortured by the anguish which she always said she had inherited from her father. This was the self for whom, as Miss Salter says, that 'growing debt under pressure from the Inland Revenue became a monster blown up into frightening proportions by the memory of her mother's disgrace'.

And, as Miss Salter had by now discovered, Edith's 'means of retreat were obvious. The brandy taken at night to calm her *angoisse*, the strong martinis beforehand and the liberal quantities of white wine. . . . Unfortunately, these were only temporary palliatives; in the long run both sickness and despair were accentuated.'

There were times when the private Edith could still persuade herself to laugh at the antics of her public self, as she made quite clear in the letter to John Lehmann from Florence at the beginning of March 1959, while sleepless nights and money-terrors were still afflicting her. That indomitable Sitwell fan, Ian Greenlees, was back in Florence at the head of the British Institute, and had persuaded Edith to give a reading of her poetry at that worthy institution. As always, Edith rose to the occasion, complete with bangles, head-dress, golden torque around her neck and aquamarines flashing on her fingers. This was the voice and public presence which had earned those standing ovations across the length and breadth of the United States: the magic worked as well as ever, and the reading was a great success. Edith described what happened afterwards.

When it was over a lot of people came onto the platform, the former Dutch ambassador, a Dean, the 'art editor of the *Daily Express*' and his protégée, a nice, pretty little goose who had been painted by that awful Annigoni, etc.

On galloped an enormous young man, and sucking down my hand into a hand like a swamp in Florida and gluing his face to mine, he enquired, 'Do you know who *I* am?' I said, coldly, that I didn't. Whereupon, he said some name I had never heard, Hannikin or Hennikin, and said, '*My father told me to ask you if you remember sitting on his knee*'!!!!

There was a deathly silence. I drew myself up and absolutely *glared* at him. The ambassador tried to look as if he hadn't heard. The Dean examined his gaiters. The pretty little goose stopped rolling her eyes, and said, 'Oo – Dame Edith!' Two young girls from Indo-China stopped tittering politely.

The silence continued. I continued to glare. The young man added as an afterthought, 'As a *child*'.

I said, 'I never sat on anybody's knee as a child, and I have never heard of your father', and turned my back. But it was a nine days' wonder in Florence.

Unfortunately it was becoming difficult for Edith to treat many slights to her reputation as a laughing matter, and from now on her public presence drove her unmercifully into battle upon demanding battle in a sort of rearguard action to defend her work. And inevitably she was the one who suffered most. Indeed, there was something almost punch-drunk in the way she seemed compelled to keep on coming back to hand out punishment to her multiple traducers. This went on through 1959, and there is now a glum monotony about her outbursts, for Edith's public reputation had become so exigent that she was even up in arms when Frank Kermode, writing quite innocently about some jazz and poetry experiment in *Encounter*, happened to mention that he thought the original of the idea had come from San Francisco. 'It did nothing of the kind,' riposted Edith. 'It derives from Sir William Walton's and my joint work, 'Façade', which was performed in public at the Aeolian Hall of the 12th June 1923, *36 years ago*. It has been performed multitudinous times since.'

It was now, in the very middle of her dog-fights and her money-worries, that the public Edith was presented with an unexpected opportunity to act out her greatest role before the largest audience of her life. The distinguished B.B.C. producer Hugh Burnett approached her to be interviewed on the programme 'Face to Face' by the grandest of inquisitors, John Freeman.

She performed superbly. Here in full public gaze was the complete façade which it had taken her a lifetime to perfect – the appearance which so many painters had attempted in their time, the dress she had carefully evolved, the manner and assurance and the wit with which she had learned to mask her inner terrors. She never faltered even when she talked about her childhood and her parents, never once allowed herself to lose her splendid dignity and regal bearing, never for one moment let the faintest doubt appear that this was indeed a great and an accomplished artist at the zenith of a life's achievement. It was an unforgettable appearance which ended as a sort of act of homage, both by the interviewer and by the nation. (Viewer research revealed later that it had been the most popular and memorable of the whole famous series.)

But, as always, the private Edith had to pay the price of her celebrity. She was prostrated by the effort and excitement and spent the next two weeks in bed recovering from the aftermath of her success. She had the inevitable shower of correspondence. Most of it was favourable, but she could not dismiss the few unpleasant letters she received. 'I had a most horrible letter yesterday,' she told Father Caraman, 'a letter of abuse brought down upon me by my television interview.'

Her 'lunatics and bores' were activated by the sudden sight of her, and there were fresh demands to cope with. 'Poor worn-out electric hare that I am,' she lamented to Miss Salter. 'I have been pursued relentlessly by greyhounds in the shape of television producers and journalists.' In the old days she could have coped with them, but now she felt defenceless and exposed before the millions who had seen her 'Face to Face'. She ate less than ever, and drank consequently more, and even the down-to-earth Miss Salter was alarmed to see that all the pressures of her life were giving her 'an alarming mixture of megalomania and inferiority complex'.

Yet even now the monstrous presence of her public image would not let her be. At times it even seemed to be taking charge of her, with quite disastrous results. For instance, early that September it led her to accept an invitation once again to read her poetry in public, this time at the Edinburgh Festival. This was a near-catastrophe. From the beginning there was trouble with her microphone, and some of the audience shouted (it would seem with reason) that they could not hear. This upset her, and she broke off her reading, 'wagged a heavily bejewelled finger at her audience, and told them – "Get a hearing aid. I am not going to shout with my voice."' Compton Mackenzie, who was acting as her chairman, found her another microphone, and the reading went ahead. But it had shaken Edith badly, and when she was interviewed shortly afterwards by Robert Muller for the London *Daily Mail* the two opposed sides of her personality both appeared. The public 'megalomaniac' Edith spoke first. 'And if I didn't write better than any woman alive I'd *shoot* myself!' she told Muller briskly, adding, 'I write in the rhythms of Sappho, though I do not have that lady's unfortunate disposition.'

That was the Edith people had almost grown to expect, but as the interview wore on the private Edith started to appear, bewailing the abuses she had suffered for her public self.

'I have had such a terrible, such an *appalling* life, with everybody clinging on to one and badgering one out of one's wits. . . . – I'd like to live just a little longer.' The interview then ended on the most pathetic note of all, with Edith begging Muller, 'You will be kind to me, won't you? Because people hate me so very much. . . .'

Edith's sense of vulnerability grew worse. After the Edinburgh fiasco she wrote to Father Caraman with her woes.

I have now, in addition to migraine, got the worst attack of sciatica I ever had, brought on I suppose by fatigue and a scurrilous attack on me in a musical comedy and by an anonymous letter telling me I am common and vulgar celà se voit, so as I cannot help it the writer should pity not blame. Most people hate me because of this and because I wear vulgar rings that the lowest barmaid would scorn to wear, that I am a third-rate poet and read poetry aloud horribly, and because I am a fake, a fraud, and (of course you will naturally know what we are coming to) a Roman Catholic.

She was coming now to doubt herself. She lacked the necessary resilience for these public shindigs, and was severely shaken by these fringe expressions of hostility which her exposure to the millions had stirred up. Alan Pryce-Jones remembers despondent sessions with her at the Sesame at which she would tell him that she knew her poetry was worthless and her life a failure. Nor did her Catholicism provide the strength or refuge that she needed. Father Caraman was still a valued friend, and Evelyn Waugh continued to observe his religious duties to his god-daughter. In August 1959, on the anniversary of her entry into the Church, he wrote to wish her every happiness in God. 'I shall never forget', he wrote, 'that sunny morning which gave August 4th a new significance – Of course don't answer. Just slip my name in with your list in your prayers – as yours is always in mine.' But now that she no longer needed an authoritative religion to subdue her rage, the zeal of her earlier devotion had subsided. As she told Jack Lindsay, apropos of Sacheverell's agnosticism, 'I oughtn't to approve as a Christian, but then I'm not a good one. I don't go to church often.'

By the late autumn of 1959 it was obvious that she was near collapse, and the inevitable accident occurred at the Sesame at 1 a.m. on 6 November. Edith herself described what happened in a letter to Father Caraman.

I had a very bad accident, because the Club *would not* mend my bed, though told millions of times that they must. It threw me out with my face against the iron of the bed opposite and I was picked up unconscious. Why I was not blinded, nobody knows. An ex-army nurse, staying in the club, a catholic was so kind. She moved into the room next to me and looked in at me once an hour for two nights. . . .

I am not telling anyone of the accident, but am saying it was influenza because the papers are such a pest and badger the life out of me.

Edith's decline appeared inevitable, for she was really so alone, and no one could provide defence against the regiment of enemies and lunatics and hidden evil-wishers and treacherous friends which she believed was constantly around her. According to Miss Salter, she was living on a diet of smoked salmon and champagne laced with milk and brandy. During the day it seemed that she could cope, but her nights had now become a time of terror, and in December 1959, when she was staying with Osbert in

Montegufoni, she had another bad fall. She described it once again to Father Caraman. Her bedside light had suddenly gone out.

Ever since my accident in London I have been terrified of being left in complete darkness. Like a fool I tried to find the window to let in some starlight, couldn't, and trying to get back to bed, I caught hold of what I thought was my bed but in fact was an extremely flimsy chair which collapsed, hurling me onto my back on the stone floor. I lay there in agony from 3 a.m. to about 8 when I was called. I have been laid up now because I could not put my feet to the floor. What absolute agony for six weeks! I got up yesterday, but not to leave my room, because of the stone stairs. What a bore!

She finally recovered, as she always seemed to. 'She was really strong as a horse, for all her sensitivity,' says Miss Andrade. But her sensitivity plagued her still, and now, on top of lunatics and journalists and anonymous letter-writers, there was another hazard – noise. After these falls she had moved to a new bedroom with a new bed when staying in London at the Sesame, and then, inevitably, building work began in the house next door. The noise was maddening. The enemy was closing in again – but Edith was not submitting even now without a fight. First she consulted the Noise Abatement Society, and when they told her there was little they could do she telephoned direct to Scotland Yard.

'If you don't tell the workmen to stop,' she informed the duty-sergeant in her most Dame-like tones, 'I shall have to go round there myself at eight o'clock on Monday morning and slap each one of their faces in turn.'

'And in that case, Madam,' said the voice of Scotland Yard, 'we shall be forced to restrain you.'

As with most of Edith's battles now, this had its funny side, but once again the private Edith suffered. The noise and tension in her life soon brought on further complications – an infection of the middle ear which left her without a sense of balance. Then in July there was kidney trouble. She was admitted to the Claremont Nursing Home. She was very ill indeed, but by the early autumn of 1960 she was sufficiently recovered once again to make the journey out to Italy with Osbert and David. There was now a real problem – who would look after her? Someone would obviously have to.

CHAPTER TWENTY-EIGHT

'Something Macabre out of Proust'

1961–1963

I am very cross because I am old
And my tales are told
And my flames jewel-cold.

'The Sleeping Beauty'

EDITH WAS now rising seventy-four and a potential burden to her family, her friends, and to herself. But, as always, the essential problem was to reconcile her private and her public selves.

The private Edith was a sick and rather frightened human being. She could now barely talk. Noise caused her agony. She scarcely ate, and she was tormented by her constant worries – about Osbert, about the unseen legion of her imaginary 'enemies', and above all about money.

But there was still the public Edith, and she was extremely difficult to cope with. For she was touchier than ever and still insisting upon her queenly dignities – her own room at the Sesame, her luncheon parties, and all the trappings of her role as the unrivalled *grande dame* of English literature.

The public Edith was apt to get bored with the private Edith. Even at Renishaw, in peace and penury, she had insisted on her trips to London, as well as her stunts and performances in public as an antidote to the tedium of private life. It was impractical for her to live full-time with Osbert any more. His condition had deteriorated over the last twelve months, and Renishaw was not the place to house two invalids. But, on the other hand, the course her doctors were suggesting – that 'for her own good' she went to live in some sort of genteel home for nice old ladies – would hardly have suited her.

Throughout her adult life Edith had always been fortunate in finding someone practical to provide the essential background to her life. First Helen Rootham, then Evelyn Wiel; and then Osbert had taken over, during and since the war. And now she found another person she could lean on – her new but increasingly devoted secretary, Elizabeth Salter.

An Australian from Sydney in her middle thirties, Miss Salter was practical, courageous and completely down to earth. An author in her own right (she writes excellent detective stories), she genuinely admired Edith's poetry, and liked her as a person. In all her dealings with her she had the great advantage of being an outsider to the London social and literary scene that had always meant so much to Edith. Thus she could be more or less immune to the snobberies and feuds and sycophany that had so long cocooned Edith in her dealings with her followers and friends. Edith valued her directness, and now, in this crisis, it was Miss Salter who alone could offer a solution. Edith took a flat near her own, from where Miss Salter could keep an eye on her, prevent her from getting lonely, and do all she could to help her live the life she wanted. It proved an admirable arrangement.

Edith's flat, at 42 Greenhill, Hampstead, was hardly Renishaw, but it was modern, comfortable and close enough to Central London for all the social life and contact with her friends that Edith wanted. She had her room, her books, her cats, and here at last the private Edith could relax and find the peace she needed.

Nor did Miss Salter's usefulness stop there. She already knew of Edith's money-worries and decided it was time that they were dealt with. Edith had always shied away from them in mortal dread. Miss Salter, on the contrary, faced them squarely, and acting for Edith marched off to Coutts Bank to discover just how bad the situation was. It was not encouraging. The overdraft had risen to £5,960, and the bank was muttering about repayment. But at Coutts Bank Miss Salter soon discovered that rarest and most admirable of mortal men – a bank manager who sympathised with writers. His name was Mr Musk, and Mr Musk became an ally. He was even introduced to Edith, and his advice and very human presence went some way to reassure her that she was not about to follow Lady Ida into Holloway for bankruptcy.

But there was still a very real need for capital to pay her debts. Sacheverell had none, and Edith still adamantly refused to 'badger' Osbert. True, she had her royalties, but they were not great, and there was still the tax man waiting to be paid. It was now that Miss Salter had a brainwave.

She had heard Edith talking several times about her life in Paris – the tiny flat in the Rue Saint-Dominique, the paintings and the sketches Tchelitchew had made of her – and how, when the war came, she had been obliged to leave behind all her possessions in the flat – her sticks of furniture, her manuscripts and notebooks, and, naturally, her Tchelitchews as well. She had never bothered to retrieve them. She was not even certain what was there. Old Evelyn Wiel might well have thrown them out, and Edith seems to have purposely dismissed from her mind these souvenirs of such a time of suffering.

But practical Miss Salter was intrigued. The value of a Tchelitchew was rising, and Americans were beginning to pay good prices for literary manuscripts. Not quite certain whether she was on a treasure hunt or a wild-goose chase, she drove off to Paris in her old green Morris Minor, and reached the Rue Saint-Dominique.

Evelyn Wiel was none too pleased at the intrusion, but Miss Salter is a forceful lady. The door of Edith's old room was unlocked, and there, beneath the dust of more than twenty years, was a treasure-trove of Sitwelliana – notebooks and manuscripts of early poems, letters from long-dead friends, and more than a dozen sketches and paintings of Edith by 'The Boyar'. Evelyn Wiel complained: Miss Salter was unstoppable. She filled the Morris and that same evening drove back to London with her booty. Next day Mr Musk was more than reassured when he heard that Edith's overdraft was now backed by something solid, although it would obviously take some time to sort and catalogue the mass of papers, and then to find a buyer.

While Edith's situation had thus magically improved, Osbert's was slowly worsening. The drugs were losing their effectiveness. His tremor had grown worse. His speech was now affected, and his face had taken on the mask-like aspect which is one of the more distressing advanced Parkinsonian symptoms. He, more than Edith now, required careful help and nursing and, ironically, D. H. Lawrence's picture of him as the wheelchair-bound Clifford Chatterley of Wragby Hall was coming true. David was still with him, and one of the incidental benefits Miss Salter bestowed upon the family was to relieve the constant tension which had inevitably existed when Edith had had to live with them at Renishaw. Relations had even been restored – superficially, at least – between Edith and David now.

But for Osbert much the most important thing that summer was that there seemed to be a chance at last of being, if not cured, at least wonderfully relieved of the most distressing symptoms of his illness. Two years earlier the devoted Bryher had been writing to him of a neurosurgeon, Professor Krayenbhül in Switzerland, who practised a brain operation for sufferers of Parkinson's Disease and was obtaining a high level of success. According to her letter, some ninety per cent were cured, and the remaining ten per cent were 'helped'.

In fact, she was being over-optimistic in her figures, but techniques of operating on the brain were being developed, and over half of the sufferers treated were being substantially improved. The only trouble was that, by its nature, it was a frightening and occasionally dangerous operation. Under local anaesthetic a small hole was drilled in the skull above the ear, and an electric probe then inserted into the brain to cauterise the small but vital site which controlled the tremor and rigidity of the limbs. An extremely high degree of surgical precision was required, and the surgeon had to have the intelligent co-operation of the patient, who remained completely conscious through the operation. Though the brain itself is insensitive to pain, the whole operation, and the knowledge of what was being attempted, called for considerable courage on the part of the patient.

Osbert consulted Dr Desmond Laurence, who was then working on a drug treatment of the disease. He put him in touch with a neurosurgeon, who explained the risks and principles involved. He told him that there was a sixty to seventy per cent chance of substantial improvement. Failing this cure, there would probably be no change in his condition, but there was also a small risk of the operation damaging the brain and even causing serious paralysis.

It was an unpleasant choice for anyone to have to take, and Osbert took some while before deciding. But his slow deterioration settled things, and at the end of September 1961 he entered the National Hospital for Nervous Diseases in Queen Square to await his fate. The operation was performed by Lawrence Walsh on 28 September 1961. For a writer who had always lived by the sharpness of his brain, the idea of having the seat of one's intelligence probed and interfered with must have been particularly abhorrent, and especially for a life-long hypochondriac like Osbert. In fact, he underwent the operation with unusual calmness and good humour.

The technique involved treating the two sides of the body separately, and on this occasion the surgeon was concerned with the right-hand side. Unlike most major operations, there was no need for a long convalescence, and within three days Osbert was back in Carlyle Square, and then went off to stay for a few more days at Weston. And at the beginning of October Edith was writing exultantly to David Horner: 'What *wonderful* news about Osbert's right hand. I feel quite stunned by it. I hardly dare say it, but now, I suppose, with *any* luck, the dear old boy will be able to write again! . . . I really bless Dr Laurence!'

When they set off together in the autumn of 1961 for their customary winter stay at Montegufoni, Edith and Osbert must have both been thinking that the nightmare of their last few years was lifting. It would take some time yet before anybody could be sure that the improvement wrought by the operation would continue. But even the chance of writing once again must have seemed miraculous. And so must Edith's liberation from the tax man, and her overdraft. Sotheby's by now had catalogued the manuscripts and drawings: if they fetched even half of what was predicted, all Edith's money troubles were behind her.

As if to underline this sudden state of grace, it was a wonderful Italian autumn, and the great house appeared an earthly paradise after the storms and tribulations of the last few months. Life there had never seemed more perfect – warm golden mornings when Osbert could sit out beneath his favourite oleander correcting the fresh proofs of *Tales My Father Taught Me*, afternoons when he and David could be driven off to visit Lucca or Siena, and evenings when they all dined out upon the terrace, with the paving-stones still giving off the faint heat of the day.

Unlike Renishaw, now dogged with servant problems, Montegufoni must have seemed a haven of simplicity. Thanks to Luigi, everything appeared to work with calm efficiency. The wine was immeasurably improved, the vineyards had yielded a record crop in the *vendemmia* that year, and Luigi's wife had proved to be a splendid cook. Osbert had always loved the food of Tuscany, which is at its best in autumn; and he particularly enjoyed *fonduto*, a dish which Norman Douglas introduced him to some thirty years before, a sort of cheese *fondue* concocted out of local ewes'-milk cheese and carbide-smelling white Italian truffles. He even had a valet for the winter, a young Sardinian called Paletta, a tactful, gentle person who made Osbert's life much easier. Miss Andrade was also there to help him with the articles he planned to write. With harmony at last restored between Edith and David, everybody in the castle could look forward to the winter with more hope than they had had for years.

But like the cracks that were now appearing in the plaster of Sir George's tower, the strains among the visitors began to show, and when the weather broke in the second week of December the atmosphere within the house began to change dramatically for the worse. Beneath their assumed civility towards one another, Edith and David clearly loathed each other still. Edith was still drinking, although she managed to avoid the falls and the disasters of the previous years. Luigi – normally the sturdiest of men – seemed out of sorts. Miss Andrade was beginning to go deaf. David was clearly very bored.

But the worst affliction once again was Osbert's, for the improvement following the operation had failed to sustain itself. The tremor in his right arm worsened once again, and with his growing disappointment the atmosphere within the castle soured. It was a gloomy Christmas. Rain came, and then the *tramontana* blew, gusting the cold, wet, mountain winds along the valley, until the castle seemed more bare and inhospitable than Renishaw in the very depth of winter. January brought *il gran' freddo*, the sudden snap of bitter cold which turned the green valley of the Pesa into an icy wilderness. The cold made Osbert's tremor worse, and he gave up all idea of working on another book. Luigi too was plainly ill by now, and in the village there was talk of the evil eye in the *castello*. Some of the peasants whispered that the strange Sardinian who looked after *Il Barone* must be a *jettatore*. Nothing but bad could come with such a man about.

As usual there was one person who seemed immune to the disaster in the air – David Horner. People who saw him now remarked that there was something just a shade uncanny in his perpetual youthfulness. The worse Osbert's health became, the better his appeared to be, and with the disappointment of the operation the contrast in their ages and appearance had never seemed so great. Osbert was now so bowed that the impression he had always given of height and dignity was all but gone. His walk was tenuous. The facial twitch was worse. His speech was often difficult to hear.

But David had scarcely changed from the *jeune premier* Rex Whistler had sketched before the war. He was sixty-one, but he still appeared to be in the very prime of life. And then, at the beginning of March 1962, whatever evil fate was hovering around the castle struck at him as well.

During the winter months, when it was not possible to sit out on the terrace after dinner, it was the custom at the castle for the guests to have their coffee in the long, blue-grey and white *salone* which looks out on the neatly planted walks and hedges of the Cardinal's Garden. They can hardly have been particularly sociable affairs, these after-dinner gatherings in murky evenings at the end of winter. It was a fine room with its chandeliers and mirrors, but it was really made for summer evenings when its rococo elegance would come as a relief from the summer heat outside. Now it seemed merely icy and, as usual in the best Italian rooms, there was no real attempt at heating – only an ancient *art nouveau* iron stove, which Luigi fed with coke, but which, like much else within the house, had not been functioning too well of late.

There was a big brown radio – of almost the same era as the stove – and Osbert liked to listen to the nine o'clock news from London every night. Sometimes the radio obliged, and when they had heard the news this strange quartet – Osbert, David, Edith and Miss Andrade – would sit on a little longer. Osbert talked less and less these days, so there were none of those celebrated anecdotes that used to rumble on and on and keep the conversation going. Miss Andrade's hearing troubled her as well, and there would be long gaps when no one spoke, and when the only sound would be the noise of Edith's knitting. At ten o'clock sharp the Sardinian would enter and silently wheel Osbert off to bed. Edith would follow, then Miss Andrade, and David would be left alone.

He usually read and drank for an hour or so before going off to bed himself. He drank a lot, usually his favourite Italian brandy, for he was a man who, as Americans would say, 'used' alcohol. The evening of 7 March was no exception, and some time around midnight he closed his book and drained his glass and then went off to bed.

He says that he is mystified by just what happened then. His room was on the first floor, down a short corridor from the *salone*. But for some reason he appeared to lose his way, and wandered, none too steadily, into another corridor that led towards the kitchens at the back of the house. Italians are notoriously stingy with electric light, and he cannot have realised that he had reached the small tiled landing at the head of a flight of ancient steps. There were thirty-eight of them leading to the basement. Not realising they were there, he went forward in the dark.

> I just don't know what happened then – or *why* it happened. I knew the house so well, and I wasn't drunk – certainly not drunk enough to tumble down a flight of steps. I still believe somebody – or something – *made* me fall, and I had a definite sensation of being pushed. I'd always known the house was haunted, and as I fell forward in the dark I had a definite sensation of something pushing firmly from behind.

Such was David's own recollection of how his evening ended, and it is not surprising that he remembers little more. The steps were very steep, the edges sharp. By the time he landed at the bottom, he had irreparably smashed his right arm, broken several ribs and cracked his skull. For the remainder of the night he lay in his own blood at the bottom of the stairs. He was not unconscious all the time. Occasionally he would emerge from his coma, but he could neither move nor speak, much less shout for help. Not that a shout would have been much use, since this part of the castle was completely uninhabited.

Luigi found him when he made his rounds at seven-thirty in the morning. The ambulance was called from Montespertoli, and within the hour David Horner, – what was left of him – was in the *ospedale ortopedico* in Florence. He was in a coma and none of the doctors seemed to think he would survive.

Italy being Italy, there were inevitably rumours in the village. One of those crazy foreigners at the *castello* had been settling old scores, possibly that villainous Sardinian with the evil eye. If he was not involved, then it was certainly a ghost. Everybody knew the place was haunted. So many dreadful things had happened at Montegufoni over the centuries that it was not surprising in the least.

The peasants shrugged as peasants do – and since it was only the *forestiere* who had suffered, and the *barone* was known to be extremely rich, the matter was allowed to rest without an investigation or an official call from the *carabinieri*. But the whole incident confirmed what everybody knew – that the *castello* was a *casa maladetta*. It was just tempting fate to live in such a place.

Miss Andrade stayed with David at the hospital – there was no one else who could. Osbert was naturally anxious, but his illness seemed to give him a curious detachment from events, and he was too sick anyhow to stay for long at David's bedside. Edith seemed torn between a certain pity and the grim certainty that this must be the hand of God. And so Miss Andrade waited on discreetly at the *ospedale*.

The surgeons still remained most pessimistic because of the coma. They talked of damage to his brain, damage so serious that if he did survive he would be bound to be an invalid for life.

But David did survive. That ever-youthful constitution somehow pulled him through, and gradually the coma lifted; his head was bandaged, and the doctors did the best they could with the smashed right arm. He finally began to speak again, but this was a painful process as his battered brain reversed the meaning of the words so that he spoke in contraries, saying 'yes' when he meant 'no', and so on. But he possessed a great will to live, and by the end of June he had recovered sufficiently to be flown back to London. Osbert soon followed to be near him.

From then on David's recovery continued. By late summer he was painfully beginning to walk again – and learning to write left-handed. But there were things that will-power and expensive doctors could not cure. His brain had slowed down. Before the accident he had been virtually bilingual in French and English. Now he could not speak a word of French. His speech and his sense of balance were affected, his right arm was useless, and he knew that he would never walk with ease again.

From now on it would be a moot point who was the more afflicted – Osbert or himself.

This suddenly created endless problems both at Renishaw and Carlyle Square. It had been bad enough attempting to run a household with one invalid. With two it was all but impossible. Both needed looking after, but suitable servants and nurses were impossible to find, and like some sick, short-tempered, ancient married couple they began to get on one another's nerves, a situation made no easier by the inconvenience of the house in Carlyle Square. It was a warren of small rooms and flights of stairs and totally unsuited to two invalids who had difficulty walking.

This settled one decision they had all but taken even before they left for Italy. The Cadogan Estate, who owned the lease, were being difficult about renewing, but offered the freehold for £20,000. Osbert decided not to purchase it, which left him until the spring of 1963 to find somewhere else to live. About this time he made another serious decision. It was still quite feasible to try the operation on the other side of his brain, and if successful this would at least relieve the tremor on the left of his body, but the risk of real deterioration now was that much greater if the operation failed. Reluctantly he decided that he could not take it. This meant that he was virtually abandoning all hope of being cured. His only hope was if some new drug or new technique were developed that could help him, but he accepted that this was most improbable.

Life appeared very grim, and his writing career was all but at an end, although he still went through the motions of observing the habits of a lifetime by forcing himself to 'work' three hours every morning in his study, but he had already completed the last new piece of writing to be published in his lifetime – an article on 'New York in the Twenties' which the *Atlantic Monthly* had published in February 1962. It was an essay of nostalgia and a picture of his favourite city which now existed only in his memory. His other fragments of memoir would remain unpublished at his death.

Superficially, at any rate, Edith's situation was more enviable than Osbert's. The flat in Greenhill suited her, as much as any mere flat ever could. She was surrounded by her books, her cats, her faithful friends, although some of them were finding the journey out to Hampstead too long. 'There is a Greenhill far away,' quipped old Sir Malcolm Bullock. Miss Salter had all but taken charge of her, and really ran her life. She was afflicted with a multitude of troubles, as she had almost always been – her eyes began to cause her pain, her back was troubling her, writing was agony. But she was still in touch with all her allies, and still entertained close friends like Father Caraman and the Clarks and the C. P. Snows to luncheon at the Sesame.

The tax man ceased to trouble her, since the sale at Sotheby's had brought her £15,000 for her manuscripts alone. (They went to the Humanities Research Center at Austin, Texas.) The Tchelitchews achieved a further £8,000, but it particularly pleased her vanity to think that Americans would pay so much for her manuscripts.

'My early manuscripts fetched me £15,000, which wasn't so dusty', she crowed to Lehmann.

But even now the public Edith would not comfortably relax: the sale had brought aspersions that annoyed Dame Edith.

Sir [she wrote to the editor of the London *Evening Standard*],

Your offensive reference to me in your issue of last night would lead the uninstructed reader to believe that I am selling certain manuscripts and pictures because otherwise I shall be reduced to sleeping on the Embankment.

You are entirely outside the Literary world, and do not know that many writers of my eminence are in the invariable habit of selling certain objects when they have become so numerous that there is no longer any room for them in one's house or flat. . . .

I must ask you to publish my letter. And you'd better apologise to me and that without any covert insolence.

It was the usual form her irritation took when she was bored. Impertinence was in the air, and the public Edith simply was not getting the applause and reverence she demanded as her due. The private Edith suffered.

Then in July 1962 the situation got considerably worse. There was another fall.

I should have written to you before [she told John Lehmann], but have been in the most dreadful pain, having slipped a disc, torn all the muscles in my back, and sprained an ankle – and torn the ligaments in that foot. I am accident prone, because I spent my childhood in irons, which has atrophied the muscles. I am in bed and must stay there until the doctor brings round a specialist.

She was, she added, absolutely 'furious' to miss the luncheon she had planned that same week at the Sesame.

But by August she was well enough to be interviewed by Kenneth Allsop, who came to her flat prepared to write a celebratory article upon a revered and famous literary figure; but instead of talking of her achievements, Edith chose to bring up her sufferings from years before, even in childhood.

My father did not value me as a daughter – he was furious that I was a girl. . . . My mother was hell. She was very beautiful and had the worst temper I have found in anyone.

Allsop was slightly puzzled, for Edith was, as he put it, 'at the summit of an elegantly full and productive life' and could look back with satisfaction on 'a life of apparent high privilege, in which she has known the most eminent and gifted of her time'. And yet she insisted on talking of her 'great grief and great pain', and ended sombrely: 'But if you ask me if I have had a happy life I must say no – I have had an extremely unhappy life.'

But even so, the public Edith should have been satisfied and gratified in the autumn of 1962 when her seventy-fifth birthday was celebrated with a triumph and a public splendour such as few poets gather in their lifetimes.

Her book on Elizabeth and Mary Stuart, *The Queens and the Hive*, as well as her final book of poems, *The Outcasts*, had been published at the end of August, and *Fanfare for Elizabeth* had been specially reprinted. That autumn she also made her private peace with one of the most unspeakable of her famous demons from the past – Noël Coward. He visited her at Greenhill, and they got on famously.

But her most impressive birthday present was the gala demonstration which her nephew Francis organised at the Royal Festival Hall on 9 October. A crammed and devoted audience heard a performance of *Façade*, and the thunderous reception of the trio made this a sort of ultimate apotheosis of the Sitwells and their work, a demonstration, if a demonstration were required, that despite the struggles and the blows and impertinences of the philistine they had won through to absolute success.

'A fine life', Vigny wrote, 'is a thought conceived in youth and realised in maturity,' and by this sort of definition Edith's life had been an enviable success. For she had laid the ghost of her appalling childhood, lived down the scandal of her mother's crime, beaten the philistine, written the poetry that formed her long life's work, and now with her fame and reputation she was receiving the applause that should have meant so much.

She and Osbert had both wanted it so badly in their time and here at last it was, to crown those years of battle – a cheering audience, a programme in their honour, distinguished admirers and friends around them on the platform. Irene Worth and Sebastian Shaw read *Façade* and afterwards Edith regally acknowledged the applause. There in her aquamarines and her extraordinary hat, the public Edith might have felt that hers had been a life well spent.

Ironically, she did not think so and her verdict on what should have been the proudest evening of her life was a bitter, almost a despairing one. She called it her 'Memorial Service'. It had been, she added, 'rather like something macabre out of Proust'.

She was quite right, although one might have hoped that she had not realised exactly how macabre it was.

To start with, there had been a painful effort to get her to the hall at all, with doctors, sedatives and wheel-chairs all required to play their part. Without Elizabeth Salter, she would not have managed it. Even with Miss Salter there it was touch and go getting Edith aboard the ambulance that brought her to the hall from Hampstead. Only her resolution not to disappoint her audience seems to have brought her through.

Afterwards there came a birthday dinner which, according to Lord Clark, was 'a terrible ordeal for Edith, who was quite bemused by then with the whole affair. We simply had to get her drunk to carry her through it, and I will never forget the appalling strain of that dinner party with

Willie Walton on one side of her and me on the other, and both of us being forced to shout across her to make conversation.'

This was all but the last appearance of the public Edith. After the cheering it was the private Edith's turn to carry on. She relapsed totally after the excitement and it was a couple of weeks before she could face anyone again (although three weeks later a bemused Edith did just manage to stagger through the ordeal of 'This is Your Life' on television). It was not merely that the evening had exhausted her, although it had. But the truth was that, far from giving her the expected sense of triumph, this public crowning of her whole career had done the opposite, and she felt only despondency and bitterness with life. It was as if this final public recognition of her fame had convinced her that eminence was not worth the price that she had paid for it. She, after all, had sacrificed her life for this, lived by and for her poetry, and passionately nurtured and defended her unique public presence. Her life and poetry, she always felt, were one: the fame that she received she claimed in her priestess role for poetry.

But now, as Miss Salter says, she was all too grimly conscious of 'the battle that her life had been and the scars that it had left'.

By the beginning of 1963 Edith had slipped into a state of almost permanent depression: her lungs were troubling her now and a virus illness showed signs of turning to pneumonia. Miss Salter often found that in the mornings 'her speech would be muzzy, her eyes owlish', and her doctors said that if she was going to survive the winter it was imperative to get her out of London – and preferably out of England too. Never one for doing things by halves, Edith decided there and then that she would go on a cruise – around the world.

The journey was dramatically disastrous. Miss Salter and Edith's nurse, Sister Farquhar, travelled with her and all went reasonably well until they reached Australia. To pass her mornings, Edith had now begun working on the memoirs she had toyed with writing since the twenties. She enjoyed the comfort of her first-class stateroom in the S.S. *Arcadia*, and at Sydney even managed a successful conference with the local press – her last. Soon after this, news arrived from London of her appointment as a Companion of Literature.

But this was not sufficient to sustain her spirits or her health; and her return aboard a less luxurious Dutch ship, the *Willem Ruys*, became a nightmare. She had been hoping friends would meet her at Miami, but when they failed to materialise Edith became upset, distressed and then apparently desperately ill. By the time the boat was due to sail she was weak and haemorrhaging and calling for a priest. The local doctor she had seen was not excessively alarmed at her condition, but by the time the

Willem Ruys had reached Bermuda it seemed that Edith was unquestionably *in extremis*, and she was swung dramatically ashore in a stretcher and raced to hospital, followed by her two distracted ladies, wondering if they would arrive in time to find her still alive.

They had reckoned without Edith's powers of survival. Despite their fears, Edith's haemorrhage subsided, and was correctly diagnosed as being caused by nothing more than a varicose vein condition at the base of her throat; she was soon considered fit to be flown back to England, along with Sister Farquhar. Faithful Miss Salter had the task of following her back by sea, along with fifteen pieces of luggage, and five days later, when she got back to Hampstead, Edith was safely in her flat. 'I found her pale. She was tired, but she was alive.'

But this was not enough for Edith. Her travels had exhausted her and in the months that followed her boredom and her bitterness grew worse. One can hear this bitterness in her voice in the last book she wrote – or, to be accurate, dictated – from her bed during the last few months she spent at Greenhill before her final illness. Even the title – *Taken Care Of* – seems to suggest the anger that the woman in her felt at the way her life had gone, and these final memoirs are the last thing anybody would expect from that treasured mother-figure of English poetry which the young had been applauding at the Royal Festival Hall that autumn.

Parts of the book are very funny, and parts are scrappy and unfair – and ancient anecdotes are told that had been better told before. But as she dictated it, each morning sitting up in bed and knowing she had not too long to live, she made it a sort of balance-sheet of life. None of her enemies had been forgotten, and her energetic anger still resounds against those old enemies like Wyndham Lewis and D. H. Lawrence who had incurred her wrath. But the real anger, one suspects, was not for them at all, but with herself and with the price that the façade of her extraordinary life had claimed from the very private woman who inhabited it. One of the things that most tormented her was the thought that, for all the praise and the success, she would die and never know what true physical love had really meant.

CHAPTER TWENTY-NINE

End Game

1963–1969

DESPITE THE pessimism of his doctors – and despite the appalling injuries of twelve months earlier – David did recover. He could now shuffle round the house. He could soon talk without slurring words. His brain, though still a little slow, was operating perfectly. He was beginning to write now with his left hand, and it seemed as if his chief affliction was quite simply irritation with his lot. He, who had always been so free, so youthful, so desired, and who had always managed to escape from Osbert when he wanted to, was suddenly shackled to the house in Carlyle Square. Predictably, his temper suffered: so did Osbert's.

It would be wrong to over-emphasise the rows that followed. They had always bickered, but they had also been together now for thirty-three years and were inevitably part and parcel of each other's life. The underlying pattern of these years persisted: Osbert was still the adoring one, David the adored.

In November 1962 they had a brief – and probably a welcome – break from one another when Osbert visited Sacheverell and Georgia at Weston: but absence soon made Osbert pine. No sooner had he arrived at Weston than he was scribbling a reassuring letter to his 'Dearest David' in his most agonising scrawl.

How are you? . . . I am better, thank Heaven.

I lead a quiet life here, with Sachie angelic in helping me. I long to see you on Tuesday. I wrote to Miss Noble asking her to find dinner for me, Tuesday or next Wednesday night.

Best love, ever dearest,
OSBERT

P.S. My watch has gone on strike. Such a warning.

Miss Noble, the housekeeper at Carlyle Square, was a loyal old lady, who was having problems feeding and coping with her two demanding invalids. So was Miss Andrade, and the succession of young males nurses who were hired – and as regularly found unsatisfactory and fired – to perform those intimate and daily tasks which Osbert was becoming incompetent to do himself.

It was a trying situation now for everyone involved, especially since everybody knew the days in Carlyle Square were numbered: indeed, these final days do form a gloomy postscript to those historic years when this large, handsome house on the King's Road corner of the square had once enshrined so much that was stylish and sparkling in the artistic life of London. Now everything had worn away to sadness and irritation as these two ancient lovers went on unhappily together. Presumably their lives would have continued thus but for a curious quirk of fate.

Frank Magro must have appeared a most unlikely candidate to break the stalemate at Carlyle Square. He was unmarried, thirty-seven, eager to be of help and rather self-effacing – a dapper, dark-haired, dark-eyed man with a deceptively gentle manner. Although of Maltese birth, he liked to regard himself as 'essentially true British' and was anxious to improve himself and lead an interesting life. So far he had not had much success. He had served several years in both the army and the R.A.F. and worked as a clerk for Thomas Cook's. This bored him, but he failed to see how he would ever lead the life he wanted. Despite this, he had started 1963 with hope. Not long before he had had his watch psychometrised by William Redmond at the headquarters of the Spiritualist Association in London – and had been assured that life would soon be looking up, with an extraordinary change of fortune round the corner. Buoyed up by this, and anxious now to give the fates their chance, Frank Magro visited the Forces Employment Bureau for a change of job.

The interview was not encouraging. No, he explained, he was not interested in industry, nor in insurance, nor, for that matter, in a position in a bank. Surely there must be something else?

The retired major behind the desk grunted.

'Well, there's an author needs a servant.'

'Really,' replied Frank Magro. 'Tell me more.'

'Titled gentleman called Sitwell. *Sir* Osbert Sitwell. Hardly your line, I would have thought.'

Frank Magro disagreed. During the *longueurs* of his air force service in Singapore he had read a lot. He had even read the Sitwells. Perhaps – who knows – this was the chance that William Redmond had foreseen.

A few days later, bright-eyed and in his best blue suit, Frank Magro rang the bell at Carlyle Square. He was interviewed by Osbert, who explained the nature of his duties. He was ill and he required a servant to look after him. He would pay seven pounds a week and offered him the post without so much as bothering to see his references, with a month's trial either way. Frank Magro made his mind up on the spot.

> On the 28th of January, 1963 [he entered in his diary], I started working for him. It was a miserable day and snow lay all over London. Sir Osbert was in bed with a cold, and my first job was to help him to get better. This wasn't easy at first, as I was suffering from a sense of awe, which was only partially overcome by my instant liking for Sir Osbert. For years I had secretly wished to work for an author, even a third-rate one, and here I was, working for none other than Sir Osbert, who had already received the Companionship of Honour from Her Majesty the Queen for his services to English literature.*

The arrival of Frank Magro almost coincided with the removal from Carlyle Square. The lease had ended with the financial year, and Miss Andrade, practical as ever, had discovered somewhere suitable for her employer to live – a second-floor apartment in York House, a florid-looking block of late-Edwardian flats off Church Street, Kensington.

David had hated it on sight. He had originally suggested they should find themselves a house in Pimlico. York House was highly inconvenient. After Carlyle Square it would be a quite abysmal come-down. Others agreed with him: one press report announced that Osbert Sitwell, 'who has lived at No. 2, Carlyle Square since 1919, is to move within a few days to a small and soulless London flat'. In fact, the York House flat was hardly the cramped garret this suggests – nor, for that matter, were its new tenants suffering much hardship from the move. There were eight good-sized rooms in the 'small soulless flat', and while Miss Andrade and the removal men got the place ready, Osbert and David pigged it at the Ritz.

But the move did bring one major difference to their lives. During the weeks at Carlyle Square Frank Magro had lived out. Now, the question of one month's trial forgotten, Osbert suggested he should move in. Although Frank Magro knew that this would naturally entail more work, he cheerfully agreed.

Do the lives of artists imitate their art? Osbert might well have wondered. For the situation that developed now might easily have come from one of his own wry, blackly cynical short stories from the past, as the whole balance in the relationship with David suddenly began to shift.

David still thoroughly disliked the flat: he also disliked the new housekeeper, an emotional Hungarian lady, who now replaced Miss Noble. Above all, he disliked Frank Magro.

* Osbert had just received the C.H. in the New Year's Honours list.

Previously, whenever David disapproved of any of Osbert's servants, the servants had departed – but not Frank Magro. Osbert would smile blandly at David's furious complaints: Frank Magro, blander still, ignored them. This continued during May and June with the situation getting more tense all the time. But throughout it all Frank did his duty, washing and shaving Osbert, helping him to bed and helping him to dress. And imperceptibly he was becoming something more to Osbert than a servant. Osbert would talk to him and seemed delighted with his company. Soon they were sharing jokes together, jokes from which David quite understandably felt excluded. By that summer Frank had already come a long way from the awestruck former clerk from Cook's who had arrived that January day, and was in the process of becoming indispensable to Osbert.

For David this development seemed threatening – and faintly sinister. But from Frank's point of view the situation was quite simple. Here, in the person of this ailing but distinguished writer, he could see somebody who needed him. After the R.A.F. and Thomas Cook's he was now being treated as a human being: and more than this, in Osbert he perceived something of a father figure. Osbert was grateful for his help, and charming now as only he could be. Frank became totally devoted in return.

David's worst mistake was his failure to appreciate the sudden weakness in his position caused by Frank Magro's coming and the advance of Osbert's illness. All Osbert now required was a devoted friend who would look after him. He had enjoyed this briefly in the company of Dan Maloney during his trips to New York and to Florida: now with Frank Magro he had this permanently. As a direct result, for the first time in the whole relationship with David, Osbert had suddenly become the stronger of the two, David the odd man out.

True, he and Osbert had the shared memories of their successful, often brilliant life together. The places they had visited, the objects they had bought, the tastes and friendships they enjoyed must have made it seem as if he and Osbert would go on united until one or other of them died. They had everything in common.

But this is not a sentimental story. Osbert and David had both changed with illness, and Osbert's new dependence was becoming greater than his ancient passion. His tender memories of David counted for little now against the fact that it was Frank who took such care to find him the food he liked, Frank who would help him walk, Frank who would take him to the bathroom if he called him in the night.

It took the months of spring and early summer for the first implications of these changes to start working out. But more and more the disagreements in this uneasy triangle seemed to take the form of Frank and Osbert in discreet alliance against a beleaguered David. One crucial row occurred in May. Osbert had been invited to a performance of the opera based upon Edith's *The English Eccentrics*. David had brusquely said he could not bear to go; and on past form Osbert would certainly have acquiesced himself. This time, however, he just nodded and said Frank would take him – which he did. The evening was a great success. Another source of trouble was the television set which Frank introduced to York House for himself. David insisted it was vulgar and refused to watch. Osbert admitted enjoying it, and there were further upsets when Osbert started viewing it in Frank Magro's room.

Lincoln Kirstein who visited York House about this time remembers the atmosphere as 'highly comic if it hadn't all been so very sad'.

The comic elements must have intensified that June when the unlikely trio duly decamped for Renishaw. Gone were the days when Osbert could boast that this household was minutely geared to his most demanding whim as artist and as master. The peace and that eccentric brand of self-indulgence he enjoyed had gone. There were no early peaches in the hot-house, the servants were suspicious of Frank Magro's status, and for these hideous weeks of summer the rows and accusations echoed and rechoed through the great haunted barracks of a house. By the end of August everyone had had more than enough and Osbert had arrived at a decision. He and David clearly had to part – for a while, at least. With Frank to help him, he could spend more time in Montegufoni and David would stay on in London at the flat.

It was more a trial separation than a final break. They were unquestionably still fond of one another, and all Osbert's old possessions from Carlyle Square remained in York House. He still regarded this as his London home, still wrote affectionately to David, and certainly intended to come back after the winter. But it was a crucial step, the first time they had ever parted for so long at Osbert's wish. Certainly Frank Magro had no doubts about the importance of what was happening.

On the 15th of September, 1963 [he recorded in his diary], Sir Osbert and I began our journey to Montegufoni, where we arrived some 24 hours later, having travelled overland via Calais, Paris and Bologna. The journey, I am glad to say, was uneventful, although it did prove somewhat difficult for me, having to look after Sir Osbert at every stage of the journey, and looking after the luggage at the same time.

On reaching the castello on a very hot September day, we were faced with a row of the household personnel, half smiling and half crying.

Thirty-eight years earlier, Sir George and Lady Ida had made the identical journey when they set out from Renishaw to settle here. Osbert had then been glad to see them go. Now in much stranger circumstances he was following them. Whether he would stay was quite another matter. When he had previously been here with David and with Edith, there had been disasters, tragedies on such a scale that it had almost seemed as if the house had turned against them. After David's accident the whole place seemed so doomed that he had thought of selling it.

Now, in his illness, he was retreating here alone with Frank. He was no Berenson, and this was no *Villa I Tatti*, that custom-built and antiseptic factory of culture with its perfect gardens, its library and its splendid pictures still lovingly maintained since B.B.'s death up on its hill at Settignano. More than ever now, Montegufoni must have seemed an isolated, cavernous, great rambling ruin of a place. Luigi had died of cancer; without him its genius had gone. But in that blazing autumn weather the emptiness and inaccessibility of the enormous house must have appealed to Osbert. Its age and its decrepitude both somehow matched his own: and for the first time in his life he had time now to enjoy its splendours. The battles and catastrophes were over. Now in old age he was finally alone there to enjoy the crumbling stone terraces, the echoing rooms, the secret gardens. Here he was *Il Barone*. No one would contradict him or embarrass him about his frightful illness. Here he could live with dignity until he died. It was his own extraordinary kingdom, and it suited him.

It suited Frank as well, which was fortunate, as Osbert could not have survived a day without him now. And Frank was revealing unexpected qualities. He was essentially a loner, so that the prospect of continuing here with Osbert did not trouble him. He was practical. He spoke fluent Italian. He could type admirably, and possessed the makings of the perfect secretary. Above all, he was desperate to learn and to improve himself. Osbert had always enjoyed the role of teacher.

And so the autumn months and then the winter were by no means the disaster people had predicted. Nor, to judge from the few letters that he wrote, was Osbert any longer missing David in the way he always had before.

One vital person in his life who could no longer trouble him was Edith. Her angry life was drawing to a close. At seventy-six she was extremely ill again. Her kidney trouble had recurred, and he and Frank had visited her in a nursing home just a few days before they left England. She had appeared so old and worn that she was timeless now, a gentler effigy of all that she had been. She was incapable of writing. Her violence had turned to a wan acceptance of her fate. She had had a full-time nurse, Sister Farquhar, for a long time, for Elizabeth Salter could not cope with her alone; and in the spring of 1964, after Edith had left the nursing home,

Miss Salter found her a little house in Keats Grove, Hampstead, which she named 'Bryher' in gratitude to her millionairess benefactor. There she could go to die in peace. Sister Farquhar tended her devotedly.

Osbert, meanwhile, was enjoying life – thanks to Frank Magro. Just before Christmas they had made an unlikely expedition to Venice. Osbert had more or less resigned himself to never seeing his favourite of all cities again, but Frank insisted they should go. They went in style in a chauffeur-driven Lancia and stayed at the Danieli. Frank pushed his master through the winter city in his wheel-chair. They saw the pictures and the sights that Osbert loved. They ate extremely well and they enjoyed themselves. Osbert was duly grateful to Frank Magro – and vice versa.

During the spring of 1964 Osbert was still at Montegufoni and still on nominally affectionate terms with David, who had spent the winter on his own in York House. From the beginning of March to the second week of April Osbert had written him four short letters in his own handwriting. They were all 'Dearest David' letters; and in the last of them he announced that he and Frank would be returning to York House in May.

Despite the upsets of the previous year, Osbert was clearly hoping that something of his former life with David could continue. Almost from the day of his arrival in York House it was obvious that it could not – at any rate, if Frank stayed too. The winter months in Italy had clearly made Osbert more, not less, dependent on Frank Magro; and David was increasingly impatient of the whole relationship. The weeks that followed in the flat seem to have been a time of torture for all concerned.

Since Osbert could no longer manage a knife and fork, Frank had to cut up all his food for him. Since Osbert's cataracts were worse, Frank started reading to him in the evenings. David objected, logically enough, that Frank was no longer what he was hired to be. Frank made it quite clear where his loyalties lay. David began to feel that Frank and Osbert were in league against him. Tempers grew. Tact declined.

The only member of this ghastly triangle who still seemed unaffected by it all was Osbert. That Buddha-like impassivity which Miss Andrade always found so disconcerting never deserted him. Illness had made him more inscrutable, and it is hard to tell exactly why he decided to prolong the agony by taking both Frank and David to Renishaw again that summer. Perhaps he was still hoping that the stalemate would finally work out: perhaps he imagined that the situation would become so insupportable that David would be driven into making the inevitable decision. Certainly the strains appear to have been magnified by Renishaw. Osbert's tremor worsened. Finally, in August, David had some sort of nervous crisis during which he fell, cutting his head badly on a mantelpiece. The fall was serious – especially for an invalid like David – and he had the sense to realise that he must get away. Frank and Osbert brought him back to York House, saw that he was comfortable, and then returned to Renishaw in mid-September.

By now Osbert had made it plain that he would be going back to Italy with Frank again in the winter of 1964–5, but he was still apparently reluctant to make a final break with David. From Renishaw he wrote him a sympathetic letter on 28 September.

Dearest David,
I am so glad the clouds are lifting a little. It must have been heavy going. . . . Life here has been very varied. Arthur Bryant invited himself and his wife over from [illegible] to see the house yesterday. . . . I shall leave for Montegufoni on November 5th, Guy Fawkes Day.

And again, on 4 October, he wrote, 'Dearest David, I hope all goes well? . . . I have no news of any sort. My nephew Francis is here. Very nice.'

Just before Frank and Osbert caught their train to Italy they lunched with Lady Aberconway at her house at 12 North Audley Street. That charming and formidable old lady had recently become concerned for her beloved Osbert. One of his oldest – and worldliest – of friends, she had never had the slightest illusions about his relationship with David. Indeed, she had been one of David's closer friends for years. But recently the rumours she had heard had worried her. She had seen both of them, realised exactly what was going on and, woman of action that she was, decided to do something definite about it.

She was in touch with Sacheverell about the situation, and at the end of that September he wrote to her strongly agreeing with the need for action. Perhaps Osbert could be persuaded to give David up, but then he would be left entirely alone – that was the problem. As for Frank Magro – 'The Maltese', as he called him – he did not care for him, but he admitted that he was an excellent servant and that he did at least look after Osbert.

A few days later Sacheverell wrote to her again. It was, he felt, a frightful situation. If only Osbert *could* be persuaded to leave David – but it was quite impossible and unfortunately legal separations just were not feasible in such bizarre relationships. All that he wanted for Osbert was the chance for him to enjoy 'a few last years of peace'.

Lady Aberconway did not share Sacheverell's hopelessness about the case. Not for nothing was she the daughter of a former head of Scotland Yard. A few days later she was writing to her old friend Tony Gandarillas, 'I think it very sensible that Osbert and David should have separate establishments, as two so greatly afflicted people as they are, are not beneficial to each other.' Having decided this, she must have discussed it at some length with Osbert, for the day after Osbert lunched with her she spoke to Frank Magro on the telephone.

'You will be pleased to know', she said, 'that Sir Osbert will soon be quite all right.'

Frank thought she was referring to some new treatment for his Parkinson's Disease. She was not.

The journey to Florence was quite uneventful. Osbert was rather quiet. Frank put this down to tiredness. They reached the castle in the late afternoon of Guy Fawkes' Day, and it was over dinner that Osbert almost casually announced that Mr Horner would be receiving a letter from his lawyer first thing tomorrow morning. It would contain an official request for him to leave York House as soon as possible and find himself somewhere else to live.

Alone with Frank again in Italy, it was as if Osbert wanted to retrace his past. The previous winter it had been Venice. Now it was Naples and Amalfi. They set off together late that November – again in the chauffeured Lancia – and when they reached the south the sun was waiting for them. They found the Hotel Cappuccini high on the cliffs above Amalfi. Osbert was anxious to show Frank where he had written what was still his favourite book – *Before the Bombardment* – in the white cell by the cloistered garden. After a few days they returned to Naples, and it was here in the Excelsior Hotel that on 12 December 1964 Sacheverell got through to Osbert on the telephone. There was bad news. Poor Edith had passed away the night before in St Thomas's Hospital. It had been a peaceful end. She had had absolution from the Church two days earlier and died peacefully. Only the nurse was with her at the time.

Osbert appeared to take the news quite calmly – as he took everything these days. Frank, who was with him, was surprised at this apparent lack of emotion. The funeral was being held near Weston, at Lois Weedon church, in two days' time. Osbert explained that he was too unwell to face the journey, let alone the service in the church. Sacheverell agreed.

Within the space of a few weeks Osbert had managed to dispose of David and had now lost Edith. But he seemed untouched by both events, almost as if they had belonged to some previous existence – which in a way they had.

In fact, he had not been able to make quite the surgical and clean-cut break it seemed, and the apparent lack of all emotion is one of the disconcerting side-effects of Parkinson's Disease. That Sunday a long, appreciative article on Edith by her old friend Kenneth Clark was published in the *Sunday*

Times. When Frank read the article to Osbert, Osbert wept: and there is no cause to doubt the words that he addressed to Father Caraman not long afterwards.

I must say I find her loss appalling and doubt whether I shall ever get used to it. Although I am not a catholic, I know how much your visits must have meant to Edith. I thought her cottage very pretty though she seemed rather far away up there. Her bedroom seemed to exist in space only. The last time I saw her she seemed well and cheerful. As you know, I too was abroad at the time of her death. Perhaps this was just as well, as I am sure I couldn't have stood the strain had I been in England.

If he felt similar strong qualms about David, he kept them hidden. Perhaps it was the effect of Parkinson's Disease again but there is something faintly chilling in the way his loving indulgence of a lifetime seemed to have changed to absolute indifference.

Not surprisingly, David had refused to vacate the York House flat. Lady Aberconway, with Osbert's full support, tried to persuade him that he should. It would be 'undignified' for him to cause a further squabble. 'Yes,' Osbert wrote to her, 'do instil into David's head that his behaviour is indeed "undignified". But he is so grasping now (or was he not always so?) that he will probably be able to stand any amount of indignity to get what he wants.' David's point of view was given in a letter which Lady Aberconway wrote to Osbert at the end of January. She had just lunched with David, and in her brisk manner tried to get things settled.

I asked him to give me a clear picture of what had happened. His great complaint is that he has never had any letter from you telling him that the time had come to part. He then said that you had been on his nerves for at least five years, and that he quite understood that he was on your nerves. But he said, 'I am terribly hurt that Osbert has not written to me personally.' I said, 'But David, how could he write when he has only got Frank with him? . . . He could not dictate to Frank such an intimate letter as he would wish to have written you.'

'No,' said David, 'but he could have dictated the letter to Miss Andrade before he left England – had he done so, I would have felt quite different.'

However I gather that when all the furniture leaves your flat he intends to get a service flat in a building in Chelsea at £4 a day – in or near Chesham Place.

I told him that I thought his behaviour was most dignified and he said, 'Yes, I suppose it is, but I shall never speak to Osbert again.'

David was far more shattered than she realised, and was in the deepest depression. Philip Frere's wife, Vera, an old friend of both Osbert and David, wrote to Montegufoni in no uncertain terms:

You unspeakable swine, Osbert – only the most unutterable cad will hit a man who is broken in mind and body – and that is what you have done to David whom you called your dearest friend; may God's curse be upon you for the rest of your days – and don't you ever dare set foot in this house again.

With Osbert's full assent Frank Magro sent the letter on to Lady Aberconway, who in turn sent it to Philip Frere, who was most apologetic and emphatic that nothing could affect their friendship. It was finally arranged that David should take over the tenancy of York House on his own account. Osbert's possessions were removed to Renishaw. The two men never met again.

If Osbert thought that he could now calmly contract out of the world that he had known he was mistaken. He was seventy-two and, thanks to the property he owned, a millionaire. His health appeared precarious and, as far as anybody knew, he had made no arrangements to dispose of his possessions when he died. Such situations generally encourage the concern of other members of the family. His was no exception.

One can see the problem; and Osbert, after all those years of bitter testamentary worries with Sir George, obviously saw it too. But it now suited him to keep everybody guessing, just as Sir George had done.

For the family back in England it was intensely worrying. From the reports they heard it seemed that Osbert's health was worsening; and if he went on being obstinate, the Sitwell patrimony, acquired so carefully across the centuries, could be at risk. With swingeing death duties, Renishaw itself might have to go. And then there was a further worry. What plans had Osbert made for Montegufoni when he died?

The question was particularly important to Sacheverell. He had now no claim on Renishaw, as he had sold the reversion to the property after the war. But Montegufoni was quite different. As with Renishaw, he had known and loved it as a child. His father purchased it with Sitwell money, and it would seem right for him and Georgia to end their days there after Osbert.

But now came the mystery. Nobody knew for certain to whom Osbert had left it: and Osbert had not said. More disquieting still, there had long been talk – he hoped it was no more – that Osbert had already left the castle and its contents to David Horner. Once again the practical, concerned old Lady Aberconway was consulted, but even Lady Aberconway was not sure:

> I think [she wrote] that David kept on trying to get Osbert to do this, but that Osbert contrived to postpone carrying this out. Then came David's accident, which of course affected his brain as well as his body, and, after he recovered enough to speak and walk he came to believe that the deed *had* been made.

This was faintly reassuring – but only faintly. Who on earth could tell how Osbert would behave? He was acting very oddly these days. Parkinson's Disease was known to affect the personality, and recently Sacheverell had had a note from David Horner in which he warned that Osbert was becoming 'very feeble-minded' and that 'The Maltese' was beginning to exert far too much influence on him. By February 1965 Sacheverell and Georgia decided that the time had come to visit Osbert, and, if possible, settle things.

Parkinson's Disease has a curious effect upon its victims – reduction to a passive state, and a levelling of activity, which in their turn produce the *appearance* of debility and feeble-mindedness. But in fact this is rarely true. With the progressive blocking of the nervous synapses, the sufferer has difficulty communicating with the outside world, and Osbert was no exception. He stayed owlishly aloof from the anxieties his behaviour was inflicting on his brother and his family. In his own quiet way he plainly relished the whole situation. Crippled, sick, virtually forgotten he might be, but he still possessed that one indubitable source of power – uncertainty about his will. He was in no hurry to relinquish it.

With his brother and his sister-in-law, he was affectionate but not over-welcoming. It was more than a quarter of a century since they had stayed at Montegufoni. Perhaps they should wait a little longer. As Osbert wrote to Lady Aberconway,

> They are to be my guests at a hotel in Florence where I will be staying for the period of their visit, as it would be quite impossible to entertain them at Montegufoni with its acute shortage of servants. I am very happy they are coming, for then they will be able to see how 'feeble-minded' I am.

They came. They stayed as Osbert's guests at the Grand Hotel beside the Arno. Relations were quite amicable, but they left little wiser about the future of the castle, or the contents of its owner's will. The visit gave them something more to worry over now – Frank Magro. He was plainly much more than a mere servant to poor Osbert. Perhaps David Horner's warning could be right.

Meanwhile, in the delicate affair of Osbert's will, the family continued to rely on Lady Aberconway. She felt that there was no time to be lost, and no point in being over-delicate with Osbert now. On 9 February she forwarded him a copy of a note from Sacheverell confirming that estate duty of something over half a million pounds was payable on Renishaw at Osbert's death – 'a lot of money to throw away', as he coolly put it. Surely, she said in tones that brooked of no delay, *something* should be done?

Osbert's reaction was dramatic. How dare the family start getting at him now when they had not been bothering with him for years? He was beginning to feel, he wrote, just like King Lear.

This brought an immediate rap-over-the-knuckles from her ladyship by return of post. 'I am distressed that you are feeling like King Lear. But in character you are not like King Lear, nor are any of the persons who may inherit Renishaw in the least like Cordelia. (She was as stupid as her father. Just a chip off the old block.)' She then reminded him about the loss that the estate would suffer if he did not make some arrangements soon. 'Your father told me that you were tenant for life at Renishaw': as such he had a clear-cut duty to see that the property was handed on as intact as when he had received it. In his reply he dropped the role of the offended Lear. He had, he grudgingly admitted, already started disentailing Renishaw and was prepared to hand over certain capital to his heir, Reresby. On the other hand, he did not feel inclined to give him the agricultural land that he had bought – presumably because it still brought in a useful income. And there, he made it plain, the whole affair would rest until he felt like doing something more about it.

As if to keep everybody guessing – and to show just how little he was worried about his heirs – Osbert chose this moment to sell off some of the treasures he possessed: the fine collection of Venetian books which he had formed through Orioli in the twenties. As he remarked, quite logically, they were too heavy for him to handle now. Rather than leave them uselessly at Renishaw for someone to inherit when he died, he sent them, unregretted, to Sotheby's. As he wrote delightedly after the sale, 'They sold magnificently' – well over £16,000 for the lot. To celebrate, he and Frank went off to Venice once again. On 6 April 1965 he wrote to Lady Aberconway, 'today Frank pushed my chair around the Giudecca, which I found fascinating'.

Then it was time for England. Since he was so determined that his life would still continue as it always had, he planned a period in London, a trip to stay with Lady Aberconway at beautiful, familiar Maenan, and finally, of course, the full seigneurial return to Renishaw for the remainder of the summer months. It was all most courageous in its way, but he was reckoning without the English climate and his depleted state of health. Instead of enjoying London, he caught influenza and the expensive suite that he had booked at the Berkeley Hotel became a sick-room as he struggled with pneumonia. His stay at Maenan was a depressing convalescence – to the accompaniment of good Welsh rain. Then at the climax of this summer of misfortunes he and Frank Magro came to Renishaw. It had never seemed less welcoming. The weather, as he wrote to Lady Aberconway, was 'very invernal, not to say infernal'. Pneumonia had left him feeling particularly weak. And then, on top of that, there was a new and unexpected hazard to contend with – the feelings of the servants.

On previous visits they had seen Frank Magro as the master's manservant: he had looked after him, waited on him at table, and eaten separately. In Italy all this had changed. Frank had become companion, secretary and friend and used to eat his meals with Osbert. In Italy this seemed civilised. At Renishaw it was deemed outrageous. The servants felt the social fabric threatened when they were asked to wait on Frank at table with Sir Osbert. There was an angry ultimatum and they finally walked out *en masse*. Osbert pretended not to notice.

For several weeks he and Frank hung on alone at Renishaw as if nothing had occurred. Frank would arrange the food, then serve Osbert and himself in Sir Sitwell Sitwell's great red dining-room. Osbert's tremor was much worse. Illness and strain made this inevitable, but he still doggedly refused to admit defeat – or recognise the extraordinary behaviour of the servants. It was the sort of situation that Sir George would have appreciated, but it was also most uncomfortable and utterly absurd. Osbert must have seen this too. He was too old and ill for gestures any more. Renishaw had beaten him. There was no point in staying.

He hired a large car – one of the little luxuries he had always loved. Frank packed his bags. He said goodbye to his agent, Peter Hollingworth, old Maynard's son. No, he would not be coming back. The big front doors were locked. The car drove off along the drive and past the Sitwell Arms. Osbert, stony-faced, did not look back.

They drove to Brighton – one of the few remaining spots in England where he could still feel happy and at ease – and there, in the Metropole Hotel, in a room on the top floor facing out to sea, he spent the next few days recovering his equanimity, and carefully considering how he would spend the remainder of his days.

Finally he telephoned his lawyers. His old friend Philip Frere had retired, and his junior partner, Hugo Southern, was now dealing with most of the Sitwell business. Osbert invited him to lunch. It was a somewhat painful meal (Osbert's tremor was so bad that he rarely lunched with strangers now). They talked of Renishaw – and then of death duties. Southern began to state the case for some sort of gradual transfer to Reresby. Osbert cut him short. 'I've decided to give up Renishaw entirely to Reresby. The whole place has become the most frightful bore. I won't be living there again.'

Later that autumn, when the transfer was complete, there was a brief announcement in *The Times*.

Sir Osbert Sitwell will be 73 in December. For several years he has suffered from Parkinson's disease, and for some years past he has been obliged to winter abroad. After a chest complaint this summer, he decided to give up completely his home in this country. He has released his interest in the family settled estates in England to his nephew, Mr. Reresby Sitwell, and will now be living permanently at the family property in Italy.

A short postscript to the announcement described Montegufoni, and added that 'The main property affected in England is Renishaw Hall, Derbyshire, and about 7,000 acres of agricultural and wooded land. English death duties will be avoided, provided Sir Osbert survives the next five years.'

But would he? People who saw him did not rate the chances very high. Hunter Davies interviewed him that October for the *Sunday Times* during a brief visit Osbert made to London. Osbert spoke very gamely about a new book he had started – a 'biographical memoir' about Edith – and mentioned he would soon be leaving on a ten-week cruise to Australia and back. But Hunter Davies's recollections of the interview are that 'he did appear at first sight very gaga and totally past it'; but appearances are deceptive, as Osbert rapidly made clear when he talked about his latest book like the professional self-publicist he was, and said, somewhat inevitably, that he spent a lot of time 'not seeing people and not answering letters'.

There is something rather splendid about him now, so ill and so alone and yet irascibly setting off on a fresh book and a fresh long journey. His life had been one of writing and of travel. He would go on with them as long as possible. And despite his illness, the voyage turned out to be a great success. The ship, the P. & O. *Orsova*, was luxurious, and he appreciated luxury as much as ever. Sir Colin Anderson, chairman of P. & O., had personally made sure that he had one of the best first-class staterooms on the ship, and he was treated as something of a celebrity. He enjoyed that too. Even the seventy-third birthday party that was held for him aboard ship at Singapore seemed to amuse him – and at Sydney there were further interviews with journalists and a visit from a writer he particularly admired, the future Nobel Prize winner, Patrick White.

By now he was corresponding – via dictated letters assiduously typed by Frank – with a few old friends, and thinking of the past with unashamed nostalgia. He wrote a special birthday letter for Lady Aberconway.

I think of so many lovely birthday parties with you in the past. Do you remember one in South Street when Violet played her clavichord and when Eltie [Lady Desborough] was there and asking numerous questions. Of course she hated music. And do you remember that occasion when somebody asked her whether she preferred Debussy or Ravel, and she replied 'I think they're neck and neck'?

The success of the Australia trip seems to have encouraged Osbert to further travel, and the new year of 1966 saw him and Frank off on a fresh adventure, this time aboard a French cruise ship (the idea of the *cuisine* appealed to him) bound for the West Indies. This was a disaster. Frank described why in a letter to Hunter Davies which the *Sunday Times* published in March.

You will be sorry to hear that it was sheer misery. The sea was very rough indeed, but even so one does not expect a boat of 20,000 tons to behave as if it were a rowing boat. We rolled for four days and four nights, consecutively. A number of people got injured, some of them *badly*. Sir Osbert was thrown out of bed but luckily only managed to injure himself slightly. . . . I, too, was thrown onto the deck and a heavy chair fell on top of me.

Shaken but still resourceful, Osbert decided they would disembark at Vigo. From there they travelled to Madrid and stayed in splendour at the Ritz. Since every day was precious now, none must be wasted. They visited the Prado – 'something I never thought I'd see again', said Osbert gratefully as Frank puffed behind him with the wheel-chair. They ate extremely well. Then they flew on to Rome, and were safely back in the *castello* by the end of March.

From that time on Osbert led the life of a near-recluse up in the great house on the hill. Yet even now the lunatics and bores who had been plaguing him for most of his life managed to reach him. As he wrote to Lady Aberconway on 23 May:

Our only diversion has been a mad manservant who came to us while on leave from an asylum apparently. One morning he came to me to say that one of his shoes was growing larger – this naturally frightened him – and soon said that he was leaving on the same day. Soon afterwards, however, he changed his mind, and told Frank that he was staying (as an invalid). . . . Eventually we called the *maresciallo* of the *carabinieri* who persuaded him to go. The *maresciallo* made enquiries and found that he was an inmate of the lunatic asylum at Volterra.

Lunatics apart, there were a few old friends in Florence he knew well enough to feel that they would not be embarrassed by his illness. One was Ian Greenlees, the erstwhile secretary and founder of the Oxford University Sitwell Society, and now the portly, gourmandising head of the British Institute in Florence. Like Osbert, he enjoyed the Montegufoni wine and the Widow Pestelli's cooking. Sometimes Osbert would tell him gloomily: 'Of course I'm only staying alive now for Reresby's sake'. But from the way he ate and laughed at the latest Florence scandal, it was quite clear that there was more to life than this. Another visitor was Christopher Pirie Gordon, the British Consul at the time.

I got the idea that he was quite resigned and not unhappy now. Certainly there was no real gloom, no sense of overwhelming melancholy at Montegufoni. He didn't like new people seeing him in the state he was in. But with those he knew and liked he could still enjoy a good story. He kept going wonderfully, although towards the end Frank virtually had to translate for him as he could barely speak.

Perhaps his best friend now was Harold Acton, that same Harold Acton who had once formed his poetic style on Edith's and partly modelled himself on Osbert to become the cynosure of Oxford in the twenties. Now in looks and manners he was more like one of those Chinese sages he admired in pre-war Peking, but he was the most loyal of friends and could offer Osbert two of the remaining things he enjoyed in life – unusual anecdotes and delicious food. Osbert and Frank would lunch with him quite regularly at La Pietra. Osbert could still consume remarkable quantities of food and here in Harold Acton's sumptuous dining-room there was a flicker of the Osbert Sitwell of his prime. One of his favourite butts was Violet Trefusis, his 'fiancée' of half a century before, and now the *monstre sacré* of Anglo-Florentine society. She had inherited her mother's villa, L'Ombrellino, on the hill of Bellosguardo, and from there she plagued her friends and enemies alike. Acton had quarrelled with her. She was, he would write, 'a law unto herself, perhaps the most selfish woman I have known, so selfish and inconsiderate that she became a joke' – and it was as a joke that Osbert still appreciated her.

Indeed, 'Aunt Vi', as he and Acton christened her, was really the last of the long line of Osbert's female caricatures, succeeding Mrs Kinfoot, poor Miss Collier-Floodgaye, and the appalling Mrs Sawnygrass. She was still indefatigably pursuing Osbert and desperately anxious to discover what was happening at Montegufoni – and who would inherit it.

By now Frank Magro had become totally indispensable, and inevitably this gave rise to rumours and to speculation. But it is not too hard to understand how grateful and attached Osbert had become to him; or why this rank outsider to the Sitwell world should now have undoubtedly become the most important person in his life. While others were talking about him being 'feeble-minded' and simply wondering when he would die, Frank was concerned with keeping him alive and helping him enjoy himself. Frank had become his one sure link with life. He fed him, read to him, washed him, clothed him, wrote letters for him, comforted him and made him laugh. He was his legs, his hands, his eyes. When visitors could not make out what he was saying, Frank would interpret.

But there was more to the relationship than this. Frank was developing himself. Although his education stopped when he left school in Malta at fourteen, he was an eager learner. After three years with Osbert he was a very different person from the awestruck character who had originally come to Carlyle Square. He was now better dressed and better read. From taking regular dictation from the master, he had picked up a definite Sitwellian style. From eating with him he had learned about food and restaurants. From travelling with him he had absorbed some knowledge of the world. Today he describes this period as 'my university' – and it was the Pygmalion element in the relationship that appealed to Osbert. If 'Aunt Vi' was his final butt, Frank Magro was his final protégé.

Two things made Osbert realise that time was passing – the changing of the seasons, and the inexorable advance of his disease. 1967 came. Spring reached the valley of the Pesa, and Frank started wheeling him out again each morning to the Cardinal's Garden and read to him beneath the oleander. But for long periods he would now become depressed, and it was a sad self-portrait that he gave in a letter on 17 April to Lady Aberconway.

About Edith, I have given up any idea of writing as my complex of maladies makes it difficult. I shake so much that I cannot do much in the line of dictation, and what I do is unsatisfactory. *Between ourselves*, my eyesight is going, due to inoperable cataract, and so on and so forth. In fact I seem to be an involuntary candidate for Job's crown.

He had been writing to Reresby and Penelope, and they had told him that they would soon be out to see him, and he looked forward to their visit. Otherwise, the chief visitors from the world outside were Dinka the cat, and Dr Piazzini, who drove down from Montespertoli to operate upon his poisoned toe.

But while it must have seemed as if the world he used to know had totally forgotten him, he still had that one source of power over it all – his will. Like Sir George before him, he was now giving it a lot of thought.

Earlier that year Violet Trefusis had written to her friend, Georgia's former brother-in-law, Sir Anthony Lindsay-Hogg on the subject.

There are lurid rumours abroad about poor Osbert who is much worse, can scarcely speak and not write at all, so it's impossible for him to make a will leaving (presumably) Montegufoni to Francis as Reresby has got Renishaw. The family appear to neglect him disgracefully. No one has been near him. He sends me pathetic little messages from time to time, but doesn't want to see me as he is practically unintelligible.

This was all nonsense, and it was absurd to suggest that he was now so utterly incapable he could not sign a will. But she had put her finger on the point that was arousing considerable interest in Florence. Who would the castle go to? She presumed Francis because it would seem fair, with Reresby getting Renishaw. (Edith had already done something for Francis by leaving him her literary estate in her will, and Osbert had helped him furnish his house in Ladbroke Grove.) But was she right? What of the claims of Georgia and Sacheverell? What about Reresby? For that matter, Montegufoni might still be left to David Horner (if he declined to change the existing will) – or Frank. It was a fascinating subject.

As the days lengthened, and the summer came, Osbert would sit, half blind and very bowed, beneath his oleander. Frank would read Proust to him – in the translation of the old enemy, Scott Moncrieff. And Osbert, still the only one who knew all the answers, would keep his secrets to himself.

Was Osbert lonely? He liked telling the story of Pachman, the great pianist who, when asked the same question in old age, would solemnly reply, 'But how can I be lonely? I have always Pachman here to keep me company.' And despite the note of near despair in a letter to Lady Aberconway, he did in fact still nurse some hope of being cured of Parkinson's Disease.

There is an American doctor in Boston who has cured a man who has been suffering from Parkinson's Disease for a number of years. Frank has written to him, inquiring whether the treatment might benefit me. At the moment, also in America, it was reported in the *Corriere della Sera*, there is another cure for Parkinson's. Apparently things are moving and discoveries are being made. But I ought to bear in mind that I am 74, and these things might not be suitable for me.

But then, when autumn came, Osbert got worse again. On top of that 'complex of maladies' he suddenly developed prostate trouble, 'which', as Frank now reported back to Miss Andrade, 'makes life for him considerably more difficult. He has had to put up with a great many difficulties for so many years, and now things are much worse.' He was confined to bed. In his state of health there was no question of an operation. For Osbert there was much discomfort: for Frank still more unpleasant work. But Frank encouraged Reresby and Penelope to come twice that autumn. Osbert was grateful and enjoyed their company. He was grateful, too, to Frank, who soon had news for Miss Andrade in England.

I am sure that you will be pleased to know that Sir Osbert has given me a flat in the castello for life. I have the whole of the rococo façade, except for one room, including the lovely and large terrace, overlooking part of the garden, the artificial lake and the surrounding hills. This was done on the 13th of September when Mr and Mrs Reresby were staying here. They were delighted with the arrangement, but I am not sure about the rest of the family. However, if they don't bother me I shan't bother them. Lady Aberconway too, is delighted and has sent me a truly wonderful letter telling me so.

Osbert was not religious but in his decline he made the most of the remaining consolations by which he had always lived. This gave the last months of his life a certain grim nobility. He was agnostic, self-indulgent an an unashamed materialist. He enjoyed scandal, food and art. He lived for literature and travel. And now, with extraordinary energy and courage, he drove his ailing body to enjoy them still. He was incontinent and well-nigh paralysed, and yet the new year saw him once again in Venice. Money helped a lot, of course. He could afford a male nurse to help relieve an overburdened Frank. They could all stay in signal comfort at the Danieli once again. And when the winter fogs had lifted from the city, and the first warm days of spring had come, Osbert, the nurse and Frank would trundle off, like something out of Thomas Mann, enjoying the piazzas, churches and picture galleries which had always formed so much of the artistic landscape of his life. Tiepolo did not change; nor did the morning view across the lagoon to San Giorgio Maggiore. And Osbert would let nothing stop him seeing what he wanted. 'Now that he has a permanent nurse,' wrote Frank to Miss Andrade, 'life is easier all round, especially when we come to cross these wonderful Venetian bridges – Sir Osbert gets hold of the person on each side of him and away we go.'

Venice delighted him but left him exhausted. On the return to Montegufoni life, as Frank now recorded in his diary, 'was on a low key', but it was Frank, not Osbert, who cracked first. The strains of the last few months had been too much, particularly the continual effort of lifting Osbert's helpless but considerable bulk. He developed painful rheumatism, and in July Osbert sent him off for a cure at Abane Termine, near Padua. During the fortnight he was absent, Reresby came out to keep his uncle company. Every evening Osbert would dictate a short note to him for Frank, for he missed him badly.

Then in August Osbert had his final contact with the world of very grand society that he had left behind him. The Snowdons were spending part of their summer with Harold Acton at La Pietra, and it was a thoughtful gesture of Princess Margaret to suggest visiting the all but forgotten old friend of the family at Montegufoni. From his wheel-chair in the *salone* Osbert somehow did the honours at this his final royal occasion. The Princess brought him personal greetings from her mother, then visited the gardens. Her husband made a point of seeing the Severini room. Briefly the castle came alive with its distinguished visitors and servants, chauffeurs, personal detectives and police. Outside, the villagers were waiting for a glimpse of *Il Barone*'s friend, the sister of the Queen of England. They saw her briefly in her green flowered dress as her car drove down the hill then sped away to Florence. Then silence was restored inside the castle. That night at dinner Osbert's verdict on the visit was, Frank noted in his diary, 'restrained but favourable'.

That November he was ill with pneumonia, but once again his powerful physique helped him recover after a fashion. But news of his illness brought out a concerned Sacheverell and Georgia. This time Osbert was too weak to entertain them at the Grand Hotel, even if he had wanted to;

so finally, and for the first time in over thirty years, Sacheverell was once more staying in his father's castle. It was a nervous visit. They both mistrusted Frank and felt his influence excessive. There was a scene, and things were said that would not be forgotten, but the two brothers still seemed to have their old affection for each other, and parted amicably.

The last visitor that year was Edith's old confessor, Father Caraman, who came at Christmas time, *en route* from Rome. Osbert had invited him, and Father Caraman is convinced that he was, as he puts it, 'now feeling his way towards the Church'. Frank, however, does not believe this, and it must remain a moot point whether Osbert invited Father Caraman as a priest, or simply as a friend.

With the new year, 1969, the routine inside the castle seemed to have reached a sort of limbo, with a completely passive Osbert carefully sustained by Frank and Pino, the male nurse. To outsiders it was slightly eerie, this living death of a distinguished writer. He was a huddled figure, all but paralysed, incapable of speaking more than in a whisper. There was just one part of his body that was still entirely alive – his brain: though whether this was a curse or blessing it is hard to know. Frank insists that he was still managing to enjoy a few small pleasures of the mind: music (Frank had recently installed a record-player in his bedroom), books (Frank still read to him each morning) and news of the outside world (the old radio in the *salone* was still dutifully tuned in to London every evening). He also loved the garden, the returning warmth of spring, the scent of the wistaria. He ate as much as ever.

Incredibly, he still had hope of being cured. Frank had been in correspondence with Dr Gurewich in New York who had been having outstanding results by treating Parkinson's Disease with the new, so-called 'miracle drug', L-Dopa, and Osbert was determined to be treated with it now as soon as possible. Dr Gurewich agreed, on condition that Osbert came out to New York.

He was excited by the prospect and for some days could think of little else. On the morning of 2 May Frank breakfasted as usual with him in his room and they discussed New York. Osbert had made up his mind. They would leave at the end of May and fly from Rome. Then Osbert started talking of the places they would visit in New York – the Museum of Modern Art, so that Frank could see the Tchelitchew that all the fuss had been about, the Metropolitan, and of course his favourite collection in the whole world of art, the Frick, with its great Rembrandt of the Polish Rider, its wonderful Bellini of St Francis, its matchless atmosphere that made it the epitome of all those Edwardian fine mansions of the very rich that he had always loved. He talked of all this as if he were already cured: as for the journey home, that would be the time for convalescence. Instead of flying, they would take a liner to Genoa, enjoy the crossing as he had done so many times before, and then be back in Florence for the autumn. That was the best time to be back in Italy. Frank agreed, then left him while the nurse washed and dressed him for the garden.

At nine-thirty Frank was summoned by the nurse. Osbert had had some sort of seizure. He could not speak, but seemed in pain and was having spasms in his limbs. When Dr Piazzini came he diagnosed a heart attack and suggested it was time to call the family. Frank telephoned Reresby in London. Reresby came that evening. Osbert was barely conscious when he saw him. Next day he was worse, and the specialist who came up from Florence said there was little hope.

It took him two more days to die, days during which the necessary calls were made, the final death-bed cast assembled. After Reresby, the first to come was Hugo Southern, the solicitor, who by chance was already on his way to stay with Osbert. Sacheverell and Georgia followed on the evening of the third, with Francis and Reresby's wife Penelope coming on the fourth. Sacheverell and Georgia stayed at the Grand Hotel in Florence.

Accordingly, it was Frank and Reresby who saw most of Osbert at this point, and on the evening of the third, at Frank's suggestion, Reresby put through a call to the Anglican Church in Florence, asking for a priest: the vicar, the Reverend George Church, came late that night. As a former chaplain to the Queen, the Vicar of St Mark's had the right to wear a scarlet cassock, so that his abrupt appearance, for all the world like some Roman Catholic cardinal, would have been doubly baffling to Osbert – if he had been in a state to be baffled by anything. The vicar read the Lord's Prayer and the Creed. Osbert made no response. He remained in a coma all that night; and the following evening – at 7.15 on 4 May – he died.

It was remarkable how death released him from the deformity of his illness, for without the frightful tremor the body was no longer that of the hunched-up invalid in the wheel-chair but of a tall, imposing figure of his youth. The face relaxed as well and reassumed its strong, commanding features.

It was felt that he should have obsequies appropriate to the lord of the *castello*; and so the body was garbed in its best dinner-jacket, Frank fixed the black bow-tie, and for the whole night of 5 May the remains of *Il Barone* Osbert Sitwell lay in state in an open coffin in the great hall by the main *cortile*; the villagers kept vigil by him.

Then on the sixth the cortège left the castle, winding its way along the Via Volterrana in blazing sunshine, through Montagnana, beneath the shadow of the great Villa of I Collazzi which Michelangelo supposedly designed, and then down the steep hill into Florence for the funeral. It was held at Anglican St Mark's in the Via Maggio, a strange place in a way for Osbert's funeral – and yet, not so strange, because St Mark's, with its smell of ancient hassocks and faded Anglican décor, is one of the last relics from that long-dead world of Anglo-Florentine society he knew so well.

The family was there, a few servants from the villa, and various old friends from Florence. The general feeling was that death had been 'a merciful release'. True to form, Violet Trefusis came, late but insistent in a dreadful hat.

The body was cremated, the ashes buried in the Allori Protestant cemetery, not far from the Certosa monastery which one passes on the way to Montegufoni. One of Osbert's final wishes was to be buried with a copy of his first and favourite novel, *Before the Bombardment*. Frank duly placed it in the urn.

In retrospect the funeral appears to have been far more bizarre than most of the mourners realised. The unspoken question naturally was: Who would succeed Osbert as the next owner of Montegufoni? And just as naturally, most of the congregation assumed it would be Sacheverell. He, after all, was now inheriting the title. He had been so close to Osbert in the past and had such strong links with Italy that it seemed probable. Georgia believed it, so did the closest Sitwell friends. But just a handful of the congregation knew the truth – Frank Magro, Reresby and Penelope, and Hugo Southern. Neither Georgia nor Sacheverell knew, and it was not until the funeral was over, and Reresby had had an interview with Osbert's lawyer, Dr Contri, that he realised, as he says, 'that I had to pull myself together and tell my father the truth about Osbert's Italian will' – and that he would not get Montegufoni after all. Sacheverell took the news 'very bravely', but it still took several days before he dared tell Georgia the unpalatable truth. This produced the awkward situation of Georgia assuming that she was the rightful chatelaine of Montegufoni, her friends and the castle servants thinking that she was as well, and the small group who knew the truth sworn to uneasy secrecy.

What in fact had happened was that, eighteen months earlier, just before his prostate trouble, Osbert had decided that the time had come to put his life in order and compose his will. Until that moment the castle *had* still been left in its entirety to David Horner. But Osbert plainly put a lot of thought into his final testament. This was when he decided he would give Frank Magro his flat for life, and it was then too that he had picked his heir for Montegufoni. Late that summer Reresby received a letter from him saying that he would inherit it as well as Renishaw. There was, however, one condition: it was to be a secret until his uncle died. Not even Hugo Southern knew until the night he reached the castle, for this was an Italian will, made by an Italian lawyer, Dr Contri.

But Osbert's will was not as simple as it seemed, and as the beneficiaries studied the details they must have realised that it was a curiously disruptive document. Perhaps it all *was* fortuitous, perhaps that invalid recluse beneath the oleander tree had been as feeble-minded as some had said he was and simply had not realised the trouble he would cause. But to say this is possibly to miss the point of what may well have been the author's last and most effective joke of all, a joke so perfectly in character that it is tempting to believe he must have meant it as a postscript to his life, as barbed and witty and baroque as one of his own best anecdotes.

Musing on the death of authors some forty years earlier in his book *The Man Who Lost Himself*, he had written that 'famous men of letters and imagination are apt to meet with an end as exceptional as themselves'. And certainly there was something exceptional in the complications he was causing. True, Reresby had got the castle in addition to Renishaw – but he had not inherited the necessary money to run the enormous house. This had all gone to Sacheverell and Georgia. They in their turn were now deprived of the Italian home they had been counting on. In some ways more wounding still, Osbert had appointed Frank, and not his brother, as his literary executor and legatee. But as a slightly hollow consolation for the loss of Renishaw and Montegufoni, Sacheverell was now well off – but he was seventy-one; the wealth had really come too late. He and Georgia would never move from Weston now. Francis got nothing more, despite the good impression he had made on Osbert.

It took some while for all the implications of the will to work out fully. For nearly five years Reresby and Penelope struggled to maintain what Osbert himself had called 'that great white elephant' of Montegufoni. They loved it, and Reresby improved the vineyards, even attempting to market 'Sir Barone Osbert Sitwell's' excellent chianti in England. But this was not enough to pay for the upkeep and repairs to the *castello*, and disagreements within the family made it impossible for them to share the place for holidays as Reresby hoped. In 1974 he sold it to a wealthy, self-made local business man, Sergio Posarelli. And so, within a space of sixty years, Sir George's Italian adventure was over, and Montegufoni had returned once more to its own people.

Sacheverell and Georgia were understandably bitter at the will and thought at one stage of disputing it. Georgia was not told the truth until some two weeks after the funeral. Her disappointment and unhappiness can be imagined. Sacheverell's misery was more profound, for as well as having lost Montegufoni – the house where he and Georgia had once believed that they would end their days – he also felt himself rejected and betrayed by Osbert after their early years of closeness and devotion and combined achievement. The Sitwells' long family association with Italy and Florence had ended, so he felt, in 'sadness and humiliation', for, as he wrote philosophically after the event, 'there are things one never thought to happen, and that do happen; and it is the same with long loyalties and affections which do not stand the stress of time'.

It took him eighteen months to recover from the shock, eighteen months in which he felt, he says, so totally discouraged that he could not work; and when he finally began to write again it was the loss of this, the castle of his childhood, that give him a theme and title for his book. He called it *For Want of the Golden City*, and within the book's account of a lifetime's appreciation of art and travel it is Montegufoni, 'bringing back pitiful memories', that becomes a symbol for the golden city of earthly happiness he had been seeking from his earliest childhood.

He compared its loss with the loss of faith a religious-minded person might experience, and, as he says, by forcing him to reconsider and recover his own personal beliefs, this loss made him appreciate the true consistency and continuity behind his life. For by looking back he realised that at thirteen or fourteen, on his early visits to Italy, he had been interested already 'in just the things, in principle, that still interest me now'.

He had achieved that earliest ambition to see and appreciate all the world's greatest works of art and architecture; and he had lived his life, as he rightly claims, as both a poet and a dilettante, in the old and real sense of that word – a private person who delights in beauty in whatever shape he sees it. It is this delight, this sense of wonder and excitement that links all his books, whether he writes on Bach or El Greco, Naples or St Petersburg, English craftsmanship or Angkor Wat. It will ensure their constant rediscovery by a discerning 'happy few', and, as he must have realised, the sixty-eight books that are his lifetime's work form a more valuable and more enduring monument than any golden city he has lost.

Not long before he died Osbert wrote to an old friend saying that he felt that Renishaw was made for Reresby and Reresby for Renishaw. Since Osbert's death the years have proved his point. Unlike the previous Sitwell generation, Reresby is a genial, good-living, country-loving man, who seems to have returned to the original strain of Sitwell squires. A good improving landlord to the local farmers, he worships Renishaw, and together he and Penelope have gradually refurbished it, so that the sense of doom has now departed from that long, low, brown house above the collieries. For the first time in over a century the house is lived in now throughout the year, and Reresby even has his vineyard there, the most northerly, he claims, in Europe. He has inherited his grandmother's skill at mimicry. One of the best performances is of his Uncle Osbert.

Frank Magro still continues his extremely private life in splendid isolation in the rococo wing of Montegufoni. He is eternally grateful to Osbert, as well he might be. 'He was my benefactor and my university and more of a father than my real one to me.' The remainder of the house is uninhabited, and while Signor Posarelli tries to decide what to do with it, Frank Magro's flat seems like a gondola moored against the deserted harbour of the great haunted castle on the hill. The flat, with its various Sitwellian relics, remains a shrine to Osbert's memory. There are the glittering brass bust of him which Dobson sculpted in the early twenties, a lot of photographs, a complete signed collection of his works, and in the *galleria* hang the framed originals of the Piper sketches which formed the illustrations for *Great Morning*. In summer evenings, when the screech-owls call to each other from the empty tower and the dogs howl back from the village down below, one wonders if it is Sir George's ghost or Osbert's that disturbs them now.

Sources and References

SOURCES OF UNPUBLISHED MATERIAL

Dame Edith Sitwell's letters to Robert Nichols are in the possession of Mrs Anne Charlton. Letters to John Sparrow and Father Philip Caraman are in their recipients' possession. Other unpublished letters and papers of hers are in the Humanities Research Center, Austin, Texas. The letters from Ian Fleming to Dame Edith are in Indiana University Library. The letter to her from Geoffrey Grigson is taken from a copy in his possession. Other unpublished letters to her are in the Humanities Research Center, Austin Texas.

Sir Osbert Sitwell's letters to Adrian Stokes are in the possession of Mrs Adrian Stokes, and those to Christabel, Lady Aberconway in the British Library. Letters to Lincoln Kirstein and Lorna Andrade are in their recipients' possession. Other unpublished letters and papers of his are in the Humanities Research Center, Austin, Texas. The letters from Lady Aberconway are in the British Library. Other unpublished letters to Sir Osbert are in the Humanities Research Center, Austin, Texas. The letter from Violet Trefusis is in the possession of Mrs Serafina Clarke.

Sir George Sitwell's unpublished papers are in the Humanities Research Center, Austin, Texas.

ABBREVIATIONS

AAS O & S: *All at Sea* (with O: 'A Few Days in an Author's Life'), 1927
AJ O: *Argonaut and Juggernaut*, 1919
AMP E: *Aspects of Modern Poetry*, 1934
AP E: *Alexander Pope*, 1930
BAC S: *British Architects and Craftsmen*, 1945
CES *A Celebration for Edith Sitwell*, ed. Jose Garcia Villa, 1948
DA O: *Dumb-Animal and Other Stories*, 1930
DG O: *Death of a God and Other Stories*, 1949
DTAL O: *Discursions on Travel, Art and Life*, 1925
E Dame Edith Sitwell
ECP E: *The Collected Poems of Edith Sitwell*, 1957
ESL E: *Selected Letters*, ed. John Lehmann and Derek Parker, 1970
EWM O: *Escape with Me!*, 1939
FC O: *The Four Continents*, 1954
GM O: *Great Morning*, 1947
GS Sir George Sitwell
ILUBS E: *I Live Under a Black Sun*, 1937
IS Lady Ida Sitwell
LHRH O: *Left Hand, Right Hand!*, 1944
LMS O: *A Letter to my Son*, 1944
LNR O: *Laughter in the Next Room*, 1948
M S: *Mozart*, 1932
MLH O: *The Man Who Lost Himself*, 1929
MWMW S: *Mauretania, Warrior, Man, and Woman*, 1940
NE O: *Noble Essences*, 1950
NGM O: *The Novels of George Meredith*, 1941
O Sir Osbert Sitwell, Bt
OC O: *On the Continent*
OCS O: *Collected Stories*, 1953
OCSP O: *Collected Satires and Poems*, 1931
OD O: *Open the Door!*, 1941
OOF O: *Out of the Flame*, 1923
PALS O: *The People's Album of London Statues*, 1928
PC E: *Poetry and Criticism*, 1925
PF O: *Penny Foolish*, 1935
PP E: *The Pleasures of Poetry, A Critical Anthology*, first series, 1930
PW O: *Pound Wise*, 1963
QMO O: *Queen Mary and Others*,
S Sir Sacheverell Sitwell, Bt
SM S: *Splendours and Miseries*, 1943
SPL S: *Sacred & Profane Love*, 1940
SS *Sacheverell Sitwell: A Symposium*
ST O: *The Scarlet Tree*, 1946
TCO E: *Taken Care of*, 1965
TF O: *Triple Fugue*, 1924
TMF O: *Tales My Father Taught Me*, 1962
TWD O: *Those Were the Days*, 1938
WC O: *Winters of Content and Other Discursions*, 1950
WKCR O: *Who Killed Cock-Robin?*, 1921
WL O: *The Winstonburg Line*, 1919
WP E: *The Wooden Pegasus*, 1920

CHAPTER ONE: THE YOUNGEST BARONET IN ENGLAND

Private information from Francis Bamford, Cyril Connolly, Reresby Sitwell, Sir Sacheverell Sitwell, Glenway Wescott

16 *of a legend* ECP 176
17 *under the earth* O: 'Portrait of Lawrence', *Week-End Review*, 7 February 1931
19 *of my own* LHRH 27
20 *such a mistake* LHRH 26
20 *with pale eyelids* TCO 22
24 *celebrated public man* ST 59
25 *known as Spiritualism* LHRH 246
25 *Cold as ice* LNR 210
28 *he was engaged* LHRH 98
29 *on God's earth* TCO 41

CHAPTER TWO: 'MY CHILDHOOD WAS HELL!'

Private information from: Sir Harold Acton, Francis Bamford, Reresby Sitwell, Sir Sacheverell Sitwell, Glenway Westcott

31 *exceedingly violent child* TCO 17
31 *most exasperating child* TCO 27
31 *to commit suicide* Lindsay, *Meetings with Poets* 68
31 *of the world* TCO 27
32 *striking personal dignity* CES 9
32 *A genius* CES 9
33 *poor little E* TCO 27
33 *experience of faithlessness* TCO 33
33 *in the nursery* LHRH 89
33 *being a female* TCO 26
33 *of his personality* LHRH 89
34 *pink or yellow* LHRH 111
34 *day had held* LHRH 112
35 *be rather fun* E: 'My Awkward Moments', *Daily Mail*, 15 October 1927
35 *frozen with nerves* Salter, *The Last Years of a Rebel* 65
36 *to melt it* TCO 40
37 *girl may become* Salter 65
37 *glory was everywhere* TCO 44
37 *a polar green* LHRH 204
38 *for a gentleman* LHRH 162
38 *to her head* SM 242
39 *child of eleven* LHRH 222–3
40 *furious, sudden rages* LHRH 98
40 *more tragic tone* LHRH 98
40 *jostling, screaming mob* ST 68
41 *victorious British Democracy* ST 69
42 *love of averageness* ST 126
41 *my football boots* O to IS, 14 January 1903
41 *beastly bothering bores* O to IS, 1 February 1903
41 *send you one* O to IS, 15 February 1903
41 *might have been* O to IS, 22 February 1903
41 *to Father yesterday* O to IS, 14 January 1903
41 *disapoitted about it* O to IS, 25 January 1903
41 *I better do* O to IS, 1 February 1903
41 *a long time* O to IS, 22 March 1903
41 *effects of overwork* GS: *The Making of Gardens* xv
42 *way in everything* ST 85
44 *like a martyr* ST 79
45 *I was four* E to Charlotte Haldane, undated
45 *of her hand* O: unpublished memoir of E
46 *in the sun* E to Haldane, undated
46 *to my own* ST 228
47 *there aren't any* GM 76
47 *castle all told* GS: unpublished essay on Montegufoni
47 *hope to keep* GM 80

CHAPTER THREE: A CLOSED CORPORATION

Private information from: Sir Harold Acton, Sir Sacheverell Sitwell.

49 *of her days* TCO 21
50 *the parallel bars* ST 18
50 *and less comfortable* PF 35
50 *rather beautiful machine* MLH 17
50 *boys and masters* ST 257
51 *differentiate from happiness* MLH 17
51 *the English Gentleman* PW 81
51 *and his Maker* AAS 17
51 *a boy's idiosyncrasies* MLH 17
51 *need of disguise* MLH 17
52 *Bores and Bored* Byron, *Don Juan*, canto 13, xcv
53 *into her notebooks* Lehmann, *A Nest of Tigers* 44
54 *when spoken to* ESL 90
54 *almost Spanish vehemence* ST 145
54 *fortune to bear* ST 146
55 *joys of humanity* ST 145–6
55 *a closed corporation* LNR 86–7
55 *Wuthering Heights country* ESL 124
56 *Queen in pantomime* E: 'Why I Look the Way I do', *Sunday Graphic*, 4 December 1955
56 *I was different* Salter 65
56 *like a Pekingese* E: Why I Look the Way I do'
57 *my own misery* GM 42
57 *relationship must rest* PW 293
57 *bore other people* GM 40
57 *disgrace at home* GM 59
57 *on an envelope* GM 59
58 *and vituperative misunderstanding* MLH 32–3
58 *for a while* GM 38
59 *a la Russe* GM 123
59 *banishment to Siberia* GM 136
59 *every cavalry officer* GM 121
59 *any primitive tribe* PF 34
59 *of the opposition* GM 136
59 *who love it* GM 136
60 *for the winter* GM 148
60 *of white grapes* GM 149

CHAPTER FOUR: 'A YORKSHIRE SCANDAL'

Private information from: Sir Harold Acton, Lady Diana Cooper, Sir Sacheverell Sitwell

62 *Yorkshire Scandal* *Daily Mail*, 12 March 1915
63 *and unnatural black* GM 161–2
64 *meaning for her* GM 165
65 *to be done* IS to Field, 2 November 1911, quoted in *Sheffield Daily Telegraph*, 12 March 1915
65 *of my family* IS to Field, 9 November 1911, quoted in *News of the World*, 14 March 1915
65 *believe my word* IS to Field, 2 December 1911, quoted in *Sheffield Telegraph*, 12 March 1915
65 *get hold of* IS to Field, December 1911, quoted in *News of the World*, 14 March 1915
66 *tongue about me* IS to Field, October 1912, quoted in *News of the World*, 14 March 1915
66 *of his clutches* IS to Field, quoted in *News of the World*, 14 March 1915
66 *debt of honour* IS to Field, 27 January 1912, quoted in *Sheffield Daily Telegraph*, 12 March 1915
67 *of a moneylender* GM 165
67 *known about it* GM 165
67 *early in September* *Sheffield Daily Telegraph*, 15 March 1915
67 *public Yorkshire scandal* *Daily Mail*, 12 March 1915
68 *of the brain* S to O, 26 December 1912
69 *sheer stupidity* TCO 77
69 *sordid tragedy* TCO 77
69 *the golf-course* TCO 77
69 *violence that prevailed* GM 237
69 *of absolute bondage* GM 237
69 *conduct and habits* E to Lehmann, 29 August 1945
69 *with my work* interview by Marjorie Proops, *Daily Mirror*, 7 September 1962
71 *hours of thought* GS to O, 4 February 1913
71 *has re-established confidence* GS to O, 6 March 1913
71 *frame of mind* ibid
71 *that could follow* ibid
71 *generous with themselves* ibid
72 *George R. Sitwell* GS to O, 15 March 1913
76 *while it lasted* GM 193
76 *of the rich* O: 'London', *Wheels* II
76 *and dusky potentates* GM 217
77 *of September splendour* GM 217–18
77 *the most remarkable* GM 217
77 *eye for character* GM 218
77 *gift of mimicry* GM 218
77 *to see again* GM 209
77 *peasant and aristocrat* GM 223

78 *you were ringing* O: unpublished notes for posthumous memoirs
78 *on the air* ibid
78 *of billiard balls* ibid
78 *bock and seltzer* ibid
78 *genial Sidney Guille* ibid
78 *at a statue* ibid
78 *old Whig Society* GM 260
79 *for her chic* GM 226
79 *and cold hardness* GM 167
80 *them the way* LNR 118
80 *that is all* GM 250
80 *Edwardian arcady* GM 253
80 *Kitchener of hostesses* GM 254
81 *for a century* GM 141
81 *could be faced* GM 141
81 *of the arts* GM 141
82 *come my way* GM 141
82 *brother and sister* GM 141

CHAPTER FIVE: DISASTERS
84 *sparkling geese* E: 'What the Goose girl said about the Dean' WP
84 *I am not* E to Jennings, undated
84 *in a dream* E: 'From an Attic Window', *Daily Mirror*, 7 January 1914
84 *cost would be* E to Elkin Matthews, 6 July 1914
84 *publisher for them* E to Elkin Matthews, 7 January 1917
85 *the twisted pillars* GS to O, 3 March 1914
86 *pair Laced Pyjamas* GM 279
86 *at that age* GM 279–80
86 *most profound depression* GM 282
87 *on the sideboards* GM 231–2
87 *thirteen at midnight* DG 6
87 *that was coming* DG 7
87 *its own death* GM 297
88 *world for you* IS to O, 4 July 1914
88 *that never heals* TWD 4
88 *us from sleeping* Brooke, 'Peace', *1914 and Other Poems*
90 *and grey discomfort* LNR 79
90 *trees and mud* LNR 78
90 *across a continent* LNR 79
90 *little more palatable* LNR 80
90 *of utmost frustration* LNR 80
91 *well as night* LNR 86
92 *the Old Bailey* *Daily Mirror*, 9 March 1915
92 *Sidelights on Society* *News of the World*, 14 March, 1915
92 *tall and slight* *Daily Mirror*, 10 March 1915
93 *black Cossack cap* *Daily Mail*, 12 March 1915
93 *of The Fork* LNR 92
94 *levity of spirit* LNR 90
94 *be more severe* *Sheffield Telegraph*, 15 March 1915
94 *the ells below* ibid

CHAPTER SIX: 'PICTURES, PALLADIO, AND PALACES'
Private information from: Geoffrey Gorer, Sir Sacheverell Sitwell
95 *our private calamity* LNR 94
95 *world of fury* LNR 93
96 *friend with her* GS to 'Ethel', 2 May 1915
97 *personally ashamed of* O to Horner, undated
97 *mother was involved* TCO 77
97 *mud and flies* TCO 77
98 *not deep enough* *The Mother*
98 *and shuddering air* *The Mother*
100 *carnal-spiritual home* Cunard, *Sublunary* 93–5
100 *scale much simplified* Holroyd, *Augustus John* ii 56
101 *enchanting child-soldier* Cooper, *The Rainbow Comes and Goes* 140
101 *fox-terrier-like phrases* LNR 95
101 *idea of it* LNR 95
101 *the chicory flowers* LNR 95
102 *farmed with father* LNR 88
102 *me look ridiculous* LNR 118–19
102 *me to paper* LNR 115
102 *Babel's direful prophecy* LNR 117
103 *and international celebrity* LNR 116
103 *found a purpose* LNR 117
103 *with their elders* Fussell, *The Great War and Modern Memory* 109
105 *and champagne orgy* Lady Cynthia Asquith, *Diaries 1915–1918* 278
105 *to talk to* Asquith, *Diaries* 267
105 *in the morning* ibid
105 *knew Italy well* LNR 107
105 *was to survive* LNR 108
106 *a juggernaut, Wheels* 'The Progress of a Poet', *Sunday Times*, 8 September 1957
106 *wrote for it* Hamnett, *Laughing Torso* 98
107 *or mine, Thais?* *Wheels* I 41
107 *much in common* *Wheels* I 89
107 *the painted world* *Wheels* I 9
107 *yet craving pleasure* *Wheels* I 21
107 *I tell you* *Wheels* I 70
109 *same genial glow* NE 101
109 *the National Gallery* NE 101
110 *kindness and encouragement* E to Ross, undated
110 *an octogenarian pirate* Clark, *Another Part of the Wood* 191
110 *coming to dinner* NE 54
111 *myself to life* LNR 27
112 *Osbert Sitwell waiting* Asquith, *Diaries* 279
112 *What a life* Morrell, *Ottoline at Garsington* 206
113 *platitudinous multitude* *Wheels* I 2nd ed. v
113 *holds in store* ibid
113 *God of Blood* *Wheels* I 2nd ed. vi
113 *treatise on Simplicity* ibid
114 *his fellow soldiers* Sassoon, *Memoirs of an Infantry Officer* 280
115 *is always happy* AJ 102
115 *tell the truth* AJ 102–3
116 *a charming type* Huxley to Julian Huxley, 3 Augus 1917, *Letters* 132
116 *a bright production* ibid
116 *to the Press* *Wheels* II v
117 *my son Saul* *Wheels* II 54
117 *as a Nubian* *Wheels* II 81
117 *your father!' twice* *Wheels* II 90
117 *but terribly nervous* Huxley to Julian Huxley, 13 December 1911, *Letters* 141
118 *very moving* Asquith, *Diaries*, 12 December 1917
118 *railing than usual* E to Nichols
119 *Osbert is route-marching* E to Nichols
119 *his flapping arms* E to Nichols
119 *grandeur in repose* E to Nichols
119 *the same again* E to Nichols
119 *of the cooking* E to Nichols
119 *whom they admired* E to Nichols
119 *all on Sundays* E to Nichols
120 *so I have* E to Nichols
120 *come back soon* E to Nichols
121 *like railway carriages* E to Nichols
121 *you cannot accomplish* E to Nichols
121 *to suffer so* E to Nichols
121 *us it lacked* O: unpublished memoir
121 *makes life endurable* Sassoon to O, 3 July 1918
121 *well some day* NE 98
122 *to frighten him* NE 103
122 *of the sensitive* NE 104
122 *the early Christians* NE 89

122 *topography of Golgotha* NE 106
122 *desert of war* NE 108

CHAPTER SEVEN: THE MACHINE STARTS UP
Private information from: Cyril Connolly, Sir Sacheverell Sitwell, Sir William Walton
123 *second Golden Age* LNR 14
123 *and Drury Lane* LNR 15
124 *at the crowds* LNR 1
124 *flower of Bloomsbury* LNR 17
124 *Infanta of Spain* LNR 15
124 *painters and writers* LNR 17
124 *from a trance* LNR 22
124 *cried and cried* E to Nichols
124 *struggle was over* LNR 24
125 *a new God* AJ vi, ix
125 *of creative art* Read, *The Contrary Experience* 257
126 *in the future* Read 139
126 *discuss the future* Read 141
126 *rich-man's gilded bolshevism* Lewis, *The Apes of God* 565
126 *Lewis and Eliot* Read 143
126 *jolly open-faced* Read 144
126 *he will "see"* Read 144
128 *c'est une femme* GM 128
128 *Van Dieren's works* E to Nichols
128 *an enormous success* E to Nichols
128 *really is intolerable* E to Nichols
129 *novel a manner* LNR 10
129 *appalled me* LNR 10
129 *in their revolt* Read 141
129 *Such a mockery* E to Nichols
130 *of sad nothingness* TCO 21
131 *distracted her existence* Peter Quennell, *The Marble Foot* 131
131 *may turn up* E to Nichols
131 *even with you* LHRH 95
131 *on my part* E to Nichols
131 *this bloody festival* E to Nichols
131 *tinged with irritabilities* E to Nichols
131 *is something definite* *New Statesman*, 24 December 1918
132 *is so unjust* E to Nichols, 26 December 1918
132 *plague, Spanish Influenza* LNR 38
132 *satyr-like Father Time* LNR 38
133 *curate's first sermon* NE 75
133 *a sauce creole* Lewis, *Blasting and Bombardiering* 225
134 *as old Maggie* Channon, *Diaries* 208
134 *eyes that blinked* Sonia Keppel, *Edwardian Daughter* 23
135 *of Italian light* LNR 46
135 *of modern pictures* LNR 51
136 *seven professions* LNR 28
136 *within its pages* Taylor, *English History 1914–45* 142
136 *But Jesus wept* OCSP 32–3
137 *my Daily Herald* WL
138 *occupation of man* Sassoon, *Siegfried's Journey* 79
138 *go our again* WL
139 *for many years* LNR 148
139 *and Sacheverell Sitwell* LNR 154
139 *of the Philistine* LNR 155
140 *secretive and crude* E to Nichols, 16 January 1918
142 *enjoyed this evening* LNR 56
142 *a dilated heart* E to Nichols
142 *but I shan't* TMF 142

CHAPTER EIGHT: MAN'S NATURAL OCCUPATION
Private information from: Geoffrey Gorer, Alan Pryce-Jones, Sir Sacheverell Sitwell, A. J. P. Taylor, Sir William Walton
144 to fight back LNR 118
145 *is up to* Fussell, *The Great War and Modern Memory* 76
145 *of their nostrils* NE 298
146 *with the multitude* Aldington, *Life for Life's Sake* 136
147 *worm, Jack Squire* Holroyd, *Lytton Strachey* 758
147 *into the bargain* Woolf to Strachey, 26 May 1919, *The Question of Things Happening* 361
147 *fun with him* E to Nichols, 12 November 1919
147 *is not returned* ibid
147 *of bright pictures* *London Mercury* i, no. 3, p. 334
148 *of the period* John Gross, *The Rise and Fall of the English Man of Letters* 262
148 *writer of verses* *London Mercury* i, no. 2, p. 204
148 *and no reason* *Daily Express*, 20 July 1920
149 *year's Silly Season* Huxley, *Letters* 188
149 *their critics mad* *Daily Express*, 22 July 1920
149 *painting and music* ibid
150 *writers of England* LNR 146
151 *School of Paris* Steegmuller, *Cocteau: A Biography* 150
151 *the Gods say* WKCR 29
151 *or the macaw* WKCR 11
151 *and cowslip-balls* WKCR 6
151 *old hag, Nature* WKCR 6
152 *fly in amber* WKCR 6
152 *home to bed* ECP 25, 31
152 *leaves of afternoon* WP
152 *genuinely feminine poetry* E to Nichols
153 *the Judgement day* OOF 71
153 *a famous name* OOF 72
153 *insists on Hell* OOF 72
153 *good – We know* OOF 73
154 *And so artistic* OCSP 64
154 *in its eyes* ibid
154 *cultivated English aristocrat* Aldington, *Life for Life's Sake* 203
155 *and Princess Bibesco* Holroyd, *Lytton Strachey* 828
155 *on Irish politics* O'Faolain, *Summer in Italy* 112
156 *to the scene* NE 11
157 *of something else* O: 'Fiume', *The Nation*, 1 January 1921
158 *there in England* NE 123
158 *moors of Devonshire* NE 123
158 *ears of absurdity* NE 123
158 *on Italian soil* NE 114
158 *or American capitalism* O: 'Fiume'

CHAPTER NINE: WORK IN PROGRESS
Private information from: Sir Harold Acton, Sybille Bedford, Sir Sacheverell Sitwell, Sir William Walton
160 *of our time* ESL 19
161 *ready for publication* ESL 20
161 *can tell you* ibid
161 *nicely balanced periods* Quennell, *The Marble Foot* 132
161 *ever carried out* Hollingworth to O, 19 February 1920
162 *decayed provincial intelligentsia* Bedford, *Aldous Huxley*, i 117
164 *of modern art* TF 189
165 *visited it altogether* DTAL 134
165 *poverty-stricken plain* DTAL 134
165 *monuments in Italy* DTAL 134
166 *could give anyone* O to M. Bennett, 17 June 1920
166 *was most interesting* Holroyd, *Lytton Strachey* 823
167 *their own verses* E to M. Bennett
167 *make use of* E to M. Bennett
167 *like it before* E to M Bennett
167 *in her place* E to M. Bennett, February 1921
168 *Yours affectionately, Edith* E to M. Bennett, 27 December 1922
168 *Yours, Edith Sitwell* E to M. Bennett, 26 January 1922
169 *species and nationality* Huxley to Julian Huxley, 24 August 1921, *Letters* 202
169 *up his traces* Bedford, *Aldous Huxley* i, 123
169 *put it high* Huxley, *Crome Yellow* 12
170 *the arts, virtuosi* Huxley, *Mortal Coils* 116
170 *rocked with laughter* Huxley, *Mortal Coils* 149–50

171 *in the oases* TF 186
171 *than the Italians* TF 188
171 *grasp his meaning* TF 188–9
171 *to his appearance* TF 186
172 *took to prose* LNR 59
172 *really quite impossible* E to Nichols
172 *felt uncomfortably rustic* Graves, *Goodbye to All That*, 395
173 *and hemming handkerchiefs* ibid 404
173 *a miraculous life* Cocteau's introduction to Radiguet, *Le Bal du Comte d'Orgel*
173 *middle-aged youth movement* Lewis, *The Apes of God* 565
173 *to Edith Sitwell* Marie-Jacqueline Lancaster, *Brian Howard* 32
174 *depends on you* ibid
174 *a 'born writer'* ibid 35
174 *you are not* ibid 36
174 *spoil your gifts* ibid
174 *is simply admirable* ibid 33
174 *Beerbohm's early works* Acton, *Memoirs of an Aesthete*
175 *an appropriate age* Lancaster, *Brian Howard* 54
175 *books, pictures, music* ibid 37
175 *I meet muscle* ibid 90
176 *a queer place* ibid 96
176 *the Eiffel Tower* ibid
176 *within his power* LNR 62
177 *match hands down* Lewis, *Blasting and Bombardiering* 237
177 *as a woman* O: unpublished memoir
177 *excessively neat woman* ibid
178 *some short stories* O to Grant Richards, November 1922
178 *really, commit suicide* O to Grant Richards, 17 November 1922
178 *of West Hampstead* TF 75

CHAPTER TEN: FACADE

Private information from Cole Lesley, Raymond Mander, Angus Morrison, Sir William Walton
179 *waltzes, polkas, foxtrots* LNR 185
179 *that I could* Salter, *The Last Years of a Rebel* 60
180 *for fun* Salter 66
180 *chances and genius* LNR 170
181 *as 'Osbert's fag'* Shead, *Constant Lambert* 87
181 *and emphasis fell* LNR 170
181 *analogies and images* LNR 183
182 *of the Guards* Lawrence to Tate Trustees, 5 February 1923, *Frank Dobson*, Arts Council Catalogue, 1966
182 *Interlude by W. T. Walton* LNR 189
183 *of the entertainment* LNR 191
183 *contempt and rage* LNR 192
183 *to smite her* Salter 166
183 *anything like it* LNR 193–4
183 *committed a murder* LNR 192
184 *could not judge* Woolf to J. Raverat, 30 July 1923, *Letters*, iii 60
184 *literature and music* LNR 192
185 *a goat's behind* Coward, *Poems by Hernia Whittlebot* (privately printed, undated)
186 *mouth is Bournemouth* ibid
186 *paroxysms of fury* Coward, *Present Indicative* 196
186 *rise above it* Nichols, *The Sweet and Twenties* 52
187 *the utmost indecency* E to Lehmann, 23 July 1947
187 *cheating at cards* O to Coward, *Star*, 28 October 1933
188 *a lady's face* O: unpublished MS
188 *are a cult* *Star*, 2 April 1923

CHAPTER ELEVEN: BREAK-UP

Private information from: Sir Cecil Beaton, Lord Clark, Cyril Connolly, Lady Lindsay, Peter Quennell, Lady Sitwell, Sir William Walton, Dame Rebecca West, Richard Wollheim
189 *complete ertably paradise* Connolly to Blakiston, 8 April 1925, *A Romantic Friendship* 66
190 *the postwar generation* Green, *Children of the Sun* 78
191 *see what happens* Beaton, *The Wandering Years* 150
191 *house in Chelsea* ibid 163
192 *large country houses* Connolly, *Enemies of Promise* 155
193 *haunted summers* ECP 176
193 *austere and elegant* ECP 174
193 *long eyelids* ECP 174
193 *for the emotions* PC 26
193 *Stravinsky and Debussy* PC 26
194 *to his friends* TF 222
194 *appearance of it* TF 226
194 *the collar-bone* TF 223
196 *outburst of spring* O: unpublished MS on seventeenth-century painting
196 *of their houses* ibid
197 *in great pain* E to Freeman, Easter 1925
197 *of delayed convalescence* ibid
197 *Osbert and Sachie* ibid
197 *so much better* ibid
197 *I most dislike* ibid
197 *I hope, apologise* E to Freeman 16/17 January 1925
198 *I'm never prepared* O to Richards, 1925
198 *of other people* AAS 15
198 *wants a lesson* O to Richards, 29 August 1924
199 *disgusting and common* E to Freeman, undated
199 *and now Turner* O to Richards, 14 January 1925
199 *do it ourselves* ibid
199 *of red cornelian* O to Richards, 14 December 1924
199 *and rather disillusioned* ibid
200 *all be rich* O to Richards, January 1925
200 *the two brothers* *Evening Standard*, 18 July 1925
200 *agree with that* ibid
200 *at the moment* ibid
201 *emigrate – to Italy* LNR 260
201 *Italian Sir George* LNR 261
201 *home for ourselves* LNR 270
202 *in those days* SS 1
202 *I could chew* SS 2
204 *to remain impassive* Acton, introduction to QMO 14
204 *most intimate friend* LNR 207
205 *malaise and ennui* Quennell, *The Marble Foot* 142
206 *of the season* Newman, *Sunday Times*, 2 May 1926
206 *merely revoltingly arty* Beaton, *The Wandering Years* 89
206 *of my life* LNR 203
207 *was now immersed* Christopher Farman, *The General Strike* 219
207 *awaited the signal* LNR 207
209 *to their jobs* *Evening Standard*, 2 September 1926
209 *the theatrical profession* *Evening Standard*, 3 September 1926
209 *Laye, Evelyn Laye* ibid
210 *very dreadful affair* ibid
210 *of trivial complaint* 'The Sitwell Trinity', *New Statesman*, 3 December 1927
210 *company of Judas* AAS 37
210 *with nervous prostration* AAS 21
210 *ultra-Conservative paper* AAS 22
210 *plotting in progress* AAS 24
210 *have been impertinent* AAS 29
211 *a great mind* AAS 26
211 *lasted for long* G. Orioli, *Adventures of a Bookseller* (1937 ed) 296

CHAPTER TWELVE: OSBERT, EDITH AND THE CHATTERLEYS

Private information from: Sir Cecil Beaton, Geoffrey Gorer, Reresby Sitwell. Acknowledgement is made to Laurence Pollinger for the quotations from D. H. Lawrence listed below and to the Estate of the late Mrs Frieda Lawrence Ravagli for the passage quoted by O. Hamilton.

214 *a summer sky* QMO 101
214 *and very unpretentious* QMO 105
214 *funny in conversation* QMO 107
214 *her grey hair* QMO 108
215 *the dear boys* QMO 108
215 *before the Fall* QMO 101
216 *and bravely sincere* E: 'Miss Stein's Stories', *Nation and Athanaeum*, 14 July 1923
216 *significance to language* ibid
216 *valuable pioneer work* ibid
216 *amount of silliness* ibid
216 *of exquisite beauty* ibid
216 *wonders of creation* Stein, *Selected Writings*, ed. C. Van Vechten, 192
216 *delicacy and completeness* ibid
216 *Easter Island idol* TCO 136
216 *immensely good humoured* TCO 136
217 *verbally very interesting* TCO 136
217 *for our language* E: 'The Work of Gertrude Stein', *Vogue*, October 1925
217 *gigantic* Woolf, *A Change of Perspective* 209
217 *half-intelligent people* *Flowers of Friendship*, Letters to G. Stein, 184
217 *many fresh admirers* ibid
217 *always must have* Stein, *The Autobiography of Alice B. Toklas*, 253
217 *swallowing a plum* Acton, *Memoirs of an Aesthete* 160
217 *in the eyes* ibid
217 *glory and excitement* Stein, *The Autobiography of Alice B. Toklas*, 253
218 *own particular elegance* E, 'Why I look the way I do', *Sunday Graphic*, 4 December 1955
218 *of sea-monsters* ibid
219 *behind her eyes* Harper, 'A Memory of Dame Edith Sitwell', unpublished
219 *a mediaeval saint* Beaton, *The Wandering Years* 148
219 *as a convolvulus* ibid
219 *Drinkwater and Squire* ibid
219 *and Tallulah Bankhead* ibid
219 *sound like poetry* ibid 147
220 *all the same* ibid
220 *of a ghost* Beaton, *The Book of Beauty* 36
221 *over the fare* Laing, unpublished note on E
221 *him as Ginger* ibid
221 *on a rock* Woolf, *A Change of Perspective* 236
221 *a drowned mermaiden* ibid 269
221 *company of Bloomsbury* TCO 86
221 *beautiful little knitter* E to G. Singleton, 11 July 1955
221 *me in Sicily* Woolf, *A Change of Perspective* 356
221 *stuck about it* ibid 352
221 *towards each other* Harper, 'A Memory'
222 *ticks on sheep* ibid
222 *tell the truth?* ibid
222 *you are, Iain* Laing, unpublished note on E
222 *some sharp slaps* TCO 13
223 *matted, dank appearance* TCO 108
223 *no literary importance* TCO 108
223 *and smelly speech* TCO 109
223 *nasty little nymphomaniac* TCO 109
223 *a famous writer* TCO 109
223 *he had suffered* TCO 109
224 *and breathe heavily* LNR 278
225 *done to it* LNR 277
225 *mattresses were soft* TCO 107
225 *Lawrence, geborene Richthofen* TCO 107
225 *and Rolf Gardiner* Lawrence, *Collected Letters*, ed. Moore, 929
226 *face of genius* O 'Portrait of Lawrence', *Week-End Review*, 7 February 1931
226 *looked so ill* ibid
226 *in his books* ibid
226 *a hen's legs* TCO 107
226 *childhood and ours* TCO 107
226 *who were "gentlemen"* TCO 107
226 *lost selves. Queer* Lawrence, *Collected Letters* 979
226 *as a' that* ibid 978
227 *disturbed by them* O. Hamilton, *Paradise of Exiles* 166
227 *of "his" Midlands* Aldington, *Portrait of a Genius, But* 319
228 *over the park* Lawrence, *The First Lady Chatterley* 2
228 *of Lady Eva* ibid 14
229 *and newspaper articles* Lawrence, *John Thomas and Lady Jane* 5
229 *a viscount's daughter* Lawrence, *Lady Chatterley's Lover* 10
230 *so sensitive about* ibid 11–12
230 *of our day* ibid 10
230 *world of chaos* ibid
230 *lot of people* ibid
230 *live together, always* ibid 12
230 *had stood for* ibid
231 *It won't last* ibid 6
231 *self-effacing, almost tremulous* ibid 15
231 *of human contact* ibid 15
231 *in his stories* ibid

CHAPTER THIRTEEN: PASTURES NEW

Private information from: Lady Sitwell, Dame Rebecca West
233 *anxious-looking young man* E: 'Hazards of Sitting for my Portrait', *Observer*, 13 November 1960
233 *seen a ghost* ibid
233 *not my affair* Parker Tyler, *The Divine Comedy of Pavel Tchelitchew*, 318
233 *Colonna and Michelangelo* ibid 338
234 *burden of both* ibid 320
234 *about Pavlik's mechancete* Tyler 318
234 *with grande-dames* ibid
234 *respect for them* ibid
235 *father's greatest friends* ibid 320
236 *a real campaign* ESL 33
237 *sincerely, Edith Sitwell* ESL 31
237 *character and outlook* O, unpublished notes on Eliot
237 *of his nature* ibid
238 *on his shoulders* ibid
238 *during many years* ibid
239 *more congenial occupation* ibid
239 *air of mystification* ibid
239 *in his unhappiness* ibid
240 *him look cadaverous* Woolf, diary, 12 March 1922
240 *paint his lips* ibid, 27 September 1922
240 *bothered to reply* ibid
242 *at Tom's face!* ibid
242 *home with it* AAS 94

CHAPTER FOURTEEN: SEPARATE LIVES

Private information from: Sir William Walton
244 *for Sachie's arrival* Whistler to O, January 1928
244 *of civilised history* MWMW 44
245 *well as scholarly* Asquith, *Diaries* 508
246 *with his piano* LNR 208
246 *on the air* LNR 208
247 *will be away* O to Horner, 3 July 1928
247 *I'm going to* O to Horner, 7 September 1928
247 *Freud Madox Fraud* QMO 105
248 *flower under glass* Tyler 338
248 *withdrawn from witnesses* ibid
248 *I had imagined* Patmore, *My Friends When Young* 129
248 *go next Monday* O to Horner, 24 September 1928
248 *the trees flowering* O to Horner, 19 November 1928
249 *have gone through* O to Horner, 17 December 1928
249 *please, at once* O to Horner, undated from Amalfi

250 *one could imagine* O to Horner, 8 April 1929
250 *excess of affection* O to Horner, undated from Athens
250 *things can be* O to Horner, April 1929
252 *clustering, clattering things* DA 95
252 *you'll be engaged* O to Horner, 25 April 1929
253 *that vaster tragedy* ECP xli
253 *from London burning* ECP 251
253 *second World War* ECP XXXV
253 *what has arisen* ECP XXXV
254 *go marching on* ECP 252–3
253 *experiments was done* TCO 152
254 *in the nose* O to Horner, November 1929
254 *thought or speech* ibid
254 *frightful boredom anyway* ibid
255 *you and fret* ibid
255 *insistent waltz lilt* WC 13
255 *of the fields* ibid
255 *man has attempted* ibid
255 *wave of sunshine* O to Horner, 14 December 1929
256 *in the armoury* Moat to O, 23 September 1929
256 *awaits the doctor* O to Horner, 19 December 1929
256 *without me, too* ibid
256 *you come here* ibid
257 *of unavoidable sadness* WC 20

CHAPTER FIFTEEN: PARTIAL ECLIPSE

Private information from: Christabel Lady Aberconway, Cyril Connolly, Sir Sacheverell Sitwell, Lady Sitwell
260 *could never face* O to Horner, 12 May 1930
260 *little friend, O* O to Horner, 15 May 1930
260 *tears – with Ginger* O to Horner 19 May 1932
261 *for the day* O to Horner, 23 May 1930
261 *Sphinx to din-din* O to Horner, undated from Montegufoni
262 *of the past* ST 209
263 *exception of Tennyson* Lynd, 'Miss Sitwell on Pope', *Westminster Gazette*, 14 March 1930
263 *of our poets* AP 1
264 *National [Portrait] Gallery* AP 11
264 *remained to us* AP 10
264 *pain-stricken mouth* AP 10
264 *administer to him* AP 12
264 *chilblained, mittened musings* AP 6
264 *of the Sur-Realists* AP 4
264 *School of Poetry* AP 2
264 *at all costs* AP 2
264 *rhetoric and formalism* AP 3
264 *the dying world* AP 2
264 *the national sports* AP 12
265 *shallowness of thought* Grigson, *Yorkshire Post*, 27 March 1930
265 *of Fantastic Inanities* *Daily Express*, 4 March 1930
265 *not got one* *Yorkshire Evening News*, 8 August 1930
265 *disease, believe me* ESL 40
266 *is better now* O to Horner, 7 August 1930
266 *heard you sing* Waugh, *The Diaries*, ed. Davie 318
266 *expression) and disappear* O to Horner, 26 June 1930
267 *wrote. Bless you* O to Horner, undated
267 *can ever imagine* O to Horner, 26 July 1930
267 *discoloured Derbyshire stone* Waugh, *Diaries*, 327
267 *with coal dust* ibid
268 *very gentle* ibid
268 *of the moonlight* O to Horner, 8 August 1930
268 *some private joke* Waugh, *Diaries* 327
268 *charwoman told her* ibid
268 *got to go* ibid 328
269 *very shy* ibid 327
269 *and Osbert's Walk* ibid 328
269 *taking him home* ibid
269 *best behaviour though* O to Horner, 17 August 1930
269 *course, it is* O to Horner, 18 August 1930
269 *they'd elope together* O to Horner, 19 August 1930
269 *all my life* O to Horner, 1 August 1930
270 *I told you* O to Horner, 3 August 1930
270 *feel about it* O to Horner, 5 August 1930
270 *and Reresby screaming* O to Horner, 10 August 1930
271 *sad and miserable* O to Horner, 2 September 1930
271 *Osmund Finnian Shaw* Lewis, *The Apes of God* 349
271 *carefully-contained obesity* ibid 350
271 *cum-soda-water* ibid 355
272 *cast was present* ibid 354–5
272 *rich-man's gilded bolshevism* ibid 565

CHAPTER SIXTEEN: A SITWELL IN PARIS

Private information from: Geoffrey Gorer
273 *unmitigated hell* TCO 135
273 *come upon us* E 'Life's Tyrannies', *Evening News*, 16 July 1931
274 *the ideal possession* ibid
274 *feeling of tenderness* Salter 108–9
274 *nobody ever will* Salter 108
274 *more than you* Salter 110
274 *is one not* E to Tanner, 11 December 1930
274 *can be prevented* ibid
274 *writes bad poetry* ibid
274 *that doesn't matter* ibid
275 *mean to us* ibid
275 *make one sadder* ibid
275 *curiously indirect way* Tyler 339
275 *was born for* E to Harper, undated
276 *a country bunch* PP 3
276 *a la douceur* E to Tanner, undated
277 *was undiluted hell* E to Horner, 28 March 1932
278 *me into trouble* O: unpublished memoir of Eliot
278 *friends I possess* E to Tanner, 28 May 1932
278 *be more peaceful* ibid
278 *good paying arrangements* ibid
278 *completely selfish people* E to Tanner, 29 April 1932
279 *become sane again* ibid
279 *much for me* E to Tanner, 15 August 1932
279 *we would have* ibid
280 *woman in London* Agate, *Ego* 1
280 *never a rage* PF 62
280 *of his ways* ibid 61
280 *quarry to escape* ibid
280 *perplexity to insult* ibid
280 *some limpid stream* ibid 62
280 *spare his feelings* ibid
280 *him for life* ibid
280 *21 Percy Street* ESL 44
281 *tease Wyndham Lewis* ESL 43
281 *than at others* E: 'Ape of God', unpublished
281 *of the gaze* ibid
282 *not return them* ibid
282 *Mr. Cyril Connolly* ibid
282 *spots from it* O to Horner, 17 October 1932
282 *back to poetry* E to Harper, undated
282 *can't afford any* E to Tanner, 8 October 1932
283 *impertinent to her* O to Horner, 28 October 1932
283 *etc., terrifies me* E to Tanner, 31 October 1932
283 *devoted to him* E to Tanner, 31 October 1932

CHAPTER SEVENTEEN: ANGKOR AND AFTER

Private information from: Lady Aberconway, Sir Sacheverel Sitwell
285 *a fashionable success* E to Tanner, 8 August 1933
285 *to be over* ESL 43
285 *poor creature's nightmares* E to Tanner, 27 June 1933
286 *unfortunate skin, Wyndham's* O to Horner, 8 August 1933
286 *talk to him* E to Tanner, 15 August 1933
286 *about telephone bills* O to Horner, 13 August 1933

286 *is very good* ibid
286 *about that £2000* O to Horner, 23 August 1933
286 *would get nothing* E to Tanner, 15 August 1933
286 *all our misdeeds* ESL 46
287 *the poem whole* E to Tanner, 15 August 1933
287 *better for it* O to Horner, 5 September 1933
287 *but without result* ibid
287 *He seemed surprised* O to Horner, 14 September 1933
287 *secretary wears trousers* O to Horner, 7 September 1933
287 *dangerously like her* E to Tanner, 4 October 1933
288 *for the hounds* ESL 46
288 *through it again* E to Tanner, 4 October 1933
288 *have ever known* E to Haldane, 5 December 1933
288 *like an animal's* Tyler 357
288 *crime-tinged amorous intrigues* Tyler 363
289 *up – and soon* Tyler 361
289 *be too late* E to Tanner, 4 October 1933
289 *be a fool* ESL 46
289 *I know them* E to Haldane, 5 December 1933
290 *behaving like this* ibid
290 *in contemplating Germany* E to Haldane, undated
290 *wrote my poetry* ESL 45–6
290 *plus-fours under everything* E to Haldane, December 1933
290 *go out together* E to Tanner, April 1933
291 *were beside him* ibid
291 *stinking in hell* E to Tanner, undated
291 *from the authorities* ibid
291 *it should perish* EWM vii
292 *seen in China* EWM 63
292 *as hairy too* O to Horner, unpublished poem
293 *call it Bloomsbury* EWM 298
294 *describe as 'mignonne'* E to O, 8 May 1934
294 *all the time* E to Tanner, 16 May 1934
294 *probably be lynched* ibid
295 *waxwork of Nijinsky* Nichols, *Fisbo* 15
295 *to leave it* ibid
295 *he's read about* ibid 67
296 *a falling leaf* O to Horner, 16 August 1934
296 *swollen French bags* ibid
296 *Mr Paul Muni* O to Horner, 17 August 1934
296 *trudges by myself* ibid
296 *that old bore* ESL 53
296 *that bloody Victoria* E to O, 8 May 1934
296 *under those numbers* O to Horner, 18 August 1934
297 *for false pretences* O to Horner, 20 August 1934
297 *put me right* O to Horner, 26 August 1934
297 *like a bear's* O to Horner, 27 August 1934
297 *howling with rage* E to Tanner, 7 December 1934
297 *tease Wyndham Lewis* ESL 43
298 *Ewart, we will!* AMP 32
298 *than of poetry* Leavis, *New Bearings in English Poetry* 73
298 *their guest Cydrax* E to O, 8 May 1934
298 *of weary camels* AMP 30
299 *certain spiritual vision* AMP 245
299 *sense are lacking* AMP 245
299 *a later volume* AMP 229
299 *always lacks interest* AMP 238
299 *touch of genius* AMP 251

CHAPTER EIGHTEEN: RAT YEARS

Private information from: Lorna Andrade, Geoffrey Grigson, John Sparrow, Stephen Spender
300 *his daily mouse* O to Horner, unpublished poem
300 *of York whisky* O to Horner, undated
300 *from missing you* O to Horner, April 1935
300 *full of jealousy* ibid
301 *champagne is allowed* O to Horner, 30 April 1935
301 *more I fancy* ibid
301 *the coal seam* O to Horner, 11 July 1935
301 *you, as usual* O to Horner, 10 July 1935
301 *etc. v. boring* ibid
301 *good that year* ibid
302 *but finest quality* O to Horner, April 1935
302 *them more bitter* PF 52
302 *of my country* PF 22
302 *reformed our food* PF 182
303 *a wicked war* PF 131–2
303 *million British deaths* PF 309
303 *and punctual trains* PF 311
303 *to his country* O: 'General Conversation', *Nash's Pall Mall Magazine*, October 1935
304 *work without acknowledgement* *New Statesman and Nation*, 15 December 1934
304 *his peculiar effrontery* ibid
304 *dead, the ghouls* E to Tanner, 6 January 1935
305 *is most amusing* ibid
305 *most unfortunate results* Sparrow, *Sense and Poetry* 132
305 *explain absolutely rationally* E to Sparrow, 14 May 1934
305 *that pigeons do* ibid
305 *protozoan in England* E to Sparrow, December 1934
305 *beauty in language* *New Statesman and Nation*, 15 December 1934
306 *of Mr W. J. Turner* E to Sparrow, December 1934
306 *Mr Grigson presides* E to Robert Herring, October 1933
306 *him so unhappy* E to Tanner, 6 January 1935
306 *money he makes* ibid
306 *at moments, achievement* ESL 53
307 *fairly good phrases* E to Sparrow, May 1934
307 *that crapulous lout* FitzGibbon, *The Life of Dylan Thomas* 188
307 *and literary oblivion* ibid 190
307 *aroused his fury* ibid 200–1
307 *and this admiration* ESL 55
307 *of his complexes* ESL 54
308 *him some advice* E to Herring, 27 January 1936
308 *in medieval costume* FitzGibbon, *Dylan Thomas* 201
309 *numskulls soundly hit* TCO 168
309 *it any longer* E to Jennings, undated
309 *awful ticking off* ibid
310 *to readdress them* E to Jennings, 3 March 1937
310 *so excited it* A Taylor 395
311 *pro-government or pro-rebel* O: 'Must we Fight for our Country?', *Sunday Referee*, 12 July 1936
311 *Salade, fromage, fruits* Daphne Fielding, *The Rainbow Picnic* 103
311 *champagne Jubloteau 1906* ibid
311 *footsore and slummy* O to Horner, 16 June 1936
311 *it once stood* Channon, *Diaries* 88
312 *rude or civil?* O: 'Mustard and Cress', *Sunday Referee*, 8 November 1936
312 *lovely for her* S to Horner, 20 August 1936
312 *A violent success* E to O, 5 March 1936
312 *head with excitement* E to O, 5 March 1936
316 *the whole nation* E to Horner, 13 December 1936
314 *they hardly knew* 'National Rat Week', *Cavalcade*, 13 February 1937
315 *even Judas queasy* ibid
315 *in the letter* Channon, *Diary* 101
316 *the whole nation* E to Horner, 13 December 1936
316 *would sound well* ibid
316 *speak about it* E to Beevers, 21 September 1936
316 *in her eyes* E to Horner, 28 October 1936
316 *go out much* E to Sparrow, undated
316 *and wonderfully uncomplaining* ibid
316 *hear the nonsense* ibid
317 *to a painter* E to Horner, 28 October 1936
317 *be very severe* E to Horner, 13 December 1936
318 *costume – flies, centipedes* E to Jennings, 29 May 1937
318 *one's nerves rent* E to Horner, 25 May 1937
318 *few years ago* ibid

318 *of the spine* ibid
318 *in its orbit* E to Jennings, 29 May 1937
318 *just lying there* E to Beevers, 10 April 1937
318 *had written it* ESL 65

CHAPTER NINETEEN: WAR COMETH

Private information from: Francis Bamford, Beverley Nichols

319 *at the weekend* O: unpublished diary
320 *to know that* ibid
320 *it fell through* E to Horner, 17 May 1937
321 *avoided!! Rather alarming* O to Horner, 18 May 1937
321 *to do so* Macmillan archives, 20 May 1937
321 *from my car* O to Horner, 19 May 1937
321 *in the ring* O to Horner, 20 May 1937
321 *Duchess of Hesse* ibid
321 *glad to say* ibid
321 *looking very cross* O to Horner, 21 May 1937
321 *up in court* ibid
321 *a horrible cough* O to Horner, undated
322 *for the summer* *Scarborough Mercury*, 16 July 1937
322 *was absolutely perfect* E to Horner, 3 October 1937
323 *and her money* ibid
324 *teeth at me* ibid
324 *of one person* ILUBS 51
324 *a good investment* ILUBS 50
324 *piratic, formidable Dago* ILUBS 231
325 *world excepting you* ILUBS 112
325 *your life itself* ILUBS 113
325 *of your nature* ILUBS 113
325 *call my friend* ILUBS 114
325 *gloutonnerie, like cleptomania* Tyler 455
325 *herself into jealousy* ILUBS 60
325 *make his life* ILUBS 62
326 *of her mouth* ILUBS 64
326 *and wide-open eyes* ILUBS 74
326 *the Carnegies, etc.* O to Horner, 18 September 1937
327 *photograph of myself* O: unpublished notes for a diary 1 November 1937
327 *train to Florence* FC 202
327 *old and frail* O to Horner, 21 November 1937
327 *return to England* O to Horner, 22 November 1937
327 *in my head* E to Tanner, 3 November 1937
328 *paper I think* O to Ratcliffe of Macmillan, February 1937
328 *put me right* O to Macmillan, 5 May 1938
329 *attacked the liver* E to S, 31 January 1938
329 *peace and enervation* MWMW 204
329 *with his proofs* O to Macmillan, 11 August 1938
329 *proofs for him* ibid
330 *contradict me please* O to Horner, 25 September 1938
330 *in the autumn* FC 237
332 *lonely and unhappy* E to Tanner, 9 June 1939
332 *been rather seedy* E to Horner, 29 October 1938
332 *me to it* E to Tanner, 9 June 1939
332 *a giant orphan* Tyler 413
332 *inflicts on one* E to Tanner, 9 June 1939
332 *angry and embittered* ibid
332 *waiting for them* E to Horner, 18 May 1939
333 *go to America* ESL 66
333 *carry with me* GS to E, 20 May 1939
333 *George R. Sitwell* ibid
333 *say for Renishaw* E to O, 16 August 1939
333 *my time sleeping* O to Macmillan, 2 June 1939

CHAPTER TWENTY: THE GREAT INTERRUPTION

Private information from: Lorna Andrade, Rache Lovat Dickson, Sir Alec Guinness, Sir Sacheverell Sitwell

335 *men no better* Tyler 419 (originally written in French)
335 *for good writing* Albert Camus, *Carnets, 1942–51* 64
336 *soon from Hardwick* O to Macmillan, 13 October 1939
336 *life is death* O: unpublished poem
337 *sketch...Oh dear* O to Andrade, 7 February 1940
338 *heavy to hold* Lehmann, *A Nest of Tigers* 103
338 *slightly cynical, smile* ibid
339 *it too dangerous* O to Macmillan, 20 April 1940
339 *Willie [Maugham] tomorrow* O to Horner, 21 May 1940
230 *all was discovered* E to Horner, 1 April 1940
340 *the habit anyway* Whistler to O, July 1940
341 *be sent here* O to Andrade, 19 August 1940
341 *than I am* O to Andrade, 5 November 1940
341 *has been caught* O to Andrade 29 August 1940
341 *fallen round here* O to Andrade, 21 August 1940
341 *an account there* ibid
341 *been badly hit* O to Andrade, 29 August 1940
341 *so far away* O to Horner, 21 August 1940
342 *down too well* O to Horner, 2 September 1940
342 *agent]. Philip Frere* O to Horner, 4 September 1940
342 *any day now* O to Horner, 6 September 1940
342 *mournful, vacant look* ibid
342 *do for us?* O to Andrade, 30 September 1940
343 *see it never* O: 'Personal Prejudice', *Life and Letters To-day*, October 1940
343 *is completely undisturbed* ESL 73
543 *point of view* ibid
343 *to one's offspring* ESL 74
343 *take so long* O to Dickson, 8 November 1940
344 *of a man* ESL 77
344 *advertise their wares* *Manchester Guardian*, 7 February 1941
344 *in this way* *Times*, 6 February 1941
345 *magpie and busybody* ESL 83
345 *Titanic was sinking* OD 1
345 *been over quicker* O to Horner, 30 October 1941
345 *of a conjuror* O to Horner, 9 March 1941
346 *Really, these patriots* O to Horner, 6 September 1941
346 *about the war* O to Horner, 1 July 1941
346 *becoming a habit* ESL 82
347 *much like now* O to Horner, 10 May 1941
347 *the question later* ibid
347 *wonderful in white* O to Horner, undated
347 *petrol in Eckington* O to Horner, 13 August 1941
348 *good Man Friday* ESL 101
348 *I promise you* E to Horner, undated
348 *painful to me* ESL 86
348 *it's warm enough* O to Dickson, 14 June 1941
349 *and rural pursuits* NGM
349 *he makes himself* ibid
349 *make sad thoughts* O to Stokes, 4 May 1941
350 *all arrive safely* Lehmann 196
351 *experiments was done* ECP xli
351 *prodigious hymns* CES 16
351 *dog-days are here* ECP 281

CHAPTER TWENTY-ONE: VICTORY

Private information from: Lorna Andrade, Major J. W. Chandos-Pole, John Piper, Sir Sacheverell Sitwell

352 *be so cold* O to Andrade, February 1942
352 *a capital offence* ibid
353 *thought for you* E to GS, 22 November 1941
353 *I should be* O to Horner, 24 September 1941
353 *want his money* O to Horner, 6 January 1942
354 *was being written* Lehmann 193
354 *carved Roquefort cheese* E to Herring, undated
354 *in by it* O to Dickson, 20 June 1942
355 *the whole time* O to Dickson, 6 July 1942
355 *description of Eden* O to Horner, 8 October 1942
355 *and ferociously well* E to Horner, 21 June 1942
355 *I am amused* O to Horner, 3 September 1942
355 *quid a year* O to Horner, 8 October 1942
355 *week of registration* O to Dickson, 8 July 1942

355 *of my powers* O to Stokes, 26 September 1942
356 *of the war* O to Andrade, 29 June 1942
356 *and amber wings* Bryher, *The Days of Mars* 64
356 *pull-over without coupons* E to Horner, 22 September 1942
356 *to a gamekeeper* Bryher 62
357 *must be righted* E to Red Cross Foreign Relations Department, 5 September 1942
357 *a lovely place* O to Mark, 19 January 1943
357 *had fallen. Nothing* O to Horner, 28 January 1943
357 *in some way* O to Dickson, 2 February 1943
358 *I ever remember* O to Andrade, 25 February 1943
358 *abandon and perfection* O to Horner, 14 March 1943
358 *to assert themselves* Bryher 83
358 *is very difficult* E to Lehmann, 6 April 1943
359 *as a poet* E to Horner, 28 March 1943
359 *stared at one* Welch, *Journals* 89
359 *of mechanised disorder* Lady Cunard to O, undated
359 *bottle of rum* O to Horner, May 1943
360 *and the reception* O: undated memorandum
360 *a great genius* Welch, *Journals* 8
361 *that bone-head Connolly* E to Horner, 4 May 1943
361 *for their art* ibid
361 *suppose I feel* ibid
361 *but I shan't* LNR 160
362 *A fearful lie* O to Horner, 12 July 1944
362 *for producing me* E to Horner, 20 July 1943
362 *his very best* ibid
363 *was most trying* ibid
363 *before he died* E to Lehmann, 20 August 1943
363 *devils with prongs* O to Mark, 10 August 1943
363 *his only grandchildren* E to Horner, 20 July 1943
363 *salamander needs flame* O to J. H. Hutchinson, 22 June 1943
363 *best possible offer* O to Price of Macmillan, 19 August 1943
364 *do you think* O to Dickson, 18 July 1943
364 *It is wonderful* E to Lehmann, 5 December 1943
364 *of her life* E to Lehmann, 23 November 1943
364 *a civilised house* E to Horner, undated
364 *like a tin-opener* O to Horner, undated
364 *on a summer day* O to Horner, 3 October 1943
364 *new and expensive* O to Horner, 14 March 1943
365 *all the windows* ibid
365 *were the thing* O to Horner, 29 September 1943
365 *in Brazil instead* O to Andrade, undated
365 *all. Do come* Fielden to O, quoted in O to Horner, 22 October 1943
365 *for my friend* O to Horner, 22 October 1943
365 *on the fire* E to Lehmann, 5 December 1943
365 *a long time* O to Horner, 13 December 1943
366 *the fact dark* E to Horner, 2 December 1943
366 *remembering his misfortunes* ibid
366 *agonies of separation* SM 243
366 *that befell her* SM 242
366 *of my childhood* E to Horner, 2 December 1943
366 *not mention this* O to Andrade, 27 July 1943
367 *most appalling bore* O to Macmillan, 2 April 1944
367 *from his mother-in-law* E to Horner, 27 February 1944
368 *bloodhounds after one* E to Horner, 4 March 1944
368 *to harass you* O to Andrade, 13 March 1944
369 *will be forthcoming* O to Andrade, undated
369 *in every way* Nell St John Montagu to E, 2 April 1944
369 *brought to justice* Nell St John Montagu to E, undated
369 *in a letter* E to Lehmann, 13 April 1944
370 *this miserable affair* E to Lehmann, 18 April 1944
370 *hope I'm wrong* O to Andrade, 11 April 1944
370 *to be made* O to Andrade, 13 April 1944
371 *and blackmailers (possibly)* E to Horner, 3 July 1944
372 *allowance had ceased* GM 296
373 *feel extra important* O to Horner, 9 July 1944
374 *my earlier poetry* ESL 121
374 *value your advice)* E to Lehmann, 13 June 1944
374 *my poetry before* ESL 89
374 *had never existed* E to Hayward, November 1944
375 *to do it* ESL 121
375 *to lunch-time, dithery* O to Andrade, undated
375 *grief to me* O to Horner, 21 July 1944
376 *look smilingly on* Agate, *Daily Express*, 2 September 1944
376 *600,000 common people* ibid
376 *mean, base persecution* E to Horner, 1 October 1944
376 *never suspects them* E to Lehmann, 10 September 1944
377 *applause was deafening* Lehmann, *I am my Brother* 281
377 *to the forces* O to Stokes, 13 July 1944
377 *been put to* O to Andrade, undated
377 *very bad tenants* quoted in O to Horner, 6 September 1944
378 *but not much* E to Lehmann, 13 June 1944
378 *way, one's language* ESL 123
378 *at the moment* ESL 124
378 *I like it* E to Lehmann, 15 November 1944
379 *heart of Man* ECP 296
379 *I praised him* E to Lehmann, 16 December 1944
379 *of my standing* quoted in E to Lehmann, 16 December 1944
379 *removed from my life* quoted in ibid
379 *to insult me* E to Lehmann, undated
380 *me horribly sad* O to Stokes, 26 March 1945
380 *will one ever?* O to Andrade, undated
380 *for a diversion* O to Andrade, 22 May 1945
380 *in my eyes* E to Lehmann, 6 May 1945

CHAPTER TWENTY-TWO: 'DIBS AND DIGNITY'
Private information from: Lord Clark, Jack Lindsay, John Piper
381 *the want of* O to Horner, 7 August 1945
382 *and even eat* O to Mark, 2 March 1946
382 *through many stages* CES 114–15
382 *just finished me* O to Andrade, undated
382 *dead with fatigue* E to Lehmann, 30 April 1946
383 *were fearfully tiring* ibid
383 *am a poet* E to Lehmann, 25 February 1945
383 *people I understand* E to Lehmann, 28 August 1946
384 *part of it* E to Horner, 30 November 1945
384 *dusty provincial theatre* E to Lehmann, 6 August 1946
385 *it is bad* E to Horner, 29 May 1946
386 *as 'Old Ugly'* E to Horner, 24 May 1946
386 *has blown over* E to Lehmann, 1 June 1946
386 *do with me* ESL 142
387 *make a row* E to Horner, 1 May 1946
387 *to the world* O to Stokes, 20 May 1946
388 *the dogs bowl* O to Andrade, undated
388 *for the future* BAC 1
389 *up a gum-tree* E to Lehmann, 12 November 1945
389 *really appalling rebuke* E to Lehmann, November 1946
389 *work to ourselves* Grigson to E, 9 December 1946
390 *my beloved Papa* E to Lehmann, May 1947
390 *my civilisation completely* E to Lehmann, July 1947
390 *meet the case* E to Lehmann, August 1947
390 *the country houses* E to Lehmann, 23 July 1947
390 *as grossly impertinent* ibid
390 *gross public insult* ibid
391 *up with them* ibid
391 *feel extremely ill* ibid
391 *of electricity, etc* O to Andrade, 14 September 1947
391 *just don't I* ibid
391 *from being paradisiacal* O to Horner, 15 August 1947
392 *Your Devoted O* ibid
392 *tall as trees* O to Andrade, undated
392 *is the garter* O to Macmillan, 5 December 1947
392 *Mr Noel Coward* O to Andrade, 16 December 1947

392 *there it is* O to Andrade, 30 December 1947
393 *this has happened* E to Lehmann, 14 December 1947
393 *obscene and filthy* ibid
393 *court of law* ibid
393 *myself Doctor Sitwell* E to Lehmann, 2 January 1948
394 *disagreeable ever since* E to Lehmann, June 1948
394 *and 'surprising' choices* ESL 155
394 *those particular circumstances* E to Plomer, undated
395 *and economic revolutions* Lindsay 121
396 *that poem's author* ibid 54
396 *true dialectical insight* ibid 59
396 *a vile age* O to Andrade, undated
397 *was not published* Rolph, *Kingsley* 197
397 *on, Miss Sitwell* E to Lehmann, 12 August 1948
397 *not a nuisance* E to Hayward, 19 May 1948
397 *of the winkle* E to Lehmann, 2 March 1948
397 *most of them* Ford to E, 2 January 1948

CHAPTER TWENTY-THREE: AMERICA

Private information from: Lincoln Kirstein, Glenway Wescott
399 *no real friends* Tyler 456
399 *important to life* Tyler 447
499 *not a destiny* Tyler 458
399 *all his letters* Tyler 460
300 *hide his irritation* Tyler 461
401 *ever be polite* E to Lehmann, 23 October 1948
401 *really be mad* ESL 164
403 *after the performance* E to Lehmann, 15 January 1949
403 *of the world* *New Yorker*, 29 January 1949
403 *give the troops* E to Lehmann, 15 February 1949
403 *and rather dopey* O to Andrade, 26 February 1949
404 *the public eye* Tyler 463
404 *can't put together* Tyler 464
404 *like a worm* ibid
405 *you call them* E to Tchelitchew, 28 March 1949
405 *behaviour towards me* ibid

CHAPTER TWENTY-FOUR: 'GID AND DIZZ'

Private information from: Lorna Andrade, Geoffrey Grigson, Lincoln Kirstein, Glenway Wescott, Monroe Wheeler
407 *and so on* O to Stokes, 22 August 1949
407 *after the U.S.A.* O to Dickson, 29 July 1949
407 *the servants make* O to Mark, 13 August 1949
407 *intuition and adaptability* Connolly, *Enemies of Promise* 150
407 *I have inherited* NE 166
408 *be no Second* ESL 164–5
408 *spoil the book* E to Lehmann, 12 April 1949
408 *curiously interesting* ESL 169
409 *my own judgement* Frere to E, 24 August 1949
409 *strong enough meat* Fleming to E, 13 September 1949
409 *of our time* TCO 162
410 *such a promise* E to O, 28 September 1949
410 *take both sides* ibid
410 *will be remembered* ibid
410 *told Lord Beaverbrook* E to Horner, 10 October 1949
411 *Manse outside Aberdeen* ibid
412 *they see them* E to Lehmann, 6 November 1949
412 *slows one up* O to Kirstein, 6 January 1950
413 *forgotten in England* Tchelitchew to O, 30 August 1949
413 *is battened on* E to Kirstein' 27 December 1949
413 *in terror financially* E to Kirstein, 14 January 1950
413 *fear and misery* ibid
413 *him against me* E to Kirstein, Good Friday [?1950]
414 *Victoria to Pisa* ESL 170
414 *a good mood* O to Kirstein, 18 May 1950
414 *there would be* O to Kirstein, 29 May 1950
414 *he can rely* E to Kirstein, 24 May 1950
414 *very great genius* ibid
415 *58 in December* Horner to Cohen, 25 July 1980
416 *very depressing look-out* Huxley, *Letters* 627
416 *old and tired* ibid 629
417 *very royal setting* Kirstein to E, August 1950
417 *you can write* ibid
417 *the right things* E to Kirstein, 30 August 1950
417 *impossible for Stravinsky* ibid
418 *with English authors* O to Dickson, 25 October 1950
418 *instead of telephoning* ibid
418 *don't exhaust yourselves* Huxley to E, 26 October 1950
420 *kind of complication* Lindsay 87
420 *I was dying* ESL 172
420 *to shape letters* O to Kirstein, 18 January 1951
420 *and Dorothy Parker* O to Dickson, January 1951
420 *a Spiritual Beacon* E to Lehmann, 17 January 1951
420 *was breathing heavily* ESL 173
421 *care for them* O to Kirstein, 3 February 1951
421 *about the same* Andrade, diary, 21 March 1951
422 *wonderful new idea* O to Macmillan, December 1950
422 *is a nuisance* O to Macmillan, 23 May 1951
422 *him over that* ESL 175
422 *a normal childhood* E to Lehmann, 21 May 1951
423 *pieces as usual* Andrade, diary, 30 June 1951
423 *remembered in future* O: 'Constant Lambert', *New Statesman*, 1 September 1951
423 *bad I am* O to Kirstein, 10 October 1951
423 *like being blind* ibid
424 *and a punchinello* O to Kirstein, 19 November 1951
424 *never quite seize* ibid
424 *is no worse* Horner to Kirstein, 2 October 1951
424 *to my bed* E to Peter Russell, 19 January 1952
424 *all the time* O to Kirstein, 20 February 1952
425 *return of OS* Andrade, diary, 17 April 1952
425 *the same place* O to Kirstein, 8 August 1952
425 *be traced through* OCS xviii
425 *write between sobs* OCS ix
425 *before I've finished* O to Dickson, March 1953

CHAPTER TWENTY-FIVE: HOLLYWOOD

Private information from: Lincoln Kirstein, Dan Maloney
427 *he was engaged* O to Kirstein, 26 January 1953
427 *of my country* ibid
427 *in the evening* Andrade, diary, 9 April 1953
427 *amiable and pale* Andrade, diary, 1 May 1953
427 *weekend in Surbiton* E to Stapleton, 5 June 1953
427 *and stay there* ibid
428 *now got gout* O to Kirstein, 16 June 1953
429 *a dreadful numbness* E to Lehmann, 29 November 1953
429 *lead simply anywhere* E to Kirstein, undated
429 *shall miss him* ESL 187
430 *never harmed them* E to Wykes-Joyce, 12 January 1954
430 *his own misdeeds* Frere to E, 5 January 1959
430 *whole collection* E to Horner, 23 January 1954
430 *belated Browning societies* E to Horner, 13 February 1954
430 *we know everything* E to Wykes-Joyce, 19 February 1954
431 *and black sandals* *Sunday Graphic*, 11 January 1954
431 *of supreme importance* E: unpublished lecture on Hollywood
431 *pleasantly shy* ibid
342 *in unlikely places* ibid
432 *on her behalf* ibid
432 *have to say* ibid
433 *raised to heaven* ESL 188–9
433 *splits one's head* E to Lehmann, 18 February 1954
433 *every fifteen months* O to Kirstein, 11 February 1954
435 *for next time* O to Mark, undated
435 *to celebrate it* Andrade, diary, 21 June
435 *she read it* E to Eric Tattersall (draft in E's hand)
436 *Yours devotedly, Osbert* O to Horner, 2 September 1954

436 *Minimum circulation, 30,000* O to Horner, 14 September 1954
436 *a red Indian* O to Horner, 20 September 1954
436 *a new atmosphere* E to Lehmann, 11 October
437 *to be over* ibid
437 *still tremble dreadfully* O to Andrade, 7 December 1954
437 *his unfortunate malady* Horner to Andrade, 26 December 1954
438 *Monroe to England* E to Braithwaite, 15 February 1955
438 *Best love, Osbert* O to Horner, 5 February 1955
438 *Mediterranean in July* O to Kirstein, 15 April 1955
438 *of his nurse* ibid

CHAPTER TWENTY-SIX: THE CONVERT
Private information from: Lorna Andrade, Philip Caraman, Martin D'Arcy, Lincoln Kirstein
439 *has made me* D'Arcy to E, 14 April 1955
439 *of my life* ibid
440 *will let me* ESL 192
441 *in this matter* Waugh to Caraman, 19 July 1955
441 *would be disastrous* Waugh, *Diaries*, 735
442 *sense in faith* ESL 194
442 *feel far away* ESL 192
442 *step years ago* E to Caraman, undated
443 *own grave faults* E to Caraman, 7 May 1955
444 *excuses for him* E to Caraman, undated
444 *of Catholic London* Waugh, *Diaries* 745
444 *terrible, blast it* O to Kirstein, 15 October 1955
445 *Monsieur Jean Cocteau* E: 'In Praise of Jean Cocteau', *London Magazine*, February 1956
446 *was greatly moved* E to Caraman, 16 December 1955
446 *and everybody else* O to Kirstein, 5 November 1955
446 *was very dangerous)* E to Lady Lovat, 20 August 1955
447 *rather a measly* S to Nancy Cunard, 8 March 1956
447 *haven't broken anything* E to Caraman, 5 February 1956
447 *making any sense* ibid
447 *flow of verse* E to Lehmann, undated
447 *particularly ferocious greyhounds* E to Caraman, 14 April 1956
448 *hate him terribly* E to Lehmann, 11 May 1956
448 *can't be possible* Andrade, diary, 11 June 1956
449 *Sitwell world there* Andrade, diary, 27 September 1956
449 *really terrifying accident* E to Caraman, 10 December 1956
450 *just up again* ibid
450 *the stone floors* ibid

CHAPTER TWENTY-SEVEN: EDITH AGONISTES
Private information from: A. Alvarez, Cyril Connolly, Peter Hollingworth, Sir Sacheverell Sitwell
451 *see Tom again* E to Lehmann, 19 January 1957
452 *of a stick* ESL 207
453 *pots look lovely* O to Horner, 27 February 1957
454 *to honour one)* ESL 215
454 *complete it seems* E to Caraman, 11 January 1957
455 *reached his mouth* Andrade, diary, 9 May 1957
455 *brother to me* E to Caraman, 4 June 1957
455 *on both sides* ibid
455 *of a reporter* ESL 216
455 *is too much* E to Caraman, 2 May 1957
456 *friend Choura Charlie* Ford to E, 1 August 1957
456 *have ever known* TCO 137
457 *a motor accident* O to Kirstein, 26 August 1957
457 *to think of* ibid
457 *old Medici Farmacia* Andrade, diary, 24 October 1957
457 *patience and goodness* E to Caraman, 17 December 1957
458 *a thousand wrinkles* OC 55
458 *The Old Scamp* OC 59
458 *stands to reason* E to Caraman, 21 January 1958
459 *about my poetry* E to Caraman, 19 August 1959
459 *Paris to Lagos* Clark SS 18
460 *really happy people* Connolly SS 5
460 *most desperate fashion* Snow SS 11
460 *Dame Edith!' Really.* E to Caraman, 9 July 1958
461 *your writing poetry* Gorer to E, 11 July 1958
461 *shan't do it* E to Lehmann, 21 July 1958
462 *of Mr Alvarez* E to Lehmann, 12 September 1958
462 *my earlier poems* E to Caraman, 19 August 1958
462 *can do anything* ESL 229
463 *Winfield Townley Scott* E to Caraman, 30 January 1959
463 *very heroic person* ibid
463 *have not got* Salter 102–3
464 *her mother's disgrace* Salter 103
464 *despair were accentuated* ibid
465 *wonder in Florence* E to Lehmann, 3 March 1959
465 *multitudinous times since* E to *Encounter*, 29 April 1959
466 *my television interview* E to Caraman, 27 May 1959
466 *producers and journalists* Salter 120
466 *and inferiority complex* Salter 120
466 *with my voice* *Daily Mail*, 10 September 1959
466 *lady's unfortunate disposition* *Daily Mail*, 10 September 1959
466 *a little longer* ibid
466 *so very much* ibid
467 *a Roman Catholic* E to Caraman, 26 September 1959
467 *to church often* Lindsay 84
467 *out of me* E to Caraman, 29 November 1959
468 *What a bore* E to Caraman, 10 January 1960
468 *to restrain you* Salter 131

CHAPTER TWENTY-EIGHT: 'SOMETHING MACABRE OUT OF PROUST'
Private information from: Lord Clark, Elizabeth Salter
469 *flames jewel-cold* ECP 55
472 *bless Dr Laurence* E to Horner, 7 October 1961
477 *wasn't so dusty* E to Lehmann, 2 July 1962
478 *any covert insolence* E to *Evening News*, undated draft
478 *round a specialist* E to Lehmann, 2 July 1962
478 *found in anyone* *Daily Mail*, 30 August 1962
478 *extremely unhappy life* ibid
480 *her eyes owlish* Salter 169
481 *she was alive* Salter 179

CHAPTER TWENTY-NINE: END GAME
Private information from: Lorna Andrade, Philip Caraman, Hunter Daves, Pirie Gordon, Ian Greenlees, Lincoln Kirstein, Frank Magro, Reresby Sitwell, Hugo Southern
482 *Such a warning* O to Horner, undated
484 *to English literature* Magro, diary, 28 January 1963
484 *soulless London flat* *Daily Telegraph*, 3 March 1963
489 *Guy Fawkes Day* O to Horner, 28 September 1964
489 *here. Very nice* O to Horner, 4 October 1964
489 *to each other* Aberconway to Gandarillas, undated
491 *been in England* O to Caraman, 19 January 1965
491 *what he wants* O to Aberconway, 21 January 1965
491 *to Osbert again* Aberconway to O, 31 January 1965
491 *this house again* Vera Frere to O, 16 February 1965
492 *had been made* Aberconway to S, 12 January 1965
493 *'feeble-minded' I am* O to Aberconway, 9 January 1965
494 *the old block)* Aberconway to O, 11 February 1965
494 *They sold magnificently* O to Aberconway, 11 March 1965
494 *I found fascinating* O to Aberconway, 6 April 1965
494 *to say infernal* O to Aberconway, 11 July 1965
495 *property in Italy* *Times*, 15 November 1965
496 *next five years* ibid
496 *not answering letters* *Sunday Times*, 17 October 1965
496 *neck and neck* O to Aberconway, 26 November 1965
497 *top of me* *Sunday Times*, March 1966
497 *asylum at Volterra* O to Aberconway, 23 May 1966
498 *became a joke* Acton, *Nancy Mitford* 96
499 *for Job's crown* O to Aberconway, 17 April 1967

499 *it practically unintelligible* Trefusis to Lindsay-Hogg, 17 February 1966
500 *suitable for me* O to Aberconway, 31 March 1967
500 *are much worse* Magro to Andrade
500 *telling me to* Magro to Andrade
501 *away we go* Magro to Andrade
501 *a low key* Magro, diary
501 *restrained but favourable* Magro, diary
505 *exceptional as themselves* MWLH 4
506 *stress of time* FWGC 303
506 *bask pitiful memories* FWGC 263
506 *interest me now* FWGC 295

Index

E = Dame Edith Sitwell GS = Sir George Sitwell IS = Lady Ida Sitwell O = Sir Osbert Sitwell RS = Reresby Sitwell S = Sir Sacheverell Sitwell

JOHN PEARSON, born in 1930 and educated at Cambridge, has worked on various newspapers, including the London *Sunday Times,* where he wrote a column with Ian Fleming, creator of James Bond. After Fleming's death Mr. Pearson wrote the bestselling *Life of Ian Fleming.* Among his other books are a history of the Colosseum, a life of the Kray brothers, the gangster twins who ran London's criminal world in the sixties, and *Edward the Rake: An Unwholesome Biography of Edward VII.*